AF699439

Ex Libris

William A. Rosché

Myxobacteria II

Myxobacteria II

Editors:

Martin Dworkin

Department of Microbiology
University of Minnesota
Minneapolis, Minnesota

Dale Kaiser

Department of Developmental Biology
Stanford University School of Medicine
Stanford, California

American Society for Microbiology
Washington, DC

Copyright © 1993 American Society for Microbiology
1325 Massachusetts Ave., N.W.
Washington, DC 20005

Library of Congress Cataloging-in-Publication Data

Myxobacteria II / editors, Martin Dworkin, Dale Kaiser.
p. cm.
Includes index.
ISBN 1-55581-060-8
1. Myxobacterales. I. Dworkin, Martin. II. Kaiser, Dale.
III. Title: Myxobacteria two.
[DNLM: 1. Myxococcales—genetics. QW 128 M9991 1993]
QR82.M95M97 1993
589.9′8—dc20
DNLM/DLC
for Library of Congress 92-48477
CIP

All Rights Reserved
Printed in the United States of America

Cover: Fruiting bodies of *Myxococcus fulvus* (photograph courtesy of Hans Reichenbach)

CONTENTS

CONTRIBUTORS

John Downard • Department of Botany and Microbiology, University of Oklahoma, Norman, OK 73019

Martin Dworkin • Department of Microbiology, University of Minnesota, Minneapolis, MN 55455-0312

Ronald E. Gill • Department of Microbiology and Immunology, School of Medicine, University of Colorado Health Sciences Center, Denver, CO 80262

J. F. Guespin-Michel • Laboratoire de Microbiologie, URA 203 CNRS, Faculté des Sciences et Techniques, Université Rouen, 76134 Mont Saint Aignan, France

Patricia Hartzell • Department of Microbiology and Molecular Genetics, University of California–Los Angeles, Los Angeles, CA 90024

David A. Hodgson • Department of Biological Sciences, University of Warwick, Coventry CV4 7AL, England

Gerhard Höfle • Chemistry of Natural Products, Gesellschaft für Biotechnologische Forschung, D-38124 Braunschweig, Germany

Masayori Inouye • Department of Biochemistry, Robert Wood Johnson Medical School, University of Medicine and Dentistry of New Jersey, Piscataway, NJ 08854

Sumiko Inouye • Departments of Biochemistry, Robert Wood Johnson Medical School, University of Medicine and Dentistry of New Jersey, Piscataway, NJ 08854

Dale Kaiser • Departments of Biochemistry and Developmental Biology, Stanford University School of Medicine, Stanford, CA 94305

Lee Kroos • Departments of Biochemistry and Microbiology, Michigan State University, East Lansing, MI 48824

Bert C. Lampson • Department of Microbiology, University of Tennessee, Knoxville, TN 37996-0845

B. Letouvet-Pawlak • Laboratoire de Microbiologie, URA 203 CNRS, Faculté des Sciences et Techniques, Université Rouen, 76134 Mont Saint Aignan, France

Mark J. McBride • Department of Biological Sciences, University of Wisconsin–Milwaukee, Milwaukee, WI 53211

Francisco J. Murillo • Departamento de Genetica y Microbiologia, Facultad de Biologia, Universidad de Murcia, 30071 Murcia, Spain

F. Petit • Laboratoire de Microbiologie, URA 203 CNRS, Faculté des Sciences et Techniques, Université Rouen, 76134 Mont Saint Aignan, France

Donald H. Pfister • Harvard University Herbaria, Cambridge, MA 02138

Hans Reichenbach • Microbial Secondary Metabolites, Gesellschaft für Biotechnologische Forschung, D-38124 Braunschweig, Germany

Eugene Rosenberg • Department of Microbiology, Tel Aviv University, 69978 Ramat Aviv, Israel

Amy Rosenbluh • Department of Membrane Research and Biophysics, The Weizmann Institute of Science, 76100 Rehovot, Israel

Hans Ulrich Schairer • Zentrum für Molekulare Biologie, Universität Heidelberg, Heidelberg, Germany

Lawrence J. Shimkets • Department of Microbiology, University of Georgia, Athens, GA 30602

David White • Department of Biology, Indiana University, Bloomington, IN 47401

David R. Zusman • Department of Molecular and Cell Biology, University of California–Berkeley, Berkeley, CA 94720

PREFACE

The first paper formally describing the myxobacteria was published by Roland Thaxter in 1892. During the early part of the 20th century, occasional papers on the myxobacteria were published, but these organisms had not yet begun to capture the imagination of microbiologists. Apart from such milestones as E. Jahn's monographic review of the myxobacteria in 1924 and the work of Helena Krzemieniewska and Seweryn Krzemieniewski (1926, 1927a, 1927b, 1930) on the myxobacteria of Poland, the myxobacteria remained an exotic and experimentally inaccessible group of bacteria. In 1950, Kühlwein began the modern era of myxobacterial research, and the work of this period crystallized with the first formal meeting on the myxobacteria, held in Cold Spring Harbor in 1974. Myxobacterial genetics was just about to be born, molecular biology was in the process of being invented, and a handful of enthusiasts were still struggling with such problems as carotenoid biosynthesis, myxospore induction and germination, bacteriophage characterization, and DNA and RNA synthesis during development. During the next 2 decades, interest and publications on the myxobacteria steadily increased, to the point where more than 40 laboratories throughout the world are now working on the myxobacteria, and hardly an issue of the *Journal of Bacteriology* does not contain a paper on the myxobacteria. The first book on the myxobacteria (Rosenberg, 1984) contained chapters on myxobacterial structure, genetics, nutrition, development, and gliding motility as well as a number of other aspects of myxobacterial physiology. The work that that book described set the stage for a deeper penetration of the problems of signaling, cell-cell interactions, and regulation of development and motility that represent the main thrust of the present volume.

We now know a great deal about the myxobacteria, and it is tempting to believe that we also understand a great deal about the phenomena of myxobacterial behavior, development, and multicellular activities that originally caught our attention and persuaded us to work with these organisms. Yet for several reasons, the tendency for self-congratulation must be tempered. First, most of this work has focused on one organism, *Myxococcus xanthus,* which has come to be viewed as a model system for the entire group of the myxobacteria. The tendency in microbiology to follow this "model organism" approach is almost irresistible. There is no question that this approach is a powerful one that allows the research community to coordinate efforts and thus accumulate a considerable amount of fundamental information that then facilitates more specialized work and leads to the deeper understanding all of us seek. However, as Reichenbach shows (see chapter 2), there is a wealth of diversity among the myxobacteria, and it is clear from Shimkets' chapter on the genome and evolution of the myxobacteria (see chapter 4) that there are at least three lines of myxobacterial descent. Moreover, the single "model organism" approach can be misleading. It can lead to improper generalizations about "the myxobacteria" based on one organism. In addition, the failure to examine other myxobacteria lessens the likelihood of discovering alternative biological strategies. Second, we know and understand almost nothing about myxobacterial ecology. Despite considerable information about the qualitative distribution of the myxobacteria, we do not understand their role in nature, their interactions with each other in natural environments, or the actual interactions between their developmental events and environmental factors. The irony is that many of us are attracted to the myxobacteria by their social and predatory behaviors.

A final reason to modulate our enthusiasm is that the facts thus far acquired about the physiological, genetic, and molecular aspects of *M. xanthus* cannot yet be assembled into a complete understanding of any one of the more complex features of myxobacterial behavior.

However, there is ample ground for optimism. The 1992 international meeting on the biology of the myxobacteria seemed to reflect a turning point in the field. A major and perhaps even necessary preoccupation of the past decade has been the development of molecular and genetic technologies appropriate for studying the myxobacteria. It is a fortunate characteristic of workers in this field that despite the preoccupation with the development of appropriate genetic and molecular technologies, the biological questions have not been forgotten. The recent meeting saw the technologies folded into the biological questions in an exciting and reassuring fashion. We are thus hopeful and cautiously optimistic that the next edition of this book will see the emergence of insights about fruiting body morphogenesis, the mechanisms of multicellular communication and coordination, the mechanism and function of rippling, and the mechanism of gliding motility, to name only a few of the fascinating aspects of myxobacterial biology.

In Thaxter's third and last paper on the myxobacteria (1904), he grumbled, "Since the appearance in this journal (June 1897) of the author's second paper on this group, comparatively little attention has been given to it, and with few exceptions one finds no mention of it in current text-books or in bacteriological literature." We are certain he would be encouraged and perhaps even pleased if he were able to read *Myxobacteria II*. Thus, on the 100th anniversary of Roland Thaxter's first paper on the myxobacteria, it is with considerable pleasure that we dedicate this book to him.

Finally, we acknowledge the contributions of Amy Vollmer, who prepared the index; of Patrick Fitzgerald, who encouraged our interaction with the ASM Books Division; and of Susan Birch, who as the responsible production editor for the book was efficient and helpful throughout the process.

Martin Dworkin
Dale Kaiser

REFERENCES

Jahn, E. 1924. Beiträge zur botanischen Protistologie. I. Die Polyangiden. Gebrüder Borntraeger, Leipzig, Germany.

Krzemieniewska, H., and S. Krzemieniewski. 1926. Miksobakterje Polski. *Acta Soc. Bot. Pol.* **4**:1–54. (With German summary.)

Krzemieniewska, H., and S. Krzemieniewski. 1927a. Miksobakterje Polski. Uzupelnienie. *Acta Soc. Bot. Pol.* **5**:79–98. (With German summary.)

Krzemieniewska, H., and S. Krzemieniewski. 1927b. Rosziedlenie miksobakteryj (über die Verbreitung der Myxobakterien im Boden). *Acta Soc. Bot. Pol.* **5**:102–139. (With German summary.)

Krzemieniewska, H., and S. Krzemieniewski. 1930. Miksobakterie Polski. Czesc trzecia (Die Myxobakterien von Polon. III. Teil) *Acta Soc. Bot. Pol.* **7**:250–273. (With German summary.)

Kühlwein, H. 1950. Beiträge zur Biologie und Entwicklungsgeschichte der Myxobakterien. *Arch. Mikrobiol.* **14**:678–704.

Rosenberg, E. (ed.). 1984. *Myxobacteria. Development and Cell Interactions.* Springer-Verlag, New York.

Thaxter, R. 1892. On the Myxobacteriaceae, a new order of Schizomycetes. *Bot. Gaz.* **17**: 389–406.

Thaxter, R. 1904. Notes on the Myxobacteriaceae. *Bot. Gaz.* **37**:405–416.

Myxobacteria II
Edited by Martin Dworkin and Dale Kaiser
© 1993 American Society for Microbiology, Washington, DC 20005

Chapter 1

Roland Thaxter and the Myxobacteria

Donald H. Pfister

Many taxonomists describe new species and new genera, and some have the opportunity to describe a new family or a new order, but few have the chance or foresight to take organisms described by their predecessors, confront common dicta, and reassign them to another kingdom. This is precisely what Roland Thaxter did with the myxobacteria. In this study, I want to outline the details of Thaxter's life and academic career and provide some background for the contribution Thaxter made to the study of the myxobacteria.

In 1892, when he published his first paper on the new family Myxobacteriaceae, Roland Thaxter had been assistant professor of botany at Harvard University for just 1 year. That paper, "On the Myxobacteriaceae, a New Order of Schizomycetes," can rightly be called the starting point for the study of the taxonomy and biology of the group. Though Thaxter published only three papers on the myxobacteria (1892, 1897, 1904), his work was so fundamental that the group can hardly be mentioned without reference to his pioneering investigations. He collected and studied these organisms in the field; recognized them for what they were, i.e., bacteria; and gave details of their biology, behavior, and activities based on laboratory studies. Although he did not anticipate it at the beginning, he would later have to defend his conclusions against several critics.

Thaxter was a multifaceted and talented scientist. One thinks of him with no small amount of respect and wonder. Within his body of published research are contributions significant enough to justify several careers. First and foremost, he was a mycologist. His doctoral work on the Entomophthorales became the standard reference on that group for almost a

Donald H. Pfister • Harvard University Herbaria, Cambridge, Massachusetts 02138.

century. Nearly our entire body of critical information on the Laboulbeniales, a group of fungi parasitic on insects, is attributable to him. In addition, he did groundbreaking work on the Zygomycetes, aquatic fungi, and Endogonaceae. Further, Thaxter was an extraordinary plant pathologist (Weston, 1933b; Horsefall, 1979) and a completely competent entomologist (Pfister, 1984). He performed one of the first experiments in North America on biological control of insects and is probably responsible for the introduction of the fungus that now seems to be attacking the gypsy moth in North America. In an age when organisms were subjected to fewer boundaries and researchers were thus less restricted in their research, Thaxter was a bacteriologist of a high order. Not only did he put the myxobacteria on the biological map, but he also studied bacterial plant pathogens.

All of his work was characterized by critical observation and detailed study of the organisms. His research took place sometimes in the laboratory and at other times in the field. Laboratory techniques that we take so much for granted were new in the 1880s, and much of the advancement in understanding microorganisms was accomplished only by using these technologies. Pasteur had published comment on using asbestos plugs in his flasks in 1876, and commercially available autoclaves appeared only in the 1880s. Petri wrote about his plate in 1887, and the van Tieghem cell, which Thaxter used routinely in his myxobacterial studies, was introduced in 1873 (Ainsworth, 1976). Thaxter was able to blend technology with a knowledge of natural history. Perhaps as important, he had the special ability to separate the important from the trivial and spend his time wisely.

The circumstances of Roland Thaxter's life (1858 to 1932) are fairly straightforward though far from conventional. The son of Celia (Laighton) Thaxter, a poet, and Levi L. Thaxter, a literary scholar and public reader of Browning, child Roland grew up in Newton (near Cambridge), the Isles of Shoals (off the coast of New Hampshire), and Kittery Point in Maine. As a youth, he traveled, hunted, and collected natural history specimens with his father. His mother, who came to reign over a select group of artists and intellectuals at her father's hotel on Appledore Island in the Isles of Shoals, was also keenly aware of the natural world. Thaxter graduated from Harvard in 1882 and began Harvard Medical School. Fortune in the form of a fellowship and misfortune in the form of his father's illness led him to the newly established graduate school at Harvard, where he received his M.A. and Ph.D. concurrently in 1888. His thesis work was done under William Gilson Farlow, who had recently been made professor of cryptogamic botany at Harvard and who had spent 2 years studying abroad, particularly in Strasbourg (then in German control) with Anton

De Bary. Farlow's connection with the German school of science undoubtedly influenced Thaxter and, in general, fostered the discipline of experimental natural science in North America. As a young man, Thaxter was interested in insects, and in fact, his first scientific papers were on insects. His dissertation on the Entomophthorales, a group of insect parasites, united his interests in insects and fungi. After obtaining his Ph.D., he moved to the Connecticut Experiment Station in New Haven and worked as a plant pathologist. His more purely scientific work was done in his leisure time. Research on insect parasites, particularly the Laboulbeniales, was his primary interest for the rest of his life.

Farlow and Thaxter remained colleagues over the years. Farlow was particularly important to Thaxter's career, since it was Farlow who brought him back to Harvard as an assistant professor in 1891. Thaxter became full professor in 1901. Complete biographical sketches were published by Weston (1933a, 1933b) and Clinton (1936). Two portraits of Thaxter are included here (Fig. 1).

As far as I can tell, Thaxter first saw a myxobacterium in about 1886 and certainly by 1888. This first sighting is recorded in his remarkable sketchbook, which begins in 1886 and carries on for several years. The book is composed of separated sheets that at some point were mounted

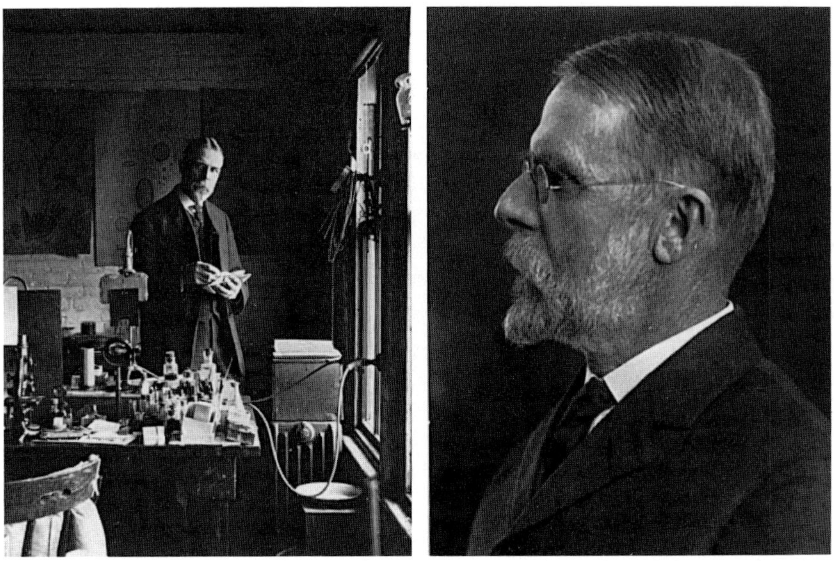

Figure 1. Portraits of Roland Thaxter in his laboratory in the 1890s (left) and in later years (right).

Figure 2. *P. aurantiacum* as illustrated in Thaxter's sketchbook, which is now in the Farlow Reference Library and Herbarium.

in a scrapbook and contains a series of colored illustrations done by Thaxter as study guides. For his myxobacterium, Thaxter used the name *Polycephalum aurantiacum* (Fig. 2) and later annotated it as *Stigmatella aurantiaca* at the recommendation of Farlow (note the "W.G.F." in the lower right-hand corner). Farlow provided the identification of what he believed was a myxomycete in a letter on 1 October 1888. In a letter to Farlow (5 October 1888), Thaxter wrote that "it is certainly a curious and disagreeable little thing." By 27 October 1888, he writes that "those spores do have a kind of oozing germination . . . and I have imagined that their contents sometimes breaks up into 'cocci'." By 1890 (letter, 13 September 1890), Thaxter wrote to Farlow that "it is certainly not a Myxomycetes and seems to be a compromise between a Schizomycetes and some acrasiaceous thing."

The locale of Thaxter's find was Kittery Point, Maine (he was to have a number of collections from New Haven as well). Kittery Point, one of the classic localities in mycology and in myxobacterial taxonomy, is a town located on a spit of land at the mouth of the Piscataqua River, which separates New Hampshire from Maine at the coast. Thaxter spent much of his time there and knew its ecology and organisms well. So many interesting collections of all kinds of organisms originated from Kittery Point that one might conclude that there was something special about it. We will never know whether that is true, since the area has been highly developed since Thaxter's day, but one suspects that the richness of the area is more likely attributable to Thaxter's facility as a collector than to the area itself. Nonetheless, Kittery Point yielded the collection of myxobacteria (as well as many other collections cited in his monographs), and it seems that this collection got Thaxter started on what would result in his paper of 1892.

Before moving further into an analysis of Thaxter's work, one should note that Thaxter was his own illustrator. Since the papers he published on myxobacteria are not readily available, some of the original plates from his first paper are reproduced here (Fig. 3 and 4). His illustrations were done with the same care he used for his observations; they were meticulous. His style is precise and characteristic in its detail and stippling. One considers his observations on the myxobacteria with amazement, since the optics he was using for cellular detail did not exceed 1,000×.

Careful examination of his first paper (Thaxter, 1892) on the myxobacteria reveals his skill and acuity in observation and his broad knowledge of organisms. The observations on life history were made in culture. On the basis of his observation that individual cells were rods similar to those of other bacteria, he concluded that these organisms were bacteria. Yet he also observed that the organisms were peculiar in their life histories, involving a phase during which they acted communally. He justified their placement in a separate family of bacteria and discussed them in relationship to the Myxomycetes, which they superficially resembled. For a group that had never been characterized, let alone accurately studied, Thaxter described in detail their locomotion, aggregation, fructification, and differences in cyst and spore formation. It was this information that allowed him to compare cogently the myxobacteria with other organisms: the fungi, the mycetozoa, and the cellular slime molds.

Given his remarkable contribution, it seems appropriate to look at the previous reports available to Thaxter. He mentioned but a few, and he notes the following genera from the literature: *Chondromyces, Stigmatella, Polycephalum,* and *Cystobacter*. Most of these were considered fungi by their describing authors; some compilers questioned the possibility that they were perhaps insect nests or eggs. Their association with fungi stems from a perception that their general morphological characters are similar to those of some of the stilbaceous fungi, i.e., those that form stalks on which conidial heads are produced. Thaxter and his contemporaries also assigned some older illustrations to the myxobacteria. I have used two of these illustrations here (Fig. 5) to provide a sense of just how sketchy information was in 1892. What was known was only the most

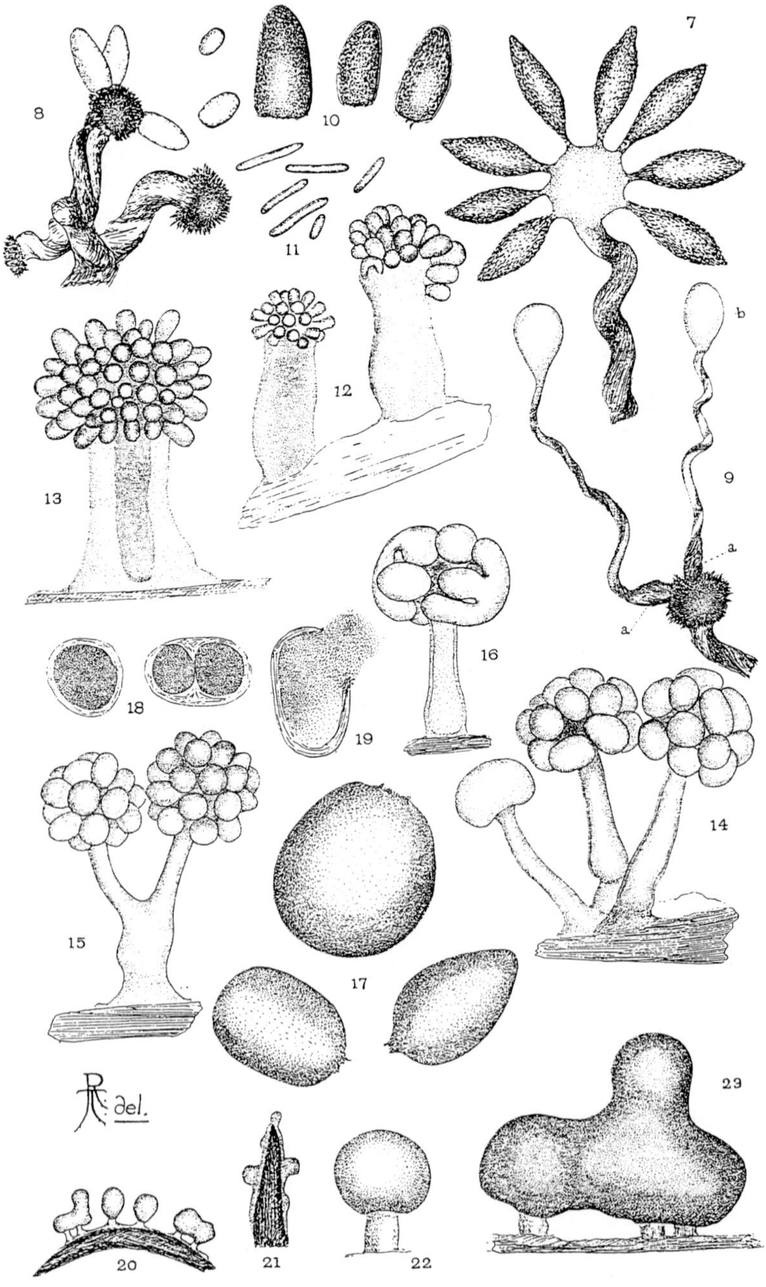
7
8
10
11
12
13
a
b
9
a
16
18
19
15
14
17
23
del.
20
21
22

general morphology. Some of the illustrations or variations on them were used over and over in the literature of the early 19th century, just as Thaxter's own illustrations appeared throughout the general literature of the early 20th century. One wonders if anyone really saw the organisms.

Thaxter's original contribution was followed by a paper in 1897. In this paper, he gave some further names, but the paper was aimed particularly at disputing the claims of Zukal (1896), who had published an account of the life cycle of myxobacteria that varied remarkably from that given in 1892 by Thaxter. By Zukal's account, the organism was a myxomycete, and although rods were described, they were said to be imbedded in a plasmodium and were additionally said to be capable of independent functions. Zukal took the position that the plasmodia had "bacteria-like energids" that were involved in their increase in size. Thaxter took a particularly dim view of this idea and of the new species Zukal had described. In 1897, Thaxter published a description of spore formation and spore germination in *Myxococcus* spp. based on his further research. The germination experiments were ingenious. Spores were placed in hanging drops in van Tieghem cells. To the bottom of the cell was added a suspension of unicellular algae, which were to supply the oxygen needed over the long period the cells would be incubated.

That the myxobacteria were not quickly accepted in the literature distressed Thaxter. In this vein, he commented (Thaxter, 1897): "The present note is therefore offered as a further contribution towards a knowledge of the species, as well as of the development of a group of organisms which, as is evidenced by its entire omission from a work like the *Pflanzen-*

Figure 3. One of the plates illustrating Thaxter's paper of 1892. The caption reads as follows: "*Chondromyces crocatus* (B. & C.) Fig. 7. Optical section of ultimate rod mass from which the rods have for the most part migrated into the immature cysts. Fig. 8. Three ultimate branches of a cystophore, one of them with three mature cysts still *in situ*. Fig. 9. Tip of an ultimate branch of a cystophore on which two cysts still *in situ* have germinated to produce secondary cystophores and cysts (a, a). Fig. 10. Five detached mature cysts showing extremes of size under ordinary conditions. Fig. 11. Vegetative rods. *Chondromyces aurantiacus* (B. & C.) Fig. 12. Young cysts budding from apex of cystophore. (Living material.) Fig. 13. A more advanced stage, a central column of ascending rods surrounded by a gelatinous layer. (Living material.) Fig. 14. Three specimens from dried material one showing terminal rod-mass from which the cysts have not yet begun to bud. Fig. 15. Specimen from dried material showing furcate habit. Fig. 16. Mature specimen from dry material in which the cysts show lateral coalescence. Fig. 17. Three mature cysts. Fig. 18. Two cysts kept on moist wood for several weeks, preparing to germinate. Fig. 19. A similar cyst germinating. *Chondromyces lichenicolus* n. sp. Fig. 20. Mature cysts on short cystophores. Fig. 21. Rod masses rising to form cysts. Figs. 22–23. Mature cysts with short cystophores, showing lobulation and coalescence."

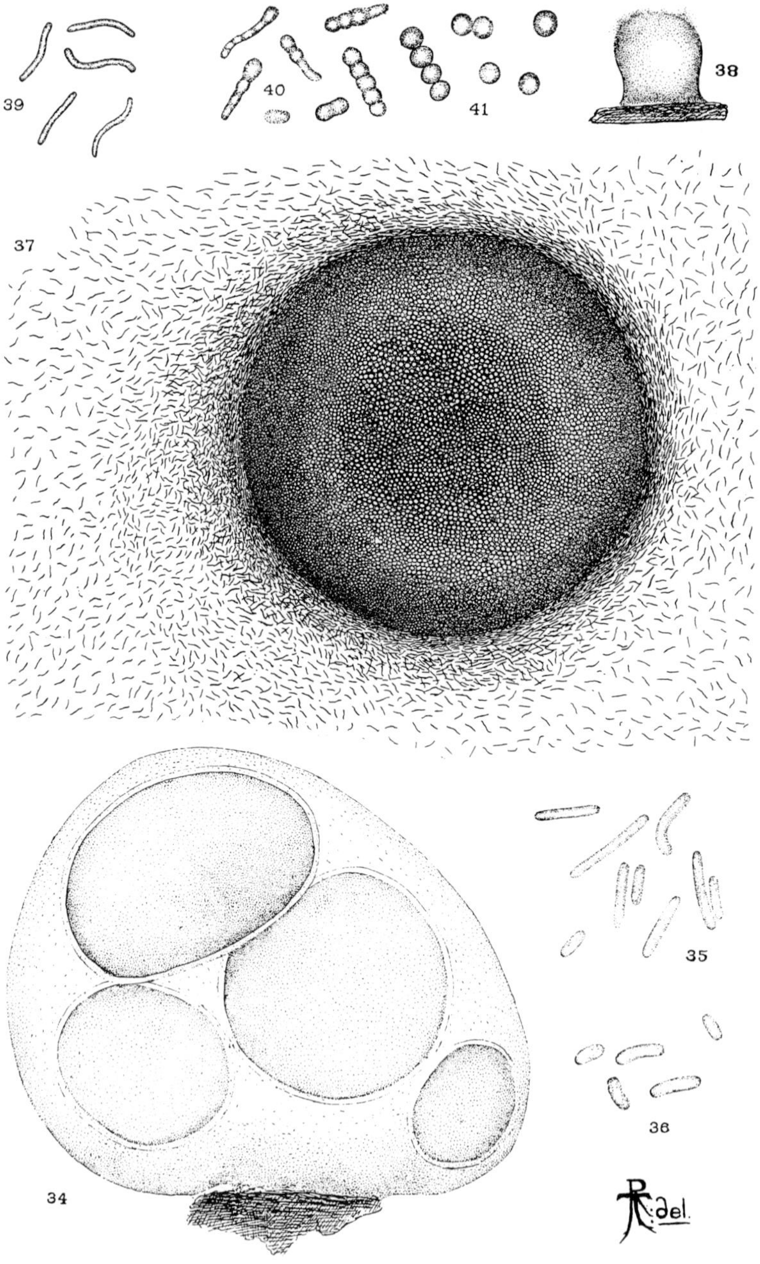
39
40
41
38
37
35
36
34

Figure 5. (A) Illustrations of *Chondromyces crocatus* and *Stigmatella aurantiaca* from M. J. Berkeley, *Introduction to Cryptogamic Botany*, 1857. (B) Illustration of *Polyangium vitellinum* from H. F. Link, *Gesell. Naturf. Freunde Mag*. **3**:3–42, 1809.

familien of Engler and Prantl, does not appear to have received the recognition it well merits. That the family, as predicted in my former paper, is likely to prove by no means numerically insignificant, is indicated by the additions enumerated below, which have for the most part made their appearance by accident, as it were, on laboratory cultures designed for other purposes; and it seems probable that any systematic attempt to enlarge our knowledge of the species could hardly fail to produce interesting results."

In 1904, 12 years after his first publication on the myxobacteria, Thaxter again wrote about the group. It had still not been included in the *Pflanzenfamilien,* and Migula's *System der Bakterien* provided but scant treatment. In this paper, Thaxter criticized the published accounts of the myxobacteria that had appeared since his 1897 paper. Zukal, who died in 1900, had finally came partly around to Thaxter's way of thinking (Zukal, 1897). A. L. Smith (1901) was probably working with a contaminant when describing *Myxococcus pyriformis,* and poor Zederbauer (1903) received the brunt of Thaxter's frustration. Zederbauer suggested that the myxo-

---

Figure 4. One of the plates illustrating Thaxter's paper of 1892. His caption was as follows: "*Myxobacter aureus* n. sp. Fig. 34. General habit showing four cysts embedded in gelatinous matrix. Fig. 35. Rods (living) from rising rod-mass. Fig. 36. Rods from cysts crushed at maturity. *Myxococcus rubescens* n. sp. Fig. 37. General appearance of young spore mass viewed from above and surrounded by vegetative rods. Fig. 38. Normal habit of spore mass viewed laterally. Deliquescence beginning at the top. Fig. 39. Vegetative rods. Fig. 40. Different stages of supposed spore formation. Fig. 41. Mature spores."

bacteria were symbiotic aggregations of ordinary bacteria and hyphomycetous molds and called them "Spaltpilzflechten." Of Zederbauer's contribution, Thaxter (1904) wrote: "This treatment of the group, though novel, seems somewhat hasty; especially in view of the fact that the figures and descriptions given in this paper show very clearly that its author is as yet unacquainted with any member of the order he discusses, having been misled by fancied resemblances and influenced no doubt by an exaggerated notion of the difficulty associated with the differentiation of rodlike bacteria from *Oedocephalum, Torula,* and similar hyphomycetous types." As stated in a footnote in the 1904 paper, Thaxter received material from Zederbauer that he determined as fungal: the conidial state of *Coryne sarcoides,* now known as *Ascocoryne sarcoides.* A look at the Thaxter archives shows that Zederbauer and Thaxter did fall into correspondence during this period. There are six letters, polite but pointed, and all revolving around Zederbauer's specimens and identifications. We have only the Zederbauer side of things. Zederbauer was determined to be right and published a later paper (1906) once more illustrating the myxobacteria he was studying. Again, these illustrations were of some fungi and some bacteria. In this paper of 1904, which would be his last published account of the members of the family, Thaxter affirmed that "further careful study of these organisms is much needed, especially in connection with the movements of the rods and with their detailed cell characters."

There is a hint of melancholy in Thaxter's last words on the group he had so much to do with characterizing. One suspects that he was pleased with his own contributions but saddened by those who tried to follow him. He was disappointed that the myxobacteria did not receive the full attention he thought they deserved. Certainly during this period many projects, not the least of which was his monograph on the Laboulbeniales, were competing for his time. One suspects that with little new information to add, he was satisfied to let his work speak for itself. Certainly, he created a climate in which critical cultural and behavioral work could be done: witness several works which soon followed (Vahle, 1909; Quehl, 1906; Baur, 1904). By the close of the first decade of the 20th century, the myxobacteria were showing up in general texts, often illustrated with redrawings of Thaxter's own illustrations. He seemed to be relinquishing the group when he wrote in 1904: "There are certainly few groups including members which, as in the present instance, have been relegated to several genera of the Hyphomycetes, to the Gasteromycetes, to Eubacteria, to insect eggs, to myxobacteria, to Myxomycetes, and finally to a group of Splatpilzflechten; and one can but look with no small interest, and perhaps with some misgivings, to such further taxonomic vicissitudes as may be in store for them."

Acknowledgments. I thank the staff of the Harvard University Herbaria Library, particularly Jean Cargill, for assisting in the location of literature and documents used in this study. All letters cited, illustrations, and photographs arc located in the archives of the Farlow Reference Library of Harvard University. Thaxter's specimens of myxobacteria are housed in the Farlow Herbarium, a part of the Harvard University Herbarbia.

REFERENCES

Ainsworth, G. C. 1976. *Introduction to the History of Mycology*. Cambridge University Press, Cambridge.

Baur, E. 1904. Myxobakterienstudien. *Protistenkunde* **5:**92–121.

Clinton, G. P. 1936. Roland Thaxter. *Natl. Acad. Biographical Mem.* **17:**55–68.

Horsefall, J. G. 1979. Roland Thaxter. *Annu. Rev. Phytopathol.* **19:**29–35.

Pfister, D. H. 1984. Roland Thaxter: mycologist and entomologist, p. 211–221. *In* Q. Wheeler and M. Blackwell (ed.), *Fungus-Insect Relationships: Perspectives in Ecology and Evolution*. Columbia University Press, New York.

Quehl, A. 1906. Untersuchungen über die Myxobakterien. *Zentralbl. Bakteriol. Parasitenkd. Abt. 2* **16:**9–34.

Smith, A. L. 1901. Myxobacteria. *J. Bot.* **39:**69–72.

Thaxter, R. 1892. Contributions from the Cryptogamic Laboratory of Harvard University. XVIII. On the Myxobacteriaceae, a new order of Schizomycetes. *Bot. Gaz.* **14:**389–406.

Thaxter, R. 1897. Contributions from the Cryptogamic Laboratory of Harvard University. XXXIX. Further observations on the Myxobacteriaceae. *Bot. Gaz.* **23:**395–411.

Thaxter, R. 1904. Contribution from the Cryptogamic Laboratory of Harvard University. LVI. Notes on the Myxobacteriaceae. *Bot. Gaz.* **26:**405–416.

Vahle, C. 1909. Vergleichende Untersuchungen über die Myxobakteriaceen und Bakteriaceen, sowie die Rhodobakteriaceen und Spirillaceen. *Zentralbl. Bakteriol. Parasitenkd. Abt. 2* **25:**178–260.

Weston, W. H. 1933a. Roland Thaxter. *Mycologia* **25:**69–89.

Weston, W. H. 1933b. Roland Thaxter (1858–1932). His influence on plant pathology. *Phytopathology* **23:**565–571.

Zederbauer, E. 1903. Myxobacteriaceae, eine Symbiose zwischen Pilzen und Bakterien. *Sitzungsber. Kaiserl. Akad. Wiss. Math.-Naturwiss. Kl. Abt. 1* **112:**447–482.

Zederbauer, E. 1906. Spaltpilzflechten. *Oesterr. Bot. Z.* **56:**213–218.

Zukal, H. 1896. *Myxobotrys variabilis* als Repräsentant einer neuen Myxomyceten-Gattung. *Ber. Dtsch. Bot. Ges.* **14:**340–347.

Zukal, H. 1897. Notiz zu meiner Mittheilung über *Myxobotrys variabilis*. *Ber. Dtsch. Bot. Ges.* **15:**17–18.

Myxobacteria II
Edited by Martin Dworkin and Dale Kaiser
© 1993 American Society for Microbiology, Washington, DC 20005

Chapter 2

Biology of the Myxobacteria: Ecology and Taxonomy

Hans Reichenbach

The myxobacteria are relatively large (0.7 to 1.2 by 3 to 12 μm), rod-shaped, gram-negative bacteria. They have several unusual features and are unique among prokaryotes by virtue of their sophisticated intercellular communication systems and morphogenetic potential.

The myxobacteria move by gliding, or creeping, along surfaces and by worming their way through porous substrates like agar. Their colonies therefore look quite different from those of ordinary bacteria; they spread film-like over the substrate and may cover the whole plate within a couple of days. These swarm colonies, or swarms, are often structured in a characteristic way, showing straight or snaking radial veins, rings, concentric ridges, and fields of tiny waves resembling the well-known ripples at the sea shore. Myxobacterial swarms usually do not move as a whole and should be clearly distinguished from pseudoplasmodia. Gliding motility is possible even in relatively dry environments and allows the myxobacteria to move in the microhabitat toward macromolecular feed materials or away from unfavorable sites. The ability to move by gliding is shared by a number of phylogenetically unrelated bacteria and thus is no argument for a relationship, at least not over large phylogenetic distances. Recently it was even found that certain flagellated bacteria may switch to gliding on surfaces (Shea et al., 1991).

The most amazing aspect of the myxobacteria is their morphogenetic potential. Under starvation conditions, the bacteria aggregate and pile up into masses consisting finally of about 10^4 to 10^6 cells. These masses may

Hans Reichenbach • Microbial Secondary Metabolites, Gesellschaft für Biotechnologische Forschung, D-38124 Braunschweig, Germany.

then differentiate into fruiting bodies, objects of a species-specific shape and structure that range in size between 10 and 1,000 μm and thus are clearly recognizable with the naked eye. They are often produced in large numbers over much of the swarm area. A fruiting body may be composed simply of cells and slime, but many myxobacteria produce sporangioles by surrounding the cell mass with a tough wall. The sporangioles may occur singly or in clusters; they may sit directly on or within the substrate or be borne on stalks. Often they are brightly colored. Fruiting bodies arise in a process of cooperative morphogenesis, which requires gliding motility as a prerequisite. Within the maturing fruiting body, vegetative cells convert into myxospores, often, but not always, by shortening and fattening substantially, i.e., by cellular morphogenesis. The myxospores are desiccation resistant and are also less sensitive than vegetative cells to a variety of other environmental stress factors, like elevated temperatures or UV light irradiation. While the myxospores obviously are responsible for the organism's survival under unfavorable environmental conditions, the function of the fruiting bodies may be to guarantee that a new cycle is started by a population rather than by individual cells. A cell population could make a more efficient use of exoenzymes for the degradation of biomacromolecules.

The latter idea is suggested by the physiology of the myxobacteria. It appears that all of them are specialized in the degradation of biomacromolecules. Most species are able to lyse whole cells of other microorganisms, in particular a wide variety of bacteria and yeasts. They can grow exclusively on the lysis products and may be cultivated on media containing, e.g., autoclaved baker's yeast or *Escherichia coli*. (They also grow on much simpler substrates, e.g., peptone or peptone-sugar media. More about isolation and cultivation methods can be found in a recent review by Reichenbach and Dworkin [1992].) Species in one group, classified in the genus *Sorangium*, decompose crystalline cellulose, e.g., in the form of filter paper. They require only the usual mineral salts in addition to cellulose or glucose for growth. They grow better if some peptone is also supplied but usually cannot be cultivated without carbohydrate.

Almost all myxobacteria can be cultivated without difficulty in pure culture on agar plates as well as in liquid media. Many species have even been grown on a large scale (5 to 50 m^3) in bioreactors.

DISTRIBUTION IN NATURE

Myxobacteria are very common in nature. The primary habitat of most species appears to be the topsoil. Myxobacteria have been demonstrated in many different types of soil, mainly in the pH range from 5 to

8, but they have also been found in soils with extreme pHs, as will be discussed below. Myxobacteria seem to prefer soils that are rich in organic material and microbial life in general (compost, garden soil, and soils from tilled fields and the forest floor). They also occur, however, in pure sand and on rocky surfaces (Krzemieniewska and Krzemieniewski, 1926, 1927, 1930; Singh, 1947; Norén, 1950, 1952; Sabadoš-Šarić, 1957; Kühlwein, 1960; Brockman and Boyd, 1963; Peterson, 1965; Peterson and Norén, 1967; Michoustine, 1968; McCurdy, 1969; Singh and Singh, 1971; Larkin and Dunigan, 1973; Rückert, 1975a, 1975b, 1976, 1978b, 1979, 1983; Brockman, 1976; Hook, 1977; Rückert and Heym, 1977; Dawid, 1978, 1979).

Another typical habitat of myxobacteria is decaying plant material, in particular rotting wood, bark, and lichens, and the dung of herbivorous animals, a classic source of myxobacteria. The myxobacteria probably colonize such substrates from the surrounding soil, for in most cases, the same species can be demonstrated in both habitats. It appears that myxobacteria are able to pass through the digestive tract of animals, as has been shown experimentally with sporangioles of *Chondromyces apiculatus* (Kühlwein, 1950). Also, myxobacteria have been found in the intestinal tracts of freshwater fish (carp; Till, 1982). Passage through animals may also be inferred from the fact that dung pellets dropped on deep layers of snow may yield myxobacteria upon incubation (Rückert, 1975c; Reichenbach, unpublished data). The organisms are probably taken up by the animal with food; since myxobacteria are strictly aerobic, it is difficult to see how they could live within the animal body, but perhaps this has never been experimentally determined. Dung may attract myxobacteria because it contains lots of debris and because the plant material in it is already more or less digested and broken down, which might be helpful for the cellulose degraders. It also harbors a rich and diversified microbial community that might supply the myxobacteria with food organisms and probably also with vitamins, e.g., B_{12}, which is required by quite a few strains of myxobacteria (e.g., *Corallococcus macrosporus, Polyangium* spp., and *Nannocystis* spp.; Reichenbach, unpublished data). Decaying wood and bark regularly host not only cellulose degraders but also species that cannot attack cellulose, like *Chondromyces apiculatus* and *Stigmatella aurantiaca*. For unknown reasons, the latter species seems to be essentially restricted to that habitat. Also, *Haploangium* sp. has been found only on wood and bark. As *Haploangium* sp. has never been cultivated, it is not known whether it is a cellulose degrader. Of course, substrates other than cellulose, e.g., xylans, are equally available in decaying wood, and it is noteworthy that most myxobacteria vigorously decompose xylans (Reichenbach, unpublished data).

It has been reported that in North America, the bark of living trees is a rich and interesting source for myxobacteria and that besides the usual species known from soil, specialized forms not found anywhere else may be isolated from tree bark (Peterson, 1957, 1959b, 1969a, 1969b). Another study, again of North American bark samples, found that indeed many myxobacteria can be obtained from that source and that species normally not as easily isolated from other sources show up in tree bark (McCurdy, 1969). Thus, conditions were clearly selective for *Melittangium* and *Stigmatella* species, but no unusual organisms were seen. The situation in Europe under comparable climatic conditions has been investigated more thoroughly (Dawid, 1979). Only half as many species of myxobacteria could be isolated from bark as from soil and rotting wood from the same area, and species were essentially the same in the three habitats. Only *Melittangium lichenicola* appeared more often on bark. Further, it became quite clear that not all species of trees are equally suitable. While bark from *Sambucus racemosa* (elder), *Fagus silvatica* (beech), and *Robinia pseudacacia* (black locust) were relatively favorable habitats, the barks of 15 other tree species only very rarely yielded myxobacteria. My own experience with North American and European bark samples is the same: in comparison with North American samples, European barks are much poorer in myxobacteria. Also, in a totally different environment, i.e., the subtropical north of India, bark samples were rich in myxobacteria; there, too, a few species were observed only on bark (Singh and Singh, 1971).

Myxobacteria are also well represented in fresh water (Geitler, 1925; Jeffers, 1964; Brauss et al., 1967, 1968; Gräf, 1975; Bernátová, 1977; Hook, 1977; Vychodilová, 1977; Tržilová et al., 1980, 1981; Kohl, 1982). This is not surprising, since it is quite common for soil organisms to be washed into bodies of fresh water and to survive there. While the survival stages of myxobacteria seem to suggest that the organisms are adapted to environments with periodic dry spells, nothing really speaks against the assumption that they may also live permanently in freshwater habitats. A *Polyangium* strain found in a pond near Vienna was reported to be parasitizing the freshwater alga *Cladophora* and forming fruiting bodies within emptied cells, obviously stimulated by nutrient depletion, thus demonstrating that fruiting body formation is possible and may even be meaningful under submerged conditions (Geitler, 1925). It would be interesting to know more about why myxobacteria show up in fresh water and how they exist there, especially since it has been suggested that myxobacteria may be useful as indicators of special kinds of water pollution (Gräf, 1975; Bernátová, 1977; Lecianová, 1977, 1978; Vychodilová, 1977; Tržilova et al., 1980, 1981). As mentioned above, myxobacteria have been demonstrated in the intestinal tracts of carp. As they are found there in low numbers, they may simply be passing through the animal (Till, 1982). Also, structures regularly observed on eggs of *Dytiscus marginalis* (water beetle) in England have been interpreted as myxobacterial fruiting bodies (Jackson, 1959).

The occurrence of myxobacteria deep in a cave in the Alps (Menne and Rückert, 1988) seems rather exotic. The myxobacteria may have entered with wind and water and perhaps subsist on organic material that has been transported there in the same way. Another unexpected habitat is the surfaces of plant leaves in central Europe (Rückert, 1981). Leaves between 0.4 and 2 m above the ground were sampled, and it appears that the incidence of myxobacteria decreased with height. It is not known whether the organisms actually grow there. Finally, myxobacteria may also be blown into the air with dust particles. They occasionally turn up as air contaminants on culture plates (Wu et al., 1968), although this is extremely rare in the microbiological laboratory.

Myxobacteria are found in all climate zones, vegetation belts, and altitudes but, of course, not everywhere with the same frequency. Myxobacterial floras that are rich in number of species and population densities are known to occur in suitable habitats from the tropics to the temperate zone (Thaxter, 1892, 1893, 1897, 1904; Quehl, 1906; Jahn, 1911, 1924; Kofler, 1913; Krzemieniewska and Krzemieniewski, 1926, 1927, 1930; Singh, 1947; Norén, 1950, 1952; Katô, 1955; Singh et al., 1958; Kühlwein, 1960; McCurdy, 1969; Peterson, 1969b; Rückert, 1972, 1975a, 1978c, 1985b; Dawid, 1978, 1979). As expected (Peterson, 1969b), the organisms also are well represented in the Southern Hemisphere. Their occurrence in South America and Southwest Africa (Namibia) has been reported (Dawid, 1978; Rückert, 1978b, 1978c); I have isolated many myxobacteria from samples from South Africa, Australia, New Zealand, and the Pacific Islands. Cold environments in arctic and subarctic areas and at high altitudes are clearly less favorable for myxobacteria. The number of species declines sharply, and more samples tend to be negative, but the organisms are still there. Soil samples from the northern coasts of Alaska and Canada yielded *Myxococcus fulvus* and *Sorangium* spp., and 4 of 17 samples were positive (Brockman and Boyd, 1963). The organisms were isolated at 24 to 26°C, and no myxobacteria were seen when the cultures were incubated at 6 to 8°C. *M. fulvus* and *Corallococcus coralloides* were found in samples from subarctic areas in northern Norway and Iceland (Rückert, 1973, 1976). In that case, the organisms were isolated at 30°C. Studies of cellulose-degrading bacteria on the Kola Peninsula in northern Russia showed *Sorangium cellulosum* to be fairly common in virgin and cultivated podzol and peat marsh soils. Isolated strains had temperature optima around 24 to 28°C. It appears, however, that the organism is not very competitive in that environment, because fertilization and neutralization of the soil with $CaCO_3$ or KOH increase its numbers. It was suggested that a low pH of the soils and antagonistic soil microorganisms control its occurrence (Zhukova, 1959a, 1959b, 1962). *S. cellulosum* was also demonstrated in Swedish and Finnish soils but virtually only in cultivated ones (Peterson and Norén, 1967). Further, myxobacteria have been obtained from soil samples from Antarctica. In one case, two of five samples were positive *(M. fulvus, Myxococcus virescens);* the strains were isolated at 30°C and did not grow at low temperatures (Rückert, 1985b). In the other case, however, true psychrophiles that could not grow at 18°C were isolated, and no mesophiles at all were observed (Dawid et al., 1988).

It seems reasonable that myxobacteria would be found in semiarid areas, since they seem well adapted to survive dry conditions (Sabadoš-Šarić, 1957; Peterson, 1965; Singh and Singh, 1971; Larkin and Dunigan, 1973; Brockman, 1976; Rückert, 1976, 1978b). However, their presence in veritable deserts with very little vegetation is surprising (Rückert and Heym, 1977). Semiarid areas and regions with seasonal rainfall often are particularly rich in myxobacteria. Another factor favoring the occurrence of myxobacteria appears to be limestone underground, although clear differences between limestone and other areas are not always seen (Norén, 1952). However, while the spectrum of myxobacteria at high altitudes in mountains is usually reduced (e.g., Katô, 1955; Kühlwein, 1960; Rückert, 1983), my experience is that many myxobacteria can be found in the northern limestone Alps in Austria up to at least 2,400 m.

The exact number of myxobacteria in any particular habitat is very difficult to estimate, because the organisms tend to stick to particles and to each other, and suspended single cells may not readily form colonies. Nevertheless, efforts toward educated guesses for the population densities of myxobacteria in various soils have been made. In English soils, between 2,000 and 75,000 bacteriolytic myxobacteria were found per g of soil, and in decomposing compost, more than 500,000 *M. fulvus* alone per g in addition to smaller numbers of *M. virescens* and *Corallococcus exiguus* were found (Singh, 1947). Between 1,500 and 80,000 bacteriolytic myxobacteria were counted per g of various types of soils in Yugloslavia, with the highest numbers in chernozem and calcareous alluvial soils and the lowest in podzol and solonetz (Sabadoš-Šarić, 1957). In Canadian soils, the *Myxococcus-Corallococcus* population was estimated to vary between 1,000 and 450,000/g (McCurdy, 1969). Up to 10×10^6 cells of *S. cellulosum* have been reported for cultivated marsh soils on the Kola Peninsula (Zhukova, 1959a). This number seems high, especially since *S. cellulosum* does not plate very well.

Even if those data are not absolutely accurate, one can still conclude that substantial numbers of myxobacteria are present in their habitats and that they consequently must have profound effects on their environment, notably on the microbial populations there. It has been well known for many years that myxobacteria attack and lyse other microorganisms, especially many kinds of bacteria (e.g., Beebe, 1941; Singh, 1947; Oetker, 1953; Kühlwein, 1955). It was speculated that myxobacteria exert a harmful or beneficial effect by controlling the environmental microbial flora. This could be done not only with excreted enzymes but also with antibiotics, of which they produce a large variety (see chapter 16). There is some experimental evidence that myxobacteria could indeed play such a role in nature. For example, *M. xanthus* inoculated into soil dramatically reduces the number of *Azotobacter* cells (Callao et al., 1966). The experiments were done with very high cell densities (approximately 10^8 cells per g of soil), which are probably unrealistic in nature, but lower population densities still may have an influence. In another study, a sequence of bacterial predations was demonstrated when soil was enriched with *Micrococcus luteus,* and *M. xanthus* was at the end of the food chain (Casida, 1980). Also, it was shown that the *Myxococcus* spp. multiplied in the soil when the *Micrococcus* sp. was present but would enter the myxospore stage when the soil was not enriched (Liu and Casida, 1983). Germination of myxospores in soil could be observed only if the feed bacterium was available (and if the soil was at a neutral pH). Myxobacteria also are strongly lytic for various cyanobacteria (Wu et al., 1968; Burnham et al., 1984). *Myxococcus* strains isolated from ditch samples were able to lyse *Phormidium luridum* very efficiently in mixed cultures. The myxobacteria formed floating spherules and, on supports, slime covers that entrapped the cyanobacteria (Burnham et al., 1981, 1984; Daft et al., 1985). Regular predator-prey cycles with a periodicity of 9 days (25°C) developed in such systems, and the myxobacterial populations still could recover when added to *Phormidium* cultures (with 10^7 cells per ml) at a ratio of 50 cells per 100 ml (Burnham et al., 1984). *Myxococcus* cells could reduce the *Phormidium* population in cultures to which the cyanobacterium was continuously added with fresh medium. Lysozyme activity was shown in the culture supernatant (Daft et al., 1985). It was concluded that myxobacteria could perhaps be used to control cyanobacteria in freshwater habitats.

The occurrence of myxobacteria obviously depends on the nutrient supply but also depends on certain physical and physicochemical factors. Myxobacteria are sometimes found under strange environmental conditions that would never allow growth in culture. This could be explained by at least three different assumptions. Their resting stages, the myxospores, may enter such places from somewhere else and are then able to survive for many years under hostile conditions; the organisms may grow in nature under circumstances totally different from ordinary culture conditions, perhaps with very long generation times; or the organisms may colonize microhabitats whose physicochemical parameters may deviate substantially from the statistical values that are usually used to characterize a habitat. Soil is in fact a very heterogeneous substrate with many microhabitats, ranging from clay particles and humic acid flocs to rotting leaves, earthworm droppings, and beetle shells.

A few examples may be sufficient to illustrate this point. All myxobacteria are strictly aerobic, so one obviously would not expect to find them in anaerobic environments. Also, in culture they usually grow only in the pH range between 6.5 and 8.0. Yet myxobacteria were isolated from water-logged samples of peat collected in bogs at a depth of 2 m, a pH of 3.7, and an oxygen content of 2.6 mg/liter, which is about 13% of normal (Dawid, 1984). However, the spectrum of observed species was very restricted, i.e., *M. fulvus, M. xanthus,* and *Polyangium* species. This could indicate that the organisms really live there, because if they had only survived there as myxospores, one would expect other myxobacteria to be there also. On the other hand, the organisms were isolated and cultivated at pH 7.2 and in a fully aerobic atmosphere, and no effort was made to grow them under conditions close to those of the habitat. In another case, myxobacteria were obtained from a slightly alkaline bog lake in Michigan (Hook, 1977). At one typical sample site, the pH was 8.1 to 8.2 and the dissolved oxygen was 0.6 to 1.3 mg/liter, yet 22% of all samples were positive, almost exclusively with *M. fulvus* and *Myxococcus stipitatus*. Again, the strains were isolated and grew under normal conditions, and it is not known whether they would also have grown at an alkaline pH. Interestingly, it was found in the same study that at the bottom of the fossa, or ditch, that surrounds the bog (pH 7.9 to 8.1, 7.0 to 7.8 mg of oxygen per liter), *Angiococcus disciformis* was rather common (22% of samples), whereas at other places, the species was rare. This is one of the few known examples of a species-specific habitat. In still another instance, truly alkaliphilic myxobacteria have been discovered. The strains were isolated from soil samples with a pH of 9.4. They grew and formed fruiting bodies only when the pH of the medium was adjusted correspondingly (Dawid, personal communication). Also, while no myxobacterium that can be cultivated at a pH below 6.0 is known, there are numerous reports of isolations from acid soils down to at least pH 2.5 (e.g., Dawid, 1979; Rückert, 1978a, 1979). The spectrum of species clearly declines when the pH drops below 5.0, with *Myxococcus* and *Corallococcus* species becoming more and more dominant. Reasoning as described above, one again concludes that the organisms live in the habitat. On the other hand, myxobacteria could be isolated from extremely acid soils (pH

3.6 to 2.5) (by the dung pellet baiting technique) only after 0.8% $NaHCO_3$ had been added to the soil (Rückert, 1978). Does this mitigation of pH extremities simply speed up growth, or are myxobacteria unable to grow at those low pHs at all? The pH effects may be modulated by other factors. A low pH appears to be better tolerated when temperature and humidity are close to optimal (Rückert, 1979). A beneficial effect of pH may also be the reason why especially rich myxobacterial floras are often found on underground limestone (the washed-out loams arising from the weathering of limestone, however, can be rather acid). In this connection, I could not demonstrate any myxobacteria at all from the acid raw-humus soils of West Virginia *Tsuga-Rhododendron* forests, in spite of the presence of abundant organic material and good temperature and humidity. This observation also casts some doubt on the hypothesis that imported myxospores would remain in the odd habitat awaiting better times. Growth of ordinary myxobacteria at rather extreme pHs has been reported (e.g., between pH 4.0 and 8.8 [Singh, 1947] and between pH 3.3 and 9.2 [Zhukova, 1962]), although I have doubts about those results. I have never seen a myxobacterium grow in cultures at a pH below 6 or above 8. Also, measurements of intracellular pH suggest that pH stasis breaks down if the external pH rises above 7.9 (Gerth, personal communication). The discrepancy may be due to difficulties with pH measurements in earlier days. Myxobacteria may also shift the pH to more favorable values by their metabolism and then start to grow.

Another important environmental factor is the concentration of inorganic ions. In general, the salt tolerance of myxobacteria is low. It has been reported that fruiting of *M. virescens* and *Archangium gephyra* is stimulated (i.e., more strains are obtained from samples) when 0.5% NaCl is added to the isolation medium or when the dung pellets used for baiting are soaked with a 1% NaCl solution (Rückert, 1978a). My own experience is that on streaks of *E. coli* on water agar containing 1% NaCl, only *M. virescens* and an occasional strain of *Nannocystis exedens* are obtained. When the salt level was reduced to 0.8% NaCl, *M. fulvus, Corallococcus* spp., and more *Nannocystis* strains also developed (unpublished observations). In these experiments, *M. virescens* sometimes was seen only on the NaCl plates and not on normal water agar. When samples from Cyprus from a sun-baked hill consisting of pure crystalline gypsum were analyzed for myxobacteria on agar with or without 1% $CaSO_4$ and at 30 and 38°C, only slight and hardly relevant differences in the spectrum of species and number of strains were observed on plates with and without gypsum. The higher incubation temperature produced greater differences, suppressing certain taxa, e.g., *Corallococcus,* completely while favoring others, such as *Archangium* and *Polyangium*. However, this effect was graded, and strains isolated at 38°C later grew as well or better at 30°C (unpublished observations). No myxobacterium that grows in culture at salt concentrations corresponding to those in the ocean has been found, yet myxobacteria could be obtained from beach sand at the high-water mark (Brockman, 1967; Rückert, 1975b), salt marshes (Rückert, 1983), marine mud flats (Rückert, 1975b), and even submarine sediments at a depth of 40 m near Helgoland, an island in the North Sea far from the German coast (Rückert, 1984). The organisms were, however, always isolated on the usual media with a low salt content, and samples with a high salt concentration gave positive results only after they had been desalted (Rückert, 1983).

Temperature seems not to play a major role in the distribution of myxobacteria. Most strains are typical mesophiles. In culture, growth can regularly be obtained between 6 and 35°C. With virtually all species, strains that still grow at 38 and even 40°C and slightly beyond, and sometimes vigorously so, can be found (Kühlwein and Reichenbach, 1964; McCurdy, 1969; unpublished results). The only myxobacterium whose cardinal growth characteristics have been carefully examined as a function of temperature is *M. xanthus* (Janssen et al., 1977). Its maximum growth rate occurred between 34 and 36°C. At 40°C, the cells were damaged and no growth took place; the absolute minimum for exponential growth was not determined, but at the lowest temperature examined (14°C), growth was still exponential but substantially reduced. Still, temperature may yet be a controlling factor through more-subtle effects. Thus, in Swedish mountains, a clear difference in the frequency of distribution of myxobacteria on southern and northern slopes was observed, with the south side being the preferred habitat (Norén, 1952). It was speculated that the difference in the summer temperatures between the slopes might be responsible for this discrepancy. When I isolated myxobacteria from forest soils in Germany, I found *Corallococcus* strains that grew reasonably well at 30°C but would form fruiting bodies only around 20°C (unfortunately, this is not a universally applicable trick to induce nonfruiting strains to fruit). No thermophilic myxobacteria have been discovered so far, but since myxobacteria can readily be obtained from soils from subtropical deserts and semideserts where no vegetation mitigates the intense environment, at least the resting stages of the organisms must be able to resist rather high temperatures (Peterson, 1965; Larkin and Dunigan, 1973; Brockman, 1976; Rückert, 1976, 1977, 1978b, 1983). The resistance of myxospores of *M. xanthus* to a variety of environmental extremes has been examined (Sudo and Dworkin, 1969). While vegetative cells were killed rapidly and exponentially at 45°C, myxospores were completely stable at 50°C and almost completely stable at 60°C. True psychrophiles have been isolated, however, from samples from Antarctica (Dawid et al., 1988). In crude cultures, those strains showed up only after 8 weeks of incubation at 4°C, and none of the isolated strains grew at 18°C. This is, incidentally, a fine example of the fact that unusual myxobacteria may indeed be found if suitable isolation conditions are applied and sufficient time is invested.

The assumptions proposed initially as explanations for the occurrence of myxobacteria in odd environments are, of course, more or less accessible to experimentation. We know that myxospores and fruiting bodies can indeed be dried and will stay alive for a long time when stored at room temperature under normal atmospheric conditions. I use dried fruiting bodies on filter paper for the preservation of strains and have found that under these circumstances, the organisms began to die after 10 to 15 years of storage. The longest survival time I have determined is 25 years (for strains of *A. disciformis*). It is likely that under different actual conditions in nature, e.g., in soil, the survival time is much longer. I regularly isolate myxobacteria from air-dried soil samples that have been stored at room temperature for 10 to 15 years. In one instance, I was able to obtain the same species after 12 years of storage that I had obtained at the beginning. Similar experiences have been reported in the literature (Rückert, 1983). The second assumption, that organisms grow in nature under conditions very different from culture conditions, is more difficult to prove, as we may first have to develop special cultivation methods that allow incubation periods of weeks and months under various temperatures, aeration conditions, humidity levels, and pHs. In principle, however, this is possible, as the example of the psychrophilic myxobacteria demonstrates. The most difficult assumption to prove is the third one, that myxobacteria colonize microhabitats very different from the overall "typical" habitat, because a very large number of different microhabitats must be expected. It would be necessary to carefully dissect the habitat and characterize individual microhabitats, which would obviously also include complex interactions with micro- and macroorganisms. A first step may be to find out exactly where myxobacteria are located, e.g., in soil. A good approach to this would be the use of fluorescence-labeled antibodies. If antibodies specific either for vegetative cells or for myxospores could be prepared, one could even decide whether the myxobacteria are simply surviving as myxospores in a specific habitat or whether they are actively growing. In any case, substantially more efforts have to be made before we really understand myxobacterial ecology.

What, then, really controls the distribution of myxobacteria in nature? It must be complex interplay between climatic, edaphic, and biological factors. This interplay can best be seen by considering the distributions of certain species that appear in very special niches in one region but not in other areas, although conditions in both niches may seem very similar at first glance. *Chondromyces crocatus* is reported to show up reliably in crude cultures started with pieces of fresh plant roots in Assam, i.e., to be a regular inhabitant of the rhizosphere (Agnihothrudu et al., 1959), yet I have never been able to isolate the organisms from that source in Europe (where the species appears to be very rare anyway). The species was isolated by other investigators from soil, decaying plant material, dung, and bark, all collected mostly in warm climates. I have obtained *C. crocatus* from the same materials collected in India, Malaysia, Brazil, and Madeira. Thus, in this case, one would conclude that *C. crocatus* favors a warm environment and habitats rich in organic material, and the occurrence in the rhizosphere is just one of many possibilities. In the central United States, *C. apiculatus* and *S. aurantiaca* can be obtained reliably from bark and rotting wood (Nellis and Garner, 1964). My own experience is that in Minnesota, at least one of three such samples produces fruiting bodies of these organisms upon incubation and sometimes in large numbers, yet in Europe, I have never been successful with the same technique despite many attempts. These organisms have only rarely been observed in Europe, although climate and soils are not so different there from those of the midlands of the United States. Thus, the controlling factor may be biological, perhaps competition by other microorganisms or perhaps tree species unsuitable as substrates. While *C. apiculatus* has also been isolated from other sources, e.g., dung, *S. aurantiaca* appears to occur only on bark and rotting wood, which is another ecological puzzle, because the organism cannot attack cellulose. The bacterium efficiently degrades chitin and xylan, however, and utilizes sugars like glucose; it may therefore rely on some other components of wood and its microbial population. *Stigmatella erecta* is very similar to *S. aurantiaca* but appears to inhabit a different niche. It is found mainly in soil and on dung and is also fairly common in Europe. That microbial competition can indeed control myxobacterial populations was demonstrated in a study of cellulose-degrading bacteria in Kola Peninsula soils (Zhukova, 1960). The chief competitors were fungi and sporeformers. The antagonism was especially severe in poor soils, because edaphic factors, particularly soil pH and organic load, obviously influence competitiveness of the cellulose decomposers. Interestingly, *S. cellulosum* was the least affected of all the cellulose-degrading bacteria, perhaps because it is less specialized and usually can also decompose other polymers, like chitin, starch, xylan, and protein. The importance of the biological factor can also be seen when the occurrence of myxobacteria on dung is studied. Here, competition by other microorganisms, the food sources and specific living conditions of the animal, peculiarities of the digestive system, the season of the year, and the degree of resistance of the various myxobacteria to pH shifts and attack by enzymes in the digestive tract may influence primary and secondary colonization of the dung. It was observed long ago that while dung of wild rabbits is an excellent source of myxobacteria, that of caged animals is not. While the dung of many herbivores is favored by myxobacteria, that of horses and cows is not. While certain myxobacteria are more likely to turn up in dung cultures than others, there seem to be only a few species that have never been observed on dung. However, the incidence may vary substantially. *Melittangium boletus* is reported to occur relatively often on rabbit and deer dung from the environs of Berlin (Jahn, 1924), but I have never found that species on dung samples from other parts of Germany.

Thus, the ecological requirements of myxobacteria are still largely obscure. It may be difficult to get clear answers, but a direct experimental approach under strictly controlled conditions is desirable.

PHYLOGENY AND CLASSIFICATION

During the past decade, the phylogenetic position of the myxobacteria has been established through 16S RNA studies. The first of these investigations, which relied on 16S RNA oligonucleotide catalogs, demonstrated quite clearly that the myxobacteria are a phylogenetically coherent group, that they belong to the *Proteobacteria* (then known as the purple bacteria), and that they are not closely related either to the *Cytophaga*-like bacteria, as had been thought at one time, or to any other gliding bacteria (Ludwig et al., 1983). Further, a deep division exists between the genera *Myxococcus, Stigmatella,* and *Cystobacter* on the one side and *Sorangium* and *Nannocystis* on the other, but genera *Sorangium* and *Nannocystis* were almost as deeply separated from each other as from the rest. This deep division nicely supported the concept of two suborders that had been proposed earlier on the basis of morphological and chemosystematic data. The picture was later corroborated and refined when the primary base sequence of 16S RNA was determined and secondary structures could be studied (Oyaizu and Woese, 1985; Shimkets and Woese, 1992). The myxobacteria were then assigned to a separate branch of the *Proteobacteria,* the δ branch, as a distinct group, with sulfate reducers and *Bdellovibrio* spp. as their nearest yet remote relatives. The two suborders and the relatively isolated position of the genus *Nannocystis* were also obvious. The proposed reclassification at the genus level on the basis of those data, however, seems premature (Shimkets and Woese, 1992), for not only do these conclusions rest on a single strain for each taxon (and such a strain could well have been misplaced taxonomically), but 16S RNA also represents only a minute fraction of the total DNA, which in myxobacteria is unusually large for prokaryotes (around 9,500 kbp; Chen et al., 1990; Neumann et al., 1992). Therefore, we are looking at the genetic constitution of the organisms through an extremely narrow window when we confine our arguments to 16S RNA.

The taxonomy of the myxobacteria rests mainly on morphological characteristics, such as the shapes of the vegetative cells and myxospores and the shape and structure of the swarm and, especially, the fruiting bodies. Some distinguishing chemosystematic markers and physiological features, e.g., the fatty acid pattern, the chemistry of the pigments and the slime, the occurrence of fluorescing compounds, and the production of specific enzymes that degrade cellulose, chitin, or yeast cells, have also been used. More information in the last category would be highly desirable and would probably be helpful in diagnostic work, for while some morphological characteristics are essentially stable and reliable, others are not. Unfortunately, shapes and structures of the fruiting bodies, which must often be known exactly for a sound classification of newly isolated strains, belong to the latter category. In crude cultures, typical fruiting bodies may often still be seen, but after purification, strains often produce only degenerate fruiting bodies or no fruiting bodies at all. In most cases, we do not know how to induce a strain to return to normal fruiting. It is therefore a good policy for taxonomically inclined microbiologists to classify a strain as early as possible or to make permanent mounts of the original fruiting bodies for later use. Also, I have found that one is often able to obtain fruiting bodies in culture from desiccated fruiting bodies preserved on filter paper prepared soon after or even before the final purification. Another difficulty with myxobacterial taxonomy is that with most species, the morphologies and colors of swarm colonies and fruiting bodies may vary substantially and overlap considerably even under seemingly identical conditions.

About 40 species of myxobacteria can presently be distinguished more or less reliably. A few more definitely exist, but it is unlikely that many more will be discovered unless there are highly specialized species that have escaped current isolation techniques. Quite a few genera and species are not very well characterized taxonomically, and thorough studies of these with modern techniques would be desirable. Thus, the delimitation of many species is more or less arbitrary, e.g., those of *M. xanthus* and *M. virescens; M. fulvus* and *M. stipitatus;* the *Corallococcus* and *Archangium* species; *Cystobacter fuscus, Cystobacter ferrugineus, Cystobacter velatus,* and *Cystobacter violaceus;* and the *Polyangium-S. cellulosum* complex, to name some of the more irritating problems.

Next, I shall discuss the taxonomy of the myxobacteria in the form of an annotated key, which could also be used to classify new strains (for more details, see McCurdy [1989] and Reichenbach and Dworkin [1992]).

THE TAXONOMY OF THE MYXOBACTERIA

Order *Myxococcales*

While it has been unnoticed by most people in the field, the name of the order has once more been changed (McCurdy, 1989). This is now the third change during my own career, but the change is made according to the rules.

The morphological descriptions given below, especially those of fruiting bodies, depend on the media used for cultivation. The reference substrates for the following observations are natural substrates, like rabbit dung, rotting wood, and bark; for the bacteriolytic species, thin streaks of living *E. coli* on water agar and VY/2 agar (=yeast agar); for the cellulose degraders, filter paper on ST21 agar (on this substrate, bacteriolytic myxobacteria in mixed culture with nonmyxobacterial cellulose decomposes often also produce typical fruiting bodies, especially in the initial crude culture), and VY/2 agar (for recipes, see Reichenbach and Dworkin [1992]).

A. Vegetative cells (Fig. 1) are slender, with more or less tapering ends; cigar, boat, or needle shaped; sometimes rather long (8 to 12 μm); and flexible. Myxospores (Fig. 2) are always clearly different in shape; i.e., they are much shorter, spheres or short fat rods, somewhat irregular in outline, optically refractile, and enclosed in a capsule that in most cases can be seen only under the electron microscope. Colonies (Fig. 3) on lean media are usually fast spreading, film-like swarms with a pattern of radial veins, often with areas of tiny oscillating waves or ripples, at the edge with delicate flame-like extensions. (The bacteria may also penetrate the usual 1.2 to 1.5% agar gels and spread within the plate. If they do so, this is a sign of degeneration, and the strain will behave atypically in other respects also; e.g., it probably will not form [typical] fruiting bodies.) The slime sheet of the swarm stains intensively with 0.01% aqueous Congo red solution (McCurdy, 1969). Fruiting bodies either are unprotected masses of slime and myxospores or consist of sporangioles with tough walls (the differentiation of sporangioles is often suppressed or lost in cultures, and much simpler forms in the shape of ridges or cushion- and sausage-like masses are then produced; still, they usually contain

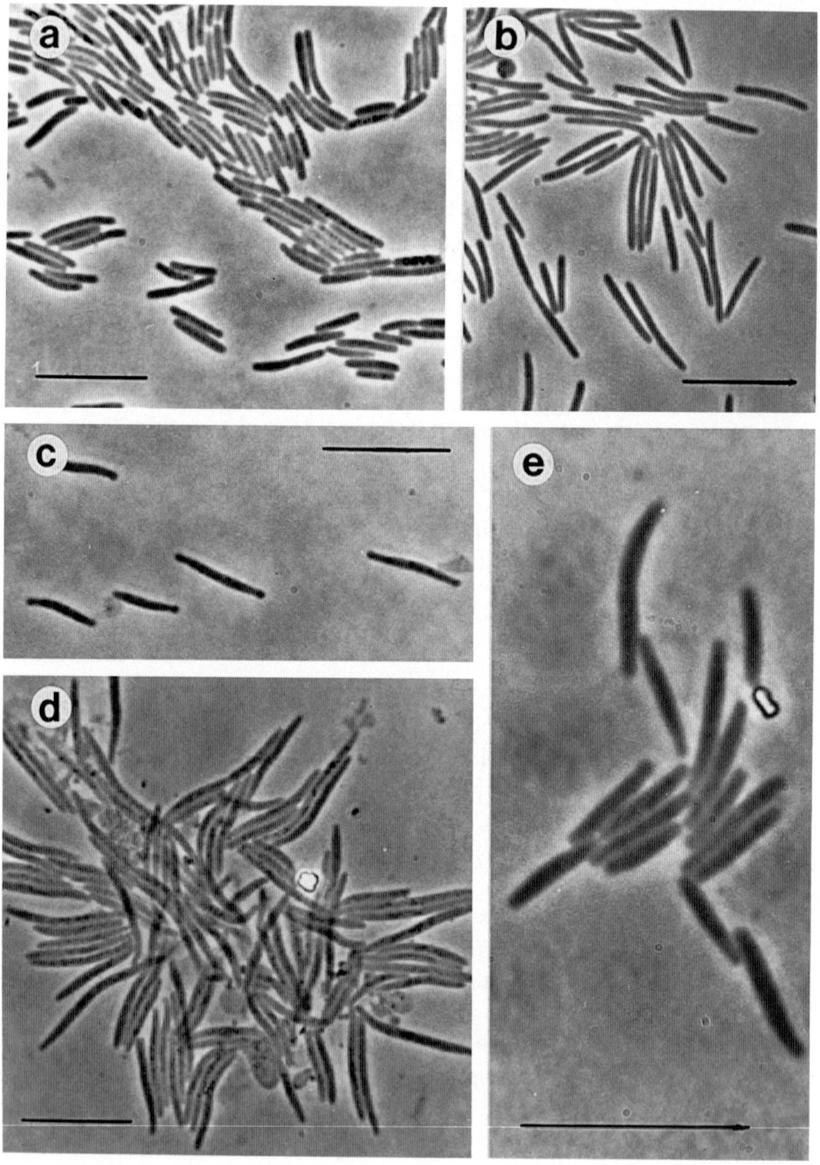

Figure 1. Vegetative cells typical of the suborder *Cystobacterineae*. (a) *C. coralloides* on yeast-VY/2 agar; (b) *M. virescens* from Casitone liquid medium; (c) *S. aurantiaca* from maize gluten liquid medium; (d) *C. fuscus* from starch agar; (e) *C. velatus* from peptone-CY agar. All pictures are in phase contrast. Bars are 10 μm.

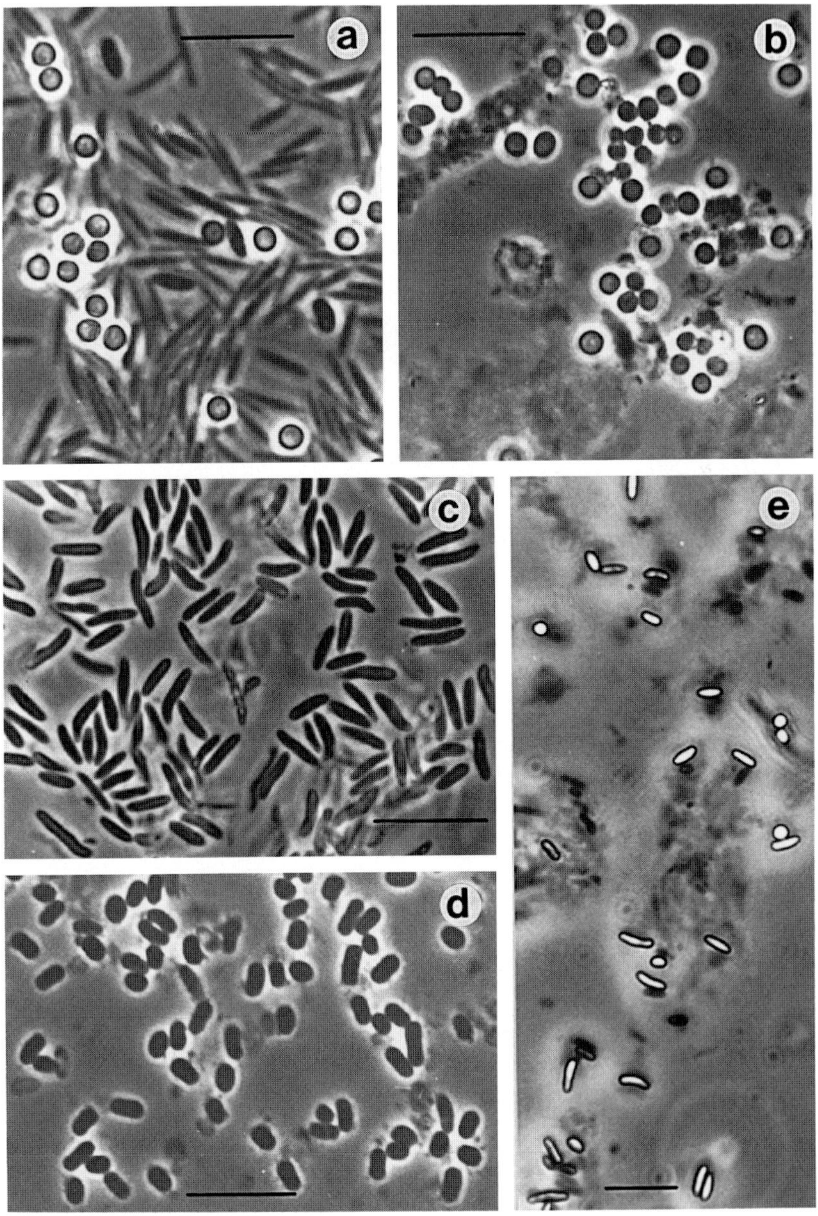

Figure 2. Myxospores of representatives of the suborder *Cystobacterineae*. (a) *M. xanthus*. In addition to myxospores, vegetative cells (out of focus) and transition forms can be seen. (b) *A. disciformis*. (c) *C. ferrugineus*. (d) *C. velatus*. (e) *S. aurantiaca*. The optical refractility of the myxospores is obscured in slide mounts under oil immersion (c and d), but it becomes apparent where the myxospores are deposited on a thin agar layer and photographed with a 40× objective (e). All myxospores are from fruiting bodies and are in phase contrast. Bars are 10 μm.

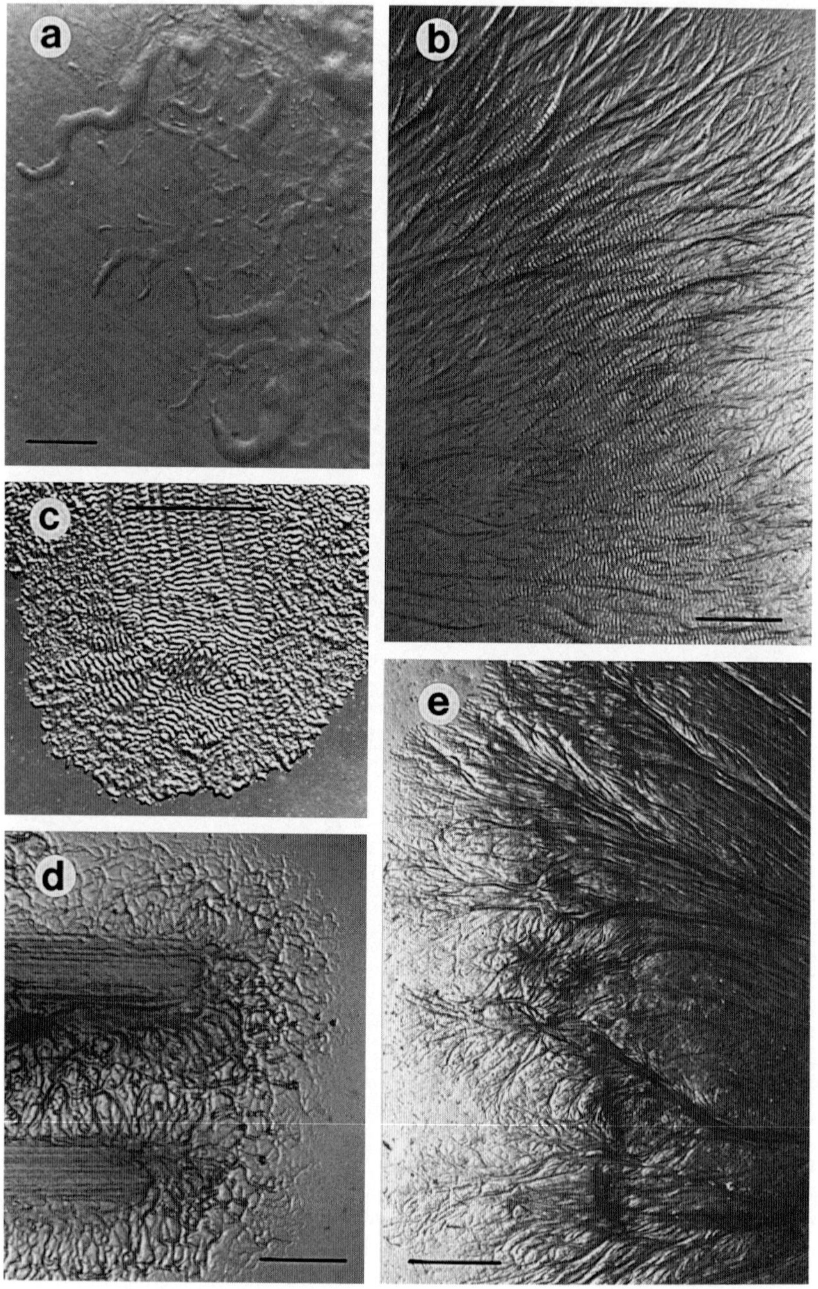
a
b
c
d
e

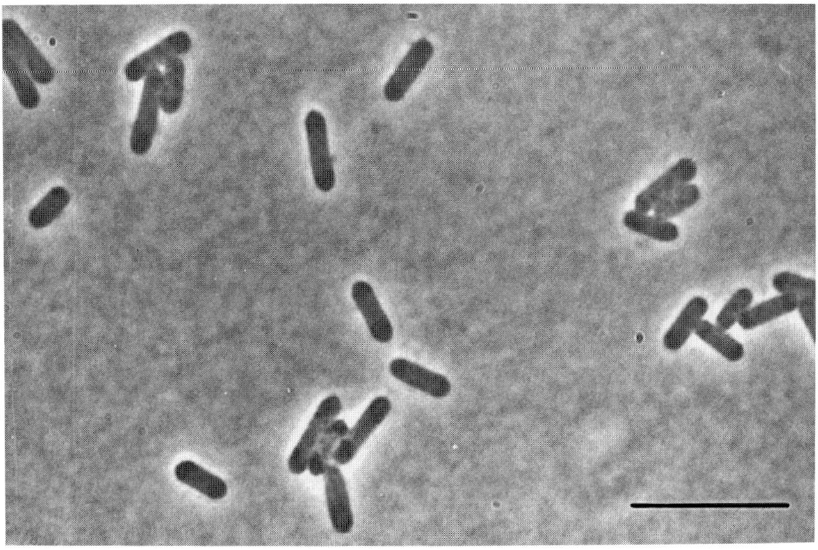

Figure 4. Vegetative cells of *S. cellulosum* showing typical cell morphology of the suborder *Sorangineae*. The cells contain dark polar granules, as are often found in older cultures; from yeast-VY/2 agar. Phase contrast; bar is 10 μm.

typical myxospores). The cell mass is usually colored in shades of yellow, orange, or red, and the main pigments are monocyclic keto-carotenoids with glucose and with a fatty acid attached to the open ends (Reichenbach and Kleinig, 1984). The fatty acid pattern is dominated by branched-chain acids and contains appreciable amounts of 2- and 3-hydroxy fatty acids (Fautz et al., 1981; Yamanaka et al., 1988). Many species degrade chitin, but none degrade cellulose. .... Suborder *Cystobacterineae*

B. Vegetative cells (Fig. 4) are cylindrical with rounded ends, relatively rigid, sometimes rather short. Myxospores usually differ

---

Figure 3. Swarm colonies characteristic of the suborder *Cystobacterineae*. (a) *C. coralloides* on yeast-VY/2 agar: edge of a swarm, with flares. (b) *A. gephyra* on xylan agar: swarm with oscillating waves. (c) *M. fulvus* on peptone-CY agar, with oscillating waves. (d) *C. coralloides* on CY agar. The swarm shows the snaking veins typical of the family *Myxococcaceae*. (e) *C. violaceus* on xylan agar. As in panel b, this swarm shows the long, straight, branching veins characteristic of the family *Cystobacteraceae*. Bars are 100 μm in panel a and 2,000 μm in all other panels.

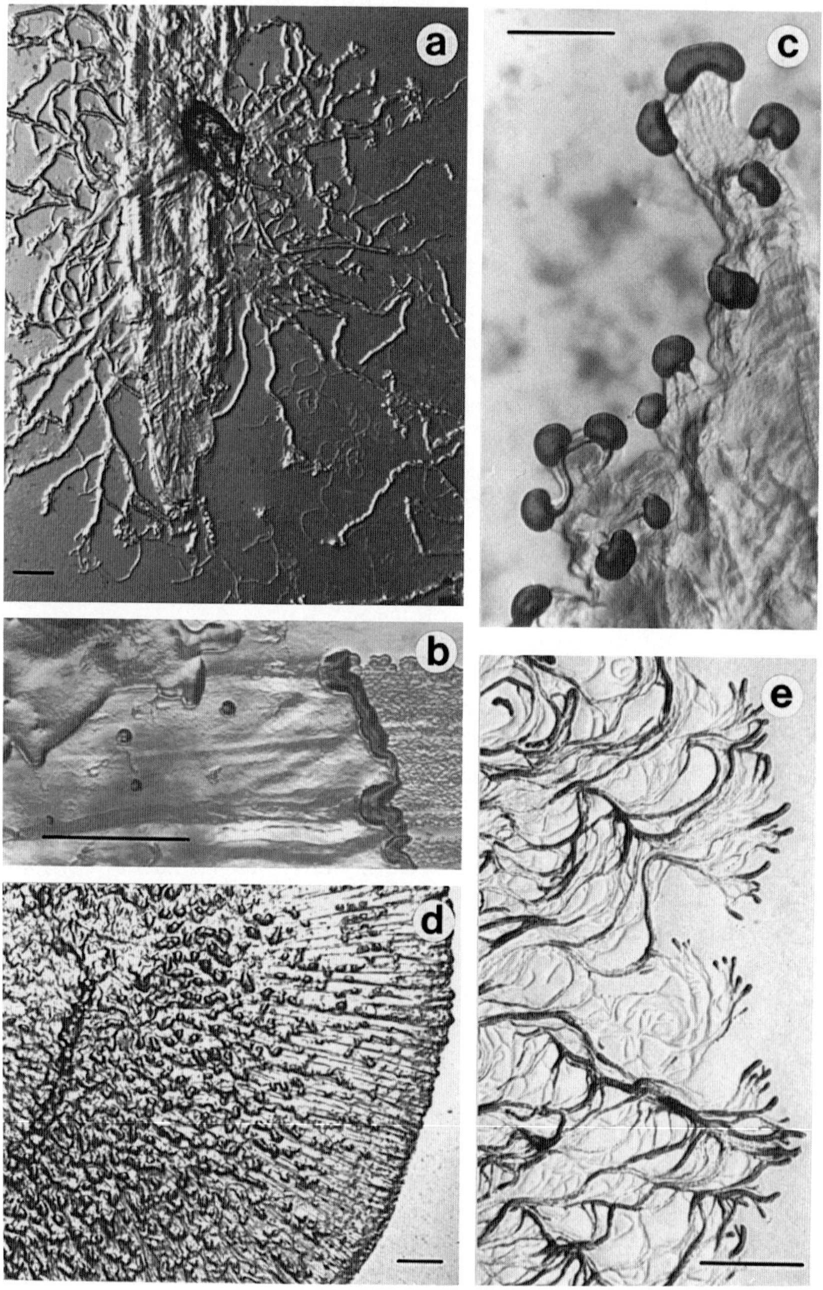
a
b
c
d
e

little from vegetative cells in shape but may be optically refractile (and are, of course, desiccation resistant); they appear never to be surrounded by a capsule. (The myxospores of *Nannocystis* spp. are ovoid and are clearly different from vegetative cells, which may, however, at times become very short, almost cube shaped.) The colonies (Fig. 5) spread and are usually sunk into the agar surface, with broad lobes or narrow radial paths; they often develop massive ridges along their margins. The organisms also tend to burrow deep into the agar and form fruiting bodies within the gel. (There are exceptions to the typical aspect of the swarm colonies, as outlined above. *Sorangium* strains may develop surface swarms with beautiful systems of branching radial veins, and there are also species that produce slime sheets on the agar surface that look exactly like those described above under A.) Congo red is not adsorbed by the slime sheet. The fruiting bodies always consist of sporangioles. (The occurrence of atypical fruiting bodies is much rarer in this group: these organisms tend, rather, to produce no fruiting bodies at all.) The cell mass is usually bright orange, yellow, or, in the case of *Nannocystis* spp., often brick red to violet; the main pigments are monocyclic hydroxy carotenoids with a sugar and with or without a fatty acid attached to the open end, and those of *Nannocystis* spp. are monocyclic aromatic carotenoids that are not glycosylated. The proportion of branched-chain fatty acids in the fatty acid pattern is low, and there are no hydroxy fatty acids. ... Suborder *Sorangineae*

A. Suborder *Cystobacterineae*

A1. Myxospores are spherical (they appear always to develop a thick capsule that during germination can sometimes be seen even in

---

Figure 5. Swarm colonies typical of the suborder *Sorangineae*. (a, b, c) *Polyangium* sp. in mixed culture on water agar with a streak of *E. coli*. In panel a, the inoculation site (a small agar block) is seen in the upper half of the picture; large clumps of cells migrate away from the streak of feed bacteria, at the same time etching deep trenches into the agar surface. In panel b, a massive ridge of myxobacteria is travelling from left to right along the *E. coli* streak. In panel c, the ridge disintegrated at the end of the streak into large cell clumps that are beginning to radiate away and eventually may differentiate into fruiting bodies. (d) *N. exedens* on peptone-glucose agar. The spreading swarm deeply corroded the agar surface. (e) *S. cellulosum* on glucose-nitrate (CA2) agar. The swarm consists of a system of isolated, branching veins; it can clearly be seen that these veins also are etched into the agar surface. Bars are 2,000 μm in panels a and d and 1,000 μm in panels b, c, and e.

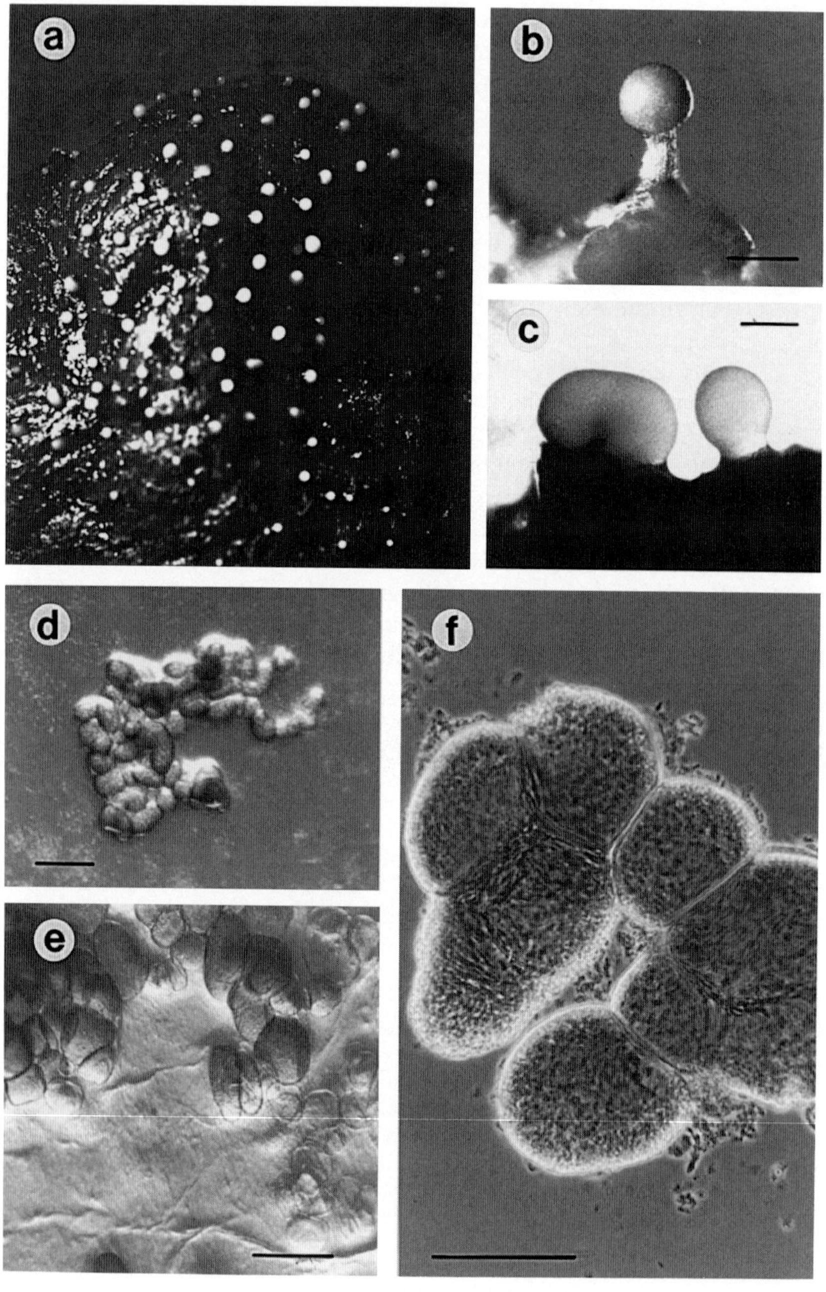
a
b
c
d
e
f

the light microscope). Vegetative cells are moderately long (3 to 8 μm) and cigar or slightly boat shaped. Swarms in the peripheral area often have characteristic snaking radial veins (usually well developed on peptone [CY] agar). Fruiting bodies in most cases are without sporangioles. .... Family *Myxococcaceae*

A2. Myxospores are short, fat rods. (Some species produce almost spherical myxospores that differ from those of the *Myxococcaceae* in their less regular and less smooth outlines. In other cases, the myxospores remain rather long and slender but still differ substantially from vegetative cells. It appears that with these organisms, the capsule always remains very thin.) Vegetative cells are usually rather long (6 to 12 μm) and needle shaped, but in some species, they are shorter and boat shaped. Swarms show long, straight radial veins that are often branching and rather delicate. (Often, a very tough slime sheet that may be almost impossible to disrupt with the inoculation loop develops.) Fruiting bodies usually contain sporangioles (but in culture, often only degenerate fruiting bodies develop). .... Family *Cystobacteraceae*

A1. Family *Myxococcaceae*

A1a. Fruiting bodies (Fig. 6) are dense clusters of spherical orange-brown sporangioles on the surface of the substrate. (The sporangioles may also be disk shaped but in culture are rarely seen in this form. In culture, the fruiting bodies usually soon degenerate into simple, *Corallococcus*-like ridges.) Vegetative cells are cigar-shaped. .... Genus *Angiococcus*

A1b. Fruiting bodies (Fig. 6) are soft, slimy globular knobs, sometimes with a constriction at the base or a stalk, and appear in shades of yellow, orange, and red or unpigmented. Vegetative cells are cigar or slightly boat shaped. .... Genus *Myxococcus*

Figure 6. Family *Myxococcaceae*, genera *Myxococcus* and *Angiococcus*. (a) *M. fulvus* fruiting bodies on a sheep dung pellet. (b) *M. stipitatus* in pure culture. (c) *M. fulvus* on a soil crumb. Two of the fruiting bodies developed close enough together to fuse, thus showing their soft slimy texture. (d, e, f) *A. disciformis* in pure culture on an agar surface (d and e). In panel e, some of the sporangioles are empty, which makes it possible to recognize the sporangiole walls more clearly. (f) Phase contrast. At a higher magnification, myxospores can be seen shining through the locally folded sporangiole walls. Bars are 100 μm in panel d and 50 μm in all other panels.

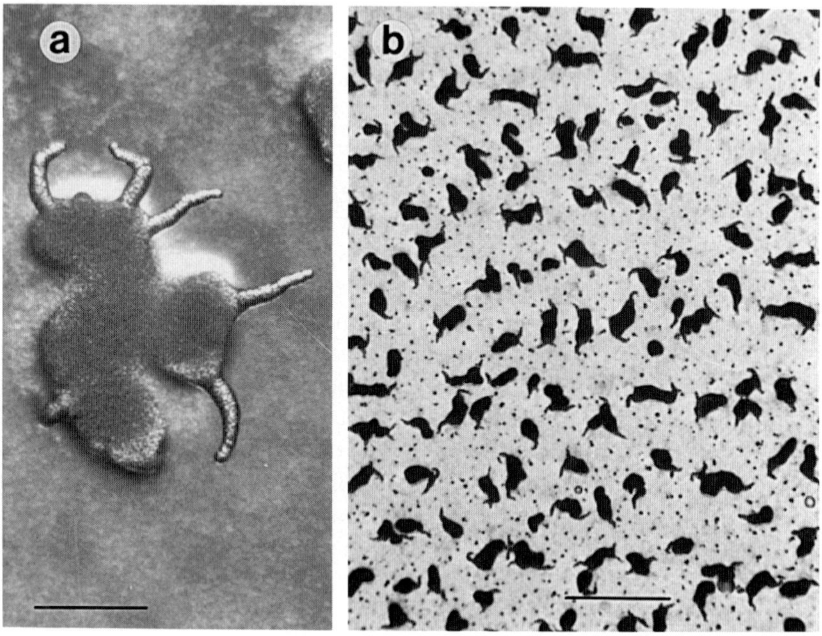

Figure 7. Family *Myxococcaceae*, genus *Corallococcus*. *C. coralloides* fruiting bodies in pure cultures on yeast-VY/2 agar. (a) The long, horn-like projections are typical of the species. Bar is 100 μm. (b) Fruiting bodies are often produced in dense populations. Here, large fruiting bodies are surrounded by tiny, dot-like ones. Bar is 1,000 μm.

A1c. Fruiting bodies (Fig. 7) are pustules or ridges, often of a bizarre shape with finger- or coral-like extensions; of a hard, cartilaginous consistency, without a sporangiole wall; in shades of yellow, orange, and red or sometimes unpigmented. Vegetative cells are cigar shaped. .... Genus *Corallococcus*

A2. Family *Cystobacteraceae*

A2a. Fruiting bodies (Fig. 8) are masses of convoluted rolls composed of slime and myxospores, of rather variable shape and size, sometimes rather large, of cartilaginous consistency, without a sporangiole wall, more or less brown to violet. (Fruiting bodies with this rather ill-defined irregular shape and structure are often also produced by other myxobacteria of this family as degenerative forms.) Myxospores are short to very short fat rods. Vegetative cells are long, slender, and needle shaped. . Genus *Archangium*

A2b. Fruiting bodies (Fig. 8) are clusters of oval or somewhat elongated sporangioles sitting directly on the surface of the substrate. With most species, the sporangioles are rather large. They may be chestnut brown, yellow- to orange-brown, or deep violet. (In culture, often only *Archangium*-like fruiting bodies are produced.) Myxospores are short and fat to relatively long (3 to 5 μm) and slender (and then often S- or C-shaped) rods. Vegetative cells are long, slender, and needle shaped. .... Genus *Cystobacter*

A2c. Fruiting bodies (Fig. 9) are small sporangioles that are spherical, kidney shaped, or like a mushroom cap; light to dark brown; and borne on a short, delicate, while pedicel (tend to degenerate to delicate, sharp ridges and *Archangium*-like structures). Myxospores are short, fat, often curved rods. Vegetative cells are rather long, delicate, and needle shaped. Cell mass is often red-orange. .... Genus *Melittangium*

A2d. Fruiting bodies (Fig. 9) are composed of moderately large, ovoid, brown sporangioles borne on a relatively robust white stalk either individually or in clusters (degenerate in culture to ridges and *Archangium*-like structures). Myxospores are short, fat, curved rods. Vegetative cells are moderately long and slightly boat shaped. All seem to be efficient chitin degraders. .... Genus *Stigmatella*

B. Suborder *Sorangineae*

B1. Swarms are more or less sunk into the agar surface in broad lobes or narrow ditches and usually also spread within the gel. In a few cases, the swarms may be film-like on the surface. Vegetative cells are short to moderately long and often rather slender. Fruiting bodies consist almost always of many clustered sporangioles, either in tight packs on or within the substrate or borne on long white slime stalks. The sporangioles often are bright orange but may also be yellow, brown, or even pitch black. Bacteriolytic or cellulose degraders. .... Family *Polyangiaceae*

B2. Swarms corrode the agar more or less intensively, often producing within the gel tunnels, holes, and caverns that may go down to the bottom of the dish. Vegetative cells are short to moderately long, relatively fat rods with round to square ends. (Particularly in old cultures, the vegetative cells may become almost cube shaped.) Cell mass is orange or brick red to deep purple. Fruiting bodies are tiny, oval or spherical sporangioles scattered through

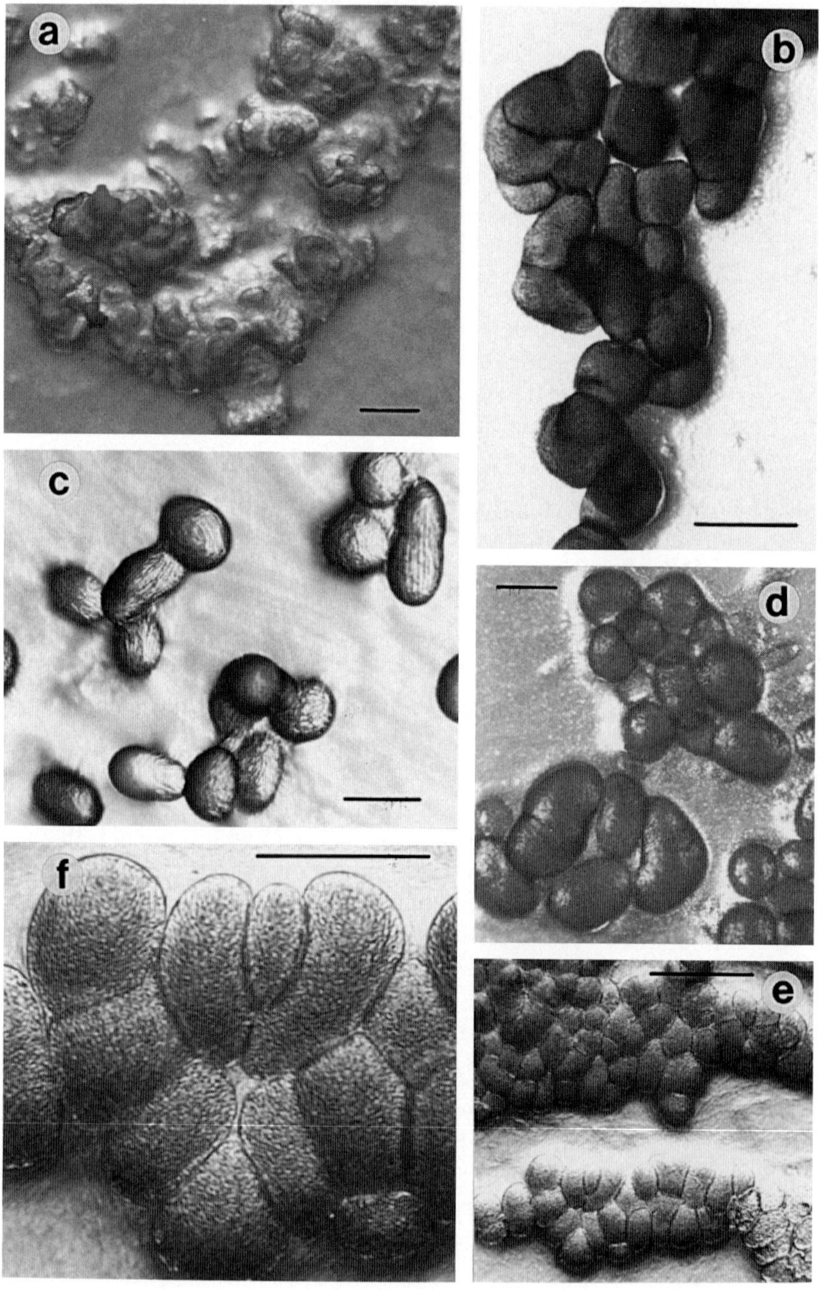
a
b
c
d
f
e

the agar gel, sometimes larger and irregular in shape on the surface, and usually completely separated from one another, but some strains also produce clusters of densely packed sporangioles. The sporangioles may be more or less brown, but often they are unpigmented. Myxospores are oval. Bacteriolytic (agar degrading). .... Family *Nannocystaceae*

(This family comprises only one genus, *Nannocystis*).

B1. Family *Polyangiaceae*

B1a. Fruiting bodies are relatively large, globular, solitary sporangioles with wrinkled, faceted surfaces; orange to orange-brown. On bark and rotting wood. (These organisms have never been cultivated.) .... Genus *Haploangium*

B1b. Fruiting bodies (Fig. 10) are composed of clusters of bright-orange sporangioles on long white stalks 500 to 1,000 μm high. .... Genus *Chondromyces*

B1c. Fruiting bodies (Fig. 11) consist of large, moderately sized, or small sessile sporangioles (diameter between 200 and 10 μm depending on the species) in loose piles or dense packs; yellow, orange, brown, or grey; on top of or within the (agar) substrate. Bacteriolytic (may be cultivated on peptone or protein media without carbohydrate). .... Genus *Polyangium*

B1d. Fruiting bodies (Fig. 12) consist of small to tiny sporangioles in densely packed bales on top of or within the substrate; often produced in tremendous numbers; yellow, orange, all shades of brown, or pitch black. Cellulose decomposers. (These appear to be the only myxobacteria that are able to grow on inorganic nitrogen compounds, like nitrate or ammonia, as the only nitrogen source. Most strains will not grow on peptone media without a carbohydrate.) .... Genus *Sorangium*

Figure 8. Family *Cystobacteraceae*, genera *Archangium* and *Cystobacter*. (a) Fruiting bodies of *A. gephyra:* pure culture on yeast-VY/2 agar. (b) *C. ferrugineus* mixed culture on water agar plus *E. coli*. (c) *C. velatus* pure culture on VY/2 agar. (d) *C. fuscus* mixed culture on water agar plus *E. coli*. (e, f) *C. minor* mixed culture on water agar plus *E. coli*. In panel e, some of the sporangioles are empty, and their walls consequently become clearly recognizable; the similarity to *A. disciformis* (Fig. 6d, e, f) is striking. Panel f is phase contrast. Bars are 100 μm in panels a through e and 50 μm in panel f.

a
b
c
e
d
f

A1a. Genus *Angiococcus*

Only one species, *A. disciformis,* is recognized (Fig. 6). It has a somewhat tortuous history. The species was described by Thaxter (1904) under the name *Myxococcus disciformis* and had been found by him twice on muskrat and deer dung in the northeastern United States. Jahn (1924) created the genus *Angiococcus* but apparently had never observed the organism. *Angiococcus* spp., however, were reported by Krzemieniewska and Krzemieniewski (1926) from Poland, where they were repeatedly obtained from dung and soil. Then for a long time, the species was not observed again, so that doubts arose as to whether the organism really existed (Peterson and McDonald, 1966). More recently, it was rediscovered in its characteristic form and in large numbers in a ditch surrounding a bog in Michigan (Hook, 1977; Hook et al., 1980). In my experience, the organism is indeed not particularly rare but may be easily overlooked, because in culture, it often forms only degenerate fruiting bodies that resemble those of *Corallococcus* spp. Also, on agar plates, the typical disk-shaped sporangioles usually become spherical, although on filter paper, the characteristic shape can often still be observed even in pure cultures. Because it produces true sporangioles, the species was recently transferred to the genus *Cystobacter* (McCurdy, 1989), but its colonies show the pattern of snaking veins that is characteristic of members of the family *Myxococcaceae,* and the vegetative cells resemble those of *Corallococcus* spp., so keeping the species in a separate genus in the family *Myxococcaceae* may be justified. In culture, the fruiting bodies of *A. disciformis* may be confused with those of *Cystobacter minor,* but the latter has a different swarm structure, rod-shaped myxospores, and somewhat longer vegetative cells.

Figure 9. Family *Cystobacteraceae,* genera *Melittangium* and *Stigmatella.* (a) Fruiting bodies of *M. lichenicola* on a piece of rotting wood. (b) *M. boletus* pure culture on yeast-VY/2 agar. (c) *S. erecta* pure culture on water agar. (d) *S. erecta* branching stalks consisting mainly of slime. The sporangioles are neatly broken off. Phase contrast. (e, f) *S. aurantiaca* pure culture; fruiting bodies are at the edge of the filter paper on water agar. Panel e is a slide mount. In panel f, the delicate pedicels supporting the sporangioles are recognizable. Bars are 100 μm in panel c and 50 μm in all other panels.

a
b
d
e
c
f

A1b. Genus *Myxococcus*

A1b-1. Myxospores are small (diameter about 1.2 μm) and often oval. Fruiting body has a long stalk, white to fawn. Swarm colony in UV light (366 nm) shows a bright-yellow fluorescence. .... *M. stipitatus*

A1b-2. Myxospores are around 1.3 to 1.6 μm in diameter, spherical. Fruiting body is globular, knob-like, often with a constriction at the base or a short pedicel; white, pink, or brick red. Vegetative cells are slightly boat shaped. .... *M. fulvus*

A1b-3. Myxospores are large (around 1.8 to 2.2 μm in diameter), spherical. Fruiting bodies are like those of *M. fulvus* but bright orange. Swarm colony is pale yellow-orange or unpigmented. Vegetative cells are cigar shaped. .... *M. xanthus*

A1b-4. Myxospores are large to very large (2.0 to 2.5 μm in diameter), spherical or slightly oval. Fruiting bodies are like those of *M. fulvus* but greenish yellow to grey; tend to become larger than those of the other species. Swarm colony is often intensively greenish yellow. Vegetative cells are cigar shaped. .... *M. virescens*

Differentiation of the *Myxococcus* species is sometimes complicated by an atypical appearance. In particular, the characteristics of *M. fulvus* and *M. stipitatus* on the one hand and *M. xanthus* and *M. virescens* on the other may substantially overlap. For example, otherwise typical *M. fulvus* strains may produce fruiting bodies with conspicuous stalks. Also, yellow fluorescence (Lampky and Brockman, 1977) does not always occur with the small oval myxospores and clearly stalked fruiting bodies, and we do not know

---

Figure 10. Family *Polyangiaceae*, genus *Chondromyces*. (a) *C. pediculatus* fruiting body from a crude culture with typical bell-shaped, drooping sporangioles; slide mount. (b) *C. crocatus* fruiting body from a crude culture; slide mount; phase contrast. (c, d) *C. lanuginosus* (small) fruiting body from a crude culture. When viewed from above, as in panel d, it can clearly be seen that the end faces of the large sporangioles are covered with many tail-like projections. Slide mount; phase contrast. (e) *C. apiculatus* fruiting bodies in induced pure culture. (f) *C. apiculatus* degenerate fruiting bodies consisting of solitary, sessile sporangioles that still show the characteristic turnip shape. Bars are 50 μm in panels c and d and 100 μm in all other panels.

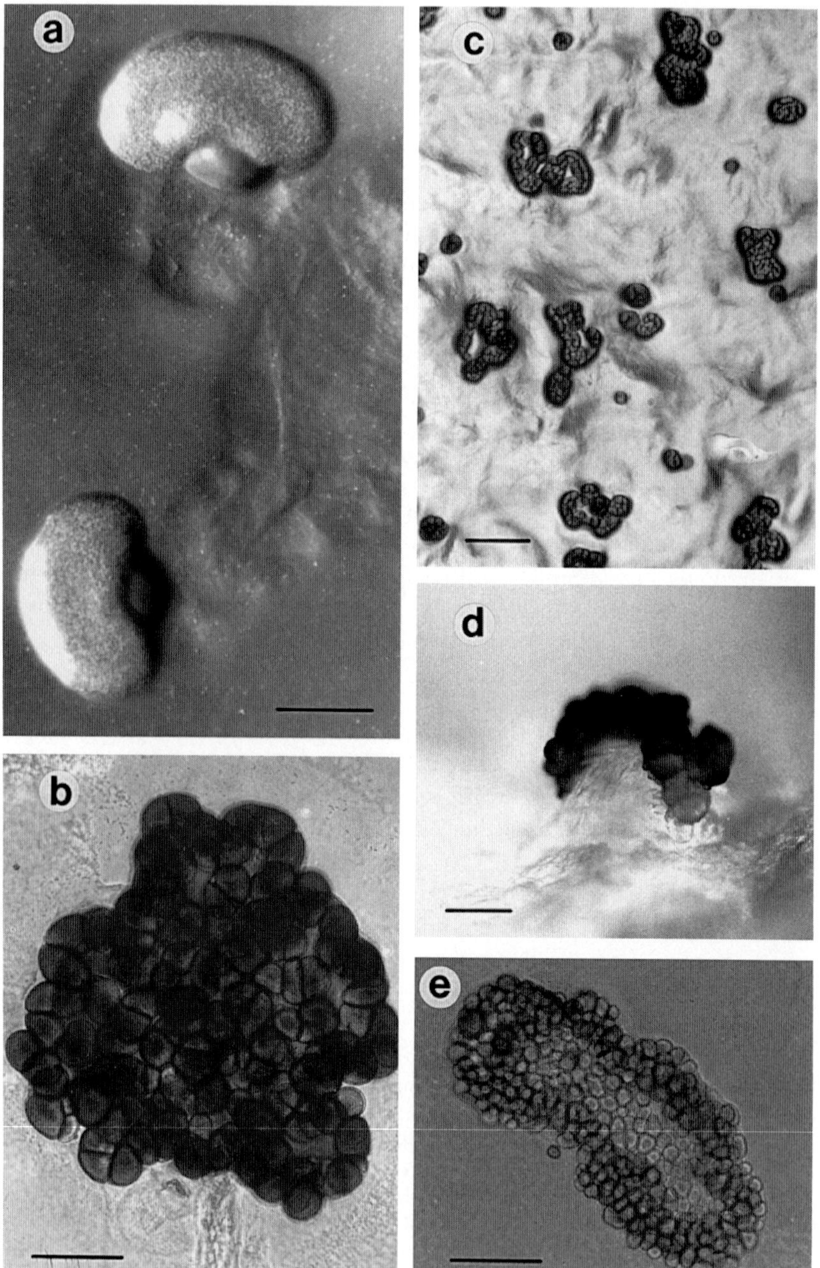
a
b
c
d
e

which of those characteristics is taxonomically relevant. In culture, *M. virescens* may become orange, and *M. xanthus* may produce greenish yellow pigments.

A1c. Genus *Corallococcus*

A1c-1. Myxospores are large to very large (1.8 to 2.5 μm in diameter). Fruiting bodies often form horn-like projections; sulfur yellow. .... *C. macrosporus*

A1c-2. Myxospores are smaller (1.3 to 1.9 μm in diameter). Fruiting bodies often form long and sometimes branched, even coral-like, projections; white, pink to brick red. .... *C. coralloides*

A1c-3. Myxospores are in the same size range as those of *C. coralloides* (1.5 to 1.8 μm in diameter). Fruiting bodies are small and brown, tend to be produced in enormous numbers, and are densely packed. .... *C. exiguus*

The genus *Corallococcus* has been under discussion since it was discovered that the old name, *Chondrococcus,* is illegitimate because an algal genus of that name has precedence. The tendency had been to lump *Corallococcus* with *Myxococcus,* but there are many substantial morphological and physiological differences that suggest that it may be wiser to keep the two apart. The fruiting bodies of *Myxococcus* and *Corallococcus* spp. can be distinguished with one glance. The vegetative cells of *Corallococcus* spp. are cigar shaped but usually shorter than those of the *Myxococcus* species with that cell shape. *C. coralloides* is an extremely variable species as far as fruiting bodies are concerned, but so far, all efforts to subdivide it have been unpersuasive. The species is probably heterogeneous, for there is considerable variation in the sizes of the myxospores among the strains.

Figure 11. Family *Polyangiaceae*, genus *Polyangium*. (a) *Polyangium* sp. migrating-cell aggregates. Their tracks are edged into the agar surface. At the rear ends, short, pedicel-like slime knobs can be seen; such masses may eventually differentiate into fruiting bodies. Mixed culture on water agar plus *E. coli*. (b) *Polyangium fumosum* fruiting body from a mixed culture on water agar plus *E. coli;* slide mount. (c) *P. spumosum* fruiting bodies on pitted agar surface; mixed culture on water agar plus *E. coli*. (d) *Polyangium thaxteri* fruiting body consisting of a cluster of large, orange sporangioles on a stalk-line slime cushion, on agar surface. (e) *P. sorediatum* fruiting body in mixed culture on water agar plus *E. coli* in situ; phase contrast. Bars are 200 μm in panels a, c, and d, and 100 μm in panels b and e.

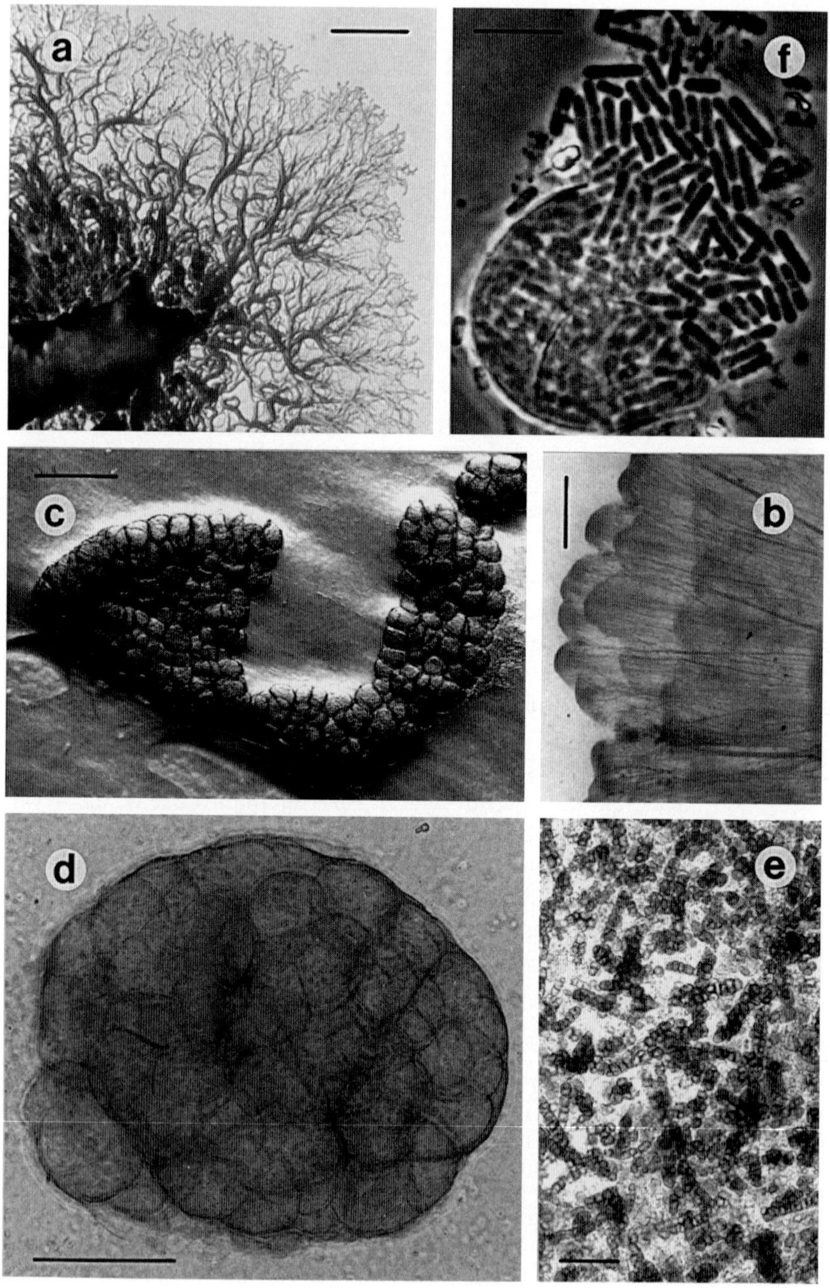
a
f
c
b
d
e

A2a. Genus *Archangium*

Presently, only one species, *A. gephyra,* is distinguished, but several more species have been described (for example, see Jahn [1924]). This happened at a time when most myxobacteria could not be cultivated and species were defined mainly from observations of fruiting bodies on rabbit dung and other natural substrates. As mentioned above, many myxobacteria produce degenerate fruiting bodies that resemble those of *Archangium* spp. in shape and structure (and some even have vegetative cells and myxospores like those of *Archangium* spp.), so that most of the earlier researchers may have been misled by such atypical forms. In some cases, it was stated that the vegetative cells of such *Archangium* species were of the *Polyangium* type, which clearly demonstrates, on the basis of our present knowledge, that they could not have been closely related to *A. gephyra.* As large, robust *Archangium*-type fruiting bodies that never differentiate further are regularly observed on natural substrates like rabbit dung, the genus *Archangium* and the species *A. gephyra* appear really to exist, but it would be very helpful if additional physiological and chemosystematic markers could be found. A DNA-DNA hybridization study would also be desirable. Another question is whether the species is identical to that described by Thaxter (1892) and perhaps should more appropriately be named *Archangium serpens.*

The myxospores of *A. gephyra* are short, fat rods, sometimes almost spherical; the vegetative cells are long, slender, flexible rods with tapering ends. In this respect, the organism resembles certain *Cystobacter* species.

A2b. Genus *Cystobacter*

I. Myxospores are relatively slender and long (3 to 7 μm), usually with tapering ends and curved in a slightly S- or C-like shape.

Figure 12. Family *Polyangiaceae,* genus *Sorangium. S. cellulosum* often produces colonies that look different from those of other representatives of the family. (a) On peptone-CY agar, massive, slimy, branching veins may develop; bar is 2,000 μm. (b) On soft carboxymethyl cellulose plates, the organism tends to grow within the substrate with a curtain-like swarm edge; bar is 1,000 μm. (c) Fruiting body on yeast-VY/2 agar; bar is 100 μm. (d) Compact fruiting body of a different type from another strain, from VY/2 agar; slide mount; bar is 50 μm. (e) Fruiting bodies from decomposed filter paper. The sporangioles developed along the vanishing cellulose fibers and are therefore aligned in chains. Bar is 100 μm. (f) Crushed sporangiole releasing myxospores that look almost like vegetative cells. Phase contrast; bar is 5 μm.

Sporangioles are large (80 to 200 μm), spherical or elongated, and chestnut brown. Efficient chitin decomposers.

A2b-1. Sporangioles are bright chestnut brown, glistening, spherical or oval, of more or less uniform size, usually in single-layered clusters, and embedded in a thick, transparent, gelatinous slime matrix. .... *C. fuscus*

A2b-2. Sporangioles are dull chestnut brown, spherical to elongated, of variable size, often only incompletely septated, sometimes piled up on top of one another, sometimes in finger-like projections. . ... *C. ferrugineus*

II. Myxospores are short, fat rods, sometimes almost irregularly spherical. Sporangioles are dark to light brown and orange, on average smaller than those of the *Cystobacter* spp. just listed.

A2b-3. Sporangioles are small (20 to 50 μm), spherical to oval, more or less orange-brown or bright orange, arising in dense clusters on top of or within the substrate (i.e., the agar plate; the other *Cystobacter* species seem never to produce fruiting bodies within the agar gel). The sporangioles may be packed so tightly that they become polygonal. Vegetative cells are shorter than those of other *Cystobacter* species. .... *C. minor*

A2b-4. Sporangioles are middle-sized (50 to 80 μm), oval but often somewhat irregular in shape, may be incompletely septated, orange-brown to fawn, in dense clusters, often in strings, and sometimes piled on top of one another. The fruiting body is covered by a thin and delicately plicated slime sheet. .... *C. velatus*

A2b-5. Sporangioles are large (50 to 140 μm), oval to elongated, dull red-brown to (rarely) violet, in dense clusters, often piled on top of one another, sometimes forming finger-like projections. The swarm may also become deep violet, at least in restricted areas. Does not decompose chitin. .... *C. violaceus*

C. fuscus is one of several myxobacterial species that had already been described in the premyxobacterial era before Thaxter, in this case by Schroeter in 1886, as Thaxter (1904) readily acknowledged. Unfortunately, however, Thaxter transferred the species to the still older genus *Polyangium,* whose members have very similar fruiting bodies but are not related to *C. fuscus* at all; in fact, we know today that the genus *Polyangium* belongs to another suborder. Thus, for many years, the *Cystobacter* species were classified with *Polyangium* until it was finally realized that the two are different, and the old genus *Cystobacter* was reestablished (McCurdy, 1970).

Cystobacter species are easy to distinguish from *Polyangium* species because of the different shapes of the vegetative cells, myxospores, and swarms. As mentioned above, *C. minor* and *A. disciformis* can be confused if only fruiting bodies are considered. (The wrong Latin term, *minus,* which is neuter in gender and was appropriate for *Polyangium* spp., is still in use, e.g., in *Bergey's Manual,* but should be replaced by *minor,* the correct diminutive of *parvus* [=small]). *C. violaceus* (Kühlwein and Gallwitz, 1958) is another case in which degenerate fruiting bodies were mistaken for a new *Archangium* species (Kühlwein and Reichenbach, 1964). The violet pigmentation regarded as typical of the species is usually not seen, and the organism resembles then *C. ferrugineus.* It can easily be distinguished, however, by its different myxospores and lack of chitinolytic enzymes. The fact that *Cystobacter* strains often quickly lose the ability to produce their typical fruiting bodies is more problematic. This is particularly true for *C. violaceus* and *C. velatus* strains, which are then hardly distinguishable from *A. gephyra.*

A2c. Genus *Melittangium*

A2c-1. Sporangiole more or less resembles a mushroom cap. Mainly in soil and on rabbit dung. .... *M. boletus*

A2c-2. Sporangiole is spherical, ovoid, or kidney shaped. Mainly on bark and rotting wood. .... *M. lichenicola*

The fruiting bodies of *Melittangium* species are small and can easily be overlooked. The sporangioles are pale yellow or light brown to red-brown and are supported by a short, delicate, white stalk. Species with longer stalks have also been described, but their status is presently uncertain. *M. lichenicola* (and the questionable *M. gracilipes*) were formerly classified together with *S. erecta* in the genus *Podangium* (Jahn, 1924), but that genus has been dissolved after *Podangium erectum* was recognized to be closely related to *S. aurantiaca.* The species *P. lichenicola* was then transferred to the genus *Melittangium,* created earlier for the species *M. boletus* (Jahn, 1924), which is very similar indeed to *P. lichenicola.* In culture, *Melittangium* strains usually fail to produce typical fruiting bodies; instead, dense fields of sharp, orange ridges and *Archangium*-like structures arise. Also, the cell mass often shows a characteristic bright-orange color.

A2d. Genus *Stigmatella*

A2d-1. Stalk bears one sporangiole only. .... *S. erecta*

A2d-2. Stalk with a cluster of sporangioles. .... *S. aurantiaca*

S. aurantiaca is another example of a myxobacterium that was described long before Thaxter. It was mistaken for a fungus (Berkeley, 1857). Following Thaxter (1892), the organism was classified in the genus *Chondromyces* until it was discovered that the similarity of the fruiting bodies is only superficial and the organism really belongs to the *Cystobacterineae* (McCurdy and Khouw, 1969; Reichenbach and Dworkin, 1969; McCurdy, 1971). The old genus *Stigmatella* was then revived, and *P. erectum* was also transferred to it.

The two *Stigmatella* species are not always easy to distinguish, because in culture, strains often produce atypical fruiting bodies. *S. erecta* fruiting bodies may arise in closely packed groups, and stalks of neighboring fruiting bodies may fuse, so that several sporangioles seem to be borne on one stalk; *S. aurantiaca* may also produce fruiting bodies with only one sporangiole on a stalk. Still, the sporangioles of *S. erecta* tend to be somewhat larger than those of *S. aurantiaca,* and the two species also appear to colonize different habitats: *S. aurantiaca* colonizes bark and rotting wood, and *S. erecta* colonizes soil and dung.

B1a. Genus *Haploangium*

B1a-1. Sporangiole is large (up to 200 μm), spherical to oval, bright orange; wall is heavily wrinkled, consisting of a smooth inner and a flaky outer layer. .... *H. rugiseptum*

B1a-2. Sporangiole is smaller (60 to 140 μm), spherical to bean shaped, dull orange-brown with a smooth surface when fresh, glistening bright orange and wrinkled when desiccated. Myxospores are packed in globose masses. .... *H. minus*

Several more species have been allocated to the genus but are of doubtful status (Peterson, 1959b). The solitary, globose, and relatively large sporangioles with more or less wrinkled surfaces are typical for the recognized species. The sporangiole is constricted at the base and may even have a short stalk (Hu et al., 1985). The species have been found so far only on bark and rotting wood. *H. rugiseptum* is rather conspicuous by its bright-orange, large fruiting bodies. While it has only rarely been recorded, it is not a rare species. I have found it regularly on rotting wood in the Black Forest in southwestern Germany. Neither organism has yet been cultivated, but vegetative cells and myxospores have been observed and are of the *Sorangineae* type. The two *Haploangium* species have been transferred to the genus *Polyangium* (McCurdy, 1974), but they are so distinct and different from the *Polyangium* species that keeping them separate appears to be justified.

B1b. Genus *Chondromyces*

I. Stalk is normally branched.

B1b-1. Dense clusters of relatively small, oval sporangioles at the ends of the branches. .... *C. crocatus*

II. Stalk is normally unbranched. Clusters of well-separated sporangioles.

B1b-2. Sporangioles have long, gradually tapering tips; turnip shaped. . ... *C. apiculatus*

B1b-3. Sporangioles are bell shaped with a flat end face, often on delicate pedicels and drooping, sometimes with a short tip in the center of the end face. .... *C. pediculatus*

III. Stalk are normally unbranched. Sporangioles are larger, almost like fused together.

B1b-4. Sporangioles are broad, bowl shaped to narrow, cylindrical, with a flat end face and a fringe of long, white, hair-like appendages around and within the end face, thus reminiscent of a hazelnut with its cupule; usually on short pedicels. .... *C. lanuginosus*

B1b-5. Sporangioles are rather small (20 to 50 μm), oval, in long chains like strings of beads, with lengths of shriveled slime in between the sporangioles; several chains arise from the tip of the stalk. . ... *C. catenulatus*

C. crocatus, too, was described before Thaxter, again as a fungus (Berkeley, 1857). This is not surprising, because the *Chondromyces* species produce the most conspicuous and beautiful fruiting bodies of all myxobacteria and simply cannot be overlooked by a collecting mycologist. Originally, several more myxobacteria with stalked and unstalked fruiting bodies of cartilaginous consistency were classified as *Chondromyces* spp. (Thaxter, 1892, 1897,

1904). Later, the genus was restricted to species with several sporangioles on one stalk (Jahn, 1924), but, as already pointed out under *Stigmatella,* the genus then was still heterogeneous. Amazingly, the myxobacteria "invented" stalked, treelet-like fruiting bodies two times independently. *C. lanuginosus* (Kofler, 1913) was described a second time under the name *Synangium thaxteri* (Faull, 1916), but the two organisms are clearly identical, unfortunately for the species epithet.

B1c. Genus *Polyangium*

I. Sporangioles are large (up to 200 μm); spherical, oval, or elongated; bright yellow to orange-brown.

B1c-1. Sporangioles are very large (75 to 200 μm), spherical to oval, golden yellow to orange, in groups of up to 20 in a common, colorless to yellowish, transparent slime envelope. Swarms are often greenish yellow. On wet rotting wood and bark, also in soil. .... *P. vitellinum*

B1c-2. Sporangioles are large to very large (80 to 180 μm), spherical to oval, bright yellow, solitary or in small groups of up to four in common, colorless, or yellowish slime envelopes. On rabbit dung and in soil. .... *P. luteum*

B1c-3. Sporangioles are more variable in size (60 to 150 μm) and shape (spherical, oval, or elongated), often incompletely septated, bright yellow to orange-brown, in large groups of 20 to 100 on top of or within the agar substrate; may be arranged in convoluted chains on top of a stout, cushion-like stalk. .... *P. thaxteri*

II. Sporangioles are very small (10 to 15 μm), yellow to orange-red.

B1c-4. Sporangioles are polygonal to spherical, tightly packed together in large numbers, thus forming flat fruiting bodies of up to 500-μm diameter that are covered by a colorless slime layer. .... *P. sorediatum*

III. Sporangioles are small to medium sized (20 to 60 μm), colorless or in shades of brown, grey, and orange.

B1c-5. Sporangioles are relatively small (20 to 35 μm), oval to slightly polygonal, colorless to light brown, in tightly packed bales of variable size. .... *P. spumosum*

B1c-6. Sporangioles are medium sized (20 to 60 μm), more or less spherical to slightly polygonal, brownish to grey, in groups of variable size, often only loosely packed and in ring-shaped strings. .... *P. fumosum*

B1c-7. Sporangioles are medium sized (30 to 40 μm), spherical, red-brown, in groups of variable sizes in a characteristic orange-yellow common slime envelope. From soil. .... *P. aureum*

B1c-8. Sporangioles are medium sized (25 to 40 μm), spherical, in groups of two to eight in a common colorless slime envelope, found in fresh water within emptied-out cells of green algae (*Cladophora* spp.), which they parasitize. .... *P. parasiticum*

P. vitellinum was the first myxobacterium to be described in a scientific journal. It was described almost a century before Thaxter by the (then) well-known botanist Heinrich Friedrich Link (1767 to 1851) in Berlin (1809) but was mistaken by him for a tiny gasteromycete. It was presented in a very exact drawing and, a few years later, on a beautiful color plate (Ditmar, 1814), so that it could be identified unequivocally.

Polyangium is a difficult genus, and its taxonomy still is unsatisfactory. Many species have been described over the years but often only from observations of fruiting bodies on natural substrates and without pure cultures. Thus, the variability of the species and their taxonomic status are sometimes difficult to evaluate, especially as descriptions of species are often so vague that the organisms are difficult or impossible to recognize. Even the species described above are not all beyond doubt. The fruiting bodies of *Polyangium* spp. resemble those of *Cystobacter, Angiococcus,* and *Sorangium* spp., and the *Cystobacter* species have indeed been classified for a long time with *Polyangium* spp.; the names in the older literature must be rather confusing for anyone not familiar with the complex history of these taxa. *Sorangium* spp. have also been classified with *Polyangium* spp., but this problem will be discussed below in the section on *Sorangium* spp. *P. thaxteri,* mentioned above, is almost certainly identical to *Archangium thaxteri* (Jahn, 1924). The aggregated cell masses often do not differentiate into sporangioles and then resemble the intestine-like, convoluted masses of the *Archangium* sp. described by Jahn, who also mentions the slime stalk. *P. parasiticum* may not be an obligate parasite and could well have been a soil bacterium that invaded the pond and attacked the algae on just that one occasion when it was observed (Geitler, 1925).

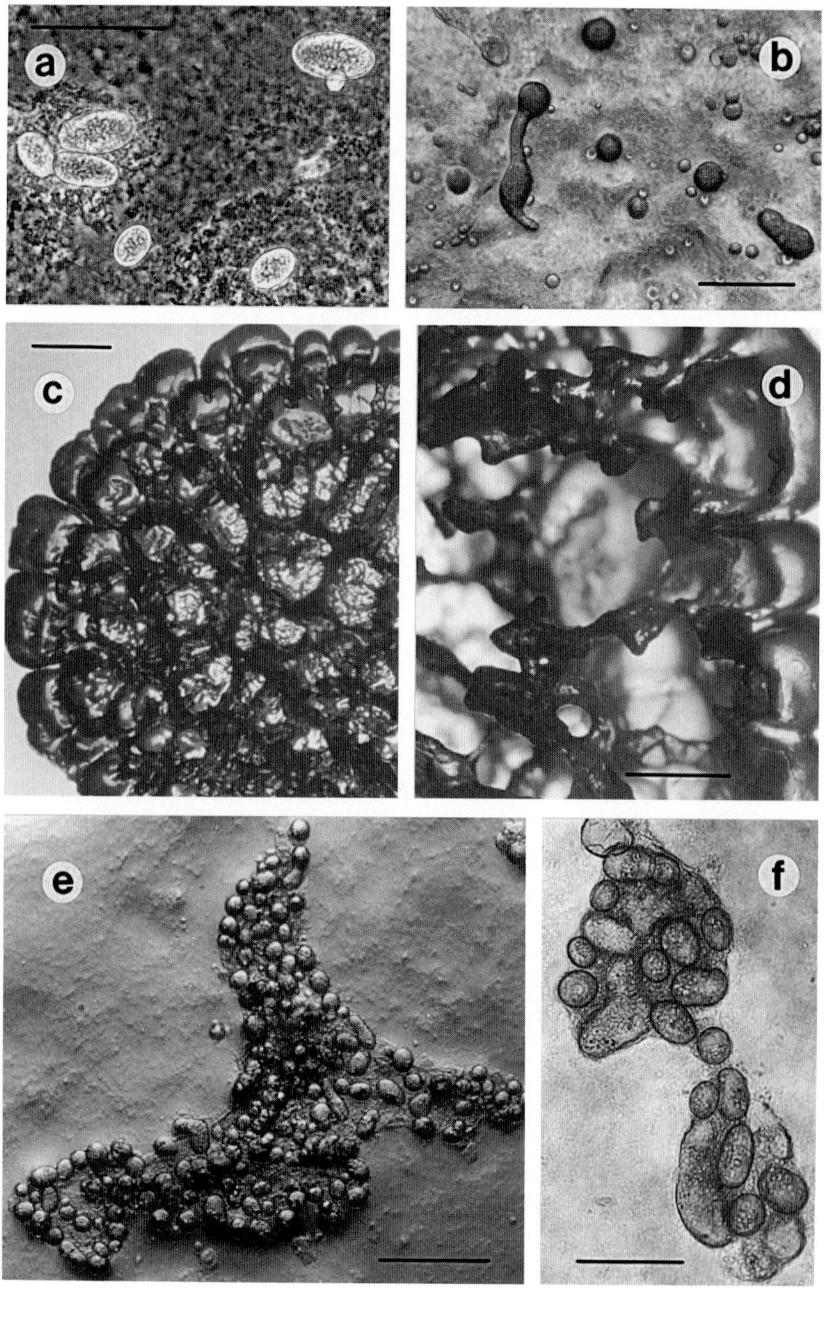
a
b
c
d
e
f

Polyangium species are somewhat fastidious but can be cultivated in pure culture. They are all bacteriolytic and proteolytic. In my experience, many strains require vitamin B_{12} and grow well on yeast (VY/2) agar. They usually penetrate deep into the gel, often to the bottom of the dish, and also tend to produce fruiting bodies within the gel. They decompose (autoclaved) yeast cells very efficiently, leaving a glass-clear lysis zone in the plate.

B1d. Genus *Sorangium*

Presently, only one species, *S. cellulosum,* can be distinguished. Several more species have been described, but their status is uncertain. Also, the original definition of the genus did not rely on cellulose decomposition but only on the presence of small, polygonal sporangioles in tight, sessile packages (Jahn, 1924). This in itself does not distinguish the genus sufficiently from *Polyangium,* and consequently, the *Sorangium* species were later classified with *Polyangium* spp. However, the cellulose-degrading species occupy a unique position among the myxobacteria with respect to their metabolism, so their classification as a separate genus appears justified. Chemosystematic differences (e.g., presence and absence of steroids [Zeggel, personal communication]) also support this position.

The strains of the *S. cellulosum* complex are extremely variable in shape and color (e.g., see Imshenetski and Solntseva [1936], Mishustin [1938], Imschenezki [1959], and Pronina [1962]). I am almost certain that several different species are hidden in this complex, but so far, all efforts to subdivide our 860 strains of that group have ended in failure. In contrast to all other myxobacteria, *S. cellulosum* attacks crystalline cellulose (filter paper) and is able to grow exclusively on cellulose with nitrate or ammonium and

---

Figure 13. Family *Nannocystaceae*, genus *Nannocystis*. (a, b) *N. exedens*. Typical fruiting bodies are tiny, solitary, oval or spherical sporangioles within the substrate, as in panel a. On the agar surface, however, much larger, irregular, though still solitary sporangioles often develop, as in panel b, which is a mixed culture on water agar plus *E. coli*. Panel a is phase contrast; bars are 50 μm in panel a and 100 μm in panel b. *N. exedens* (c, d) pure culture on peptone-CY agar. The developing swarm corroded the plate almost to the bottom, producing deep holes and large caverns. Bars are 2,000 μm in panel c and 1,000 μm in panel d). (e, f) Certain strains produce aggregates of sporangioles in addition to solitary ones and are distinguished as *N. aggregans;* mixed culture on water agar plus *E. coli*. Bars are 100 μm in panel c and 50 μm in panel f.

the usual mineral salts. Cellulose may be replaced by a sugar such as glucose or fructose, but most strains will not grow in the absence of a carbohydrate on, e.g., peptone. However, peptone and certain amino acids can serve as the sole nitrogen source.

Several more cellulolytic myxobacteria have been described, e.g., *Angiococcus cellulosae* (Mishustin, 1938) and *Myxococcus cellulosus* and *Podangium cellulosum* (Pronina, 1962), but these species have never been verified and must be regarded as very doubtful. Also, as *Cytophaga* and *Sporocytophaga* spp. were classified with the myxobacteria for some time, they of course appear in the older literature as cellulose-decomposing myxobacteria.

B2a. Genus *Nannocystis*

Only one species, *N. exedens* (Fig. 13), has been described; it is the latest accession to the myxobacteria. Probably there are more species, but as the strains are very variable in appearance, it has not been possible to define them clearly. One may be *N. aggregans,* with dense packages of small sporangioles in addition to the usual scattered ones. The tiny sporangioles scattered throughout the agar plate are typical for *Nannocystis,* as is a deep corrosion of the agar gel in the swarm area. *Nannocystis* sp. is also peculiar in that it is able to synthesize true steroids (Kohl et al., 1983).

N. exedens is a very common soil organism, certainly one of the most widely distributed myxobacteria; it was (and still is) overlooked by most people interested in myxobacterial taxonomy and ecology because of its very small and inconspicuous fruiting bodies and relatively slow development.

It is plain from this discussion that there are many open questions in the field of myxobacterial taxonomy and ecology. While an increasing number of studies of myxobacteria have been published in recent years, these deal almost exclusively with two organisms, *M. xanthus* and, to a lesser extent, *S. aurantiaca,* and mainly with genetics and molecular biology. Interesting and fascinating as these results are, myxobacteria are obviously no more all like *M. xanthus* than all bacteria are like *E. coli,* so there is still a wide and exciting field for enterprising microbiologists.

REFERENCES

Agnihothrudu, V., G. C. S. Barua, and K. C. Barua. 1959. Occurence of *Chondromyces* in the rhizosphere of plants. *Indian Phytopathol.* **12:**158–160.

Beebe, J. M. 1941. Studies on the myxobacteria. 2. The role of myxobacteria as bacterial parasites. *Iowa State College J. Sci.* **15:**319–337.

Berkeley, M. J. 1857. *Introduction to Cryptogamic Botany*, p. 313–315. H. Bailliere Publishers, London.

Bernátová, V. 1977. Occurrence of myxobacteria in surface water. *Folia Microbiol.* **22:**468.

Brauss, F. W., I. Heyne-Katzenberger, and W. Heyne. 1968. Beiträge zur Mikrobiologie von Binnengewässern (II). *Arch. Hyg. Bakteriol.* **152:**346–349.

Brauss, F. W., I. Heyne-Katzenberger, H. Pech, and H. Barth. 1967. Beiträge zur Mikrobiologie von Binnengewässern (I). *Arch. Hyg. Bakteriol.* **150:**716–724.

Brockman, E. R. 1967. Fruiting myxobacteria from the South Carolina coast. *J. Bacteriol.* **94:**1253–1254.

Brockman, E. R. 1976. Myxobacters from arid Mexican soil. *Appl. Environ. Microbiol.* **32:** 642–644.

Brockman, E. R., and W. L. Boyd. 1963. Myxobacteria from soils of the Alaskan and Canadian arctic. *J. Bacteriol.* **86:**605–606.

Burnham, J. C., S. A. Collart, and M. J. Daft. 1984. Myxococcal predation of the cyanobacterium *Phormidium luridum* in aqueous environments. *Arch. Microbiol.* **137:**220–225.

Burnham, J. C., S. A. Collart, and B. W. Highison. 1981. Entrapment and lysis of the cyanobacterium *Phormidium luridum* by aqueous colonies of *Myxococcus xanthus* PCO2. *Arch. Microbiol.* **129:**285–294.

Callao, V., R. Alvarado, A. Sedano, J. Olivares, and E. Montoya. 1966. Efecto antagonico del *Myxococcus xanthus* sobre los *Azotobacter*. *Microbiol. Espan.* **19:**45–51.

Casida, L. E. 1980. Bacterial predators of *Micrococcus luteus* in soil. *Appl. Environ. Microbiol.* **39:**1035–1041.

Chen, H., J. M. Keseler, and L. J. Shimkets. 1990. Genome size of *Myxococcus xanthus* determined by pulsed-field gel electrophoresis. *J. Bacteriol.* **172:**4206–4213.

Daft, M. J., J. C. Burnham, and Y. Yamamoto. 1985. Lysis of *Phormidium luridum* by *Myxococcus fulvus* in continuous flow cultures. *J. Appl. Bacteriol.* **59:**73–80.

Dawid, W. Personal communication.

Dawid, W. 1978. Fruchtkörper-bildende Myxobakterien in Böden Brasiliens. *Z. Allg. Mikrobiol.* **18:**83–93.

Dawid, W. 1979. Vorkommen und Verbreitung Fruchtkörper-bildender Myxobakterien im Siebengebirge. *Z. Allg. Mikrobiol.* **19:**705–719.

Dawid, W. 1984. Myxobakterien in ungestörten Hochmooren des Hohen Venn (Hautes Fagnes, Belgien). *Syst. Appl. Microbiol.* **5:**555–563.

Dawid, W., C. A. Gallikowski, and P. Hirsch. 1988. Psychrophilic myxobacteria from Antarctic soils. *Polarforschung* **58:**271–278.

Ditmar, L. P. F. 1814. Die Pilze Deutschlands, p. 55–56, plate 27. *In* J. Sturm (ed.), *Deutschlands Flora,* III, Abteilung 1, Bändchen, Heft 2. Nuremberg, Germany.

Faull, J. H. 1916. *Chondromyces thaxteri,* a new myxobacterium. *Bot. Gaz.* **62:**226–232.

Fautz, E., L. Grotjahn, and H. Reichenbach. 1981. 2- and 3-hydroxy fatty acids as valuable chemosystematic markers in gliding bacteria, p. 127–133. *In* H. Reichenbach and O. B. Weeks (ed.), *The Flavobacterium-Cytophaga Group*. Verlag Chemie, Weinheim, Germany.

Geitler, L. 1925. Über *Polyangium parasiticum* n.sp., eine submerse, parasitische Myxobacteriacee. *Arch. Protistenkd.* **50:**67–88.

Gerth, K. Personal communication.

Gräf, W. 1975. Myxobakterien der Gattung *Myxococcus* als indirekte Fäkalstoffindikatoren in Oberflächengewässern. *Zentralbl. Bakteriol. Abt. 1 Orig. Reihe B* **160:**28–39.

Hook, L. A. 1977. Distribution of myxobacters in aquatic habitats in an alkaline bog. *Appl. Environ. Microbiol.* **34:**333–335.

Hook, L. A., J. M. Larkin, and E. R. Brockman. 1980. Isolation, characterization, and emendation of description of *Angiococcus disciformis* (Thaxter 1904) Jahn 1924 and proposal of a neotype strain. *Int. J. Syst. Bacteriol.* **30:**135–142.

Hu, H. L., J. E. Peterson, E. R. Brockman. 1985. Stalked sporangia of *Polyangium rugiseptum. Int. J. Syst. Bacteriol.* **35:**362–363.

Imschenezki [also Imshenetski], A. A. 1959. *Mikrobiologie der Cellulose.* Akademie Verlag, Berlin. (Translated from Russian edition of 1953.)

Imshenetski, A. A., and L. Solntseva. 1936. On aerobic cellulose-decomposing bacteria. *Izvestiia Akad. Nauk SSSR Classe Sci. Math. Nat. Ser. Biol.* **1936:**1115–1172. (In Russian, with English summary.)

Jackson, D. J. 1959. The association of a slime bacterium with the inner envelope of the egg of *Dytiscus marginalis* (Coleoptera), and the less common occurrence of a similar bacterium on the egg of *D. semisulcatus. Q. J. Microsc. Sci.* **100:**433–443.

Jahn, E. 1911. Myxobacteriales, p. 187–206. *In Kryptogamenflora der Mark Brandenburg,* vol. 5. Gebrüder Borntraeger, Leipzig, Germany.

Jahn, E. 1924. Beiträge zur botanischen Protistologie. I. Die Polyangiden. Gebrüder Borntraeger, Leipzig, Germany.

Janssen, G. R., J. W. Wireman, and M. Dworkin. 1977. Effect of temperature on the growth of *Myxococcus xanthus. J. Bacteriol.* **130:**561–562.

Jeffers, E. E. 1964. Myxobacters of a freshwater lake and its environs. *Int. Bull. Bacteriol. Nomencl. Taxon.* **14:**115–136.

Katô, H. 1955. Notes on myxobacteria. I. The myxobacterial population in alpine soils. *Ecol. Rev.* **14:**25–28.

Kofler, L. 1913. Die Myxobakterien der Umgebung von Wien. *Sitzber. Kais. Akad. Wiss. Wien Math. Naturw. Klasse Abt. 1* **122:**845–876.

Kohl, W. 1982. Die Rolle der Mikroorganismen beim Abbau von Abwasserinhaltstoffen. *Schweiz. Z. Hydrol.* **44:**204–215.

Kohl, W., A. Gloe, and H. Reichenbach. 1983. Steroids from the myxobacterium *Nannocystis exedens. J. Gen. Microbiol.* **129:**1629–1635.

Krzemieniewska, H., and S. Krzemieniewski. 1926. Miksobakterje Polski. *Acta Soc. Bot. Polon.* **4:**1–54. (In Polish, with German summary.)

Krzemieniewska, H., and S. Krzemieniewski. 1927. Rozsiedlenie miksobakteryj (Über die Verbreitung der Myxobakterien im Boden). *Acta Soc. Bot. Polon.* **5:**102–139. (In Polish, with German summary.)

Krzemieniewka, H., and S. Krzemieniewski. 1930. Mikrobakterje Polski. Czesc trzecia (Die Myxobakterien von Polen. Teil III.) *Acta Soc. Bot. Polon.* **7:**250–273. (In Polish, with German summary.)

Kühlwein, H. 1950. Beiträge zur Biologie und Entwicklungsgeschichte der Myxobakterien. *Arch. Mikrobiol.* **14:**678–704.

Kühlwein, H. 1955. Untersuchungen über die lytische Aktivität von *Chondromyces apiculatus* Thaxter (?). *Zentralbl. Bakteriol. Abt. 1 Orig.* **162:**296–303.

Kühlwein, H. 1960. Zur Systematik und Verbreitung der Myxobakterien. *Zentralbl. Bakteriol. Abt. 2* **113:**480–490.

Kühlwein, H., and E. Gallwitz. 1958. *Polyangium violaceum* nov. spec. Ein Beitrag zur Kenntnis der Myxobakterien. *Arch. Mikrobiol.* **31:**139–145.

Kühlwein, H., and H. Reichenbach. 1964. Ein neuer Vertreter der Myxobakteriengattung *Archangium* Jahn. *Arch. Mikrobiol.* **48:**179–184.

Lampky, J. R., and E. R. Brockman. 1977. Fluorescence of *Myxococcus stipitatus. Int. J. Syst. Bacteriol.* **27:**161.

Larkin, J. M., and E. Dunigan. 1973. Myxobacteria from southwestern USA soils. *Soil Sci. Soc. Am. Proc.* **37:**808–809.

Lecianová, L. 1977. Occurrence and significance of myxobacteria in waste waters. *Folia Microbiol.* **22:**468–469.

Lecianová, L. 1978. Myxobakterien im Wasser, ein neuer Indikator in der Hydromikrobiologie. *Wasser Abwasser* **21:**43–65.

Link, H. F. 1809. Observationes in Ordines plantarum naturales. Dissertatio prima, complectens Anandrarum ordines Epiphytas, Mucedines Gastromycos et Fungos. *Der Gesellschaft Naturforschender Freunde zu Berlin Magazin für die neuesten Entdeckungen in der gesamten Naturkunde* **3:**3–42. (In Latin.)

Liu, K. C., and L. E. Casida. 1983. Survival of myxobacter strain 8 in natural soil in the presence and absence of host cells. *Soil Biol. Biochem.* **15:**551–555.

Ludwig, W., H. H. Schleifer, H. Reichenbach, and E. Stackebrandt. 1983. A phylogenetic analysis of the myxobacteria *Myxococcus fulvus, Stigmatella aurantiaca, Cystobacter fuscus, Sorangium cellulosum* and *Nannocystis exedens. Arch. Microbiol.* **135:**58–62.

McCurdy, H. 1970. Studies on the taxonomy of the Myxobacterales. II. *Polyangium* and the demise of the Sorangiaceae. *Int. J. Syst. Bacteriol.* **20:**283–296.

McCurdy, H. 1971. Studies on the taxonomy of the Myxobacterales. III. *Chondromyces* and *Stigmatella. Int. J. Syst. Bacteriol.* **21:**40–49.

McCurdy, H. 1974. Order 1. Myxobacterales Thaxter *emend.* Stanier 1957, 854, p. 76–98. *In* R. E. Buchanan and N. E. Gibbons (ed.), *Bergey's Manual of Determinative Bacteriology,* 8th ed. The Williams & Wilkins Co., Baltimore.

McCurdy, H. D. 1969. Studies on the taxonomy of the Myxobacterales. I. Record of Canadian isolates and survey of methods. *Can. J. Microbiol.* **5:**1453–1461.

McCurdy, H. D. 1989. Order Myxococcales Tchan, Pochon and Prévot 1948, 398 (with contributions of E. R. Brockman, H. Reichenbach, and D. White), p. 2139–2170. *In* J. T. Staley, M. P. Bryant, N. Pfennig, and J. G. Holt (ed.), *Bergey's Manual of Systematic Bacteriology,* vol. 3. The Williams & Wilkins Co., Baltimore.

McCurdy, H. D., and B. T. Khouw. 1969. Studies on *Stigmatella brunea. Can. J. Microbiol.* **15:**731–738.

Menne, B., and G. Rückert. 1988. Myxobakterien (Myxobacterales) in Höhlensedimenten des Hagengebirges (Nördliche Kalkalpen). *Höhle* **4:**120–131.

Michoustine (also Mishustin), E. 1968. Microorganismes cellulolytiques des sols de l'U.S.S.R. *Ann. Inst. Pasteur* **115:**596–603.

Mishustin, E. N. 1938. Cellulose decomposing myxobacteria. *Mikrobiologiya* **7:**427–444. (In Russian.)

Nellis, L. F., and H. R. Garner. 1964. Methods of isolation and purification of *Chondromyces. J. Bacteriol.* **87:**230–231.

Neumann, B., A. Pospiech, and H. U. Schairer. 1992. Size and stability of the genomes of the myxobacteria *Stigmatella aurantiaca* and *Stigmatella erecta. J. Bacteriol.* **174:** 6307–6310.

Norén, B. 1950. Notes on myxobacteria in Sweden. *Sven. Bot. Tidskr.* **44:**108–112.

Norén, B. 1952. Further notes on the distribution of myxobacteria in Swedish soils. *Sven. Bot. Tidskr.* **46:**446–453.

Oetker, H. 1953. Untersuchungen über die Ernährung einiger Myxobakterien. *Arch. Mikrobiol.* **19:**206–246.

Oyaizu, H., and C. R. Woese. 1985. Phylogenetic relationships among the sulfate respiring bacteria, myxobacteria and purple bacteria. *Syst. Appl. Microbiol.* **6:**257–263.

Peterson, J. E. 1957. The bark of living trees as a substrate for representatives of the order Myxobacterales. Ph.D. thesis. Michigan State University, East Lansing.

Peterson, J. E. 1959a. A monocystic genus of the Myxobacterales (Schizomycetes). *Mycologia* **51:**1–8.

Peterson, J. E. 1959b. New species of myxobacteria from the bark of living trees. *Mycologia* **51:**163–172.

Peterson, J. E. 1965. A group of strongly cellulolytic myxobacteria previously unreported in North American soils. *Am. J. Bot.* **52:**636.

Peterson, J. E. 1969a. Isolation, cultivation and maintenance of the myxobacteria, p. 185–210. *In* J. R. Norris and D. W. Ribbons (ed.), *Methods in Microbiology,* vol. 3B. Academic Press, London.

Peterson, J. E. 1969b. The fruiting myxobacteria: their properties, distribution and isolation. *J. Appl. Bacteriol.* **32:**5–12.

Peterson, J. E., and J. C. McDonald. 1966. The demise of the myxobacterial genus *Angiococcus. Mycologia* **58:**962–965.

Peterson, J. E., and B. Norén. 1967. The occurrence of the cellulose-decomposing myxobacterium, *Sorangium cellulosum,* in Scandinavian soils. *Am. J. Bot.* **54:**648.

Pronina, N. I. 1962. A description of new species and varieties of cellulose decomposing myxobacteria. *Mikrobiologiya* (translated) **31:**384–390.

Quehl, A. 1906. Untersuchungen über die Myxobakterien. *Zentralbl. Bakteriol. Abt. 2* **16:** 9–34.

Reichenbach, H. Unpublished data.

Reichenbach, H., and M. Dworkin. 1969. Studies on *Stigmatella aurantiaca* (Myxobacterales). *J. Gen. Microbiol.* **58:**3–14.

Reichenbach, H., and M. Dworkin. 1992. The myxobacteria, p. 3416–3487. *In* A. Balows, H. G. Trüper, M. Dworkin, W. Harder, and K. H. Schleifer (ed.), *The Prokaryotes,* 2nd ed. Springer Verlag, New York.

Reichenbach, H., and H. Kleinig. 1984. Pigments of myxobacteria, p. 127–137. *In* E. Rosenberg (ed.), *Myxobacteria. Development and Cell Interactions.* Springer Verlag, New York.

Rückert, G. 1972. Vergleichende Untersuchungen über die Verbreitung einiger Fruchtkörperbildender Myxobakterien-Arten. *Z. Allg. Mikrobiol.* **12:**655–665.

Rückert, G. 1973. Biogeographische Untersuchungen über die Verbreitung Fruchtkörperbildender Myxobakterien in Island. *Acta Bot. Isl.* **2:**56–63.

Rückert, G. 1975a. Zur Verbreitung bakteriotropher Myxobakterien in Waldböden. *Mitteilungen des Vereins für forstliche Standortskunde und Forstpflanzenzüchtung* **24:**43–47.

Rückert, G. 1975b. Zur Verbreitung von Fruchtkörper-bildenden Myxobakterien in europäischen Strand- und Dünenböden. *Zentralbl. Bakteriol. Abt. 2* **130:**343–347.

Rückert, G. 1975c. Koprochorie einiger Myxobakterien-Arten (Myxobacterales). *Z. Allg. Mikrobiol.* **15:**565–567.

Rückert, G. 1976. Myxobakterien (Myxobacterales) in natürlichen und naturnahen Substraten aus der europäischen Subarktis. *Acta Bot. Isl.* **4:**4–9.

Rückert, G. 1977. Bakteriotrophe Myxobakterien (Myxobacterales) in ariden Substraten. *Karlsruher Geogr. Hefte* **8:**101–111.

Rückert, G. 1978a. Förderung der Fruchtkörper-Bildung von *Myxococcus virescens* Thaxter (Myxobacterales) in Rohkulturen durch Salzzusatz. *Z. Allg. Mikrobiol.* **18:**69–71.

Rückert, G. 1978b. Myxobakterien in Substraten aus ariden Gebieten unter besonderer Berücksichtigung Südwestafrikas. *Namib Meer* **8:**53–66.

Rückert, G. 1978c. Zur Verbreitung der Myxobakterien (Myxobacterales) im gemäßigten Südamerika (Raum Valdivia/Chile). *Zentralbl. Bacteriol. Abt. II* **133:**45–49.

Rückert, G. 1979. Myxobakterien-Artenspektren von Böden in Abhängigkeit von bodenbildenden Faktoren unter besonderer Berücksichtigung der Bodenreaktion. *Z. Pflanzenernahr. Bodenkd.* **42:**330–343.

Rückert, G. 1981. Myxobakterien (Myxobacterales) auf Blattoberflächen. *Z. Allg. Mikrobiol.* **21:**761–763.

Rückert, G. 1983. Myxobakterien in Böden extremer Biotope. *Mitt. Dtsch. Bodenkundl. Ges.* **38:**355–360.

Rückert, G. 1984. Untersuchungen zum Vorkommen von Myxobakterien in von Meerwasser beeinflußten Substraten unter besonderer Berücksichtigung der Insel Helgoland. *Helgoländer Meeresunters* **38:**179–184.

Rückert, G. 1985a. Myxobakterien (Myxobacterales) in kleinstandörtlich aufgegliederten Teillebensräumen der südwestdeutschen Rheinauen. *Tuexenia* **5:**455–459.

Rückert, G. 1985b. Myxobacteria from Antarctic soils. *Biol. Fert. Soils* **1:**215–216.

Rückert, G., and G. Heym. 1977. Bakteriotrophe Myxobakterien (Myxobacterales) in ariden Substraten. *Karlsruher Geogr. Hefte* **8:**101–111.

Sabadoš-Šarić, A. 1957. Istrazivanja o miksobacterijama u thina Jugoslavije (Myxobacteria in different types of soils in Yugoslavia). *Yugoslavenska Akademija Znanosti i Umjetnosti Zagreb* **312:**5–29. (In Serbo-Croatian, with English summary.)

Schroeter, J. 1886. Pilze, p. 170. *In* F. Cohn (ed.) *Kryptogamen-Flora von Schlesien*, vol. 3, part 1. J. U. Kern's Verlag, Breslau, Poland.

Shea, C., J. W. Nunley, and H. E. Smith-Somerville. 1991. Variable expression of gliding and swimming motility in *Deleya marina*. *Can. J. Microbiol.* **37:**808–814.

Shimkets, L., and C. R. Woese. 1992. A phylogenetic analysis of the myxobacteria: basis for their classification. *Proc. Natl. Acad. Sci. USA* **89:**9459–9463.

Singh, B. N. 1947. Myxobacteria in soils and composts; their distribution, number and lytic action on bacteria. *J. Gen. Microbiol.* **1:**1–10.

Singh, B. N., S. Mathew, and M. Screenivasaya. 1958. Distribution of higher myxobacteria in Indian soils, p. 1–4. *In Symposium on Antibiotics.* Sree Saraswaty Press, Calcutta, India.

Singh, B. N., and N. B. Singh. 1971. Distribution of fruiting myxobacteria in Indian soils, bark of trees and dung of herbivorous animals. *Indian J. Microbiol.* **11:**47–92.

Sudo, S. Z., and M. Dworkin. 1969. Resistance of vegetative cells and microcysts of *Myxococcus xanthus*. *J. Bacteriol.* **98:**883–887.

Thaxter, R. 1892. On the Myxobacteriaceae, a new order of Schizomycetes. *Bot. Gaz.* **17:** 389–406.

Thaxter, R. 1893. A new order of Schizomycetes. *Bot. Gaz.* **18:**29–30.

Thaxter, R. 1897. Further observations on the Myxobacteriaceae. *Bot. Gaz.* **23:**395–411.

Thaxter, R. 1904. Notes on the Myxobacteriaceae. *Bot. Gaz.* **37:**405–416.

Till, M. 1982. Das Vorkommen von Myxobakterien im Darminhalt von Fischen, p. 84–86. 23. Arbeitstagung des IAD, Vienna.

Tržilová, B., L. Miklošovičová, E. Golaisová, and M. Bobková. 1980. Indikatoren der Verunreinigung des Donauwassers im tschechoslowakischem Abschnitte. *Acta Ecol.* (Bratislava) **8:**91–102.

Tržilová, B., L. Miklošovičová, G. Morháčová, and E. Golaisová. 1981. Die Wasserqualität der Zuflüsse des tschechoslowakischen Donauabschnittes von limnologischer und hygienischer Sicht. *Biologa* (Bratislava) **36:**765–774.

Vychodilová, M. 1977. Occurrence of myxobacteria in drinking water. *Folia Microbiol.* **22:** 468.

Wu, B., K. Hamdy, and H. B. Howe. 1968. Antimicrobial activity of a myxobacterium against blue-green algae. *Bacteriol. Proc.* **1968:**48.

Yamanaka, S., R. Fudo, A. Kawaguchi, and K. Komagata. 1988. Taxonomic significance of hydroxy fatty acids in myxobacteria with special reference to 2-hydroxy fatty acids in phospholipids. *J. Gen. Appl. Microbiol.* **34:**57–66.

Zeggel, B. Personal communication.

Zhukova, R. A. 1959a. Prevalence of aerobic cellulose bacteria in soils of the Kola Peninsula. *Mikrobiologiya* (translated) **28:**62–68.

Zhukova, R. A. 1959b. Toxicity of Kola Peninsula soils for aerobic cellulose bacteria. *Mikrobiologiya* (translated) **28:**844–850.
Zhukova, R. A. 1960. Role of the biological factor in toxicity of Kola Peninsula soil for aerobic cellulose bacteria. *Mikrobiologiya* (translated) **29:**165–170.
Zhukova, R. A. 1962. Aerobic cellulose bacteria in northern soils. *Microbiologiya* (translated) **31:**855–860.

Myxobacteria II
Edited by Martin Dworkin and Dale Kaiser
© 1993 American Society for Microbiology, Washington, DC 20005

Chapter 3

Cell Surfaces and Appendages

Martin Dworkin

The cell surface is the interface between the cell and its environment; as such, it is the point at which the cell first interacts with its surroundings. Like all cells, the myxobacteria must be able to perceive and respond to the external environment. However, because of the social nature of the myxobacteria, their perceptions must also reflect their dynamic interactions with other cells of their clone. Thus, the myxobacteria must be able to perceive signals from other cells and transmit these perceptions so that they can be transduced into a cellular response. Since it is likely that these signal receptions and transmissions occur via various macromolecules on the cell surface, a study of the cell surface and appendages of the myxobacteria is of central importance.

For this discussion, the cell surface comprises adhering extracellular polysaccharides, pili-fimbriae, fibrils, outer membrane (OM), and the spore coat or capsule.

FIMBRAE-PILI

Myxobacterial fimbriae were first reported by MacRae and McCurdy (1975), who described them for negatively stained cells of *Chondromyces crocatus* and ended their discussion with the following prescient comment: ". . . it is tempting to speculate that the slime and the pili, by providing cell-to-cell interconnections, influence the organized gliding movement of *C. crocatus*." Subsequently, Dobson and McCurdy (1979) isolated and purified pili from surface-grown cells of *Myxococcus xanthus*. They did so by simple mechanical depiliation followed by differential and

Martin Dworkin • Department of Microbiology, University of Minnesota, Minneapolis, Minnesota 55455-0312.

isopycnic centrifugation in CsCl. They found that the cells contained two types of pili, which they designated R (rigid) and F (flaccid). These pili had different diameters (6.0 versus 8.5 nm, respectively) but were similar in all other respects, i.e., resistance to hydrolytic enzymes, amino acid composition, molecular weight, carbohydrate composition, and antigenic determinants. Dobson et al. (1979) showed electron micrographs of *M. xanthus* cells presumably connected to each other via their pili and concluded that the pili were playing a direct role in cell-cell communication. This conclusion, however, was based on a single published micrograph and was therefore not persuasive. A more convincing approach was based on the fact that polyclonal antiserum directed against purified pili inhibited swarming motility but had no effect on the motilities of individual cells. On the basis of these experiments and a variety of other, more indirect data, Dobson et al. concluded that pili function ". . . to establish and maintain intercellular contacts . . . during the coordinated movement of cell aggregates. . . ." (1979). At about the same time, in a genetic study, Hodgkin and Kaiser (1979a, 1979b) discovered and described the A (adventurous) motility of single cells and the S (social) motility of groups of *M. xanthus* cells. Kaiser then showed (1979) that the presence of polar pili in *M. xanthus* was completely correlated with social motility of the cells. A^-S^+ cells separated from each other by more than one cell length were unable to glide (Hodgkin and Kaiser, 1979b). Furthermore, when social motility mutants that lacked pili were brought into close apposition with other such mutants of a different complementation group, the former transiently regained piliation as well as social motility (Kaiser, 1979).

Recently, Rosenbluh and Eisenbach (1992) obtained direct evidence for a role of pili in social motility by showing that mechanical depiliation of *M. xanthus* resulted in the loss of social motility but had no effect on the movement of individual cells. Figure 1 shows the polar pili of *M. xanthus*.

Thus far, various laboratories have made a number of unsuccessful attempts to use the Dobson and McCurdy (1979) technique to recover and analyze the pili of *M. xanthus*. It is not clear why these attempts have not been successful. In addition, neither Kaiser (1979) nor Rosenbluh and Eisenbach (1992) reported the presence of the two types of pili (rigid and flaccid) described by Dobson and McCurdy, nor were they able to confirm the peritrichous arrangement of the pili depicted in one of the electron micrographs of Dobson et al. (1979). It is possible that the McCurdy group was working with a strain of *M. xanthus* whose piliation differed in a number of important ways from that of the DK1622 strain used by Kaiser's group. In fact, Dobson and McCurdy referred to their strain of *M. xanthus*

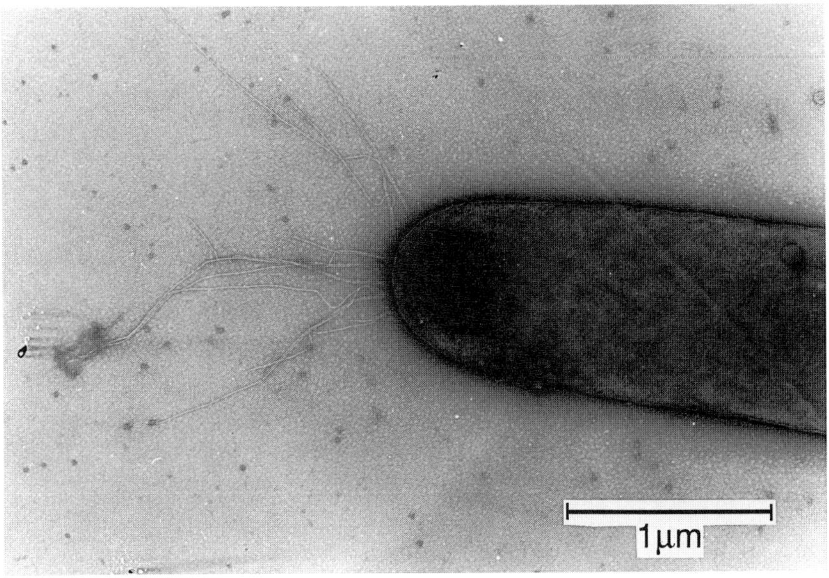

Figure 1. Electron micrograph of the polar pili of vegetative cells of *M. xanthus* DK1622 grown in liquid culture. Cells were fixed and negatively stained by the method of Kaiser (1979). Photo courtesy of Amy Rosenbluh.

(M36-F^+) as ". . . a spontaneous tan mutant that . . . produces large amounts of fimbriae on solid media" (1979).

While the evidence that pili are required for social motility is quite convincing, their precise role is not at all obvious. Rosenbluh and Eisenbach (1992) have suggested that the pili act as tactile sensors that indicate the close proximity of other cells and that such sensing may be the signal for the cells to synthesize or export the extracellular fibrils (Arnold and Shimkets, 1988a, 1988b) that actually hold the cells together during their social movement (Behmlander and Dworkin, 1991).

FIBRILS

Wild-type cells produce extracellular organelles referred to as fibrils that are filaments attached around the entire cell and are often as long as 50 µm. The diameters of the fibrils falls into two approximately equal size distributions, 15 and 30 nm. Fibrils were originally described by Arnold and Shimkets (1988a, 1988b) during their investigation of the effects of Congo red on cells *M. xanthus*. (Congo red is a diazo dye that binds

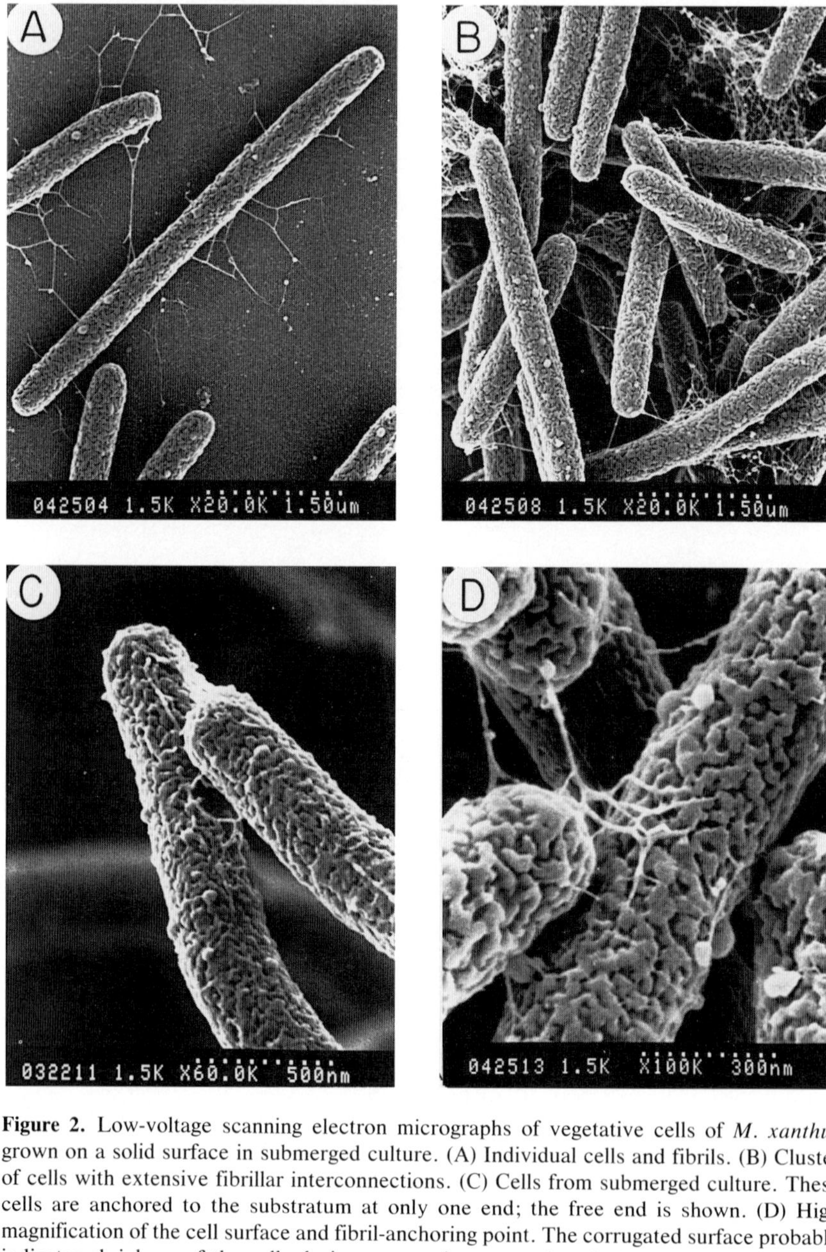

Figure 2. Low-voltage scanning electron micrographs of vegetative cells of *M. xanthus* grown on a solid surface in submerged culture. (A) Individual cells and fibrils. (B) Cluster of cells with extensive fibrillar interconnections. (C) Cells from submerged culture. These cells are anchored to the substratum at only one end; the free end is shown. (D) High magnification of the cell surface and fibril-anchoring point. The corrugated surface probably indicates shrinkage of the cells during preparation, even though they had been fixed and critical point dried (Behmlander and Dworkin, 1991).

most strongly to polysaccharides that contain contiguous β-1,4-linked D-glucopyranosyl units.) When cells were treated with Congo red, cohesion, social motility, and fruiting body formation were inhibited. These effects were similar to the properties of an A^+S^- mutant, dsp (Shimkets, 1986). The dsp mutant was shown to lack fibrils, as did wild-type cells that had been treated with Congo red. Since both the wild-type cells that have been treated with Congo red and the dsp mutant were able to form pili but were unable to cohere or manifest social motility, Arnold and Shimkets concluded that the fibrils supplied the cohesive force that held the cells together in the swarm (1988b).

Subsequently, Behmlander and Dworkin (1991) examined the properties of fibrils more closely. Using a low-voltage, high-resolution scanning electron microscope, they obtained a series of high-resolution photomicrographs of fibrils and the associated cells (Fig. 2). (Since these cells were not negatively stained, pili, if present, would not have been observed.)

These photos confirmed that the cells were connected to each other by the fibrils and that the Congo red-treated cells and the dsp mutant (Fig. 3) lacked fibrils. Behmlander and Dworkin also obtained data (1991) that suggested that fibril formation was cell density dependent (Table 1 and

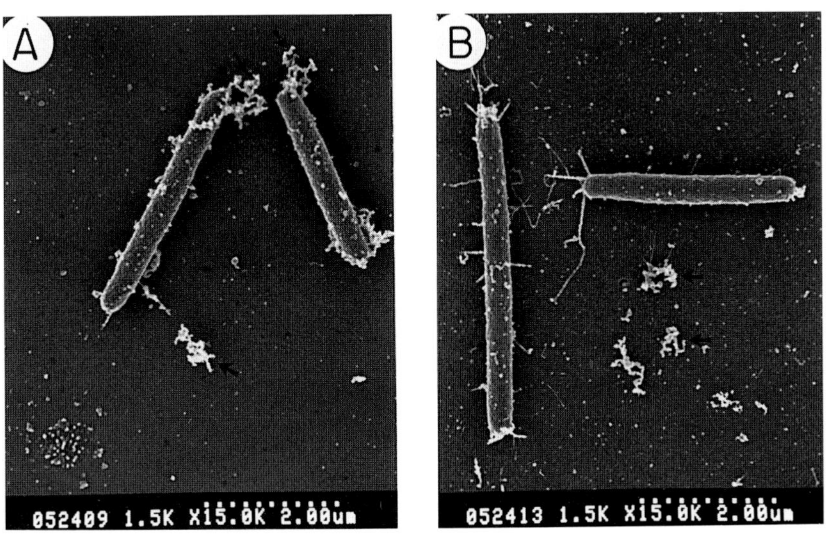

Figure 3. Low-voltage scanning electron micrographs of dsp mutants displaying what appear to be aberrant fibrils. Only rarely were such fibrils observed to connect two cells. The arrows point to debris that could be aberrant fibril material that had become detached from the cells (Behmlander and Dworkin, 1991).

Table 1. Fibril expression in socially (grouped) motile and adventurously (ungrouped) motile cells of *M. xanthus*

Type of cells	No. of cells[a]	% of total
In groups[b]		
With fibrils	441	85
Without fibrils	78	15
Individual[c]		
With fibrils	0	0
Without fibrils	21	100

[a] Six fields were counted from two different sample sets.
[b] Cells not separated from other cells by more than one cell length.
[c] Cells separated from other cells by more than one cell length.

Fig. 4) (this is consistent with the suggestion by Rosenbluh and Eisenbach [1992] that the pili serve as tactile sensors of cell proximity and subsequently of fibril synthesis). Behmlander and Dworkin (1991) isolated fibrils and showed by immunogold labeling that one of the monoclonal antibodies (MAb 2105) against a cell surface protein of *M. xanthus* (Gill and Dworkin,

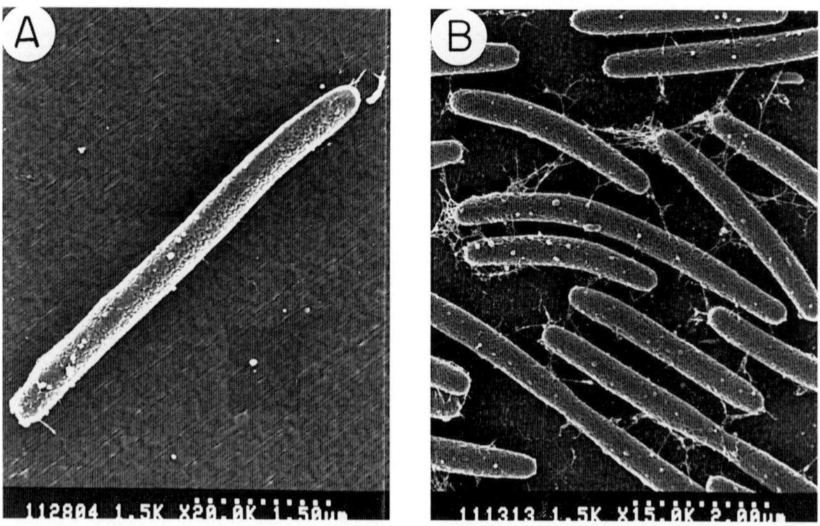

Figure 4. Cell density influences fibril production. Cells were deposited on dialysis membrane strips on nutrient agar at 1×10^7 (A) and 5×10^8 (B) cells per ml and incubated at 32°C for 12 h. Cells at high density began to produce fibrils and apparently began to align themselves into swarms. Cells at low density produced very few fibrils (Behmlander and Dworkin, 1991).

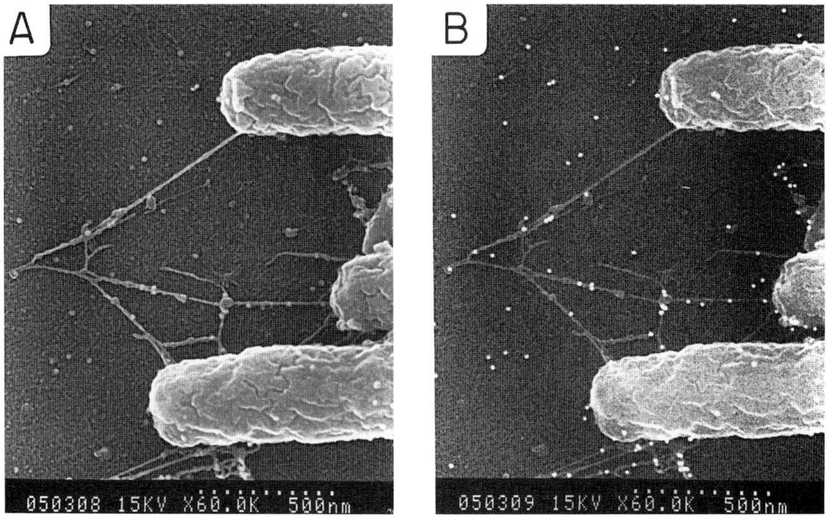

Figure 5. Immunogold localization of fibril antigen FA-1. Cells were labeled with MAb 2105 and a 15-nm-diameter gold conjugate (protein A). (A) Secondary electron image; (B) backscatter electron image. Note that gold particles in the secondary electron image that are not attached to the main fibril strands are nevertheless attached to fibril fragments (Behmlander and Dworkin, 1991).

1986) reacts with the fibrils (Fig. 5). Western immunoblots showed that MAb 2105 was reacting with several bands on electrophoretic gels of isolated fibrils (Fig. 6).

Suva (1978) reported that structures similar to fibrils could be observed by dark-field microscopy of living cells of *M. xanthus*. I have confirmed this observation. It thus seems clear that the fibrils are not an artifact of preparation for the scanning electron microscope. Shimkets and Rafiee (1990) have recently shown by immunogold labeling that C signal is not a membrane-bound signal but rather is a part of the extracellular matrix. It is thus possible that C signal is attached to fibrils and that the fibrils are also involved in signal presentation. If the fibrils are also responsible for holding the cells together during their social behavior, then an interesting problem emerges. Since the cells often alternate between social and individual motility, it is necessary that they be able to break and reform these fibrillar connections. This process may be based on an easily reversible release-reattachment mechanism or on a mechanism that involves reversible assembly-disassembly of the fibrils. Are the fibrils self-assembling structures like pili and flagella? The fact that they are extracellular organelles suggests that they might be.

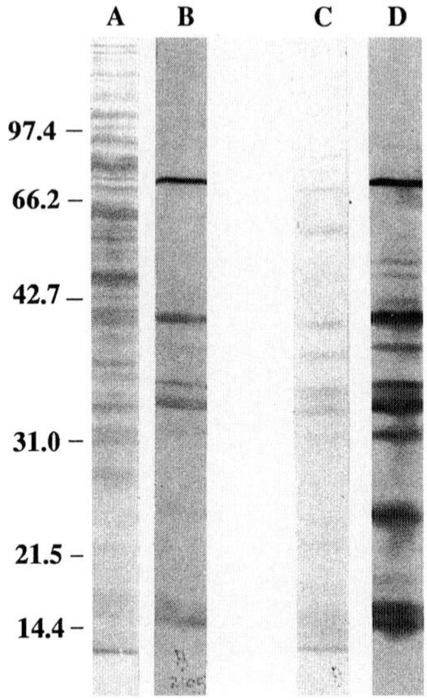

Figure 6. Western immunoblots probed with MAb 2105 (lanes B and D) or stained with Ponceau S (lanes A and C). Lanes A and B contained whole cells, and lanes B and D contained fibrils (Behmlander and Dworkin, 1991).

It is formally possible that the fibrils play a role in any or all of three aspects of contact-mediated cell-cell interactions. They may be involved in establishing the physical associations between cells required for social motility, cell-cell cohesion, and aggregation as a prelude to fruiting body formation. They may serve as antennae or tactile sensors of the presence of other cells or of particulate substrates at a distance. And finally, they may serve as vehicles for the presentation of cell-cell signals. In any case, attempts to understand the composition, structural properties, and biosynthesis of the fibrils and whether there are fibril receptors will help clarify their role in the contact-mediated social behavior of *M. xanthus*.

EXTRACELLULAR POLYSACCHARIDE

The myxobacteria produce large amounts of extracellular polysaccharide, often referred to as slime. While it has been tacitly assumed that slime exists in an amorphous form, the recent discovery of fibrils (Arnold and Shimkets, 1988a, 1988b; Behmlander and Dworkin, 1991), which consist largely of polysaccharide, suggests that this may not be the case. A fair amount of informal and anecdotal evidence suggests that excreted slime plays a role in fruiting body morphogenesis, the laying down of trails, and cell-cell connections. If so, characterization of the myxobacterial polysaccharide and its possible roles in developmental behavior and cell-cell interactions represents an important and neglected aspect of myxobacterial biology.

A few studies of the monosaccharide components of hydrolyzed, bulk polysaccharide in vegetatively growing and developing myxobacteria have been done. The generalizations that have emerged are that the major monosaccharide components of the hydrolyzed extracellular polysaccharide are glucose, galactose, mannose, and rhamnose. The components are approximately the same for a number of different genera representing the two major groups of the myxobacteria, and the values are roughly the same for material excreted by growing cells and for that adhering to cells undergoing development (Sutherland and Thomson, 1975; Sutherland, 1979). However, there have been no descriptions of the nature of the unhydrolyzed polysaccharide. We do not know how many classes of polysaccharide are synthesized by *M. xanthus* or what the sequence and composition of any class of polysaccharides are. We do not know which polysaccharides are excreted by growing vegetative cells and which are associated with the fruiting body, motility trails, or fibrils.

MAbS DIRECTED AGAINST CELL SURFACE ANTIGENS OF *M. XANTHUS*

As part of a program to characterize the cell surface macromolecules of *M. xanthus,* my laboratory has isolated over 40 MAbs directed randomly against various antigens on the surfaces of growing and developing *M. xanthus* cells (Gill and Dworkin, 1986, 1988). These antigens include proteins and the O antigen and core polysaccharide of the lipopolysaccharide (LPS) of the OM. Thirty-one of these MAbs are listed in Table 2. (Those MAbs indicated as reacting in a smear in the Western immunoblot are directed against the O-antigen portion of the LPS. MAb 2254 is directed against the core-lipid A portion of the LPS.)

The MAbs have served a number of purposes. They have been used to determine whether or not a particular cell surface antigen is involved in some aspect of the cell's behavior or development. Gill et al. (1987) showed that cells treated with MAb 1604 were blocked in development but that treatment with another MAb directed against a different cell surface protein had no effect on development. Fink et al. (1989) claimed that this effect of MAb 1604 was a reflection either of nonspecific binding of the MAb to the cell surface or of the MAb blocking development by serving as a nutrient source. This difference has not been resolved. However, it is interesting that the same strategy has been used successfully by Shimkets and Rafiee (1990), who showed that an affinity-purified polyclonal antibody directed against the C signal of *M. xanthus* blocked development.

Table 2. MAb immunoglobulin types and Western blot analysis with MAbs

Hybridoma	Immunoglobulin	Mol wt of material in Western blot analysis[a]
480	M	NR
529	G2A	NR
1070	G2A	NR
1246	G2A	NR
1992	G2B	NR
3095	G	NR
G69	G2A	NR
G357	G2B	NR
G645	G2B	NR
G999	G2B	NR
G1007	G2B	NR
G1172	G2B	NR
2886	G2B	ND
2952	G2A	ND
2982	G	ND
783	G2A	Smear
1412	M	Smear
1514	G2A	Smear
1733	G	Smear
2600	G	Smear
3061	M	Smear
302	G2A	97K, 64K
1604	G2A	200K, 170K, 140K
2105	G1	60K, 25K, 24K
2254	G3	12.5K
2788	G2B	19K, 18K
3012	G2B	41.5K
3079	M	56.5K
3112	G2A	70K
3148	G	29.5K
4000	G2A	24.5K

[a] NR, no reaction; ND, not determined; 97K, 97,000 molecular weight.

The MAbs have been used by Zissler's group to select mutants defective in cell surface antigens and then to examine the effect of those mutations on the behavior and development of the mutants (Fink and Zissler, 1989b; Fink et al., 1989). This work is described below in the discussion of LPS. The MAbs are also useful as immunolabels for determining the locations of cell surface antigens. Using this approach, Clemans et al. (1991) showed that antigen dFA-1, which was absent from vegetatively growing cells but appeared during development, was located on the extracellular fibrils of *M. xanthus*. Behmlander and Dworkin (1991) showed that antigen FA-1 was absent from the fibril$^-$ dsp mutant (Arnold and Shimkets, 1988a) and was localized on the fibrils of wild-type cells (Fig. 4).

The possibility of using MAbs for detailed topographic mapping of the cell surface is quite feasible; the combination of low-voltage field emission scanning electron microscopy (Behmlander and Dworkin, 1991) and the use of Fab fragments substantially increases the resolving power of the process, and the simultaneous use of three sizes of gold markers (Horisberger and Vonlanthen, 1978) allows determination of locations of cell surface antigens in relation to each other. Finally, MAbs have been useful as immunoaffinity reagents for assisting in the isolation of cell surface antigens dFA-1 and 1604 (Clemans et al., 1991; Jarvis and Dworkin, 1989a, 1989b).

NON-LPS SUGARS ASSOCIATED WITH THE CELL SURFACE

Johnson and White (1972) found that lysozyme-solubilized fragments of vegetative cell walls of *M. xanthus* contain a carbohydrate consisting largely of glucose. They suggested that the glucose comprised a polymer that was probably covalently linked to a modified murein fragment. The linkage with murein suggests that the glucose polymer may be located in the periplasm. If so, it may be analogous to the membrane-derived oligosaccharide described for other gram-negative bacteria (Kennedy, 1987). Sutherland and Mackenzie (1977) subsequently showed that the fruiting body myxospores of a number of myxobacteria contain α-1,3-linked glucans on their surfaces. White also showed that over 51% of the dry weight of the capsule surrounding the glycerol-induced myxospore of *M. xanthus* consists of galactosamine (1975).

THE OM OF *M. XANTHUS*

The OM of *M. xanthus* is similar to that of other gram-negative bacteria in that it consists of phospholipid, LPS, and protein (Orndorff and Dworkin, 1980). When the OM and inner membrane are separated from each other, however, several interesting features of the OM emerge. The buoyant density of the OM is considerably lower than that of the inner membrane, which is the reverse of what is usually observed in gram-negative bacteria. When measured by isopycnic density centrifugation, the buoyant density of the OM is 1.166 g/ml, and that of the cytoplasmic membrane is 1.221 g/ml. The atypically low buoyant density of the OM was attributed by Orndorff and Dworkin (1980) to the high phospholipid content of the OM; phosphatidylethanolamine constituted 76% of the phospholipids of the OM. (Phosphatidylethanolamine has also been shown to be the major phospholipid in the membranes of *Myxococcus fulvus* [Kleinig, 1972].)

M. xanthus had earlier been shown to be more sensitive than other gram-negative bacteria to lipophilic antibiotics such as actinomycin D (Dworkin, 1969), rifampin, novobiocin, and thiostrepton (Manoil, personal communication) and to detergents such as sodium dodecyl sulfate (Dworkin, unpublished data). Nikaido and Vaara (1985) have suggested that in the enterobacteria, three hypotheses may explain such an increased permeability to lipophilic agents. (i) There may be a reduction in the hydrophilic carbohydrate chains in the LPS; (ii) there may be a reduction in the amount of normal, lateral LPS-LPS interaction; or (iii) there may be increased amounts of phospholipid in the outer layer. The last possibility has already been shown to be the case in *M. xanthus;* whether the other possibilities also play a role will be determined by a more detailed examination of the LPS of *M. xanthus*. The functional significance of the increased permeability to lipophilic agents of *M. xanthus* also remains to be understood.

A subsequent paper by Orndorff and Dworkin (1982) described the patterns of appearance and disappearance of membrane proteins during development. (It is interesting that the technique devised to separate inner membranes and OM from vegetative cells was not effective with cells that had begun to undergo development.) Using one-dimensional polyacrylamide gel electrophoresis (but with a 27-cm-long gel to maximize the resolution), Orndorff and Dworkin (1982) found six protein bands whose patterns changed reproducibly during development. While this approach has been superseded by far more sophisticated strategies for looking at proteins that change during development, it was clear that while some of the changes required that all the conditions for development be present, some of the protein changes could be induced by varying either the cell density or the presence or absence of a solid surface.

The reorganization of cell surface proteins of *M. xanthus* (presumably OM proteins) during development was also demonstrated by Maeba (1983). Cells were labeled with ^{125}I-lactoperoxidase, and the labeled surface proteins were separated by one- and two-dimensional electrophoreses and autoradiography. A number of proteins appeared only during development, and some proteins in vegetative cells disappeared during development. In this latter group, one of the most heavily labeled proteins was isolated and shown to be a glycoprotein with an apparent molecular mass of 74 kDa and an isoelectric point of 3.2 to 3.3 (Maeba, 1986). (The carbohydrate moiety comprised approximately 13.5% of the dry weight of the protein, but Maeba pointed out that he was unable to determine whether the protein was a true glycoprotein or a covalently linked heteropolysaccharide-protein complex.) This protein, referred to as VGP, for vegetative glycoprotein, was shown to be released by the cells into the medium during development (Glufka and Maeba, 1991). A polyclonal antibody prepared against VGP inhibited both adventurous and social motility, and Maeba suggested that VGP may be involved in the motility machinery.

LPS

The work of Luderitz's group represents the first attempt to characterize the LPS of the myxobacteria. Weckesser et al. (1971) demonstrated the presence of the lipophilic sugar 3-*O*-methyl-D-xylose in the LPS of *M. fulvus,* and later, Rosenfelder et al. (1974) did a thorough chemical study of the components of the LPS of *M. fulvus*. (In a later study, Sutherland [1976] was unable to detect 3-*O*-methyl-D-xylose in vegetative cells of *M. xanthus*.) Rosenfelder et al. (1974) found that the main sugar components were mannose, D-galactose, D-glucose, rhamnose, glucosamine, and galactosamine. There were relatively small amounts of ribose, xylose, arabinose, and 2-keto-3-deoxyoctonate. Heptose, a common component of the LPS of other gram-negative bacteria, was absent. Among the fatty acids, isobranched fatty acids predominated, with an average chain length of 16 (compared with 14 for the enterobacteria). Lipid A could be demonstrated, with glucosamine as its principal sugar component. Rosenfelder et al. (1974) concluded that the LPS of *M. fulvus* was not fundamentally different from those of other gram-negative bacteria.

The interesting question as to the fate of the LPS during cell morphogenesis was first addressed by Sutherland (1976). He found that at least the polysaccharide portion of the LPS present in vegetative cells of *M. xanthus* was absent from fruiting body spores and suggested that the LPS was replaced by a capsule containing the α-1,3-glucan and a galactosamine-containing polysaccharide.

In an attempt to determine whether covalent modifications played a role in the regulation of development in *M. xanthus*, Panasenko (1985) looked at the patterns of methylation during development. She found that a fraction that behaved like LPS became methylated during the early stages of development. This interesting finding has not been pursued.

The nature of the LPS of *M. xanthus* and its role in behavior and development has been further examined in a series of three papers from Zissler's group. These papers described the isolation and electrophoretic examination of the LPS of *M. xanthus* (Fink and Zissler, 1989a), the use of MAbs directed against LPS to isolate a series of LPS mutants (Fink et al., 1989), and the properties of those mutants (Fink and Zissler, 1989b).

Phenol-extracted LPS was characterized electrophoretically and showed a ladder configuration consisting of about 30 higher-molecular-mass bands within a range of 15 to 70 kDa plus a broad, low-molecular-mass band at about 5 kDa (Fink and Zissler, 1989a). These bands were then further characterized by Western immunoblot with a set of MAbs directed against the LPS of *M. xanthus* (Gill et al., 1987). Five of the MAbs reacted with the laddered bands but not the broad, low-molecular-mass band; a sixth MAb (MAb 2254) reacted with both the laddered bands and the low-molecular-mass band. Fink and Zissler (1989a) suggested that the five MAbs were reacting with the O antigen of the LPS, while MAb 2254 was reacting with the core-lipid A portion. Fink et al. (1989) then used these MAbs to isolate Tn*5* mutants of *M. xanthus* defective in O antigen and core polysaccharide. All of these mutants were resistant to myxophage Mx8; phage resistance was then used as the primary selection to isolate two additional classes of mutants, one class resistant to phage Mx8 and defective in O antigen and a second class resistant to myxophage Mx1 and defective in core antigen (i.e., nonreactive with MAb 2254) or O antigen.

All of the mutants initially selected for on the basis of nonreactivity with MAbs directed against O antigen (and later shown to be resistant to phage Mx8) were defective in adventurous motility, normal in social motility, and delayed in developmental activities. On the other hand, many of the mutants that were isolated on the basis of their Mx8 resistance and found to be defective in O antigens 1514 and 783 still showed normal adventurous and social motilities as well as normal development (Fink and Zissler, 1989b). It should be emphasized that this latter group of mutants was not subjected to a rigorous genetic analysis to show that each phenotype was a result of a single mutation.

Mutants blocked in their abilities to react with MAb 2254, directed against core LPS, showed a different phenotype. These mutants were capable of adventurous and social motilities but had smaller colonies, reflecting a reduced motility. These mutants were completely blocked in their development (Fink and Zissler, 1989b).

Another group of previously characterized A^- mutants (Hodgkin and Kaiser, 1979a, 1979b) were screened and found to contain all of the LPS epitopes that were absent in the mutants isolated by Zissler's group. They thus concluded that their LPS mutants constitute a new group of mutants with mutations in adventurous motility (Fink and Zissler, 1989b).

The precise structure of the LPS of *M. xanthus* and its possible role in shape change during myxospore development and in adventurous motility remain interesting and unresolved.

MUREIN OF *M. XANTHUS*

An earlier hope was that an examination of the chemical changes in the murein of *M. xanthus* cells undergoing morphological change would provide some insights into the nature of the morphogenetic process (White et al., 1968). White et al. (1968) did examine the murein of *M. xanthus* and found that it was fundamentally similar to those of other gram-negative bacteria; the molar ratios of glutamate, diaminopimelate, alanine, glucosamine, and muramic acid were 1.0, 1.0, 1.7, 0.75, and 0.75, respectively, but these components were accompanied by substantial amounts of glycine, serine, and glucose. There were a number of other interesting features of the *M. xanthus* murein. When isolated, intact murein ghosts were treated with trypsin, which had no effect on the murein ghosts of *Pseudomonas* spp. (Heilmann, 1972), the murein was unexpectedly fragmented into smaller pieces and most of the glycine was solubilized. This suggested to White et al. that the murein of *M. xanthus* was actually patches of murein separated by trypsin-sensitive interpatch material rather than a "bagshaped macromolecule," the term used to describe conventional murein (Weidel and Pelzer, 1964). It is still not clear whether this patchiness is designed to facilitate the shape change that cells undergo when they convert from vegetative cells to myxospores or whether it is a general feature of gliding bacteria.

When these smaller fragments were solubilized with lysozyme, one of the solubilized fragments consisted of a moiety containing muramic acid and alanine as its major component and considerable amounts of covalently bound glucose (Johnson and White, 1972). The serine remained bound to the insoluble fraction of the murein. The amount of glucose in the murein was substantial, as evidenced by the fact that the molar ratio of glucose to diaminopimelic acid in the murein was 2.6.

During conversion of the vegetative rods to glycerol-induced myxospores, the amount of cross-linking between diaminopimelic acid and alanine transiently decreased (White et al., 1968) and then increased in the mature myxospore to a value 11% greater than that in the vegetative cells (Johnson and White, 1972). This suggested to the latter authors that there was autolysis and murein turnover during myxospore morphogenesis, a proposal that was subsequently verified by Dawson and Jones (1979a) for both glycerol-induced myxospores and fruiting body myxospores (Dawson and Jones, 1979b).

THE MYXOSPORE COAT

The myxospore of *M. xanthus* is the only myxobacterial resting cell that has been characterized. Early studies of the spore surface focused on the properties of myxospores induced by exposure to high concentrations of glycerol or to one of a number of other inducing compounds (Dworkin and Gibson, 1964; Kottel et al., 1975).

While glycerol-induced myxospores share a number of important physiological and morphological properties with fruiting body myxospores (see chapter 14), there are some equally important differences (Zusman, 1984a, 1984b), and attention has now shifted back to myxospores formed during normal development.

Each fruiting body spore is surrounded by a thick coat whose major component is protein S. The synthesis of protein S begins early in development, and soluble protein S accumulates inside the cell until sporulation begins. Synthesis of protein S reaches a maximum at late aggregation, at which time it represents 15% of the total protein synthesis of the cell; it is ultimately found on the surface of the spore (Inouye et al., 1979).

Protein S can be removed from the cell surface by extraction with 1 M NaCl or EDTA, which removes the entire outer layer of the spore coat; in the presence of Ca^{2+}, protein S will then reassemble on the spore surface (Inouye et al., 1979).

There are several interesting properties of protein S and its regulation. The protein, which has a molecular mass of 18,792 Da, has no signal sequence; there are two high-affinity Ca^{+}-binding domains whose amino acid sequences have a striking similarity to those of calmodulin (Inouye et al., 1983a) and an even more striking resemblance to the β and γ bovine crystalline lens proteins (Wistow et al., 1985).

Protein S and the closely related protein S1 are coded for by the *tps* and *ops* genes, respectively, two adjacent, highly similar genes located in tandem on the *M. xanthus* chromosome and separated by a short spacer sequence (Inouye et al., 1983a, 1983b). The genes are expressed at different developmental stages: *tps* codes for protein S, is induced by starvation, and is expressed during the preaggregation stage; *ops* codes for protein S1, and expression begins at the onset of sporulation. Protein S1 accumulates in the cytoplasm along with small amounts of protein S and represents about 1/10 of the amount of protein S plated on the outside of the cell (Inouye et al., 1979). The facts that the two closely related genes are expressed at different times during development and that their products are located in different parts of the cell suggest that their functions may be different. Deletion mutants with deletions of the *tps* or *ops* gene have been constructed, and while the absence of the *tps* gene had no effect on the behavior or development of the cells, deletion of *ops* caused a slight delay in development. However, when both genes were deleted, development was substantially delayed (Komano et al., 1984; Teintze et al., 1985). Thus, an alternative explanation is that the *tps* and *ops* genes function as backups for each other. The most recent review on protein S can be found in Teintze et al. (1991).

Protein U is another spore coat protein that is produced at a late stage of development (Inouye et al., 1979). However, unlike protein S, it has a signal sequence and is synthesized as a pro-protein precursor. Its molecular mass is 15,555 Da; like protein S, it has a high content of β structure but bears no sequence similarity to any known protein (Gollop et al., 1991).

In view of the fact that protein S is excreted during sporulation, it has been puzzling that neither of the protein S genes contains information that codes for a signal peptide (Inouye et al., 1983b). Teintze et al. (1985) have suggested that protein S is released during developmental autolysis (Wireman and Dworkin, 1977) and is taken up and plated on the surface of the surviving cells. They have suggested that that may, in fact, be the function of developmental autolysis. The fact that protein U, likewise a spore coat protein, is excreted via a signal sequence indicates that there are at least two pathways of developmental protein secretion but does not indicate why the two pathways exist.

MBHA

Myxobacterial hemagglutinin (MBHA) is a developmental lectin with a molecular mass of 27,920 Da that appears on the cell surface of *M. xanthus* during late aggregation and before myxospore formation (Cumsky and Zusman, 1979). (The first demonstration of lectin-hemagglutinin activity in *M. xanthus* was by Dobson et al. [1979], who showed that the fimbriae [pili] of *M. xanthus* were able to agglutinate guinea pig erythrocytes and that the agglutination was strongly inhibited by mannose and *N*-acetyl galactosamine.)

Receptors for MBHA are present on the surfaces of both vegetative and developing cells, even though MBHA itself is synthesized only by developing cells. On developing cells, the receptors are bipolar (Nelson et al., 1981); their locations on vegetative cells have not been determined. Binding of MBHA to vegetative-cell receptors but not to those of developing cells was inhibited by the glycoprotein fetuin and specifically by its penultimate sugar, galactose (Cumsky and Zusman, 1981).

The MBHA receptors are more numerous on developing cells and have a substantially higher K_a than do the receptors on vegetative cells (Cumsky and Zusman, 1981). The presence of MBHA receptors on vegetative cells, which are unable to synthesize MBHA and which have properties different from those of developmental cells, suggests that these vegetative-cell receptors may function in lectin-mediated interactions that do not involve MBHA. It would indeed be interesting to see the properties of cells with mutations in both vegetative and developmental MBHA receptors.

While mutations in the structural gene for MBHA had no effect on development under normal conditions, development was severely delayed in the absence of added Mg^{2+}. Wild-type cells, however, developed normally in the absence of added Mg^{2+} (Romeo and Zusman, 1987). This suggests that the cells require either MBHA or Mg^{2+} for some sort of cell-cell interaction during aggregation. The function of lectins in general suggests that MBHA may play a role in establishing or maintaining physical cell-cell interactions during development. What could be the nature of this interaction? A few clues indicate that this interaction may involve social motility. Nelson et al. (1989) made the statement that "the level of MBHA production is also correlated with the S-motility system," and Cumsky and Zusman (1979) quoted a personal communication from LaRossa and Kaiser that ". . . an SglA1 mutant and several other S^- strains have less hemagglutinating activity than a genetically comparable S^+ strain." Thus, the relationship between MBHA and S motility may be worth reexamining.

Relatively little attention has been paid to the polysaccharide components of the cell surface of *M. xanthus*. Thus, the extent of functional, structural, and biosynthetic overlap among the fibrils, extracellular polysaccharide, and LPS remains undefined. Nevertheless, work is now proceeding in these areas and will undoubtedly lead to an improved understanding of the role of these surface components in growth, development, motility, and cell-cell and cell-substratum interactions.

REFERENCES

Arnold, J. W., and L. J. Shimkets. 1988a. Inhibition of cell-cell interactions in *Myxococcus xanthus* by Congo red. *J. Bacteriol.* **170:**5765–5770.

Arnold, J. W., and L. J. Shimkets. 1988b. Cell surface properties correlated with cohesion in *Myxococcus xanthus*. *J. Bacteriol.* **170:**5771–5777.

Behmlander, R. M., and M. Dworkin. 1991. Extracellular fibrils and contact-mediated cell interactions in *Myxococcus xanthus*. *J. Bacteriol.* **173:**7810–7821.

Clemans, D. L., C. M. Chance, and M. Dworkin. 1991. A development-specific protein in *Myxococcus xanthus* is associated with the extracellular fibrils. *J. Bacteriol.* **173:** 6749–6759.

Cumsky, M., and D. R. Zusman. 1979. Myxobacterial hemagglutinin, a development-specific lectin of *Myxococcus xanthus*. *Proc. Natl. Acad. Sci. USA* **76:**5505–5509.

Cumsky, M. G., and D. R. Zusman. 1981. Binding properties of myxobacterial hemagglutinin. *J. Biol. Chem.* **256:**12596–12599.

Dawson, H., and M. V. Jones. 1979a. Cell wall turnover during myxospore formation in *Myxococcus xanthus*. *J. Gen. Microbiol.* **112:**143–148.

Dawson, H., and M. V. Jones. 1979b. Use of membrane filters to study the development of fruiting bodies of *Myxococcus xanthus*. *J. Gen. Microbiol.* **114:**217–220.

Dobson, W. J., and H. D. McCurdy. 1979. The function of fimbriae in *Myxococcus xanthus*. I. Purification and properties of *M. xanthus* fimbriae. *Can. J. Microbiol.* **25:**1152–1160.

Dobson W. J., H. D. McCurdy, and T. H. MacRae. 1979. The function of fimbriae in *Myxococcus xanthus*. II. The role of fimbriae in cell-cell interactions. *Can. J. Microbiol.* **25:** 1359–1372.

Dworkin, M. 1969. Sensitivity of gliding bacteria to actinomycin D. *J. Bacteriol.* **98:**851–852.

Dworkin, M. Unpublished data.

Dworkin, M., and S. M. Gibson. 1964. A system for studying microbial morphogenesis: rapid formation of microcysts in *Myxococcus xanthus*. *Science* **146:**143–144.

Fink, J. M., M. Kalos, and J. F. Zissler. 1989. Isolation of cell surface antigen mutants of *Myxococcus xanthus* by use of monoclonal antibodies. *J. Bacteriol.* **171:**2033–2041.

Fink, J. M., and J. F. Zissler. 1989a. Characterization of lipopolysaccharide from *Myxococcus xanthus* by use of monoclonal antibodies. *J. Bacteriol.* **171:**2028–2032.

Fink, J. M., and J. F. Zissler. 1989b. Defects of motility and development of *Myxococcus xanthus* lipopolysaccharide mutants. *J. Bacteriol.* **171:**2042–2048.

Gill, J. S., and M. Dworkin. 1986. Cell surface antigens during submerged development of *Myxococcus xanthus* examined with monoclonal antibodies. *J. Bacteriol.* **168:**505–511.

Gill, J. S., and M. Dworkin. 1988. Isolation of additional monoclonal antibodies directed against cell surface antigens of *Myxococcus xanthus* cells undergoing submerged development. *J. Bacteriol.* **170:**5953–5955.

Gill, J. S., B. W. Jarvis, and M. Dworkin. 1987. Inhibition of development in *Myxococcus xanthus* by monoclonal antibody 1604. *Proc. Natl. Acad. Sci. USA* **84:**4505–4508.

Glufka, R., and P. Maeba. 1991. Release of a cell surface protein during development of *Myxococcus xanthus*. *J. Bacteriol.* **173:**7988–7991.

Gollop, R., M. Inouye, and S. Inouye. 1991. Protein U, a late-developmental spore coat protein of *Myxococcus xanthus*, is a secretory protein. *J. Bacteriol.* **173:**3597–3600.

Heilmann, H. D. 1972. On the peptidoglycan of the cell walls of *Pseudomonas aeruginosa*. *Eur. J. Biochem.* **31:**456–463.

Hodgkin, J., and D. Kaiser. 1979a. Genetics of gliding motility in *Myxococcus xanthus* (Myxobacterales): genes controlling movements of single cells. *Mol. Gen. Genet.* **171:**167–176.

Hodgkin, J., and D. Kaiser. 1979b. Genetics of gliding motility in *Myxococcus xanthus* (Myxobacterales): two gene systems control movement. *Mol. Gen. Genet.* **172:**177–191.

Horisberger, M., and M. Vonlanthen. 1978. Multiple marking of cell surface receptors by gold granules: simultaneous localization of three lectin receptors on human erythrocytes. *J. Microsc.* **115:**97–102.

Inouye, M., S. Inouye, and D. R. Zusman. 1979. Biosynthesis and self-assembly of protein S, a development specific protein of *Myxococcus xanthus*. *Proc. Natl. Acad. Sci. USA* **76:**209–213.

Inouye, S., T. Franceschini, and M. Inouye. 1983a. Structural similarities between the developmental-specific protein from a gram negative bacterium, *Myxococcus xanthus*, and calmodulin. *Proc. Natl. Acad. Sci. USA* **80:**6829–6833.

Inouye, S., Y. Ike, and M. Inouye. 1983b. Tandem repeat of the genes for protein S, a development specific protein for *Myxococcus xanthus*. *J. Biol. Chem.* **258:**38–40.

Jarvis, B. W., and M. Dworkin. 1989a. Purification and properties of *Myxococcus xanthus* cell surface antigen 1604. *J. Bacteriol.* **171:**4655–4666.

Jarvis, B. W., and M. Dworkin. 1989b. Role of *Myxococcus xanthus* cell surface antigen 1604 in development. *J. Bacteriol.* **171:**4667–4673.

Johnson, R. J., and D. White. 1972. Myxospore formation in *Myxococcus xanthus:* chemical changes in cell wall during cellular morphogenesis. *J. Bacteriol.* **112:**849–855.

Kaiser, D. 1979. Social gliding is correlated with the presence of pili in *Myxococcus xanthus*. *Proc. Natl. Acad. Sci. USA* **76:**5952–5956.

Kennedy, E. P. 1987. Membrane-derived oligosaccharides, p. 672–679. *In* F. C. Neidhardt, J. L. Ingraham, K. B. Low, B. Magasanik, M. Schaechter, and H. E. Umbarger (ed.), *Escherichia coli and Salmonella typhimurium: Cellular and Molecular Biology*, vol. 1. American Society for Microbiology, Washington, D.C.

Kleinig, H. 1972. Membranes from *Myxococcus fulvus* (Myxobacterales) containing carotenoid glycosides. I. Isolation and composition. *Biochim. Biophys. Acta* **274:**489–498.

Komano, T., T. Furuichi, M. Teintze, M. Inouye, and S. Inouye. 1985. Effects of the deletion of the gene for the development-specific protein S on differentiation in *Myxococcus xanthus*. *J. Bacteriol.* **158:**1195–1197.

Kottel, R. H., K. Bacon, D. Clutter, and D. White. 1975. Coats from *Myxococcus xanthus:* characterization and synthesis during myxospore differentiation. *J. Bacteriol.* **124:** 550–557.

LaRossa R., and D. Kaiser. Personal communication.

MacRae, T. H., and H. D. McCurdy. 1975. Ultrastructural studies of *Chondromyces crocatus* vegetative cells. *Can. J. Microbiol.* **21:**1815–1826.

Maeba, P. 1986. Isolation of a surface protein from *Myxococcus xanthus*. *J. Bacteriol.* **166:** 644–650.

Maeba, P. Y. 1983. Iodination of *Myxococcus xanthus* during development. *J. Bacteriol.* **155:**1033–1041.

Manoil, C. Personal communication.

Nelson, D. D., M. G. Cumsky, and D. R. Zusman. 1981. Localization of the myxobacterial hemagglutinin in the periplasmic space and on the cell surface of *Myxococcus xanthus* during developmental aggregation. *J. Biol. Chem.* **256:**12589–12595.

Nikaido, H., and M. Vaara. 1985. Molecular basis of bacterial outer membrane permeability. *Microbiol. Rev.* **49:**1–32.

Orndorff, P. E., and M. Dworkin. 1980. Separation and properties of the cytoplasmic and outer membranes of vegetative cells of *Myxococcus xanthus*. *J. Bacteriol.* **141:**914–947.

Orndorff, P. E., and M. Dworkin. 1982. Synthesis of several membrane proteins during developmental aggregation in *Myxococcus xanthus*. *J. Bacteriol.* **149:**29–39.

Panasenko, S. M. 1985. Methylation of macromolecules during development in *Myxococcus xanthus*. *J. Bacteriol.* **164:**495–500.

Romeo, J. M., and D. R. Zusman. 1987. Cloning of the gene for myxobacterial hemagglutinin and isolation and analysis of structural gene mutations. *J. Bacteriol.* **169:**3801–3808.

Rosenbluh, A., and M. Eisenbach. 1992. The effect of mechanical removal of pili on gliding motility in *Myxococcus xanthus*. *J. Bacteriol.* **174:**5406–5413.

Rosenfelder, G., O. Luderitz, and O. Westphal. 1974. Composition of lipopolysaccharides from *Myxococcus fulvus* and other fruiting and nonfruiting myxobacteria. *Eur. J. Biochem.* **44:**411–420.

Shimkets, L. J. 1986. Role of cell cohesion in *Myxococcus xanthus* fruiting body formation. *J. Bacteriol.* **166:**842–848.

Shimkets, L. J., and H. Rafiee. 1990. CsgA, an extracellular protein essential for *Myxococcus xanthus* development. *J. Bacteriol.* **172:**5299–5306.

Sutherland, I. W. 1976. Novel surface polymer changes in development of *Myxococcus* spp. *Nature* (London) **259:**46–47.

Sutherland, I. W. 1979. Polysaccharides produced by *Cystobacter, Archangium, Sorangium* and *Stigmatella* species. *J. Gen. Microbiol.* **74:**259–266.

Sutherland, I. W., and C. L. Mackenzie. 1977. Glucan common to the microcyst walls of cyst-forming bacteria. *J. Bacteriol.* **129:**599–605.

Sutherland, I. W., and S. Thomson. 1975. Comparison of polysaccharides produced by *Myxococcus* strains. *J. Gen. Microbiol.* **89:**124–132.

Suva, R. 1978. Surface filaments of *M. xanthus,* p. 25. *In Proceedings of the Fifth International Conference on the Biology of the Myxobacteria.* Spring Hill Conference Center, Wayzata, Minn.

Teintze, M., T. Furuichi, R. Thomas, M. Inouye, and S. Inouye. 1985. Differential expression of two homologous genes coding for spore-specific proteins in *Myxococcus xanthus,* p. 253–260. *In* J. Hoch and P. Setlow (ed.), *Spores IX: The Molecular Biology of Microbial Differentiation.* American Society for Microbiology, Washington, D.C.

Teintze, M., M. Inouye, and S. Inouye. 1991. A development-specific Ca^{2+}-binding protein from *Myxococcus xanthus,* p. 437–446. *In* C. W. Heitzmann (ed.), Novel calcium-binding proteins. Springer-Verlag, Berlin.

Weckesser, J., G. Rosenfelder, H. Mayer, and O. Luderitz. 1971. The identification of 3-*O*-methyl-D-xylose and 3-*O*-methyl-L-xylose as constituents of the lipopolysaccharides of *Myxococcus fulvus* and *Rhodopseudomonas viridis,* respectively. *Eur. J. Biochem.* **24:** 112–115.

Weidel, W., and H. Pelzer. 1964. Bagshaped macromolecules: a new outlook on bacterial cell walls. *Adv. Enzymol.* **26:**193–232.

White, D. 1975. Myxospores of *Myxococcus xanthus,* p. 44–51. *In* P. Gerhardt, H. I. Sadoff, and R. N. Costilow (ed.), *Spores VI.* American Society for Microbiology, Washington, D.C.

White, D., M. Dworkin, and D. J. Tipper. 1968. Peptidoglycan of *Myxococcus xanthus:* structure and relation to morphogenesis. *J. Bacteriol.* **95:**2186–2197.

Wireman, J. W., and M. Dworkin. 1977. Developmentally induced autolysis during fruiting body formation by *Myxococcus xanthus. J. Bacteriol.* **139:**796–802.

Wistow, G., L. Summers, and T. Blundell. 1985. *Myxococcus xanthus* spore coat protein S may have a similar structure to vertebrate lens βγ-crystallins. *Nature* (London) **315:**771.

Zusman, D. R. 1984a. Developmental program of *Myxococcus xanthus,* p. 185–213. *In* E. Rosenberg (ed.), *Myxobacteria: Development and Cell Interactions.* Springer-Verlag, New York.

Zusman, D. R. 1984b. Cell-cell interactions and development in *Myxococcus xanthus. Q. Rev. Biol.* **59:**119–138.

Myxobacteria II
Edited by Martin Dworkin and Dale Kaiser
© 1993 American Society for Microbiology, Washington, DC 20005

Chapter 4

The Myxobacterial Genome

Lawrence J. Shimkets

The myxobacterial genome encodes the marvelous morphogenic behaviors of fruiting body formation and rippling that appear to be unique in the prokaryotic world. The evolutionary emergence of myxobacteria more than 650 million years ago appears to be correlated with a large increase in genome size (Shimkets, 1990). Where did the genes encoding these behaviors originate? Perhaps the secrets to the ancient changes in genome structure that accompanied the evolution of these morphogenic behaviors can be discerned by comparing myxobacterial genes and genomes with those of related organisms. This chapter summarizes what is known about genome structure and evolution in myxobacteria and their closest relatives.

DNA BASE COMPOSITION

Myxobacteria have G+C values of 67 to 71 mol% (Johnson and Ordal, 1968; Mandel and Leadbetter, 1965; McCurdy and Wolf, 1967) and thus lie near one extreme of the range of 24 to 74 mol% G+C observed in bacteria (Normore, 1976). The G+C content of *Myxococcus xanthus* is 67.5 mol% according to high-performance liquid chromatography of enzymatically digested DNA (Mesbah et al., 1989). Bacterial genes that code for proteins utilize codons whose base compositions are characteristic of the overall DNA base compositions of the DNA. Because of the degeneracy of the genetic code, changes in the third codon position can be used to adjust the average base composition of the codon without altering the amino acid composition of the protein. Organisms with high

Lawrence J. Shimkets • Department of Microbiology, University of Georgia, Athens, Georgia 30602.

G+C contents tend to use codons with G or C residues in the third position of the codon (Bibb et al., 1984). The combined codon usage of 22 *M. xanthus* genes revealed 70% G+C residues in the first codon position, 47% in the second position, and 91% in the third position, for an average G+C composition of 69% (Table 1). This G+C distribution is typical of other unrelated bacteria with similar base compositions (Bibb et al., 1984). Several developmental genes are preceded by long intercistronic regions that appear to play a role in the regulation of gene expression. The intercistronic regions preceding the *csgA* gene (Li et al., 1992), the *tps* and *ops* genes (Inouye et al., 1983b), and the *socA* operon (Lee and Shimkets, unpublished data) have an average G+C content of 68%. It appears, then that the base compositions of noncoding regions are maintained at levels approximately equal to those of the protein-coding regions. Unlike the protein-coding regions, however, the noncoding regions show no obvious pattern in the distribution of G and C residues. These results allow reliable prediction of the reading frame and coding strand of DNA sequences.

PHYLOGENY AND AGE

One way to investigate the phylogeny of the myxobacteria is a comparison of the rRNA genes, which are the most highly conserved genes in the cell. Oligonucleotide cataloging (Ludwig et al., 1983) and sequencing (Shimkets and Woese, 1992) of the 16S rRNA molecules from a variety of myxobacteria have suggested that all myxobacteria are closely related but that they can be divided into three distinct subgroups, one of which contains the lone species *Nannocystis exedens* (Fig. 1). Unique secondary structural features of the 16S rRNA molecule also define these subgroups (Shimkets and Woese, 1992). Other features differentiate and support this subgrouping. Members of the subgroup containing *Myxococcus* spp. have long, slender vegetative cells that change their shapes considerably during myxospore formation (Reichenbach and Dworkin, 1991; McCurdy, 1989). Their colonies also absorb the diazo dye Congo red. These cells tend to contain monocyclic carotenoids with reddish keto pigments (Reichenbach and Kleinig, 1984), 3-hydroxy and 2-hydroxy fatty acids, and antibiotics of different chemical structures than those of the other two groups (Reichenbach et al., 1988). These are the only bacteria known to move in pulsating waves called ripples (Reichenbach, 1986). Members of the other two subgroups, which include *Chondromyces* spp. and *N. exedens,* have much shorter vegetative cells that change their shapes little during myxospore formation. Colonies of members of this group do not ripple or absorb Congo red. There are also substantial differences in the chemistry of pigments and secondary metabolites between *N. exedens* and mem-

Table 1. *M. xanthus* codon usage[a]

First position	Second position											
	U			C			A			G		
	Amino acid	Codon	No. of codons	Amino acid	Codon	No. of codons	Amino acid	Codon	No. of codons	Amino acid	Codon	No. of codons
U	Leu	UUG	64	Ser	UCG	87	End	UAG	5	Trp	UGG	56
	Leu	UUA	0	Ser	UCA	13	End	UAA	0	End	UGA	17
	Phe	UUU	9	Ser	UCU	11	Tyr	UAU	13	Cys	UGU	10
	Phe	UUC	185	Ser	UCC	142	Tyr	UAC	106	Cys	UGC	32
C	Leu	CUG	406	Pro	CCG	161	Gln	CAG	221	Arg	CGG	160
	Leu	CUA	4	Pro	CCA	16	Gln	CAA	13	Arg	CGA	20
	Leu	CUU	13	Pro	CCU	29	His	CAU	19	Arg	CGU	41
	Leu	CUC	202	Pro	CCC	153	His	CAC	107	Arg	CGC	329
A	Met	AUG	57	Thr	ACG	175	Lys	AAG	283	Arg	AGG	26
	Ile	AUA	4	Thr	ACA	12	Lys	AAA	10	Arg	AGA	4
	Ile	AUU	34	Thr	ACU	8	Asn	AAU	20	Ser	AGU	9
	Ile	AUC	225	Thr	ACC	148	Asn	AAC	172	Ser	AGC	103
G	Val	GUG	370	Ala	GCG	418	Glu	GAG	388	Gly	GGG	84
	Val	GUA	5	Ala	GCA	35	Glu	GAA	72	Gly	GGA	33
	Val	GUU	10	Ala	GCU	37	Asp	GAU	54	Gly	GGU	52
	Val	GUC	190	Ala	GCC	375	Asp	GAC	319	Gly	GGC	395

[a] This table is derived from the coding sequences of *mbhA* (Romeo et al., 1986); *tps* and *ops* (Inouye et al., 1983b); *vegA* (Komano et al., 1987); *mglA* (Stephens et al., 1989); *csgA* and *fprA* (Hagen and Shimkets, 1990); *frzA, frzB,* and *frzCD* (McBride et al., 1989); *sigA* (Inouye, 1990); *sigB* (Apelian and Inouye, 1990); *ndk* (Munoz-Dorado et al., 1990); *frzE* (McCleary and Zusman, 1990); *uraA* and its upstream protein-coding region (Kimsey and Kaiser, 1992); *pkn-1* (Munoz-Dorado et al., 1991); the protein-coding region upstream of *csgA* (Li et al., 1992); the three genes in the *socA* operon; Mx162 reverse transcriptase (Inouye et al., 1989); and Mx65 reverse transcriptase (Inouye et al., 1990).

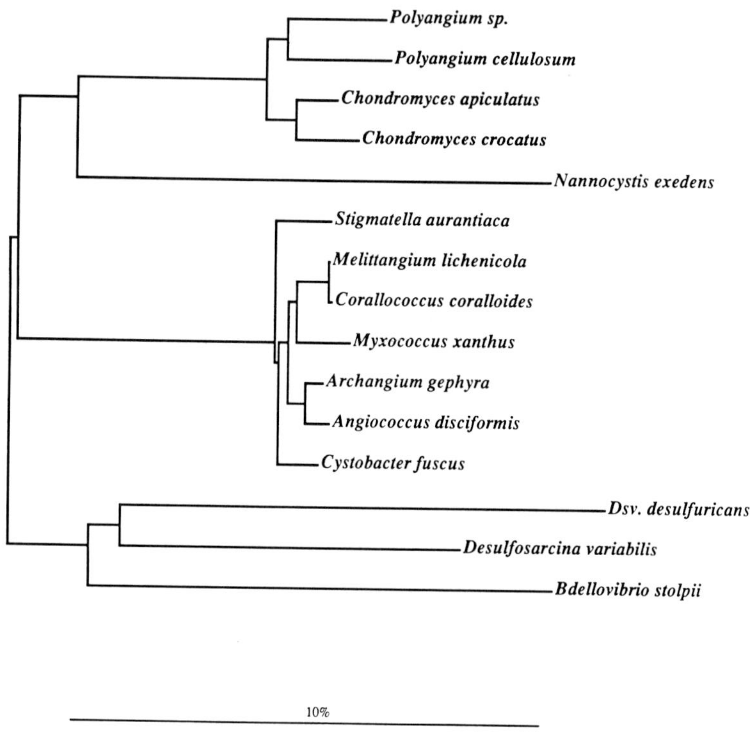

Figure 1. Phylogenetic tree for the myxobacteria. Tree was derived from evolutionary distances of the 16S rRNA molecules (courtesy of Shimkets and Woese, in press).

bers of the subgroup containing *Chondromyces* spp.: the carotenoid glucosides of *Chondromyces* spp. lack the keto groups found in the carotenoid glucosides of members of the *Myxococcus* subgroup, while *Nannocystis* sp. forms monocyclic aromatic compounds that are unique to that species (Reichenbach and Kleinig, 1984). *Nannocystis* sp. is also the only myxobacterium to synthesize sterols and squalene (Kohl et al., 1983). While the taxonomic rank based on 16S rRNA sequences is currently in a state of flux and may change in the future as the phylogenetic tree expands, the phylogenetic distances between these three subgroups are consistent with their classification as families (Shimkets and Woese, 1992).

Studies of 16S rRNA have suggested that the closest relatives of the myxobacteria are *Bdellovibrio* spp. and the sulfur- and sulfate-reducing bacteria that together make up the delta subgroup of the purple bacteria (Fig. 1; Fowler et al. 1986; Oyaizu and Woese, 1985; Shimkets and Woese, 1992). Sequence comparisons of 5S rRNA also support a phylogenetic relationship between the myxobacteria and *Desulfovibrio vulgaris,* one of the sulfate reducers (Van den Eynde et al., 1990). These results suggest that myxobacteria and *Bdellovibrio* spp. are aerobic descendants of an ancestral sulfur- or sulfate-reducing bacterium (Woese, 1987).

Although the rate of 16S rRNA evolution is not the same in all lineages, estimates of the average rate of change can be used to approximate the ages of bacterial groups. The 16S molecule appears to acquire nucleotide substitutions at a rate of about 1% per 50 million years (Ochman and Wilson, 1987). The minimum age of the myxobacteria can be estimated from the point at which the three major myxobacterial subgroups diverged. These calculations ignore *Nannocystis* sp., which is evolving at a highly accelerated rate relative to all other myxobacteria (Shimkets and Woese, 1992). The common ancestor of the myxobacterial subgroups has about 13% nucleotide sequence divergence, which corresponds to a divergence time of about 650 million years before present (BP). The maximum age of the myxobacteria can be estimated from the point at which the myxobacteria diverged from the most closely related sulfate-reducing bacterium, which corresponds to a divergence time of about 800 million years BP. The minimum age of the delta subgroup can be estimated from the deepest branch within the subgroup (Devereux et al., 1989) and corresponds to a divergence time of about 750 million years BP. The maximum age of the delta subgroup can be estimated from the 22% sequence divergence between *Escherichia coli* and the sulfate-reducing bacteria (Devereux et al., 1989), which corresponds to a divergence time of about 1,100 million years BP.

Nucleotide substitutions in the 5S rRNA molecule can be similarly used to calculate the age of myxobacteria, although there are fewer 5S sequences with which to make comparisons, particularly within the sulfate-reducing bacteria. 5S rRNA appears to accumulate nucleotide substitutions twice as fast as 16S rRNA, for a divergence rate of about 1% per 25 million years (Ochman and Wilson, 1987). The common ancestor of the myxobacteria has about 26% nucleotide dissimilarity between members of the myxobacterial subgroups (Van den Eynde et al., 1990), for a minimum divergence time of about 650 million years BP. The common ancestor of the myxobacteria and *D. vulgaris* is about 30% dissimilar, for a maximum divergence time of about 750 million years BP. The maximum age of the entire delta subgroup appears to be about 1,070 million years judging by a 42% dissimilarity between the common ancestor and members of the group.

Together, these results suggest that the myxobacteria are 650 to 800 million years old and that the delta subdivision is 750 to 1,100 million years old.

GENOME SIZE

The genomes of *M. xanthus, Stigmatella aurantiaca,* and *Stigmatella erecta* have been measured by using pulsed-field gel electrophoresis of DNA digested with rare-cutting restriction endonucleases. The *M. xanthus* genome, which is 9,454 kbp (Chen et al., 1990); the S. *aurantiaca* genome, which ranges in size from 9,200 to 9,870 kbp among different isolates (Neumann et al., 1992); and the *S. erecta* genome, which is 9,710 to 10,010 kbp (Neumann et al., 1992), are among the largest bacterial genomes known (Shimkets, 1990). In marked contrast, genomes of the other members of the delta group are considerably smaller. The genomes of *Bdellovibrio* species range from 1,950 to 2,550 kbp (Seidler et al., 1972; Torella et al., 1978) and appear to be slightly larger than those of the sulfate-reducing bacteria *D. vulgaris* (1,720 kbp) and *Desulfovibrio gigas* (1,630 kbp) (Postgate et al., 1984). Therefore, the evolution of the myxobacteria from their progenitor in the delta subgroup appears to have been accompanied by a quadrupling of genome size.

Related bacteria tend to have similar genome sizes, as if major changes in genome size accompanied the evolution of new groups of bacteria (Wallace and Morowitz, 1973; Herdman, 1985). The environmental and physiological factors controlling genome size are largely unknown, but there are two striking correlations. The largest bacterial genomes are found in organisms with both efficient respiratory systems and the ability to sporulate. The myxobacteria have the largest genomes, while *Anabaena* spp. have a genome size of 6,400 kbp (Bancroft et al., 1989). *Bacillus cereus* has a genome size of 5,700 kbp (Kolsto et al., 1990), and *Streptomyces coelicolor, Streptomyces lividans* (Kieser et al., 1992), and *Streptomyces ambofaciens* (Leblond et al., 1990) have genomes sizes of about 8,000 kbp. These results suggest the importance of an efficient respiratory system in generating the energy needed to replicate a large genome.

Although genome size expansion accompanied the evolution of development, in no case has it been demonstrated that the additional DNA has an essential or even accessory role in development. On the contrary, estimates of the number of *M. xanthus* developmental genes suggest that only a small portion of the genome is devoted to development. By screening over 2,300 independent Tn*5 lac* insertions for β-galactosidase activity during development, it was observed that only 8% of the genome increases transcription during development and only 0.3 to 0.6% of the genome is essential for development (Kroos et al., 1986). An average gene size of about 1 kbp means that about 740 genes increase expression during development, and only about 55 of these genes are essential for development. A similar situation is observed with endospore formation by *Bacillus subtilis,* in which mutations at only 28 loci disrupt sporulation (Piggot, 1973).

Perhaps the acquisition of the resources to replicate a large genome gave these organisms the flexibility to evolve new genetic functions that led to increased morphological and developmental complexity; still, much of the genome could be used for other functions.

PHYSICAL AND GENETIC MAPS OF *M. XANTHUS*

The organization of *M. xanthus* genomic DNA was examined by hybridization of the 16 *Ase*I restriction fragments with a genomic library prepared in yeast artificial chromosomes (YAC) and *Spe*I-digested genomic DNA (Chen et al., 1991). Many of the YAC clones and *Spe*I restriction fragments contain an *Ase*I site and consequently hybridize with two different *Ase*I restriction fragments, providing evidence that these *Ase*I fragments are linked to each other in the intact chromosome. Without exception, the YAC clones and the *Spe*I restriction fragments identified the same *Ase*I linkages. The results are consistent with a single, circular chromosome that is shown in Fig. 2 for DK1622 and in Fig. 3 for other FB derivatives. The only difference between the two maps appears to be a deletion of about 222 kbp in DK1622 that removed one *Ase*I site, resulting in the fusion of *Ase*I fragments M and D′, and one *Spe*I site, resulting in the fusion of *Spe*I fragments U′ and U″ (Chen et al., 1991). The 222 kbp of deleted DNA appears to consist of tandem copies of the genome of the phage-like element Mx alpha (Chen et al., 1991; Starich and Zissler, 1989).

Myxobacteria also contain multicopy single-stranded DNA molecules (msDNA) that are present at more than 500 copies per haploid genome (Yee et al., 1984). These molecules are synthesized from chromosomal loci by using a novel synthesis mechanism involving reverse transcription of a primary transcript (see chapter 5). Since msDNA is found in most myxobacterial strains (Dhundale et al., 1985) and since the reverse transcriptase gene has a codon usage typical of myxobacteria (Inouye et al., 1989, 1990), it appears that reverse transcriptase was present in an ancestral myxobacterium. Since msDNA is found in only some *E. coli* strains and since its codon usage in these strains suggests recent transfer from some other source (Lampson et al., 1989), this element may have originated within the delta subgroup. The distribution of msDNA in *Bdellovibrio* spp. and sulfate-reducing bacteria has not been assessed. Deletion of the two *M. xanthus* msDNA loci, Mx65 and Mx162, had no detectable effect on cell growth, morphogenesis, or motility (Inouye et al., 1990).

The resolution of the physical map was improved by aligning the *Ase*I physical map with a genomic YAC clone library containing ordered *Eco*RI restriction sites (Kuspa et al., 1989). The YAC clones were sorted by hybridization with individual *Ase*I restriction fragments or portions thereof into groups that are located in the same region of the chromosome.

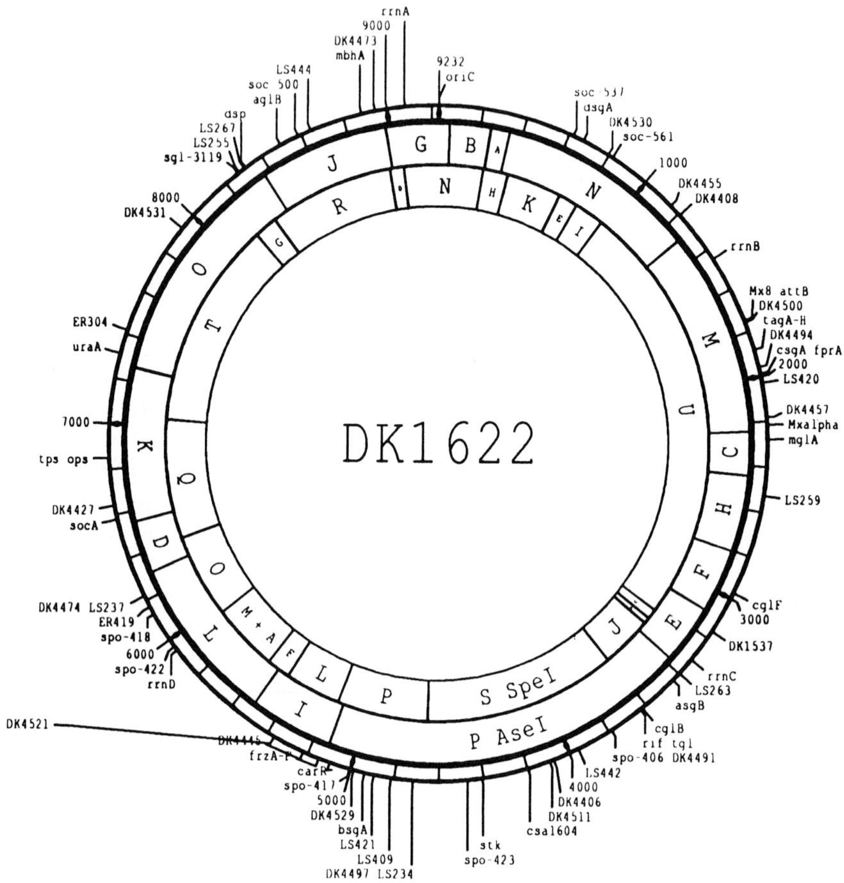

Figure 2. Physical and genetic map of the *M. xanthus* DK1622 chromosome.

Then, the clones were aligned with each other into continuous arrays of overlapping clones known as contigs based on the order of their internal *Eco*RI restriction sites. Finally, the contigs were aligned with the *Ase*I-*Spe*I physical map by common elements such as internal *Ase*I and *Spe*I restriction sites and by hybridization with gene-specific probes. The YAC clones span over 92% of the chromosome and provide the basis for the construction of a detailed genetic map.

Transposon tagging (Kuner and Kaiser, 1981) and generalized transduction (Campos et al., 1978; Martin et al., 1978) have been used in strain construction for a large number of developmental genes. Transposons shown to be linked to genes of interest by generalized transduction were used to superimpose a genetic map on the physical map. Certain Tn5

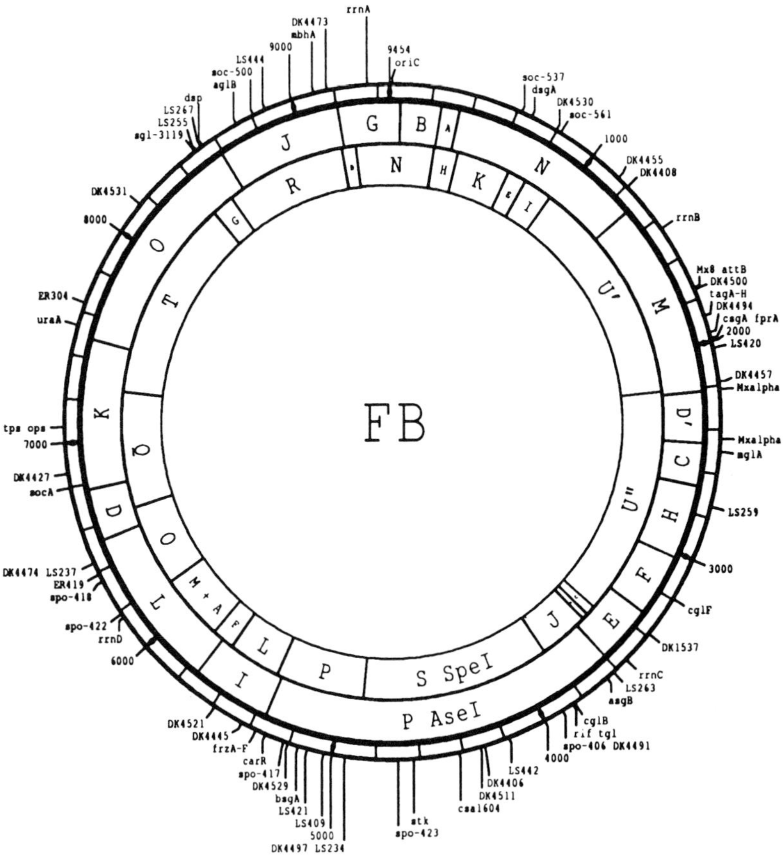

Figure 3. Physical and genetic map of the *M. xanthus* FB chromosome. Strains DZF1, DK101, and MD2 (also known as YS) have similar maps (Chen et al., 1991).

derivatives such as Tn*5*-132 (Jorgenson et al., 1979) and Tn*5 lac* (Kroos and Kaiser, 1984) contain *Ase*I sites that were used to determine the location of the transposon on the *Ase*I restriction map (Chen et al., 1990, 1991). This technique was used to identify the *Ase*I fragment containing the transposon and the distance of the transposon from one of the two ends. Many of the transposons have since been more precisely located by using the inverse polymerase chain reaction to prepare probes that were then hybridized with the ordered YAC library (Shimkets, unpublished data). On the map are more than 70 genetic markers, primarily genes that are essential for morphogenesis or genes whose expression increases during fruiting body formation (Fig. 2 for DK1622; Fig. 3 and Table 2 for other FB derivatives).

Table 2. Genetic markers of *M. xanthus* FB and derivatives DK101 and DZF1[a]

Gene symbol	Mnenomic	Homologous YAC clone(s)	Map position (kbp)[b]	Phenotypic trait
ade	Adenine		4335 or 4614	Adenine biosynthesis
aglB	Adventurous gliding	1280, 1702	8720	Adventurous motility
asgA	A signal		6775–7490	A signal
asgB[c]	A signal	799	3659	A signal
asgC	A signal		6775–7490	A signal
bsgA[c]	B signal	1465	5158	ATP-dependent protease
carR[c]	Carotenoid synthesis	1305	5319	Regulation of carotenoid synthesis
cglB1	Contact gliding	1120	3841	Contact-stimulated A motility
cglF1, cglF2	Contact gliding	452, 786, 1468, 1520	3195	Contact-stimulated A motility
csa-1604	Cell surface antigen		4418	Cell surface antigen
csgA[c]	C signal	1409	1987	C signal
DK1537			3386	Unknown
DK4401			2825 or 2871	Developmentally regulated promoter
DK4406		470, 803, 912	4298	Developmentally regulated promoter
DK4408		336, 458, 1341	1208	Developmentally regulated promoter
DK4414			8597 or 9167	Developmentally regulated promoter
DK4427		1400	6862	Developmentally regulated promoter
DK4435			4020 or 4929	Developmentally regulated promoter
DK4445		797	5444	Developmentally regulated promoter
DK4455			1150	Developmentally regulated promoter
DK4457			2182	Developmentally regulated promoter
DK4459			4157 or 4774	Developmentally regulated promoter
DK4469			9337 or 9353	Developmentally regulated promoter
DK4473		198, 1135, 1751	9168	Developmentally regulated promoter
DK4474			6444	Developmentally regulated promoter

DK4480			4010 or 4939	Developmentally regulated promoter
DK4491		776, 1183	3911	Developmentally regulated promoter
DK4492			3052 or 3299	Developmentally regulated promoter
DK4494		423	1960	Developmentally regulated promoter
DK4497			4939	Developmentally regulated promoter
DK4500			1757	Developmentally regulated promoter
DK4506			4175 or 4774	Developmentally regulated promoter
DK4511		470, 803, 912	4318	Developmentally regulated promoter
DK4514			5505 or 5590	Developmentally regulated promoter
DK4521[c]		812, 1038	5589	Developmentally regulated promoter
DK4529		617, 1465	5208	Developmentally regulated promoter
DK4530			796	Developmentally regulated promoter
DK4531			8149	Developmentally regulated promoter
dsgA[c]	D signal	1080, 1683	670	D signal (translation IF-3)
dsp	Dispersed		8535	Social motility
ER419		263	6389	Antibiotic TA biosynthesis
ER304		1306	7639	Autocide-dependent development
ER1310			5074	Antibiotic TA biosynthesis
ER1912			3915 or 5034	
ER6118			5080	Antibiotic TA biosynthesis
fprA	Flavoprotein		1987	Flavin-associated protein
frzA–frzF[c]	Frizzy	797, 1305	5369	Motility regulation
igl	Independent gliding		3875 or 5074	Motility regulation
lps-1, lps-2, lps-5	Lipopolysaccharide		6075 or 6196	Lipopolysaccharide biosynthesis
lps-3	Lipopolysaccharide		3675 or 5274	Lipopolysaccharide biosynthesis
lps-4	Lipopolysaccharide		6938 or 7325	Lipopolysaccharide biosynthesis
LS234			4948	Developmentally regulated promoter
LS237		263	6444	Developmentally regulated promoter
LS255			8499	Developmentally regulated promoter
LS257			4450 or 4499	Developmentally regulated promoter
LS259		406, 1343	2748	Developmentally regulated promoter
LS261			3960 or 4989	Developmentally regulated promoter

(continued)

Table 2. *(continued)*

Gene symbol	Mnenomic	Homologous YAC clone(s)	Map position (kbp)[b]	Phenotypic trait
LS263		461	3618	Developmentally regulated promoter
LS265			113 or 188	Developmentally regulated promoter
LS267			8527	Developmentally regulated promoter
LS409		617, 1465	5038	Developmentally regulated promoter
LS411			5798 or 6473	Developmentally regulated promoter
LS420		1409	2030	Developmentally regulated promoter
LS421			5115	Developmentally regulated promoter
LS441			8876 or 8918	Developmentally regulated promoter
LS442		819	4179	Developmentally regulated promoter
LS444		1416	8870	Developmentally regulated promoter
LS445			4180 or 4769	Developmentally regulated promoter
mbhA[c]	Myxobacterial hemagglutinin	209	9106	Lectin
mglA[c]	Mutual gliding	105, 984	2500	Required for A and S motility
Mx8 *attB*	Mx8 attachment		1748	Bacteriophage Mx8 integration site
Mx alpha	Mx alpha prophage		2217–2439	Tandem copies of Mx alpha
ops	Other protein S		7072	Developmentally regulated gene
oriC[c]	Origin	1619	1	Chromosomal origin of replication
rif	Rifampin resistance		3858	High-level rifampin resistance
rrnA	rRNA	1142	9354	rRNA operon
rrnB	rRNA	1569	1448	rRNA operon
rrnC	rRNA	461, 799, 1242	3539	rRNA operon
rrnD	rRNA	131, 1049	6165	rRNA operon
sgl-3119	Social gliding	921	8491	Social motility
sglA1	Social gliding		6773–7490	Social motility
soc-500	Suppressor of C		8809	C-signal suppressor
soc-537	Suppressor of C	1683	620	C-signal suppressor

soc-561	Suppressor of C	907	820	C-signal suppressor
socA	Suppressor of C	535	6799	C-signal suppressor
spo-75	Sporulation		9270 or 9440	Essential for sporulation
spo-87	Sporulation		4040 or 4909	Essential for sporulation
spo-133	Sporulation		1316 or 2190	Essential for sporulation
spo-147	Sporulation		8627 or 9167	Essential for sporulation
spo-406	Sporulation		3908	Essential for sporulation
spo-416	Sporulation		6091 or 6180	Essential for sporulation
spo-417	Sporulation	617	5228	Essential for sporulation
spo-418	Sporulation	454	6354	Essential for sporulation
spo-422	Sporulation		6185	Essential for sporulation
spo-423	Sporulation		4698	Essential for sporulation
spo-510	Sporulation		6626 or 6672	Essential for sporulation
spo-782	Sporulation		3875 or 5124	Essential for sporulation
spo-792	Sporulation		5979 or 6292	Essential for sporulation
stk	Sticky		4629	Cell adhesion
tagA–tagH[c]	Temp-sensitive aggregation	1690	1882	34°C aggregation system
tgl[c]		557, 1183	3858	Contact-stimulated S motility
tps[c]	True protein S	504, 798	7072	Spore coat protein S
uraA[c]	Uracil	1306, 1566	7534	Orotidine-5′-phosphate decarboxylase

[a] Data are from Chen et al. (1991) unless otherwise indicated. A transposon linked to the gene of interest was mapped to a particular *Ase*I restriction fragment by physical mapping. Where data on hybridization to the ordered YAC library are available, a more precise map location is given along with the names of the homologous YAC clones.

[b] Location is relative to *oriC*, which is located at 1 kbp in the 9,454-kbp map; average range, ±30 kbp.

[c] Kuspa et al., 1989.

GENOME EVOLUTION

A large increase in genome size appears to have occurred during the evolution of the myxobacteria, and many of the mechanisms reported to increase genome size in other bacteria, including phage integration, plasmid integration, and tandem duplication, operate in *M. xanthus* as well. Reverse transcription of an RNA molecule into a complementary DNA molecule is believed to have played a major role in the evolution of the eukaryotic genome (Weiner et al., 1986). The discovery of reverse transcriptase in myxobacteria and other prokaryotes (Inouye et al., 1989, 1990) suggests that similar mechanisms of gene duplication may operate in some prokaryotes. Template switching by DNA polymerase at inverted-repeat structures can also result in gene duplication (Ohshima et al., 1992).

Phage Insertion

Two examples of bacteriophage insertion have been described for *M. xanthus*. Bacteriophage Mx8 is a temperate phage that integrates into the *M. xanthus* chromosome at a site referred to as *attB* by analogy with the lambda phage attachment site in *E. coli* (Orndorff et al., 1983; Stellwag et al., 1985; Fig. 2 and 3 and Table 2). As with bacteriophage lambda, the integration mechanism appears to involve efficient site-specific recombination; plasmids containing the restriction fragment with Mx8 *attP* integrate much more efficiently than do plasmids relying on homologous recombination (Li and Shimkets, 1988). DNA molecules isolated from Mx8 phage particles are 56 kbp (Martin et al., 1978), and integration of a single Mx8 genome would increase the total genome size less than 1%.

The *M. xanthus* genome also contains multiple copies of a cryptic prophage referred to as Mx alpha. The virus was originally detected following transposon mutagenesis, when strains containing certain Tn*5* insertions were discovered to transduce the transposon through culture supernatant fluids (Starich et al., 1985). The transducing activity was purified by isopycnic centrifugation in cesium chloride and found to comigrate with 35-nm icosahedral particles whose size, shape, density, and resistance properties are suggestive of a phage. DNA from purified particles hybridized with discrete restriction fragments that were cloned and studied by physical mapping (Starich and Zissler, 1989). In *M. xanthus* YS, a spontaneous variant of FB, physical-mapping studies studies suggested the presence of multiple phage genomes of about 80 kbp each. All but one of these units was deleted in DK1622, an FB derivative subjected to UV mutagenesis, for a net deletion of 221 kbp. Later, the physical maps of DK1622 and the FB derivatives DK101, DZF1, and MD2 were determined with two enzymes, *Spe*I and *Ase*I, and found to differ in only one region, where DK1622 had suffered a 222-kbp deletion (Chen et al., 1991; compare Fig. 2 and 3). It appears as if UV irradiation caused deletion of all but one copy of Mx alpha through either prophage induction or homologous recombination. The multiple copies of Mx alpha in strain FB make up over 3% of the total genome.

Plasmid Insertion

Bacteria readily transmit plasmids to members of even distantly related species (Mazodier and Davies, 1991). The recipient may allow the incoming DNA either to replicate autonomously, if a functional replication system is available, or to integrate into the chromosome, if a suitable recombination system is present. Attempts to identify freely replicating plasmids in *M. xanthus* have been unsuccessful. However, the broad-host-range plasmid RP4 can be transferred from *E. coli* to *M. xanthus* by conjugation where it integrates into the chromosome (Breton et al., 1985). *E. coli* containing RP4 or its derivatives was also able to transfer DNA to *Myxococcus stipitatus, Myxococcus fulvus* (Saulnier et al., 1988), and *S. aurantiaca* (Glomp et al., 1988). When RP4 derivatives containing Tn*5* were transferred from *E. coli* to *S. aurantiaca* with selection for Tn*5*-encoded kanamycin resistance, transposons and in some cases fragments of the plasmid were integrated into the chromosome of the transconjugants (Glomp et al., 1988). Integration of RP4 into the genomes of gram-negative bacteria is generally mediated by transposon Tn*1* (Harayama et al., 1980), IS*21* (Haas and Holloway, 1978), or sequence homology with the host chromosome (Julliot and Boistard, 1979). Recombination between RP4 and the *M. xanthus* chromosome generally occurs in a 3-kbp region of the plasmid near IS*21* (Jaoua et al., 1987). However, deletion of the 3-kbp RP4 region undergoing recombination greatly reduced but did not completely eliminate the integration of RP4, suggesting that the plasmid may have multiple sites capable of serving as substrates for integration. RP4 insertion does not appear to be due to IS*21*, since deletion of IS*21* increased the frequency of integration.

Integrated plasmids were transferred back to *E. coli* by conjugation and examined for changes by physical-mapping techniques. With DZ1 as the donor, 80% of the plasmids had the same physical map and the same *E. coli*-to-*M. xanthus* transfer efficiency (Jaoua et al., 1990). When structural changes occurred in the plasmid, an additional DNA fragment of 0.3 to 3.7 kbp was observed in the area undergoing recombination. These modifications may result from imprecise excision of the plasmid from the *M. xanthus* chromosome.

Given the long evolutionary history of the myxobacteria and their ecological association with soil bacteria containing conjugal plasmids, plasmid insertion may have played some role in genome evolution of myxobacteria.

Tandem Duplications

A plot of the genome size frequency of 605 prokaryotes shows peaks at 0.75, 1.7, and 3.9 Mbp (Herdman, 1985). This result raises the possibility that larger genomes arose from smaller genomes by genome duplication (Hopwood, 1989; Stahl, 1967; Wallace and Morowitz, 1973). If the genome duplication hypothesis were true, one might expect to see genes with related functions located 180° from each other on the chromosome. The *E. coli* genes encoding enzymes involved in glucose metabolism appear to have this distribution (Riley et al., 1978). A more striking example is seen in *Streptomyces coelicolor,* in which genes controlling different steps of the same metabolic pathway are located 180° from each other and the intervening regions have no obvious functions (Hopwood, 1989). While only a portion of the genes involved in motility have been mapped, an interesting spatial distribution has been observed. The *dsp, sgl-3119,* and *aglB* loci on *Ase*I fragments O and J are located about 180° from the *cglB* and *tgl* loci on fragment P, while the *sgl* genes on fragment K are located about 180° from *mgl* (Fig. 3). DNA sequencing of these genes should help determine whether they are derived from a common ancestor. In addition, the rRNA operon *rrnB* is located directly across from *rrnD*. The *csgA* gene is also located about 180° from *socA,* a locus in which suppressor mutations restore development to *csgA* mutants (formerly *csp-560;* Rhie and Shimkets, 1989). The first gene in the *socA* operon has amino acid sequence similarity with the gene encoding CsgA (Lee and Shimkets, unpublished data).

One method of duplicating large regions of the chromosome involves formation of tandem genetic duplications, which can arise by homologous recombination between identical or nearly identical sequences located in different regions of the chromosome. Tandem duplications occur spontaneously with a frequency of 10^{-5} to 10^{-3} in the enteric bacteria (Anderson and Roth, 1977). These duplications can be very large; duplications as large as one-fourth of the chromosome have been reported in *Salmonella typhimurium*. Tandem duplications result in precise duplication of the genetic information and maintenance of the same gene order within the duplicated segment. The only mutated area of a tandem duplication is the novel joint, the area between the two duplicated regions, but since intact copies of the two genes forming the joint are located at the two ends of the duplication, there is generally no loss of genetic information. Since strains containing two copies of a particular gene should produce approximately twice as much gene product as the haploid strain, strains containing tandem duplications are frequently selected when overproduction of a gene product confers a selective advantage. Tandem duplications can be lost by a single homologous crossover anywhere within the duplicated region, and they generally revert at a high frequency in the absence of selective pressure for their maintenance.

A unique feature of tandem duplications is that regardless of how large they are, transfer of the novel joint to a wild-type cell followed by a homologous recombination on either side of the novel joint results in reconstruction of the entire duplication in the recipient bacterium. This means that megabase duplications can be faithfully reconstructed following transduction or transformation of DNA fragments less than 10 kbp in length. Spontaneous tandem duplications have been identified in *M. xanthus* by using a clever selection mechanism similar to that used in *S. typhimurium* (Anderson and Roth, 1978; Avery and Kaiser, 1983a). Transduction of the novel joint from a strain containing a duplication to the wild type led to reconstruction of the duplication in the recipient strain (Avery and Kaiser, 1983a). The duplications were unstable and segregated haploid progeny at a frequency of 0.1% per generation. The methods have been further refined to allow construction of tandem duplication with defined endpoints (Avery and Kaiser, 1983b) that could be used to examine phenomena thought to be important to genome evolution, such as gene dosage and gene order.

It is becoming increasingly apparent that *M. xanthus* has functionally redundant systems for accomplishing the tasks associated with its life cycle. There are two motility systems (Hodgkin and Kaiser, 1979a, 1979b) and two aggregation systems (O'Conner and Zusman, 1990). However, genetic analysis of these systems has not yet progressed to point that we can determine whether these systems are derived from single ancestral systems by a duplication process. DNA-DNA hybridization was used extensively during the construction of the physical and genetic maps, and it was not obvious that large repeated regions were present at the stringency used in these experiments. However, such duplications could have occurred hundreds of millions of years ago and then so diverged in nucleic acid sequence as to be unrecognizable by DNA-DNA hybridization.

GENE EVOLUTION

The source(s) of the genes specifying social and developmental behaviors of the myxobacteria is presently unknown. These genes could have been transferred to the myxobacteria from other organisms, or they could have been recruited to their current functions from genes already present in the host. Recruitment is generally believed to involve duplication of the ancestral gene so that one copy is free to evolve with little selective pressure to retain current functions (Kimura and Ohta, 1974). A small duplication of potential interest in this regard is the spore coat protein S, the product of the *tps* gene, which is downstream of and in the same orientation as a second copy of the gene referred to as *ops* (Inouye et al., 1983a, 1983b). While the two gene products have 88% amino acid identity, they have different expression patterns during development and different cellular locations (Downard et al., 1984; Teintze et al., 1985). Expression of *tps* is induced by starvation and begins about 5 h into the developmental program (Downard and Zusman, 1985). Protein S is one of the most abundant developmental proteins, reaching about 15% of the total amount of protein at the peak of its synthesis (Inouye et al., 1979). Expression of *ops* begins much later in development, and its gene product accumulates in much smaller quantities inside the spore (Downard and Zusman, 1985). The regulatory regions of *ops* and *tps* are unusually large and appear to include both positive and negative regulatory factors (Downard et al., 1988).

Perhaps the most striking example of gene duplication followed by evolution of the duplicates in *M. xanthus* concerns a group of genes with homology to the serine-threonine kinase genes of eukaryotic cells (Munoz-Dorado et al., 1991; see chapter 9). This multigene family appears to contain about 26 members (Zhang et al., 1992). The most extensively studied member, *pkn-1,* produces a protein that is autophosphorylated at serine and threonine residues (Munoz-Dorado et al., 1991). Expression of this gene starts just prior to initiation of sporulation, and mutations in this gene result in poor spore yields. The mechanism(s) by which this gene family became amplified is not yet known.

CONCLUSIONS AND SPECULATIONS

The nucleotide sequence divergence of 5S and 16S rRNA molecules suggests that the delta group of the purple bacteria emerged between 750 and 1,100 million years ago, while the myxobacteria appeared 650 to 800 million year ago. Since neither aerobic sulfate reducers nor anaerobic myxobacteria have been discovered, the ancestral myxobacterium probably acquired an aerobic respiratory system during the same time frame. This seems to correlate with the appearance of oxygen tensions high enough to support aerobic growth. The origin of oxygenic photosynthesis by cyanobacteria about 2.8 billion years ago was the first step in the evolution of an aerobic atmosphere (Schopf et al., 1983). The O_2 concentration in the atmosphere reached about 1% of present-day levels about 1.7 billion years ago and then slowly accumulated to present-day levels. Obligate aerobic microorganisms began to appear at about 1.7 billion years ago and became widespread somewhat later (Schopf et al., 1983).

The myxobacteria have much larger genomes than other members of the delta group, which suggests that chromosome size expansion accompanied the evolution of the myxobacteria. Replication of the large myxobacterial genome would place on the cell an increased metabolic burden that could have been balanced by a shift from anaerobic sulfate respiration to the more bioenergetically favorable aerobic respiration. The nature of the forces coupling genome evolution with developmental evolution has not yet been elucidated, but phage integration, plasmid integration, tandem genetic duplication, and polymerase-mediated gene duplication may have played some role in genome evolution.

Acknowledgments. This work was supported by National Science Foundation grant DCB-9001755.

Many individuals have made contributions to the gene mapping work: Qian He, Yvonne Cheng, Dale Kaiser, Hongwu Chen, and Adam Kuspa.

REFERENCES

Anderson, R. P., and J. R. Roth. 1977. Tandem genetic duplications in phage and bacteria. *Annu. Rev. Microbiol.* **31:**473–505.

Anderson, R. P., and J. R. Roth. 1978. Gene duplication in bacteria: alteration of gene dosage by sister-chromosome exchanges. *Cold Spring Harbor Symp. Quant. Biol.* **42:**1083–1087.

Apelian, D., and S. Inouye. 1990. Development-specific sigma-factor essential for late-stage differentiation of *Myxococcus xanthus. Genes Dev.* **4:**1396–1403.

Avery, L., and D. Kaiser. 1983a. *In situ* transposon replacement and isolation of a spontaneous tandem genetic duplication. *Mol. Gen. Genet.* **191:**99–109.

Avery, L., and D. Kaiser. 1983b. Construction of tandem genetic duplications with defined endpoints in *Myxococcus xanthus. Mol. Gen. Genet.* **191:**110–117.

Bancroft, I., C. P. Wolk, and E. V. Oren. 1989. Physical and genetic maps of the genome of the heterocyst-forming cyanobacterium *Anabaena* sp. strain PCC 7120. *J. Bacteriol.* **171:**5940–5948.

Bibb, M. J., P. R. Findlay, and M. W. Johnson. 1984. The relationship between base composition and codon usage in bacterial genes and its use for the simple and reliable identification of protein coding regions. *Gene* **30:**157–166.

Breton, A. M., S. Jaoua, and J. Guespin-Michel. 1985. Transfer of plasmid RP4 to *Myxococcus xanthus* and evidence for its integration into the host chromosome. *J. Bacteriol.* **161:** 523–528.

Campos, J. M., J. Geisselsoder, and D. R. Zusman. 1978. Isolation of bacteriophage MX4, a generalized transducing phage for *Myxococcus xanthus. J. Mol. Biol.* **119:**167–178.

Chen, H.-W., I. M. Keseler, and L. J. Shimkets. 1990. The genome size of *Myxococcus xanthus* determined by pulsed-field gel electrophoresis. *J. Bacteriol.* **172:**4206–4213.

Chen, H.-W., A. Kuspa, I. M. Keseler, and L. J. Shimkets. 1991. Physical map of the *Myxococcus xanthus* chromosome. *J. Bacteriol.* **173:**2109–2115.

Devereux, R., M. Delaney, F. Widdel, and D. A. Stahl. 1989. Natural relationships among sulfate-reducing eubacteria. *J. Bacteriol.* **171:**6689–6695.

Dhundale, A. R., T. Furuichi, S. Inouye, and M. Inouye. 1985. Distribution of multicopy single-stranded DNA among myxobacteria and related species. *J. Bacteriol.* **164:**914–917.

Downard, J. S., S.-H. Kim, and K.-S. Kil. 1988. Localization of the *cis*-acting regulatory DNA sequences of the *Myxococcus xanthus tps* and *ops* genes. *J. Bacteriol.* **170:** 4931–4938.

Downard, J. S., D. Kupfer, and D. R. Zusman. 1984. Gene expression during development of *Myxococcus xanthus:* analysis of the genes for protein S. *J. Mol. Biol.* **175:**469–492.

Downard, J., and D. R. Zusman. 1985. Differential expression of protein S genes during *Myxococcus xanthus* development. *J. Bacteriol.* **161:**1146–1155.

Fowler, V. J., F. Widdel, N. Pfenning, C. R. Woese, and E. Stackebrandt. 1986. Phylogenetic relationships of sulfate-sulfur-reducing eubacteria. *Syst. Appl. Microbiol.* **8:**32–41.

Glomp, I., P. Saulnier, J. Guespin-Michel, and H. U. Schairer. 1988. Transfer of IncP plasmids into *Stigmatella aurantiaca* leading to insertional mutants affected in spore development. *Mol. Gen. Genet.* **214:**213–217.

Haas, D., and B. W. Holloway. 1978. Chromosome mobilization by R plasmid R68.45 in *Pseudomonas aeruginosa* PAO. *Plasmid* **9:**42–52.

Hagen, T. J., and L. J. Shimkets. 1990. Nucleotide sequence and transcriptional products of the *csg* locus of *Myxococcus xanthus*. *J. Bacteriol.* **172:**15–23.

Harayama, S., M. Tsudo, and T. Lino. 1980. High frequency mobilization of the chromosome of *Escherichia coli* by a mutant of plasmid RP4 temperature-sensitive for maintenance. *Mol. Gen. Genet.* **180:**47–56.

Herdman, M. 1985. The evolution of bacterial genomes, p. 37–68. *In* T. Cavalier-Smith (ed.), *The Evolution of Genome Size*. John Wiley & Sons, Inc., New York.

Hodgkin, J., and D. Kaiser. 1979a. Genetics of gliding motility in *Myxococcus xanthus* (Myxobacterales): genes controlling movements of single cells. *Mol. Gen. Genet.* **171:**167–176.

Hodgkin, J., and D. Kaiser. 1979b. Genetics of gliding motility in *Myxococcus xanthus* (Myxobacterales): two gene systems control movement. *Mol. Gen. Genet.* **172:**177–191.

Hopwood, D. A. 1989. *Streptomyces* genetics in pure and applied research, p. 12–20. *In* C. L. Hershberger, S. W. Queener, and G. Hegeman (ed.), *Genetics and Molecular Biology of Industrial Organisms*. American Society for Microbiology, Washington, D.C.

Inouye, M., S. Inouye, and D. R. Zusman. 1979. Biosynthesis and self-assembly of protein S, a development specific protein of *Myxococcus xanthus*. *Proc. Natl. Acad. Sci. USA* **76:**109–213.

Inouye, S, 1990. Cloning and DNA sequence of the gene coding for the major sigma factor from *Myxococcus xanthus*. *J. Bacteriol.* **172:**80–85.

Inouye, S., T. Franceschini, and M. Inouye. 1983a. Structural similarities between the developmental-specific protein S from a Gram-negative bacterium, *Myxococcus xanthus,* and calmodulin. *Proc. Natl. Acad. Sci. USA* **80:** 6829–6833.

Inouye, S., P. J. Herzer, and M. Inouye. 1990. Two independent retrons with highly diverse reverse transcriptases in *Myxococcus xanthus*. *Proc. Natl. Acad. Sci. USA* **87:**942–945.

Inouye, S., M.-Y. Hsu, S. Eagle, and M. Inouye. 1989. Reverse transcriptase associated with the biosynthesis of the branched RNA-linked msDNA in *Myxococcus xanthus*. *Cell* **56:** 709–717.

Inouye, S., Y. Ike, and M. Inouye. 1983b. Tandem repeat of the genes for protein S, a development-specific protein of *Myxococcus xanthus*. *J. Biol. Chem.* **258:**38–40.

Jaoua, S., J. F. Guespin-Michel, and A. M. Breton. 1987. Mode of insertion of the broad-host-range plasmid RP4 and its derivatives into the chromosome of *Myxococcus xanthus*. *Plasmid* **18:**111–119.

Jaoua, S., B. Letouvet-Pawlak, C. Monnier, and J. F. Guespin-Michel. 1990. Mechanism of integration of the broad-host-range plasmid RP4 into the chromosome of *Myxococcus xanthus*. *Plasmid* **23:**183–193.

Johnson, J. L., and E. J. Ordal. 1968. Deoxyribonucleic acid homology in bacterial taxonomy: effect of incubation temperature on reaction specificity. *J. Bacteriol.* **95:**893–900.

Jorgenson, R. A., S. J. Rothstein, and W. S. Reznikoff. 1979. A restriction enzyme cleavage map of Tn*5* and location of a region encoding neomycin resistance. *Mol. Gen. Genet.* **177:**65–72.

Julliot, J. S., and P. Boistard. 1979. Use of RP4-prime plasmids constructed in vitro to promote a polarized transfer of the chromosome in *E. coli* and *Rhizobium meliloti*. *Mol. Gen. Genet.* **173:**189–198.

Kieser, M. H., T. Kieser, and D. M. Hopwood. 1992. A combined genetic and physical map of the *Streptomyces coelicolor* A3(2) chromosome. *J. Bacteriol.* **174:**5496–5507.

Kimsey, H. H., and D. Kaiser. 1992. The orotidine-5′-monophosphate decarboxylase gene of *Myxococcus xanthus*. *J. Biol. Chem.* **267:**819–824.

Kimura, M., and T. Ohta. 1974. On some principles governing molecular evolution. *Proc. Natl. Acad. Sci. USA* **71:**2848–1852.

Kohl, W., A. Gloe, and H. Reichenbach. 1983. Steroids from the myxobacterium *Nannocystis exedens*. *J. Gen. Microbiol.* **129:**1629–1635.

Kolsto, A.-B., A. Gronstad, and H. Oppengaard. 1990. Physical map of the *Bacillus cereus* chromosome. *J. Bacteriol.* **172:**3821–3825.

Komano, T., T. Franceschini, and S. Inouye. 1987. Identification of a vegetative promoter in *Myxococcus xanthus:* a protein that has homology in histones. *J. Mol. Biol.* **196:**517–524.

Kroos, L., and D. Kaiser. 1984. Construction of Tn*5lac*, a transposon that fuses *lacZ* expression to endogenous promoters, and its introduction into *Myxococcus xanthus*. *Proc. Natl. Acad. Sci. USA* **81:**5816–5820.

Kroos, L., A. Kuspa, and D. Kaiser. 1986. A global analysis of developmentally regulated genes in *Myxococcus xanthus*. *Dev. Biol.* **117:**252–266.

Kuner, J. M., and D. Kaiser. 1981. Introduction of transposon Tn*5* into *Myxococcus* for analysis of developmental and other nonselectable mutants. *Proc. Natl. Acad. Sci. USA* **78:**425–429.

Kuspa, A., D. Vollrath, Y. Cheng, and D. Kaiser. 1989. Physical mapping of the *Myxococcus xanthus* genome by random cloning in yeast artificial chromosomes. *Proc. Natl. Acad. Sci. USA* **86:**8917–8921.

Lampson, B. C., J. Sun, M.-Y. Hsu, J. Vallejo-Ramirez, S. Inouye, and M. Inouye. 1989. Reverse transcriptase in a clinical strain of *Escherichia coli:* production of branched RNA-linked msDNA. *Science* **243:**1033–1038.

Leblond, P., F. X. Francou, J.-M. Simonet, and B. Decaris. 1990. Pulsed-field gel electrophoresis analysis of the genome of *Streptomyces ambofaciens* strains. *FEMS Microbiol. Lett.* **72:**79–88.

Lee, K., and L. J. Shimkets. Unpublished data.

Li, S.-F., B.-U. Lee, and L. J. Shimkets. 1992. *csgA* expression entrains *Myxococcus xanthus* development. *Genes Dev.* **6:**401–410.

Li, S.-F., and L. J. Shimkets. 1988. Site-specific integration and expression of a developmental promoter in *Myxococcus xanthus*. *J. Bacteriol.* **170:**5552–5556.

Ludwig, W., K.-H. Schleifer, H. Reichenbach, and E. Stackebrandt. 1983. A phylogenetic analysis of the myxobacteria *Myxococcus fulvus, Stigmatella aurantiaca, Cystobacter fuscus, Sorangium cellulosum* and *Nannocystis exedens*. *Arch. Microbiol.* **135:**58–62.

Mandel, M., and E. R. Leadbetter. 1965. Deoxyribonucleic acid base composition of myxobacteria. *J. Bacteriol.* **90:**1795–1796.

Martin, S., E. Sodergren, T. Masuda, and D. Kaiser. 1978. Systematic isolation of transducing phages for *Myxococcus xanthus*. *Virology* **88:**44–53.

Mazodier, P., and J. Davies. 1991. Gene transfer between distantly related bacteria. *Annu. Rev. Genet.* **25:** 147–171.

McBride, M. J., R. A. Weinberg, and D. R. Zusman. 1989. "Frizzy" aggregation genes of the gliding bacterium *Myxococcus xanthus* show sequence similarities to the chemotaxis genes of enteric bacteria. *Proc. Natl. Acad. Sci. USA* **86:**424–428.

McCleary, W. R., and D. R. Zusman. 1990. FrzE of *Myxococcus xanthus* is homologous to both CheA and CheY of *Salmonella typhimurium. Proc. Natl. Acad. Sci. USA* **87:** 5898–5902.

McCurdy, H. D. 1989. Fruiting gliding bacteria: the myxobacteria, p. 2139–2168. *In* J. T. Staley, M. P. Bryant, N. Pfennig, and J. G. Holt (ed.), *Bergey's Manual of Systematic Bacteriology,* vol. 3. The Williams & Wilkins Co., Baltimore.

McCurdy, H. D., and S. Wolf. 1967. Deoxyribonucleic acid base composition of fruiting Myxobacterales. *Can. J. Microbiol.* **13:**1707–1708.

Mesbah, M., U. Premachandran, and W. B. Whitman. 1989. Precise measurement of the G + C content of deoxyribonucleic acid by high-performance liquid chromatography. *Int. J. Syst. Bacteriol.* **39:**159–167.

Munoz-Dorado, J., M. Inouye, and S. Inouye. 1990. Nucleoside diphosphate kinase from *Myxococcus xanthus.* I. Cloning and sequencing of the gene. *J. Biol. Chem.* **265:** 2702–2706.

Munoz-Dorado, J., S. Inouye, and M. Inouye. 1991. A gene encoding a protein serine/threonine kinase is required for normal development of *M. xanthus,* a gram-negative bacterium. *Cell* **67:**995–1006.

Neumann, B., A. Pospiech, and H. U. Schairer. 1992. Size and stability of the genomes of the myxobacteria *Stigmatella aurantiaca* and *Stigmatella erecta. J. Bacteriol.* **174:** 6307–6310.

Normore, W. M. 1976. Guanine-plus-cytosine (GC) composition of the DNA of bacteria, fungi, algae, and protozoa, p. 65–232. *In* G. D. Fasman (ed.), *CRC Handbook of Biochemistry and Molecular Biology,* 3rd ed., vol. II. CRC Press, Cleveland.

Ochman, H., and A. C. Wilson. 1987. Evolution in bacteria: evidence for a universal substitution rate in cellular genomes. *J. Mol. Evol.* **26:**74–86.

O'Connor, K. A., and D. R. Zusman. 1990. Genetic analysis of the *tag* mutants of *Myxococcus xanthus* provides evidence for two developmental aggregation systems. *J. Bacteriol.* **172:**3868–3878.

Ohshima, A., S. Inouye, and M. Inouye. 1992. *In vivo* duplication of genetic elements by the formation of stem-loop DNA without an RNA intermediate. *Proc. Natl. Acad. Sci. USA* **89:**1016–1020.

Orndorff, P., E. Stellwag, T. Starich, M. Dworkin, and J. Zissler. 1983. Genetic and physical characterization of lysogeny by bacteriophage MX8 in *Myxococcus xanthus. J. Bacteriol.* **154:**772–779.

Oyaizu, H., and C. R. Woese. 1985. Phylogenetic relationships among the sulfate respiring bacteria, myxobacteria and purple bacteria. *Syst. Appl. Microbiol.* **6:**257–263.

Piggot, P. J. 1973. Mapping of asporogenous mutations of *Bacillus subtilis:* a minimum estimate of the number of sporulation operons. *J. Bacteriol.* **114:**1241–1253.

Postgate, J. R., H. M. Kent, R. L. Robson, and J. A. Chesshyre. 1984. The genomes of *Desulfovibrio gigas* and *D. vulgaris. J. Gen. Microbiol.* **130:**1597–1601.

Reichenbach, H. 1986. The myxobacteria: common organisms with an uncommon behavior. *Microbiol. Sci.* **3:**268–274.

Reichenbach, H., and M. Dworkin. 1991. The myxobacteria, p. 3416–3487. *In* A. Balows, H. G. Truper, M. Dworkin, W. Harder, and K. H. Schleifer (ed.), *The Prokaryotes,* 2nd ed. Springer-Verlag, New York.

Reichenbach, H., K. Gerth, H. Irschik, B. Kunze, and G. Holfe. 1988. Myxobacteria: a new source of antibiotics. *Bio/Technology* **6:**115–121.

Reichenbach, H., and H. Kleinig. 1984. Pigments of myxobacteria, p. 127–137. *In* E. Rosenberg (ed.), *Myxobacteria: Development and Cell Interactions*. Springer-Verlag, New York.

Rhie, H.-G., and L. J. Shimkets. 1989. Developmental bypass suppression of *Myxococcus xanthus csgA* mutations. *J. Bacteriol.* **171:**3268–3276.

Riley, M., L. Solomon, and D. Zipkas. 1978. Relationship between gene function and gene location in *Escherichia coli*. *J. Mol. Evol.* **11:**47–56.

Romeo, J. M., B. Esmon, and D. R. Zusman. 1986. Nucleotide sequence of the myxobacterial hemagglutin gene contains four homologous domains. *Proc. Natl. Acad. Sci. USA* **83:** 6332–6336.

Saulnier, P., J. Hanquier, S. Jaoua, H. Reichenbach, and J. F. Guespin-Michael. 1988. Utilization of IncP-1 plasmids as vectors for transposon mutagenesis in myxobacteria. *J. Gen. Microbiol.* **134:**2889–2895.

Schopf, J. W., J. M. Hayes, and M. R. Walter. 1983. Evolution of earth's earliest ecosystems: recent progress and unsolved problems, p. 361–384. *In* J. W. Schopf (ed.), *Earth's Earliest Biosphere: Its Origin and Evolution*. Princeton University Press, Princeton, N.J.

Seidler, R. J., M. Mandel, and J. N. Baptist. 1972. Molecular heterogeneity of the bdellovibrios: evidence of two new species. *J. Bacteriol.* **109:**209–217.

Shimkets, L. J. Unpublished data.

Shimkets, L. J. 1990. Social and developmental biology of the myxobacteria. *Microbiol. Rev.* **54:**473–501.

Shimkets, L. J., and C. Woese. 1992. A phylogenetic analysis of the myxobacteria: basis for their classification. *Proc. Natl. Acad. Sci. USA* **89:**9459–9463.

Stahl, F. W. 1967. Circular genetic maps. *J. Cell Physiol.* **70**(Suppl. 1)**:**1–12.

Starich, T., P. Cordes, and J. Zissler. 1985. Transposon tagging to detect a latent virus in *Myxococcus xanthus*. *Science* **230:** 541–543.

Starich, T., and J. Zissler. 1989. Movement of multiple DNA units between *Myxococcus xanthus* cells. *J. Bacteriol.* **171:**2323–2336.

Stellwag, E., J. M. Fink, and J. Zissler. 1985. Physical characterization of the genome of the *Myxococcus xanthus* bacteriophage MX-8. *Mol. Gen. Genet.* **199:**123–132.

Stephens, K., P. Hartzell, and D. Kaiser. 1989. Gliding motility in *Myxococcus xanthus: mgl* locus, RNA, and predicted protein products. *J. Bacteriol.* **171:**819–830.

Teintze, M., R. Thomas, T. Furuichi, M. Inouye, and S. Inouye. 1985. Two homologous genes coding for spore-specific proteins are expressed at different times during development of *Myxococcus xanthus*. *J. Bacteriol.* **163:**121–125.

Torella, F., R. Guerrero, and R. J. Seidler. 1978. Further taxonomic characterization of the genus Bdellovibrio. *Can. J. Microbiol.* **24:**1387–1394.

Van den Eynde, H., Y. Van de Peer, H. Vandenabeele, M. van Bogaert, and R. de Wachter. 1990. 5S rRNA sequences of myxobacteria and radioresistant bacteria and implications for eubacterial evolution. *Int. J. Syst. Bacteriol.* **40:**399–404.

Wallace, D. C., and H. J. Morowitz. 1973. Genome size and evolution. *Chromosoma* **40:** 121–126.

Weiner, A. M., P. L. Deininger, and A. Efstratiadis. 1986. Nonviral retroposons: genes, pseudogenes, and transposable elements generated by the reverse flow of genetic information. *Annu. Rev. Biochem.* **55:**631–661.

Woese, C. R. 1987. Bacterial evolution. *Microbiol. Rev.* **51:**221–271.

Yee, T., T. Furuichi, S. Inouye, and M. Inouye. 1984. Multicopy single-stranded DNA isolated from a gram-negative bacterium, *Myxococcus xanthus*. *Cell* **38:**203–209.

Zhang, W., J. Munoz-Dorado, M. Inouye, and S. Inouye. 1992 Identification of a putative eukaryotic-like protein kinase family in the developmental bacterium. *Myxococcus xanthus*. *J. Bacteriol.* **174:**5450–5453.

Myxobacteria II
Edited by Martin Dworkin and Dale Kaiser
© 1993 American Society for Microbiology, Washington, DC 20005

Chapter 5

Retron Elements of the Myxobacteria

Bert C. Lampson

Retrons are a new class of retroelements found in the prokaryotes that encode a reverse transcriptase (RT) similar in structure to the polymerase found in retroviruses. They were first discovered in the myxobacteria *Myxococcus xanthus* and *Stigmatella aurantiaca* and have also been found in *Escherichia coli*. These elements also code for the production of an unusual satellite DNA known as multicopy, single-stranded DNA (msDNA), which is synthesized by the retron-encoded RT.

The discovery of retroelements in a prokaryotic organism raises a number of important questions about the biological functions of msDNA and RT, the origin and divergence of retroelements, and the role of reverse transcription in the evolution of prokaryotes and eukaryotes. Several reviews describing the msDNA-producing retron elements have recently appeared (Inouye and Inouye, 1991a, 1991b, 1992; Lampson et al., 1991b; Lim and Maas, 1989). This chapter emphasizes the retron elements found in the myxobacteria and recent discoveries concerning the retrons of *E. coli*. Although research efforts have largely shifted to the *E. coli* elements in recent years, myxobacterial retrons have provided many of the key findings about these retroelements, particularly regarding the structure and synthesis of msDNA. In addition, because some of the unique features observed among the retrons of *M. xanthus* are not seen in the *E. coli* elements, the myxobacterial retrons may provide important information about the origin and evolution of RT and retroelements.

Bert C. Lampson • Department of Microbiology, University of Tennessee, Knoxville, Tennessee 37996-0845.

REVERSE TRANSCRIPTION AND EVOLUTION OF THE EUKARYOTIC GENOME

The process of reverse transcription, or the reverse flow of genetic information from RNA back into DNA, has been a powerful force at work on the structure and organization of eukaryotic genomes. A large number of pseudogenes and repetitive sequence elements, for example, exist in higher eukaryotic genomes and appear to have been generated by reverse transcription of cellular RNAs. The highly repeated sequences known as SINEs, which are dispersed throughout the mammalian genome, are derived from the reverse transcription of tRNA sequences (Brosius, 1991; Weiner et al., 1986). Likewise, the Alu family of repetitive DNA elements found in primate genomes is derived from 7SL RNA (Howard and Sakamoto, 1990). These and many other repeat sequences can serve as recombination sites that may result in dramatic genome rearrangements (Finnegan, 1989; Weiner et al., 1986). In addition, numerous mobile elements that transpose via reverse transcription can act as insertional mutagens and promote large-scale changes in chromosome structure (Boeke and Corces, 1989; Bucheton, 1990). Although the source of RT activity for many of these events is largely unknown, this process clearly has had a profound effect, with as much as 5 to 15% of some eukaryotic genomes being composed of small repetitive sequences generated by reverse transcription.

DISCOVERY OF RT IN BACTERIA

RTs resembling the polymerases found in retroviruses have recently been found in *M. xanthus* (Inouye et al., 1989) and *E. coli* (Lampson et al., 1989b; Lim and Maas, 1989). However, the key piece of evidence that led to the discovery of bacterial RT occurred a number of years earlier with the discovery of a small satellite DNA, known as msDNA, in *M. xanthus* DZF1 (Yee et al., 1984). Indeed, it was soon shown that the gene encoding bacterial RT is closely linked to the locus encoding msDNA and is required for the synthesis of this unusual molecule. This extrachromosomal DNA is present at a high copy number, estimated at 500 to 700 copies per cell genome of *M. xanthus,* and was thus designated multicopy, single-stranded DNA, or msDNA (Furuichi et al., 1987a). Subsequent work established that msDNA is a complex of DNA and RNA and has an unusual if not unique structure. The msDNA molecule from *M. xanthus* is composed of a 162-base single strand of DNA joined to an internal residue of a 77-base single-stranded RNA (Fig. 1) (Dhundale et al., 1987). Branching out from the 2′ position of a specific internal guanosine residue on the RNA strand (rG), a unique 2′-5′ phosphodiester bond joins thc RNA strand to the 5′ end of the DNA strand.

Several different msDNA molecules from both the myxobacteria and *E. coli* have now been characterized. msDNAs are identified by two letters representing the genus and species of the host cell and a number corresponding to the size of the DNA strand in nucleotide bases. Although these msDNAs show little primary sequence homology in either the DNA or the RNA strand, they all display common secondary structural features (Fig. 1). These include (i) secondary folding of DNA and RNA strands into stable stem-loop structures; (ii) a 2′-5′ branch linkage always to a specific internal rG; and (iii) a short region of base pairing (a hybrid region) between the 3′ ends of the RNA and DNA chains. The structure of msDNA is, no doubt, more complex than current models depict. For example, msDNA may exist in the cell as a complex with protein(s), presumably RT, since msDNA has been shown to copurify with RT activity (Viswanathan et al., 1989; Lampson et al., 1990).

Locus Encoding RT

When msDNA is labeled with ^{32}P and used as a hybridization probe, it cross-hybridizes with a single restriction fragment from the host cell chromosome, indicating that a single-copy locus encodes msDNA (Yee et al., 1984). This locus contains the genes *msr* and *msd,* which correspond in sequence to the RNA strand and DNA strand, respectively, of the msDNA molecule. Located a short distance upstream of *msd* (downstream of *msr*) is a large open reading frame (ORF) encoding the bacterial RT (Fig. 2). With two exceptions, all known retrons have a similar gene organization (see below). For Mx162 of *M. xanthus,* the ORF encoding RT lies only 77 bp from the *msd* gene.

The ORF of Mx162 potentially encodes an RT of 485 amino acids, and this amino acid sequence can be aligned with the amino acid sequences of over 70 RTs found in eukaryotic retroelements including retroviruses (Xiong and Eickbush, 1990). Using the conserved-residue alignment method, which is based on common structural domains, the myxobacterial RT sequence shows significant similarities to the sequences of eukaryotic RTs; more specifically, it shares all but 2 of the 17 strictly conserved amino acids found in eukaryotic RTs (Inouye and Inouye, 1991b).

RT Is Required for Production of msDNA

Several lines of evidence strongly indicate that the product of the RT ORF is required for the production of msDNA. First, mapping experi-

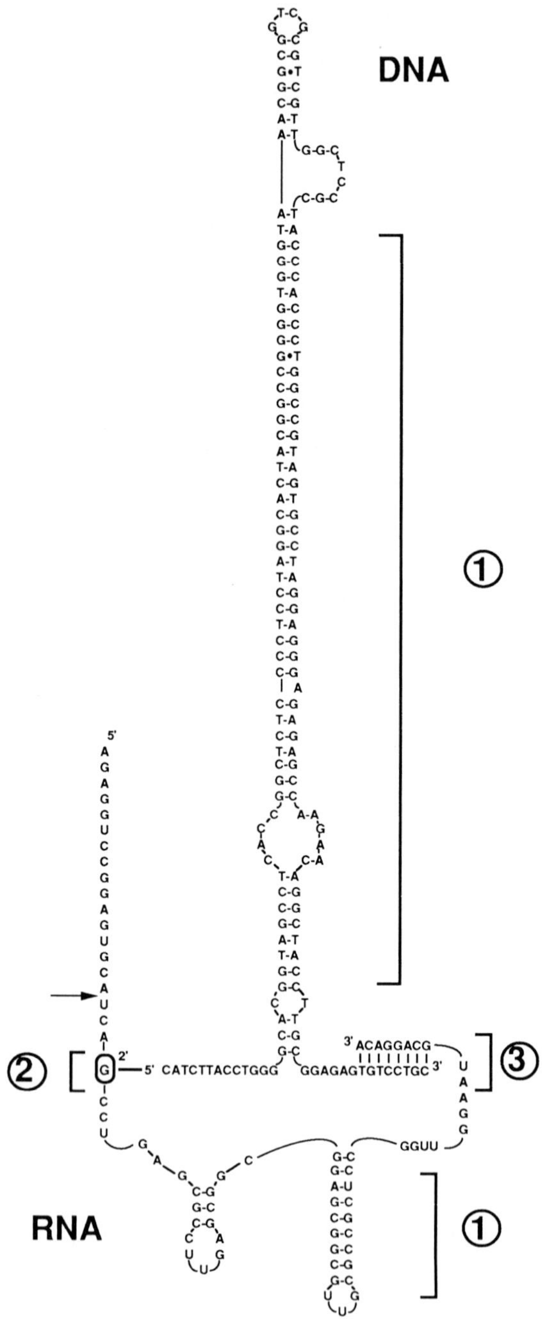

DNA
①
5'
2'
②
5' CATCTTACCTGGG
GGAGAGTGTCCTGC 3'
3' ACAGGACG
③
GGUU
RNA
①

ments with S1 nuclease show that transcription of the locus for Mx162 msDNA of *M. xanthus* initiates at about 75 bp 5′ from the start of the *msr* gene and extends through and beyond the *msd* gene (Dhundale et al., 1987) (Fig. 2). Transcription probably extends beyond this point to include the RT ORF, since similar mapping studies of *E. coli* retrons indicate that transcription begins at or near the 5′ start of the *msr* gene and goes on to include the RT ORF. In addition, *lacZ* fusion constructions to the RT ORF of the Ec86 retron (*E. coli* B) showed that deletion of the region 5′ to the start of the *msr* gene reduced both β-galactosidase activity and msDNA production, suggesting that one long transcript encompasses *msr*, *msd*, and the RT ORF (Lim and Maas, 1989; Lampson et al., 1989b).

Second, deletion of the RT ORF generally blocks production of msDNA. For example, a deletion mutation as well as a simple disruption mutation (insertion of a kanamycin resistance gene) of the RT ORF of Mx162 severely reduced the level of msDNA produced by *M. xanthus* (Dhundale et al., 1988a). Interestingly, deletion of the amino-terminal residues of the Mx162 ORF does not appear to completely destroy RT activity, as a small amount of msDNA is still produced. More-extensive deletions of some of the *E. coli* RTs show a complete block of msDNA synthesis (Lampson et al., 1989b; Lim and Maas, 1989).

HOW msDNA IS SYNTHESIZED

The model presented here describes the synthesis of msDNA based on a mechanism of reverse transcription from an RNA that serves as both template and primer (Dhundale et al., 1987; Lampson et al., 1989a). The reasons for strongly favoring this model over alternative mechanisms that have recently been proposed (Lease and Yee, 1991) can be summarized as follows. (i) The genes encoding msDNA are closely linked to an ORF coding for bacterial RT. (ii) The amino acid sequence of this ORF clearly resembles those of eukaryotic RTs, including those from retroviruses. (iii)

Figure 1. Primary and proposed secondary structures for Mx162 msDNA produced by *M. xanthus*. Mx162 msDNA is composed of a 77-base single-stranded RNA joined to the 5′ end of a 162-base single-stranded DNA. Although other species of msDNA differ considerably in primary sequence, all msDNAs share certain secondary structural features. Corresponding to the numbered brackets, these secondary structures are stable stem-loops in both the DNA and RNA strands as a result of complementary base pairing (1), an rG that forms the 2′-5′ phosphodiester linkage to the 5′ end of the DNA strand (2), and a short region of base pairing between the 3′ ends of the RNA and DNA chains (3). The arrow indicates a late processing event that removes part of the 5′ end of the RNA strand to yield a smaller, mature form of Mx162 msDNA.

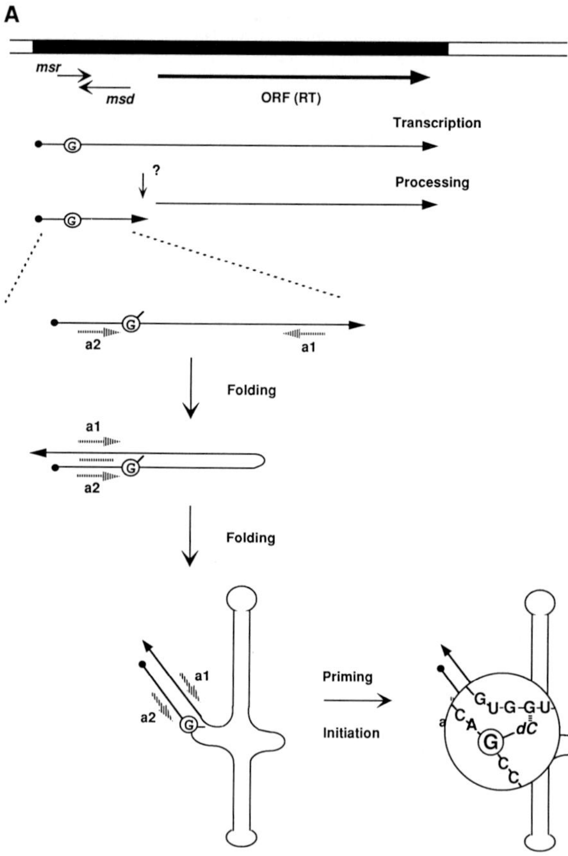

Figure 2. Proposed reverse transcription model for the synthesis of msDNA. msDNA is synthesized from a precursor RNA that serves as both primer and template by reverse transcription that is primed by an internal 2′-OH. See the text for details. (A) Transcription and folding of the RNA template, priming, and initiation of DNA synthesis.

The genes for msDNA are coexpressed, that is, cotranscribed, with the RT gene, and deletion of the RT ORF blocks production of msDNA. (iv) The msDNA molecule copurifies with RT activity from cell extracts. (v) The overlapping, convergent arrangement of the genes *msr* and *msd* (encoding msDNA) are consistent with the reverse transcription model. (vi) A mutational change in the *msr* gene at the rG residue that will form the branch linkage to msDNA blocks synthesis of msDNA. (vii) Intermediate forms of msDNA that are observed when synthesis of msDNA is interrupted in an in vitro system are also consistent with the model described

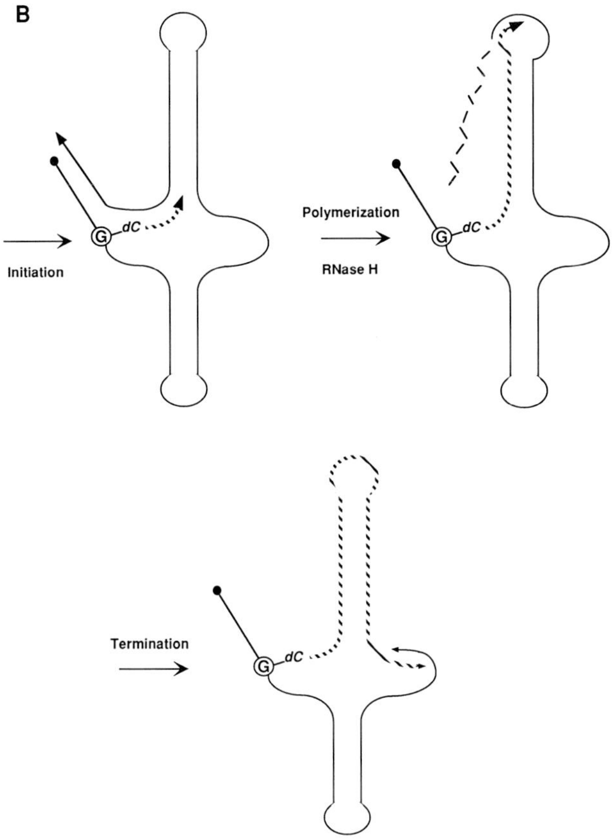

Figure 2. (*continued*) (B) Polymerization of DNA, removal of the RNA template, and termination of synthesis.

here. Proposed alternative mechanisms (Lease and Yee, 1991; Inouye et al., 1987) do not adequately account for, interpret, or incorporate all of these findings.

Transcription, Folding, and Processing of Template RNA

Synthesis of msDNA begins with transcription of a long RNA; for Mx162 msDNA of *M. xanthus,* transcription starts upstream of *msr* (encodes the RNA strand) and extends to include *msd* (encodes the DNA strand) and the RT ORF (Fig. 2). At some time, the long RNA is cleaved down to a smaller RNA precursor, but how or when this event occurs during the synthesis of msDNA is unknown. The RNA template then folds up into a complex secondary structure. Two inverted repeat sequences, designated a1 and a2, are critical for proper folding of the RNA into a stable stem structure. This secondary structure positions the internal rG residue (circled G in Fig. 2), which will contain the 2′ branch linkage in msDNA, in such a way as to prime initiation of DNA synthesis by RT. Correct positioning of the branched rG in the folded RNA is critical for the synthesis of msDNA. If mutations are introduced into either inverted repeat (a1 or a2) of the Mx162 retron, causing a base pair mismatch, the stem structure immediately adjacent to the branched rG is disrupted, preventing the synthesis of msDNA (Hsu et al., 1989). The inverted repeats that form this important stem structure, although different in base sequence, are found in the same location in all known retrons: a2 is situated immediately upstream from the branched rG residue, and a1 is immediately 5′ to the start of the DNA coding region (Fig. 2).

Priming and Branch Formation

After the precursor RNA template has folded into its proper secondary structure, a priming reaction initiates synthesis of the DNA strand of msDNA. Here, a phosphodiester linkage is formed between the 2′-OH of the internal rG and the 5′ end of the first deoxyribonucleotide. Analysis of intermediate structures formed during the synthesis of Ec86 msDNA (from *E. coli* B) in an in vitro system (Lim and Maas, 1989) indicates that formation of the linkage between RNA and DNA occurs very early, most likely as the first step in the synthesis of msDNA. Indeed, recent in vitro experiments seem to suggest that the retron-encoded RT can utilize the 2′-OH of the internal rG de novo as a primer to initiate synthesis of msDNA (Hsu et al., 1992a). Partially purified Ec67 RT can incorporate, in a template-directed event, the first deoxynucleotide (dTTP) into a 132-base precursor RNA derived from RNA isolated from cells containing only the *msr-msd* genes of the Ec67 retron. However, it is not clear whether this precursor RNA template has in some way been structurally modified or processed prior to the initiation reaction with RT, since in vitro-synthesized RNA does not serve as a competent template primer for this reaction.

Although the full details of the priming mechanism remain to be determined, it is clear that the 2′ position of a specific rG is required to prime the synthesis of msDNA. For Mx162 msDNA of *M. xanthus,* it is residue 20 of the RNA strand, since replacement of this rG with rA or rC completely blocks the synthesis of msDNA (Hsu et al., 1989). This mutation is significant because it provides important supporting evidence for the proposed priming mechanism described here. In an alternative mechanism recently suggested, the DNA strand is first synthesized from a conventional (3′-OH) RNA primer and then later ligated to the 2′ position of the RNA strand of msDNA. However, this mechanism does not explain the mutation described above, since replacement of the rG (residue 20) with rC should not affect synthesis of the DNA strand according to the conventional primer model.

Polymerization and Termination of DNA Synthesis

Judging from intermediate structures observed in a permeabilized cell system developed to study the synthesis of msDNA in *M. xanthus*, polymerization of DNA by RT begins after formation of the branch structure (Lampson et al., 1989a). In addition, as elongation of the DNA strand continues along the RNA template, there appears to be a rapid removal of the RNA template within the growing DNA-RNA duplex by RNase H activity. The RNase H activity is probably derived from the *M. xanthus* host cell, since the Mx162 RT gene does not appear to encode an RNase H domain (Inouye and Inouye, 1991b).

For Mx162 msDNA, elongation of the DNA strand continues along with removal of the RNA template until termination at 162 deoxyribonucleotides, leaving a 77-base template RNA remaining with the molecule (Fig. 2B). It is not clear how this precise termination of msDNA synthesis occurs in vivo. Termination signals may be a combination of secondary structure in the template RNA and the complexed protein.

STRUCTURE OF THE RETRON ELEMENT

Mapping Ends of the Retron

Retrons exist as single-copy elements in the bacterial chromosome, and thus, the ends of a retron can be defined as the junction site at which sequences unique to the retron end and sequences homologus to the bacterial chromosome begin. For example, retron Ec107 is found in several natural strains of *E. coli*, including strain ECOR 70. By comparing the nucleotide sequences 5′ and 3′ to the retron genes with homologous sequences found in the *E. coli* K-12 chromosome (which lacks a retron), the precise ends of the retron as well as its insertion site can be determined. Retron Ec107 contains only 1.3 kb of unique sequence with 70 bp of untranscribed sequence at the 5′ end followed by the genes for msDNA, a 50-bp spacer, the RT gene, and finally, a 47-bp untranslated sequence at the 3′ end (Fig. 3). The retron element has inserted between the *pyrE* and *ttk* genes at 82 min on the *E. coli* chromosome map. Unlike transposons or prophages, the retron has no flanking direct or inverted DNA sequence repeats. However, as a consequence, presumably, of its insertion into the

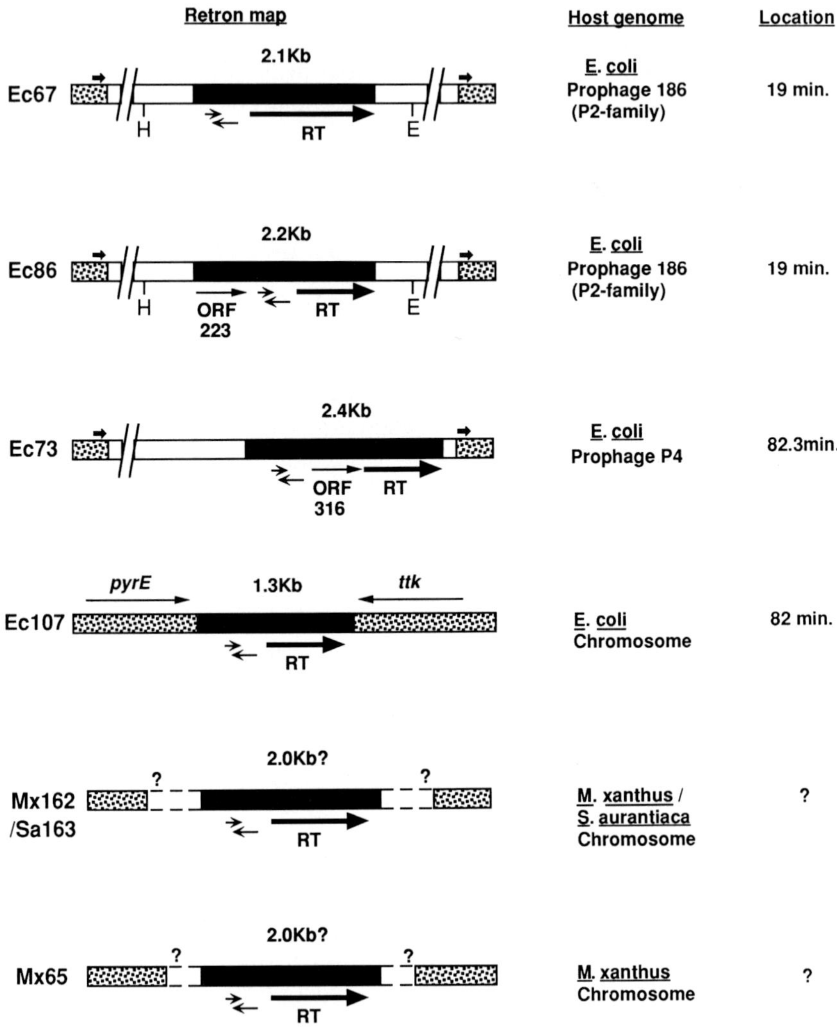

Figure 3. Insertion sites and gene organizations of the retron elements of *M. xanthus* and *E. coli*. The black bar in each genetic map represents the entire retron element, defined as unique sequences associated with the genes for msDNA and RT. Approximate sizes are given in kilobase pairs. Open bars represent prophage DNAs, and stippled bars represent bacterial chromosomes. Retron-encoded genes are RT (thick arrow), msDNA (two small convergent arrows), and in some cases a second ORF of unknown function (thin arrow). For retrons Ec67 and Ec86, the restriction sites *Hin*dIII (H) and *Eco*RI (E) are shown to indicate that these two retrons have inserted into the same site. The prophage genomes are flanked by direct sequence repeats (small arrows). However, retrons are not flanked by either direct or inverted repeat sequences. Question marks indicate that neither the ends nor the insertion sites have been determined for the retron elements of the myxobacteria.

chromosome, the element has replaced a 34-bp intergenic sequence at the insertion site (Herzer et al., 1992).

Genetic Organization and Insertion Sites

A comparison of the genetic maps and insertion sites for the known retrons of *E. coli* and *M. xanthus* is shown in Fig. 3. A retron element is defined here as all unique sequences associated with the genes encoding msDNA and RT. The Ec107 retron is the smallest element at only 1.3 kb, and thus, the minimum genetic unit constituting a retron element is simply the genes for msDNA, an ORF encoding RT, and short terminal sequences. Some retrons (Ec73 and Ec86), however, are larger and contain a second ORF of unknown identity in addition to that for RT (Lim and Maas, 1990). For the Ec73 retron, ORF 316 lies between the genes encoding msDNA and RT, and a comparison of its amino acid sequence with the sequences of all known proteins contained in GenBank (as of 1990) revealed no significant identity (Sun et al., 1991). ORF 316 was also deleted from the Ec73 retron, but this deletion had no effect on production of msDNA.

Another interesting finding is that some *E. coli* retrons have inserted into a prophage genome of the P2 family or into the P2 parasitic phage P4. For example, mapping the ends of retron Ec73 showed the element to be associated with a much larger segment of foreign DNA inserted into the chromosome of the host *E. coli* strain Cl-23 (Sun et al., 1991). Indeed, when the DNA sequence of the entire 12.7-kb foreign DNA was determined, it showed significant homology to that of the *E. coli* phage P4 (Halling et al., 1990). The foreign segment has a gene organization similar to that of P4 except for an additional 2.4-kb region encoding the Ec73 retron that is incorporated into the right end of the prophage. However, this P4-like prophage also differs from wild-type P4 in several other ways. For example, the element encodes an integrase gene different from that of P4. Also, the 29-bp direct repeats that flank the prophage are significantly different from the P4 *att*L and *att*R sequences, and the site of insertion in the chromosome of *E. coli* Cl-23 (82.3 min) is not the same as the P4 insertion site at 97 min (Bertani and Six, 1988; Ghisotti et al., 1990). In addition, this P4-like prophage containing the Ec73 retron can excise from the chromosome, form an infectious virion, and transduce the ability to produce msDNA Ec73 to a new host strain if phage P2 is provided as a helper phage (Inouye et al., 1991).

In addition to P4, phage 186, a member of the P2 family, has also apparently incorporated retron elements into its genome. As a result of two independent events, two different retrons, Ec67 (*E. coli* Cl-1) and Ec86 (*E. coli* B), have inserted into precisely the same site of the 186 prophage (Hsu et al., 1990; Lim, 1991). This phage 186-like prophage, containing either retron, is about 34 kb in size and resides at 19 min on the *E. coli* host chromosome.

Coresiding with the Mx162 retron in *M. xanthus* DZF1 is a second retron element, Mx65, having an entirely different nucleotide sequence, producing a smaller msDNA, and located at a different position on the DZF1 chromosome (Dhundale et al., 1988b). However, the exact boundaries or ends of these retron elements from *M. xanthus* have not been determined. One major problem is that no strains of *M. xanthus* lack the Mx162 retron; thus, the ends of the element cannot be mapped by comparison with chromosome sequences (as was done in *E. coli*).

COMPARING RETRONS WITH EUKARYOTIC RETROELEMENTS

A large variety of different genetic elements from eukaryotic organisms utilize RTs for their replication and mobilization. These elements include RNA viruses such as Moloney murine leukemia virus and human immunodeficiency virus and DNA viruses such as hepatitis B virus and cauliflower mosaic virus. In addition, transposons such as the Ty elements of yeast, Copia and Gypsy of *Drosophila melanogaster,* some mitrochondrial introns, and various long interspersed repetitive-sequence (LINE) elements like L1 and I factor all encode RT (McClure, 1991; Xiong and Eickbush, 1990).

Functional domains or regions of proposed enzymatic activity encoded by the ORF(s) of a retroelement can be inferred by detecting amino acid sequence similarity with a known protein. For example, in addition to RT, retroviruses encode RNase H, protease, and endonuclease activities (Doolittle et al., 1989). A comparison of the inferred structural and functional domains of the ORF encoding RTs from various eukaryotic retroelements with retron Mx162 of *M. xanthus* (Fig. 4) reveals the bacterial elements to be the retroelements perhaps simplest in structural organization. Except for msDNA, a retron appears to encode only RT activity. Eukaryotic elements vary considerably in structural organization, with the viral or virus-like elements (Ty, Gypsy) being the most complex and the nonviral elements or those lacking long terminal repeats being much simpler (McClure, 1991; Xiong and Eickbush, 1990).

ORIGIN AND DIVERGENCE OF RETRON ELEMENTS

One of the most interesting questions concerning the discovery of retrons in bacteria concerns where these RT-encoding elements came from: are they prokaryotic in origin, and what is their relationship to the

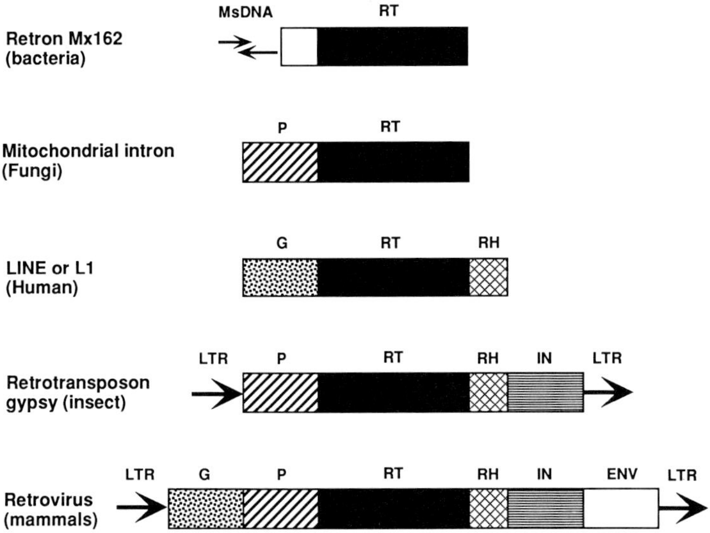

Figure 4. Functional domains encoded by eukaryotic retroelements and myxobacterial retrons. Retron elements such as Mx162 appear to be among the simplest retroelements in terms of functional domains associated with the RT ORF. Retrons appear to lack such domains as a capsid protein (G), protease (P), RNase H (RH), integrase (IN), envelope proteins (ENV), and long terminal repeats (LTR), which are encoded by other eukaryotic retroelements. The open bar associated with the Mx162 RT gene indicates that the amino acid sequence contains an N-terminal extension that has no sequence similarity to any known activities.

eukaryotic retroelements? As discussed below, the retrons of the myxobacteria, in particular, may provide important clues to the answers to these questions.

Prevalence and Diversity

Seven different retron elements (Ec67, Ec73, Ec86, and Ec107 from *E. coli* and Mx162, Mx65, and Sa163 from myxobacteria) have been studied in some molecular detail. Except for Mx162 and Sa163, these elements have almost no similar nucleotide sequences. The extreme heterogeneity of retrons is reminiscent of the highly diverse eukaryotic retroelements and may reflect the possible replication of these bacterial elements by RT, with its poor fidelity of DNA polymerization (Preston et al., 1988).

In addition to *M. xanthus* and *E. coli,* a number of other bacterial groups have recently been shown to contain msDNA-producing elements. These include *Salmonella* spp., *Proteus mirabilis, Klebsiella pneumoniae, Rhizobium* spp., *Bradyrhizobium* spp., and the myxobacterium *Nannocystis exedens* (Rice and Lampson, 1992).

Several studies have examined collections of *E. coli* strains for the presence of msDNA and have found that only a small number (10 to 13%) of strains appear to harbor a retron element in their chromosome (Sun et al., 1989; Lim et al., 1990; Herzer et al., 1990). There are at least seven different types of retrons, including the four mentioned above, and probably more from these *E. coli* strains. Quite a different result was obtained when a similar survey of a collection of *M. xanthus* strains was done. In that study, 20 strains of *M. xanthus* from sites in the United States and Europe were all found to produce msDNA essentially the same as Mx162 according to size and DNA hybridization (Dhundale et al., 1985; Lampson et al., 1991a). Also all of these *M. xanthus* strains cross-hybridize with a probe containing the region encoding the Mx162 RT gene, and most (but not all) of the strains appear to contain a second retron species similar to Mx65.

Origin and Evolutionary Relationships

The large variety of retrons and their limited distribution among a few strains of *E. coli* suggest that retron elements found today are not native to this species. Indeed, a recent study of retron elements in the ECOR collection supports this idea. The ECOR collection is a standard reference collection of 72 *E. coli* isolates that have been defined by electrophoretic analysis of 20 enzyme-encoding loci as representative of the broad genetic diversity of this species (Ochman and Selander, 1984). From this study, it was concluded that msDNA-producing retroelements were acquired independently from some outside source in several different clonal lineages of *E. coli* (Herzer et al., 1990). Additional evidence indicates that the transmission of retrons into the genome probably occurred during the recent evolution of *E. coli*. A comparison of the amino acid codons used among RT genes from *E. coli* retrons shows a pattern of usage rarely observed among *E. coli* genes (Inouye et al., 1989). For example, in the Ec67 RT gene, 65% of arginine codons are AGA or AGG, while these codons are rarely used (2.7%) among 199 *E. coli* genes surveyed. The atypical codon usage found in RT ORFs of *E. coli* retrons implies that these elements were probably recently introduced into the genome of certain *E. coli* strains. Another indication of how long retron elements have resided in the *E. coli* genome is the degree of nucleotide sequence diversity among several individual (homologous) retrons. For example, the *E. coli* retron Ec107 is found to occur quite frequently among strains of *E. coli* producing msDNA, a fact that allowed 12 individual retrons of the Ec107 type to be isolated and sequenced (Kawaguchi et al., 1992). Although these 12 retrons were found in strains from three different phylogenetic branches of the ECOR reference population of *E. coli*, the observed nucleotide sequence variability was very small (9 of 12 were identical), indicating a recent acquisition or spread among the different *E. coli* strains.

Whereas retron elements presently found in *E. coli* appear to be recent acquisitions in the genome, evidence suggests that the retrons of *M. xanthus* may be quite ancient. Unlike *E. coli* populations, in a genetically diverse collection of *M. xanthus*, every strain, without exception, contains a retron essentially the same as the Mx162 element originally found in strain DZF1. Also the codon bias found in the RT genes of *M. xanthus* is similar, that is, typical, for codons used in other known *M. xanthus* proteins (Inouye et al., 1989). In addition, *S. aurantiaca*, a myxobacterium related to *M. xanthus*, also contains a retron element, designated Sa163 (Furuichi et al., 1987b). This retron is very similar to Mx162, having more than 81% sequence identity in the genes coding for msDNA and about 77% sequence identity in the gene coding for RT (Hsu et al., 1992b). All of these unusual features imply that retrons and their RT genes existed in the myxobacteria at least before the divergence of *M. xanthus* and *S. aurantiaca* from a common ancestor some 100 million years ago.

However strong the evidence is for the postulated ancient lineage of RT genes in the myxobacteria, this idea is contrary to the findings of a study that compared bacterial and eukaryotic RTs (Xiong and Eickbush, 1990). The amino acid sequences of 4 retron-encoded RTs were compared with those of more than 70 RTs from various eukaryotic elements. Evolutionary relationships derived from these comparisons showed that the bacterial RTs appeared to be most closely related to certain organelle RTs, such as the RTs encoded by mitochondrial introns and plasmids of fungi, and were probably recently acquired from some unknown source. However, one of the difficulties in deriving these relationships and comparisons is the extreme variation among RT sequences. This variability may be because many of these eukaryotic elements are replicated by reverse transcription, a process highly error prone.

We have continued to investigate the fascinating questions of the origin and evolution of RT by seeking to identify retrons homologous to Mx162 among related myxobacteria. Indeed, a retron element very similar to Mx162 was recently discovered in the myxobacterium *N. exedens* (Rice and Lampson, 1992). The phylogenetic relationship of *N. exedens* to *M. xanthus* and other myxobacteria is presented in Fig. 5. According to 16S rRNA sequences, *M. xanthus* and *S. aurantiaca* appear to be closely related, having diverged from a common relative as recently as 100 million year ago. However, *N. exedens* appears to be more distantly related, perhaps emerging as a separate lineage about 900 million years ago (Ludwig et al., 1983). Comparing the nucleotide sequence of the retron from *N. exedens* with sequences of homologous examples from other myxo-

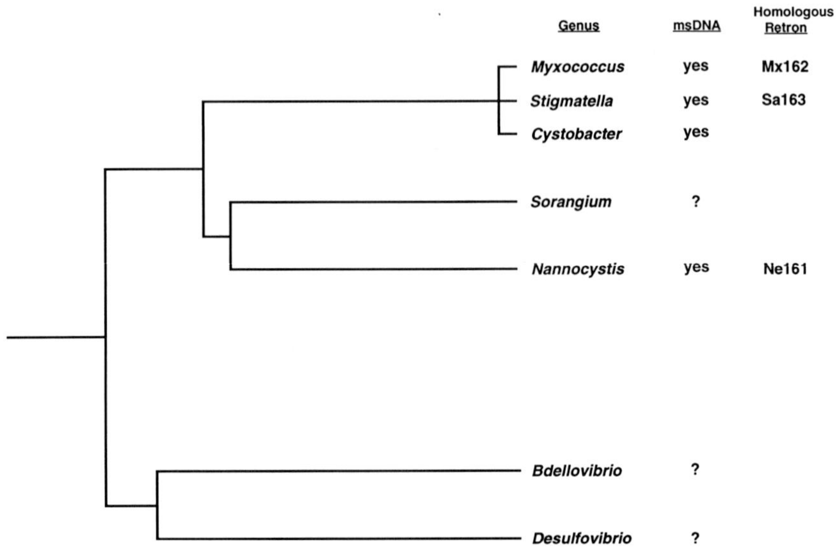

Figure 5. Phylogenetic relationships among the myxobacteria (genera *Myxococcus, Stigmatella, Cystobacter, Sorangium,* and *Nannocystis*) and their closest relatives (bdellovibrios and the sulfate and sulfur reducers). The length of each line in the dendrogram reflects the relative genetic distance or time from divergence for each group. The prevalences of msDNA and retron elements homologous to Mx162 are also indicated.

bacteria (Mx162 and Sa163) should provide important information about phylogenetic relationships among bacterial RTs and how long these genes have existed in the myxobacterial genome.

CONCLUDING REMARKS

The msDNA molecule was the first crucial clue leading to the novel discovery of retroelements and RTs in the prokaryotic world, yet this peculiar molecule remains the most mystifying character in the retron element story. The function of msDNA and its effects on the *M. xanthus* host cell are unknown. What purpose does the 2′-5′ branch linkage between RNA and DNA serve? Why is this satellite molecule produced in such large quantities? A double deletion mutation of both retrons (Mx162 and Mx65) from strain DZF1 produced no apparent effect on cell growth, motility, fruiting body formation, spore formation, or germination (Inouye et al., 1990).

However, circumstantial evidence suggesting that these bacterial retroelements are mobile continues to accumulate. On separate occasions, the same prophage genome in *E. coli* appears to have picked up two different retrons (Ec67 and Ec86), and another retron, Ec107, appears to have recently disseminated among different population groups of *E. coli*. It is interesting to note some resemblance between retrons and the structurally simplistic LINE elements of eukaryotes (Fig. 4). Indeed, a number of different members of the LINE element family have recently been shown to be active transposable elements (Dombroski et al., 1991; Heidmann and Heidmann, 1991). If retrons are eventually observed to transpose to a new site, then these elements may prove to be attractive tools for investigating a novel mechanism of transposition.

Although the function of msDNA remains obscure, continued investigation of its structure and mode of synthesis could reveal new activities and functions for an RNA molecule and for RT. In particular, what is the nature of the rG residue of the precursor RNA that serves as a primer and of the interaction of RT to form the 2′-5′ branch structure? The unusual structure of msDNA is made more intriguing by the recent recognition of the important role certain 2′ hydroxyl groups play in a number of catalytic RNAs (Williams et al., 1992; Pyle and Cech, 1991). Even more interesting is the recent report of a debranching activity in yeast cells that hydrolyzes 2′-5′ phosphodiester bonds. Mutant yeast cells containing the Ty element and lacking the debranching activity show a pronounced reduction in transposition frequency of this retrotransposon (Chapman and Boeke, 1991). This suggests that formation of a 2′-5′ linkage must play a role in retrotransposition of the Ty element, perhaps because a 2′-OH group functions as a primer for the reverse transcription reactions. Perhaps msDNA also serves some role in regulating the transposition of retron elements.

Further investigation of the msDNA-producing retroelements of the myxobacteria is likely to provide new insights into the mechanism of retrotransposition, how RT initiates reverse transcription, and the origin and divergence of RT and retroelements.

Acknowledgments. My sincere thanks to Masayori and Sumiko Inouye for their critical evaluation of this manuscript. I also thank Gary Stacey for his helpful comments and Scott Rice for help with the figures and drawings.

ADDENDUM IN PROOF

The reader is alerted to some recent papers published on retron elements: D. Lim, *Mol. Microbiol.* **6:**3531–3542, 1992; J. Kirchner et al., *Mol. Microbiol.* **6:**2815–2824, 1992; S. Miyata et al., *Proc. Natl. Acad. Sci. USA* **89:**5735–5739, 1992; and T. Shimamoto et al., *J. Biol. Chem.* **268:**2684–2692, 1993.

REFERENCES

Bertani, L. E., and E. W. Six. 1988. The P2-like phages and their parasite, P4, p. 73–143. *In* R. Calendar (ed.), *The Bacteriophages*. Plenum Press, New York.

Boeke, J. D., and V. G. Corces. 1989. Transcription and reverse transcription of retrotransposons. *Annu. Rev. Microbiol.* **43:**403–434.

Brosius, J. 1991. Retroposons—seeds of evolution. *Science* **251:**753.

Bucheton, A. 1990. I transposable elements and I-R hybrid dysgenesis in *Drosophila*. *Trends Genet.* **6:**16–21.

Chapman, K. B., and J. D. Boeke. 1991. Isolation and characterization of the gene encoding yeast debranching enzyme. *Cell* **65:**483–492.

Dhundale, A., T. Furuichi, M. Inouye, and S. Inouye. 1988a. Mutations that affect production of branched RNA-linked msDNA in *Myxococcus xanthus*. *J. Bacteriol.* **170:**5620–5624.

Dhundale, A., T. Furuichi, S. Inouye, and M. Inouye. 1985. Distribution of multicopy single-stranded DNA among myxobacteria and related species. *J. Bacteriol.* **164:**914–917.

Dhundale, A., M. Inouye, and S. Inouye. 1988b. A new species of multicopy single-stranded DNA from *Myxococcus xanthus* with conserved structural features. *J. Biol. Chem.* **263:** 9055–9058.

Dhundale, A., B. Lampson, T. Furuichi, M. Inouye, and S. Inouye. 1987. Structure of msDNA from *Myxococcus xanthus:* evidence for a long self-annealing RNA precursor for the covalently linked, branched RNA. *Cell* **51:**1105–1112.

Dombroski, B. A., S. L. Mathias, E. Nanthakumar, A. F. Scott, and H. H. Kazazian, Jr. 1991. Isolation of an active human transposable element. *Science* **254:**1805–1807.

Doolittle, R. F., D.-F. Feng, M. S. Johnson, and M. A. McClure. 1989. Origins and evolutionary relationships of retroviruses. *Q. Rev. Biol.* **64:**1–30.

Finnegan, D. J. 1989. Eukaryotic transposable elements and genome evolution. *Trends Genet.* **5:**103–107.

Furuichi, T., A. Dhundale, M. Inouye, and S. Inouye. 1987a. Branched RNA covalently linked to the 5′ end of a single-stranded DNA in *Stigmatella aurantiaca:* structure of msDNA. *Cell* **48:**47–53.

Furuichi, T., S. Inouye, and M. Inouye. 1987b. Biosynthesis and structure of stable branched RNA covalently linked to the 5′ end of multicopy single-stranded DNA of *Stigmatella aurantiaca*. *Cell* **48:**55–62.

Ghisotti, D., S. Finkel, C. Halling, G. Deho, G. Sironi, and R. Calendar. 1990. Nonessential region of bacteriophage P4: DNA sequence, transcription, gene products, and functions. *J. Virol.* **64:**24–36.

Halling , C., R. Calendar, G. E. Christie, E. C. Dale, G. Deho, S. Finkel, J. Flensburg, D. Ghisotti, M. L. Kahn, K. B. Lane, C.-S. Lin, B. H. Lindqvist, L. S. Pierson III, E. W. Six, M. G. Sunshine, and R. Zierman. 1990. DNA sequence of satellite bacteriophage P4. *Nucleic Acids Res.* **18:**1649.

Heidmann, O., and T. Heidmann. 1991. Retrotransposition of a mouse IAP sequence tagged with an indicator gene. *Cell* **64:**159–170.

Herzer, P. J., S. Inouye, and M. Inouye. 1992. Retron-Ec107 is inserted into the *Escherichia coli* genome by replacing a palindromic 34 bp intergenic sequence. *Mol. Microbiol.* **6:** 345–354.

Herzer, P. J., S. Inouye, M. Inouye, and T. S. Whittam. 1990. Phylogenetic distribution of branched RNA-linked multicopy single-stranded DNA among natural isolates of *Escherichia coli*. *J. Bacteriol.* **172:**6175–6181.

Howard, B. H., and K. Sakamoto. 1990. Alu interspersed repeats: selfish DNA or a functional gene family? *New Biol.* **2:**759–770.

Hsu, M.-Y., S. G. Eagle, M. Inouye, and S. Inouye. 1992a. Cell-free synthesis of the branched RNA-linked msDNA from retron Ec67 of *Escherichia coli*. *J. Biol. Chem.* **267:** 13823–13829.

Hsu, M.-Y., S. Inouye, and M. Inouye. 1989. Structural requirements of the RNA precursor for the biosynthesis of the branched RNA-linked msDNA of *Myxococcus xanthus*. *J. Biol. Chem.* **264:**6214–6219.

Hsu, M.-Y., C. Xu, M. Inouye, and S. Inouye. 1990. Retron for the 67-base branched RNA-linked multicopy DNA from *Escherichia coli:* a potential transposable element encoding both reverse transcriptase and Dam methylase functions. *Proc. Natl. Acad. Sci. USA* **87:** 9454–9458.

Hsu, M.-Y., C. Xu, M. Inouye, and S. Inouye. 1992b. Similarity between the *Myxococcus xanthus* and *Stigmatella aurantiaca* reverse transcriptase genes associated with multicopy, single-stranded DNA. *J. Bacteriol.* **174:**2384–2387.

Inouye, M., and S. Inouye. 1991a. Retroelements in bacteria. *Trends Biochem. Sci.* **16:**18–21.

Inouye, M., and S. Inouye. 1991b. msDNA and bacterial reverse transcriptase. *Annu. Rev. Microbiol.* **45:**163–186.

Inouye, M., and S. Inouye. 1992. Retrons and multicopy single-stranded DNA. *J. Bacteriol.* **174:**2419–2424.

Inouye, S., T. Furuichi, A. Dhundale, and M. Inouye. 1987. Stable branched RNA covalently linked to the 5′ end of a single-stranded DNA of Myxobacteria, p. 271–284. *In* M. Inouye and B. S. Dudock (ed.), *Molecular Biology of RNA: New Perspectives*. Academic Press, Inc., San Diego, Calif.

Inouye, S., P. J. Herzer, and M. Inouye. 1990. Two independent retrons with highly diverse reverse transcriptases in *Myxococcus xanthus*. *Proc. Natl. Acad Sci. USA* **87:**942–945.

Inouye, S., M.-Y. Hsu, S. Eagle, and M. Inouye. 1989. Reverse transcriptase associated with the biosynthesis of the branched RNA-linked msDNA in *Myxococcus xanthus*. *Cell* **56:** 709–717.

Inouye, S., M. G. Sunshine, E. W. Six, and M. Inouye. 1991. Retronphage R73: an *E. coli* phage that contains a retroelement and integrates into a tRNA gene. *Science* **252:**969–971.

Kawaguchi, T., P. J. Herzer, M. Inouye, and S. Inouye. 1992. Sequence diversity of the 1.3 kb retron (retron-Ec107) among three distinct phylogenetic groups of *Escherichia coli*. *Mol. Microbiol.* **6:**355–361.

Lampson, B. D., M. Inouye, and S. Inouye. 1989a. Reverse transcriptase with concomitant ribonuclease H activity in a cell-free system of msDNA of *Myxococcus xanthus*. *Cell* **56:** 701–707.

Lampson, B. C., M. Inouye, and S. Inouye. 1991a. Survey of multicopy single-stranded DNAs and reverse transcriptase genes among natural isolates of *Myxococcus xanthus*. *J. Bacteriol.* **173:**5363–5370.

Lampson, B. C., S. Inouye, and M. Inouye. 1991b. msDNA of bacteria. *Prog. Nucleic Acid Res. Mol. Biol.* **40:**1–24.

Lampson, B. C., J. Sun, M.-Y. Hsu, J. Vallejo-Ramirez, S. Inouye, and M. Inouye. 1989b. Reverse transcriptase in a clinical strain of *E. coli:* its requirements for the production of branched RNA-linked msDNA. *Science* **243:**1033–1038.

Lampson, B. C., M. Viswanathan, M. Inouye, and S. Inouye. 1990. Reverse transcriptase from *Escherichia coli* exists as a complex with msDNA and is able to synthesize double-stranded DNA. *J. Biol. Chem.* **265:**8490–8496.

Lease, R. A., and T. Yee. 1991. Early events in the synthesis of the multicopy single-stranded DNA-RNA branched copolymer of *Myxococcus xanthus*. *J. Biol. Chem.* **266:** 14497–14503.

Lim, D. 1991. Structure of two retrons of *Escherichia coli* and their common insertion site. *Mol. Microbiol.* **5:**1863–1872.

Lim, D., T. A. T. Gomes, and W. K. Maas. 1990. Distribution of msDNA among serotypes of enteropathogenic *Escherichia coli* strains. *Mol. Microbiol.* **4:**1711–1714.

Lim, D., and W. K. Maas. 1989. Reverse transcriptase dependent synthesis of a covalently linked, branched DNA-RNA compound in *E. coli* B. *Cell* **56:**891–904.

Lim, D., and W. K. Maas. 1990. Mapping of the msDNA operon in the chromosome of *Escherichia coli* B. *Mol. Microbiol.* **4:**2201–2203.

Ludwig, W., K. H. Schleifer, H. Reichenbach, and E. Stackebrandt. 1983. A phylogenetic analysis of the myxobacteria *Myxococcus fulvus, Stigmatella aurantiaca, Cytobacter fuscus, Sorangium cellulosum,* and *Nannocystis exedens. Arch. Microbiol.* **135:**58–62.

McClure, M. A. 1991. Evolution of retroposons by acquisition or deletion of retrovirus-like genes. *Mol. Biol. Evol.* **8:**835–856.

Ochman, H., and R. K. Selander. 1984. Standard reference strains of *Escherichia coli* from natural populations. *J. Bacteriol.* **157:**690–693.

Preston, B. D., B. J. Poiesz, and L. A. Loeb. 1988. Fidelity of HIV-1 reverse transcriptase. *Science* **242:**1168–1171.

Pyle, A. M., and T. R. Cech. 1991. Ribozyme recognition of RNA by tertiary interactions with specific ribose 2′-OH groups. *Nature* (London) **350:**628–631.

Rice, S., and B. C. Lampson. Diversity of retron elements among a population of rhizobia and other gram-negative bacteria. Submitted for publication.

Sun, J., P. J. Herzer, M. P. Weinstein, B. C. Lampson, M. Inouye, and S. Inouye. 1989. Extensive diversity of branched-RNA linked multicopy single-stranded DNAs in clinical strains of *Escherichia coli. Proc. Natl. Acad. Sci. USA* **86:**7208–7212.

Sun, J., M. Inouye, and S. Inouye. 1991. Association of a retroelement with a P4-like cryptic prophage (retronphage R73) integrated into the selenocystyl tRNA gene of *Escherichia coli. J. Bacteriol.* **173:**4171–4181.

Viswanathan, M., M. Inouye, and S. Inouye. 1989. *Myxococcus xanthus* msDNA-Mx162 exists as a complex with proteins. *J. Biol. Chem.* **264:**13665–13671.

Weiner, A. M., P. L. Deininger, and A. Efstratiadis. 1986. Nonviral retroposons: genes, pseudogenes, and transposable elements generated by the reverse flow of genetic information. *Annu. Rev. Biochem.* **55:**631–661.

Williams, D. M., W. A. Picken, and F. Eckstein. 1992. Function of specific 2′-hydroxyl groups of guanosines in a hammerhead ribozyme probed by 2′ modifications. *Proc. Natl. Acad. Sci. USA* **89:**918–921.

Xiong, Y., and T. H. Eickbush. 1990. Origin and evolution of retroelements based upon their reverse transcriptase sequences. *EMBO J.* **9:**3353–3362.

Yee, T., T. Furuichi, S. Inouye, and M. Inouye. 1984. Multicopy single-stranded DNA isolated from a gram-negative bacterium, *Myxococcus xanthus. Cell* **38:**203–209.

Myxobacteria II
Edited by Martin Dworkin and Dale Kaiser
© 1993 American Society for Microbiology, Washington, DC 20005

Chapter 6

Genetic Approaches for Analysis of Myxobacterial Behavior

Ronald E. Gill and Lawrence J. Shimkets

The isolation and characterization of mutants with altered behavior are the basis for a genetic analysis of myxobacteria. Understanding such mutants and the relationships between them should reveal the regulatory networks that coordinate myxobacterial behavior.

By comparison with other bacteria, such as the enteric bacteria, for which sophisticated and elegant genetic systems have been devised, the repertoire of genetic tools available for genetic analysis of the myxobacteria is modest. Through the creative and imaginative application of the available techniques and the very successful integration of "classic" genetics with molecular biology, all essential genetic manipulations are now possible.

The techniques and strategies currently used to isolate and characterize mutants and the methodology available for cloning and characterizing *Myxococcus xanthus* genes are described in this chapter.

ISOLATION OF MUTANTS

Conventional mutagens such as UV light, nitrosoguanidine, ethyl methanesulfonate, transposon insertions, and in vitro constructions have all been used successfully to isolate a wide variety of *M. xanthus* mutants. Each method has limitations, yet taken together, the various methods afford considerable flexibility in the isolation and subsequent characterization of mutants.

Ronald E. Gill • Department of Microbiology and Immunology, School of Medicine, University of Colorado Health Sciences Center, Denver, Colorado 80262. **Lawrence J. Shimkets** • Department of Microbiology, University of Georgia, Athens, Georgia 30602.

The advantage of conventional mutagenesis is that a wide variety of types of mutations, some of which may not arise by methods such as transposon mutagenesis, are isolated. Although all of the methods can be used to isolate loss-of-function mutations, conventional mutagens causing single-base-pair alterations can also be expected to result in conditional mutations or alterations in essential genes that alter function yet allow survival of the bacterium. For example, Rudd and Zusman (1979) found that approximately 5% of the mutations that conferred resistance to rifampin also resulted in a developmental defect. Although these mutants were shown to have altered RNA polymerases, most of the mutations had no discernible effect on cell growth. The advantages of conventional mutagenesis over transposons, for example, are also exemplified by the *dsg* mutations. These map to the same gene, which is believed to be vital, because it has been impossible to isolate viable *dsg* transposon insertion mutations in haploid cells. However, point mutations that affect developmental functions of the *dsg* gene product yet allow apparently normal rates of vegetative growth have been isolated after UV irradiation. Given these results, it is unlikely that *dsg,* a gene required for cell viability and independently for normal developmental gene expression, would have been isolated from a collection of transposon insertion mutants.

Transposon mutagenesis with Tn*5* has been used effectively for the isolation of *M. xanthus* mutants and the mutagenesis of cloned *M. xanthus* DNA in *Escherichia coli* (see below). Transposon insertion mutations are particularly useful when complete loss of gene product function is sought. In one study of mutants (Rhie and Shimkets, 1989), the frequency of a particular mutant class was 2×10^{-4} per cell when transposon mutagenesis was used and only 1×10^{-5} per cell when UV mutagenesis was used. This enhanced frequency would be highly advantageous when there was no direct selection for the mutant phenotype. In addition, the availability of several specialized derivatives of Tn*5* has greatly enhanced the utility of transposon mutagenesis.

In practice, transposon mutagenesis requires a means of introducing the transposon into the cell, identification of cells in which transposition has taken place, transposition of the element to numerous (if not random) sites, and stable integration after transposition. Introduction of Tn*5* into *M. xanthus* is accomplished with P1::Tn*5*, a variant of coliphage P1 into which Tn*5* has transposed (Kuner and Kaiser, 1981). This virus is a nearly ideal suicide vector in *M. xanthus*. Infection of *Myxococcus* cells with P1::Tn*5* forms stable kanamycin-resistant (Km^r) transductants, yet there is no production of progeny virus or any maintenance of P1 viral DNA by the infected cells. Some cell killing may result, however. Kuner and Kaiser (1981) showed that Km^r transductants resulted from the transposition of Tn*5* from P1::Tn*5* to many perhaps random sites on the bacterial chromosome. Since Tn*5* encodes resistance to kanamycin, to which *M. xanthus* is naturally sensitive, cells into which Tn*5* has transposed can be selected by the stable inheritance of kanamycin resistance. Once isolated, the frequency of further transposition of Tn*5* from its initial chromosomal site to a new site has been estimated to be less than 10^{-3} per cell (Avery and Kaiser, 1983a; Sodergren and Kaiser, 1983). Each of the kanamycin-resistant colonies produced by infection with P1::Tn*5* resulted from a single independent transposition into the *Myxococcus* chromosome. This is significant for two reasons. First, one is assured that each Km^r colony that is screened has in fact suffered a mutation at some position in the chromosome. Second, one is reasonably certain that each strain contains only a single mutation, a supposition that can be confirmed by restriction enzyme analysis for any particular mutant identified. Efficient mutagenesis is thus attained without the complication, often encountered with conventional mutagens, of multiple mutations and undoubtedly contributes to the enhanced frequency of isolating certain classes of mutants.

Tn*5* Derivatives for the Production of Gene Fusions

Gene fusion technology has become increasingly valuable and indispensable for genetic analysis. The fusion of an endogenous promoter to a reporter gene can be used to detect genes expressed under a specific set of circumstances, to detect a protein that has not been identified or for which no assay exists, to analyze gene regulation, and to analyze the mechanism of localization of a protein to an intracellular compartment or to the extracellular space (Silhavy and Beckwith, 1985). The two basic types of fusions are operon fusions and translational fusions. In operon fusions, the promoterless reporter gene coding sequence is complete with its own sequences for translation initiation. It constitutes a separate gene in a multigene operon that is under the transcriptional control of the same endogenous promoter. The translational efficiency, specific activity, and posttranslational fate (such as cellular localization or secretion) of the protein are all determined by the reporter gene. In translational fusions, a portion of the reporter gene coding sequence is fused in the correct reading frame to the endogenous gene. The product of the gene fusion is a chimeric polypeptide in which the amino-terminal portion is derived from the endogenous gene and the carboxy-terminal portion is derived from the reporter gene. Therefore, transcription and translation are generally controlled by the endogenous gene. However, the specific activity and posttranslational fate of the chimeric protein are often controlled by both portions of the chimeric protein.

Two Tn*5* derivatives, Tn*5lac* (Kroos and Kaiser, 1984) and Tn*phoA*

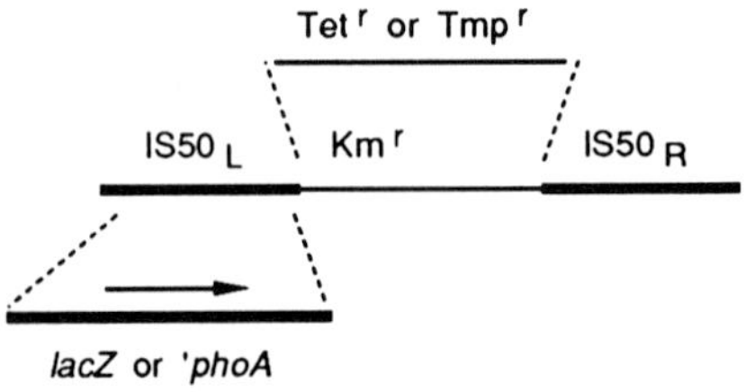

Figure 1. Structures of Tn*5* and some of its derivatives. Tn*5* (center line) is composed of three segments: two IS*50* elements arranged in inverted orientation and a unique center segment encoding resistance to kanamycin, streptomycin, and bleomycin. Only kanamycin resistance is routinely utilized in *M. xanthus*. Replacement of the unique segment of Tn*5* with genes encoding resistance to tetracycline (Tet^r) or trimethoprim (Tmp^r), schematically shown above Tn*5*, yields Tn*5*-132 and Tn*5*-Tp, respectively. Replacement of a segment of the left IS*50* element (IS50$_L$) with reporter genes for formation of operon or protein fusions is schematically shown below Tn*5*. The *lacZ* fragment contains the entire protein-coding sequence but lacks a transcription promoter and is used for generating operon fusions. The *'phoA* fragment contains only a portion of protein-coding sequence; it lacks the translational start and signal sequences at the amino-terminal end of the protein that are required for export of the protein to the periplasmic space.

(Manoil and Beckwith, 1985), each of which contains a promoterless reporter gene, provide a simple yet powerful in vivo means of producing operon or protein fusions, respectively (Fig. 1). (In vitro-constructed gene fusions to cloned *Myxococcus* genes are also widely used and are discussed below.) Of these two Tn*5* derivatives, Tn*5lac* is more widely used and has been particularly valuable for the identification of large numbers of developmentally regulated genes. Tn*5lac* is transposition proficient, encodes resistance to kanamycin, and has been modified by the addition of a promoterless *lacZ* gene from *E. coli* to the left IS*50* element. Transcription of *lacZ* is dependent on and under the regulatory control of a properly oriented endogenous promoter adjacent to the site into which the element transposes. Once transcribed, the *lacZ* reporter mRNA contains the sequences for translational initiation, and wild-type β-galactosidase enzyme is produced. The Tn*5lac* transposon is introduced into *M. xanthus* on bacteriophage P1::Tn*5lac*, wherein it transposes to many perhaps random sites on the chromosome. The resulting kanamycin-resistant colonies can then be screened for developmentally regulated β-galactosidase production. Many developmentally regulated genes have been identified by screening for the production of β-galactosidase upon nutrient downshift (Kroos et al., 1986). These developmentally regulated genes were grouped into classes based on time and level of expression of β-galactosidase. Furthermore, by introducing representative fusions into other developmental mutant backgrounds, the effect on the transcription of the *lac* fusions was assessed (Kroos and Kaiser, 1987; Kuspa et al., 1986). The results of these experiments suggest a program of differential gene expression during fruiting body development in which development-specific transcription is dependent on intercellular signaling.

It is a curious and yet to be adequately explained observation that many of the Tn*5lac* insertions resulting in developmentally regulated *lacZ* fusions do not themselves result in developmental defects (Kroos et al., 1990). One would expect a significant proportion of these fusion strains to have lost the function of the developmental gene. Among the possible explanations for the unexpectedly low frequency of developmental mutants is that many of the essential developmental functions either are redundant or are also essential for cell viability. Such genes would not be readily identified by transposon mutagenesis. Alternatively, the transposon may have a higher level of insertion site specificity than it seems to have.

A second reporter gene, useful in *M. xanthus* for the detection of cell envelope proteins, is that for alkaline phosphatase *(phoA)*. Alkaline phosphatase requires export to the periplasm or beyond to be enzymatically active (Michaelis et al., 1983). Tn*phoA* is a derivative of Tn*5* in which a portion of the *phoA* gene is inserted into the left end of the transposon; however, this portion lacks the sequences needed for transcription, translation, or export (Manoil and Beckwith, 1985). Insertion of Tn*phoA* in the proper orientation and correct reading frame yields alkaline phosphatase activity if the *phoA* gene is fused to sequences that allow export. In a screen of over 10,000 Tn*phoA* insertions in *M. xanthus*, 36 strains containing insertions were found to produce alkaline phosphatase, and 3 of these insertions disrupted *cgl* cell motility genes that are required for adventurous motility (Kalos and Zissler, 1990). Hodgkin and Kaiser (1977) had previously shown that the *cgl* class of motility mutant is transiently stimulated to move when in contact or close proximity to cgl^+ cells and have suggested that the *cgl* genes may encode signals passed between cells that are required for motility. On the basis of PhoA enzyme activity, the protein products of the *cgl-phoA* gene fusions were judged to be localized in the cell envelope and extracellular space. The protein product of each of the wild-type loci was apparently transferred between cells, since contact with cgl^+ cells restored adventurous motility to the mutants. These observations demonstrate that Tn*phoA* can detect genes whose products are secreted in *M. xanthus*.

Strategies for Identification of Mutations

By far the most common strategy for identification of mutations affecting myxobacterial behavior is simply screening for mutants in which the behavior of interest is altered. In this manner, mutations that alter motility, fruiting body formation, sporulation, carotenoid biosynthesis, and antibiotic production have been identified.

The intercellular interactions that serve to coordinate myxobacterial multicellular behaviors are intriguing. The challenge in studying the genetics of these interactions is to differentiate mutations that cause defective intercellular interactions governing a particular behavior from mutations that result in structural or mechanical defects in the same behavior. Both types may have the same primary phenotype, e.g., a nonmotile or developmental mutant. Mutations that alter signaling may act either at the initiation of such a signal or at the level of perception and response to the signal. Extracellular complementation has been used successfully to identify mutants that appear to be defective in the production of a signal by applying the premise that when mixed with wild-type cells, these mutants will still be able to respond to the signal produced by the wild-type cells. In this way, "stimulatable" mutants that have altered single cell or social motility have been identified (Hodgkin and Kaiser, 1977). Likewise, developmental mutants able to undergo extracellular complementation were identified (Hagen et al., 1978). The developmental signaling mutants were subsequently divided into four extracellular complementation groups on the basis of the behavior of pairs of mutants when they were mixed. Pairs of mutants that did not exhibit extracellular complementation were presumed to be defective in production of the same signal, whereas pairs of mutants that underwent extracellular complementation were presumed to be defective in different signals so that when they were mixed together, both signals were produced. It was determined for both the stimulatable motility mutants and the developmental signaling mutants that extracellular complementation results in transient responses but no genetic alteration in the responding cells.

Kroos and Kaiser (1987) and Kuspa et al. (1986) have shown that each of the four classes of developmental signaling mutants (*asg, bsg, csg,* and *dsg,* for A signal, B signal, etc.) has defects in transcription of distinctive sets of developmental genes. The isolation of extragenic bypass suppressor mutations has been used to begin to identify the regulatory steps intervening between requisite cell-cell signaling during development and induction of genes whose expression is dependent on those signals (Rhie and Shimkets, 1989; Kaplan et al., 1991). Two similar strategies have been used. Rhie and Shimkets (1989) isolated extragenic suppressors of *csgA* by direct selection for *csgA* mutant cells capable of forming heat- and sonication-resistant spores. Six distinct groups of extragenic bypass suppressors were identified, and five of these groups appeared to restore a normal program of gene expression as judged by their effects on the expression of nine developmentally regulated promoters. It seems likely that these mutations define additional regulatory steps that occur near the time of CsgA action. A similar strategy was used by Kaplan and coworkers (1991) to isolate extragenic suppressors of *asg*. The *asg* (A-signal) mutants are blocked early in development and fail to express several *lac* fusions to developmentally induced genes. Those workers isolated a series of mutants mapping to one of two loci that allow expression of an *asg*-dependent gene in an *asgB* mutant background. Therefore, these new mutations appear to bypass the requirement for the production of A signal for the expression of at least certain developmental genes and will likely identify components in the regulatory network that respond to the signal.

Schemes to isolate mutations targeted to specific genes or gene products have also been employed. The most common strategy, which will be discussed in more detail later in this chapter, is to introduce insertion mutations into specific *Myxococcus* genes cloned in *E. coli* and then recombine the mutant allele into the *M. xanthus* chromosome, thereby replacing the wild-type gene. This method is of particular value in determining the effects of mutations in genes cloned on the basis of nucleotide or amino acid sequence homology for which the function in *Myxococcus* is only inferred or is unknown.

A second strategy for isolating targeted mutations made use of monoclonal antibodies directed against cell surface epitopes. Dworkin and coworkers (Gill and Dworkin, 1986, 1988) have isolated a series of such monoclonal antibodies that react with surface antigens found on developing *M. xanthus* cells. The cell surface antigens implicated by this criteron included both lipopolysaccharide and cell surface proteins. Some of these antibodies were shown to block development when they were added to developing wild-type cells and were inferred to identify cell surface components required for development (Gill et al., 1987). To test the hypothesis that these structures play a role in development, Tn*5* or ethyl methanesulfonate-induced mutants that no longer expressed the antibody-reactive epitope on the cell surface were isolated (Fink et al., 1989). These mutants were then tested for defects in development and related behaviors such as motility and cohesion (Fink and Zissler, 1989). For certain mutations affecting lipopolysaccharide biosynthesis, motility and/or development was indeed impaired or abolished.

Epistasis

Epistasis is the masking of the action of one gene by the action of another gene. It is often used to test whether two genes function in the same pathway. If two genes are part of the same pathway, then a strain containing two mutations with distinguishable phenotypes will have the phenotype of the mutation occurring earlier in the pathway (Hereford and Hartwell, 1974). This principle has been applied to analysis of the developmental pathway of *M. xanthus* by using strains containing both a mutation that disrupts development and a reporter for a developmental gene. For example, when an *asg* mutation was transferred to 21 strains containing developmentally regulated transcriptional *lacZ* fusions, β-galactosidase activity was reduced or eliminated in 18 strains (Kuspa et al., 1986). These results suggest that the block imposed by the *asg* mutation precedes the expression of roughly 85% of the developmentally regulated genes and is consistent with results obtained with biochemical markers that suggest that *asg* mutants are blocked early in development (LaRossa et al., 1983). While the results of this approach have not yet been extended to very many developmental mutations, a pathway is apparent with the few that have been tested. The *asg* mutation appears to block developmental gene expression at about 2 h after initiation and *csgA* at about 6 h (Kroos and Kaiser, 1987).

GENETIC MANIPULATION AND SHORT-RANGE GENETIC MAPPING OF MUTATIONS

Generalized transduction between strains of *M. xanthus* is readily accomplished with bacteriophages Mx4 (Campos et al., 1978; Geisselsoder et al., 1978) and Mx8 (Martin et al., 1978) and provides tremendous flexibility in genetic analysis of mutations. Generalized transducing virus particles are thought to be formed during lytic infection, when segments of bacterial DNA are, on rare occasions, packaged in place of viral DNA. When such particles infect a recipient host, the encapsidated bacterial DNA is injected into the cell and is free to undergo homologous recombination with the host cell chromosome. Individual genes may be transduced at various frequencies, but all are within the range of 10^{-5} to 10^{-7} per particle. There are no reports of loci that cannot be transduced, which suggests that virtually any portion of the bacterial chromosome may be packaged into transducing particles.

Bacteriophages Mx4 and Mx8 normally package 56 and 62 kbp of DNA, respectively (Kaiser, 1984). This length is also believed to represent the length of chromosomal DNA (approximately 0.5% of the *M. xanthus* genome [Chen et al., 1990]) packaged in the transducing virus particle. If two genetic markers lie close enough together on the bacterial chromosome, they may be packaged together in a single segment of transducing DNA and subsequently coinherited by the recipient. The cotransduction frequency of two markers is inversely proportional to the distance between them (Wu, 1966). The cotransduction frequencies for two different pairs of Tn*5* insertions were used to estimate the distance between them in *M. xanthus* (Sodergren et al., 1983). The physical distance predicted by the Wu equation based on cotransduction frequency (see below) agrees with the physical distance determined by mapping with restriction enzymes, which suggests that physical distance can be estimated from the cotransduction frequency by the Wu equation as follows:

$$\text{cotransduction frequency} = \left(1 - \frac{\text{distance between two markers}}{\text{length of DNA in transducing particle}}\right)^3$$

Tn*5* as a Linked Marker in Transduction

Genetic analysis by cotransduction is limited by the requirement that at least one of the genetic markers be selectable. The frequency of generalized transduction (usually less than 10^{-5} transductants per recipient cell) makes such an analysis impractical in the absence of selection. In *M. xanthus*, the number of selectable markers is limited. To address this problem, Kuner and Kaiser (1981) developed a method to identify Tn*5* insertions that are genetically linked by cotransduction to any mutation for which there is a distinguishable phenotype. First, a library of approximately 1,000 independent Tn*5*-containing (Km^r) transductants is isolated following infection of wild-type *M. xanthus* cells with P1 :: Tn*5*. To identify the particular Tn*5* from this library that is linked to a specific mutation of interest, the library is divided into pools of 100 strains, and transducing phage are grown on each pool. Each phage stock is used to transduce the Tn*5* insertions from the pool into the mutant strain by selection for kanamycin resistance. These Tn*5*-containing transductants are then tested for coinheritance of the wild-type allele of the original mutation, which occurs if the two loci are sufficiently close to each other to be located on the same fragment of DNA in the transducing particle. The linkage of the Tn*5* to the mutation of interest is finally verified by growing a new transducing phage stock on the cotransduced strain and repeating the transduction into the original mutant.

Tn*5* appears to transpose to many perhaps random sites in the chromosome. Thus, assuming a genome size of 9.5×10^3 kbp (Chen et al., 1990), this collection of 1,000 Tn*5*-containing strains would be expected to contain an average of one insertion every 10 kbp. Since this frequency is well within the limits of what can be transduced by either Mx4 or Mx8 (according to the Wu formula, two markers separated by 10 kbp would be cotransduced at a frequency of approximately 50% by these phages and so should be observed at a frequency of about 0.5% when transduction is from the pool of 100 independent Tn*5*-containing strains), it is likely that this library of strains is large enough to contain at least one Tn*5* insertion that is sufficiently close to some (any) chromosomal locus for the two to be cotransduced. Indeed, virtually every attempt to find a Tn*5* linked to a new genetic locus has been successful. In the very rare cases in which no Tn*5* could be identified (Gill, unpublished data), it is entirely possible that the initial Tn*5* library was not large enough or that too few transductants were screened.

Once a genetically linked Tn*5* has been identified, it can then be treated as any other stable linked genetic marker for subsequent analysis of the original mutation. For example, when several mutants with similar phenotypes are isolated, the linkage relationships between the respective mutations and particular Tn*5* insertions can be determined. In this way one can determine whether the mutations map to a single genetic locus, and are all linked to the same Tn*5* insertion, or to multiple, genetically unlinked loci. Zusman (1982) used two- and three-factor crosses in which a Tn*5* was one of the markers to establish that the "frizzy" aggregation-defective mutations *(frz)* lie in a single cluster covering approximately 7.5 kbp and to determine the relative gene order within the *frz* locus. In similar fashion, *asg* mutations that abolish production of a required developmental cell interaction known as A signal have been mapped by using linked Tn*5* insertions to three genetically unlinked loci. In addition, the linked Tn*5* serves as a convenient reference point from which to clone the gene(s) of interest and to determine its position on the physical map of the *M. xanthus* chromosome.

A Tn*5* linked to a particular genetic locus is useful for various aspects of strain construction. For example, it is used to transduce mutations from the strains in which they are originally isolated into an otherwise wild-type genetic background. Because of the relatively small size of the DNA that is transduced, this backcross helps to ensure that all aspects of the original mutant phenotype are due to a single mutation rather than to secondary mutations acquired during isolation of the original strain. Similarly, cotransduction of Tn*5* and the mutant allele is used to construct double-mutant strains in order to make epistasis determinations.

In principle, this strategy is generally applicable to other organisms for which extensive genetic maps are not yet available. It is of particular value since it permits routine genetic manipulation of interesting genes for which no selection exists. The general application of this scheme to other organisms requires that relatively few conditions be met. First, the transposon must insert in a random or nearly random distribution in the bacterial chromosome. Second, the element must be relatively stable at its new chromosomal site. A high frequency of transposition within the bacterium in which it was isolated or upon transduction to a new recipient strain would certainly complicate the analysis. Finally, the transposon must not result in a high frequency of mutations affecting the phenotypic property under study.

Tn*5* Derivatives Encoding Resistance to Other Antibiotics

The versatility of Tn*5* as a linked genetic marker is greatly enhanced by the availability of Tn*5* derivatives encoding resistance to antibiotics other than kanamycin (Fig. 1). The Tn*5* derivatives Tn*5*-132 (Rothstein et al., 1981; Avery and Kaiser, 1983a) and Tn*5*-Tp (Sasakawa and Yoshikawa, 1987; Shimkets, unpublished data) have each been used successfully in *M. xanthus*. These two transposons retain most of the right and left IS*50* elements of Tn*5*. However the unique central region of Tn*5*, which normally encodes resistance to kanamycin, has been removed and replaced with genes encoding resistance to tetracycline and trimethoprim, respectively. These elements are most frequently used to replace the kanamycin resistance of an existing Tn*5* insertion (the technique for this procedure was described by Avery and Kaiser [1983a]). This replacement is accomplished by infecting a Tn*5*-containing strain with P1::Tn*5*-132 or P1::Tn*5*-Tp and selecting for resistance to tetracycline or trimethoprim, respectively. Because the modified Tn*5* elements retain the basic structure of Tn*5* and in particular most of the flanking IS*50* elements, one of the possible fates of the incoming modified Tn*5* is homologous recombination with the endogenous Tn*5*. Recombination will be restricted to the IS*50* sequences, since these are the only regions of shared homology. In those cells in which a double recombination occurs, i.e., one in each of the two flanking IS*50* elements, the kanamycin resistance of the endogenous element will be replaced by resistance to either tetracycline or trimethoprim. This replacement results in a strain having a single transposon insertion at precisely the same chromosomal location as the original Tn*5* yet being distinguishable on the basis of its antibiotic resistance. Although transposition of the incoming modified Tn*5* is known to occur, it is relatively infrequent. Furthermore, cells in which transposition has occurred can be distinguished from those in which homologous recombination has led to a replacement event on the basis of their antibiotic resistance: whereas transposition results in cells that retain kanamycin resistance, replacement results in cells that have become kanamycin sensitive.

The construction of isogenic strains having a Tn*5* located at identical sites in the chromosome yet differing by the antibiotic resistance encoded by the Tn*5* is useful in a number of situations. For example, since two Tn*5* insertions can be differentially marked and therefore distinguished from one another, cotransduction analysis can be used to determine linkage relationships (and therefore an estimate of physical distance) between pairs of independent Tn*5* insertions. In addition, this capability enables one to utilize more than one Tn*5* in the construction of multiply marked strains.

Long-Range Mapping

Long-range mapping deals with physical distances of >20 kbp and, in the context referred to here, determines relative positions of genes on the chromosome. In many bacteria, long-range mapping is performed by conjugation; however, the development of a conjugation system for mapping *M. xanthus* genes is still in its infancy. Using an alternate strategy developed by Kohara and coworkers for *E. coli* mapping (1987), Kuspa et al. (1989) developed a long-range gene-mapping technique based on DNA-DNA hybridization by using a series of partially overlapping *M. xanthus* clones contained in yeast artificial chromosomes. The clone library was aligned with the *M. xanthus* physical map (Chen et al., 1991), and hybridization of a cloned gene to the library can determine the location of the gene of interest. A physical and genetic map of the *M. xanthus* chromosome is shown in chapter 4 for DK1622 (Fig. 2 and Table 2 in that chapter) and FB (Fig. 3 and Table 3 in that chapter).

CLONING AND ANALYSIS OF *M. XANTHUS* GENES

Several conventional strategies have been successfully adapted for cloning *M. xanthus* genes in *E. coli*. Among these strategies are hybridization using degenerate oligonucleotide probes based on limited peptide sequence of the *M. xanthus* protein product, hybridization to cloned genes from other organisms, and identification of the cloned *M. xanthus* gene in *E. coli* either by complementation of auxotrophic mutants or by the antigenic specificity of the cloned gene product. These strategies are not unique to *M. xanthus* and are described in detail in most manuals on molecular cloning.

More of a challenge is to clone an *M. xanthus* gene that has been identified simply by mutation or for which one has little if any information about the gene product. Because many aspects of myxobacterial behavior have no counterpart in *E. coli,* genes governing these behaviors cannot be identified by complementation of *E. coli* mutants. In these cases, a linked Tn*5* is often used as a reference point from which to clone (e.g., Shimkets et al., 1983; O'Connor and Zusman, 1983). If the chromosomal DNA of the Tn*5*-containing strain is digested with an appropriate restriction enzyme, then the fragment containing the kanamycin resistance gene and its flanking *M. xanthus* sequences can be cloned in a suitable vector by selecting for kanamycin resistance in *E. coli*. The cloned fragment can then be transferred back into *M. xanthus* (see below) to determine, usually by complementation of a known mutant, whether the gene of specific interest is present on the cloned fragment. Furuichi et al. (1985) have devised a clever variation of this scheme by constructing Tn*V*, a modified Tn*5* containing an origin of replication derived from the plasmid pSC101 and encoding tetracycline resistance in both *M. xanthus* and *E. coli*. When Tn*V* is used, chromosomal DNA is merely cut with an appropriate enzyme, and the fragments are circularized in a very efficient, unimolecular ligation reaction. The ligated DNA is then transformed into *E. coli* with selection for tetracycline. The particularly inefficient, bimolecular ligation step of joining the cloning vector to the fragment to be cloned has been eliminated. Using a related strategy, Gill and coworkers (1988) described a cloning method in which the vector is first integrated into the IS*50* of a Tn*5* insertion located near the gene of interest. Integration of the vector in vivo serves to physically join the vector to the sequences to be cloned. DNA is then prepared from this strain and digested with appropriate enzymes that generate a single fragment containing all essential portions of the integrated vector, IS*50*, and the adjacent *M. xanthus* sequences. As in the case of Tn*V*, this fragment is merely circularized and recovered by transformation of *E. coli* with selection for kanamycin resistance encoded by the vector.

Transfer of Recombinant Plasmids to *M. xanthus*

Currently, no transformation procedure is generally applicable for *M. xanthus*. Although electroporation has been described, the reported efficiency of one transformant per μg of supercoiled DNA is less than that necessary for cloning directly into *M. xanthus*. Nevertheless, quite reliable means of transferring DNA from *E. coli* to *M. xanthus* that are dependent on bacteriophage P1 have been developed. In *E. coli*, this virus may either grow lytically or enter a lysogenic state, in which the viral DNA is maintained as an extrachromosomal plasmid. As discussed above, the virus also mediates the entry of viral DNA into *Myxococcus* cells in what amounts to an abortive infection. The P1 DNA is not maintained by the *Myxococcus* cells, and no progeny virus are produced (Kuner and Kaiser, 1981).

O'Connor and Zusman (1983) have shown that P1 is able to mediate the transfer of plasmids from *E. coli* to *M. xanthus*, presumably by generalized transduction. Since at least 40 kbp of linear DNA is required for encapsidation in P1 phage particles (Yarmolinsky, 1977), it is likely that multimeric forms of the plasmids are the substrates for packaging. It was found that although transductants could be isolated with a number of different plasmids, the efficiency of transfer was generally higher for larger plasmids yet could be variable even for plasmids of the same size.

A somewhat more efficient and reliable scheme for transferring plasmids from *E. coli* to *M. xanthus* utilizes a form of specialized P1 transduction. Advantage is taken of the fact that bacteriophage P1, which lysogenizes *E. coli* as an extrachromosomal self-replicating plasmid, exhibits the phenomenon of incompatibility (Sternberg et al., 1981). That is, it cannot be stably maintained in a cell containing a plasmid of the same incompatibility type. Therefore, cloning vectors modified by the addition of the *inc* region of P1, which is carried on a 6.7-kbp *Eco*RI fragment, can be stably maintained in a lysogenic cell only by the formation of a cointegrate molecule by recombination between the shared P1 sequences. Upon induction of the cointegrate P1 lysogen to lytic growth, packaging of linear DNA into virions initiates from a unique *pac* site in the P1 portion of the cointegrate and proceeds by a sequence-independent mechanism which sequentially fills the virions with DNA. Analysis of the DNAs contained in the mature virions produced from such a strain showed that a very large proportion contain the entire vector plasmid flanked by P1 viral DNA (Gill et al., 1988). Upon subsequent infection of *M. xanthus* by such a virus stock, these packaged linear DNA molecules likely undergo homologous recombination between the P1 *inc* sequences to regenerate the original circular plasmid and subsequent loss of all other P1-derived sequences.

This scheme is generally reliable. However, it requires the use of specially modified cloning vectors, which, owing to their larger size and fewer unique restriction sites, are somewhat more cumbersome to use. Although most *inc* vectors contain a 6.7-kbp P1 fragment, smaller fragments will suffice (Gill, unpublished data). As yet, there has been no systematic study to determine the smallest fragment that would represent a balance between providing the necessary incompatibility functions and still providing sufficient DNA for the recombination between plasmid and P1 leading to the formation of the cointegrate molecule.

Integration of Plasmid Sequences

Most cloning studies of *M. xanthus* DNA have used pBR322-based plasmids. However, these plasmids are apparently unable to replicate in *M. xanthus*. In fact, there are no published reports of naturally occurring plasmids that replicate in *M. xanthus,* including those that typically have a broad host range. In the absence of a functional plasmid origin for use in *M. xanthus,* plasmids transduced from *E. coli* are readily lost from the cell unless a means of incorporating all or a portion of the plasmid sequences into the host chromosome is provided. Three strategies have been used to integrate plasmid DNA into the *M. xanthus* chromosome (Fig. 2): (i) homologous recombination between cloned *Myxococcus* DNA on the incoming plasmid and its homolog on the bacterial chromosome, (ii) site-specific integration at the bacteriophage Mx8 attachment site, and (iii) Tn*5*-mediated transposition of plasmid sequences to numerous sites in the bacterial chromosome.

Homologous recombination

Homologous recombination between cloned *M. xanthus* DNA and its chromosomal homolog can have two primary outcomes. A single recombination event taking place between a circular form of the incoming plasmid and the chromosome will lead to integration of the entire plasmid into the bacterial chromosome at the site of shared homology (homologous integration; Fig. 2A). Alternatively, double recombination replaces the chromosomal sequences with those on the plasmid, and all plasmid vector sequences are lost (gene replacement; Fig. 2D).

Homologous integration of plasmid DNA is predicted to result in a merodiploid strain with a tandem duplication of the sequences corresponding to the cloned DNA separated by vector sequences at the novel joint. The structures of duplications formed in this manner, first documented by Shimkets et al. (1983) and O'Connor and Zusman (1983), have since been verified by genetic and restriction enzyme analysis with numerous different cloned *M. xanthus* DNA fragments. Since recombination frequency is generally proportional to the length of the interval through which recombination may occur, it should not be surprising that very small fragments are particularly inefficient at mediating integration. Downard (1988) reported that he was unable to detect plasmid integration when he used fragments of less than about 400 bp.

Homologous integration has proven to be an invaluable tool for assigning genes to a particular cloned fragment by complementing known *M. xanthus* mutations. If the entire locus is present on the cloned DNA, then one expects the resulting merodiploid to contain two intact copies of the locus, one derived from the donor and one derived from the recipient. Provided the mutation in the recipient chromosome is recessive to the cloned wild-type locus, genetic complementation will result in phenotypically wild-type merodiploids. This strategy is used routinely for the initial characterization of cloned DNA fragments, especially for fragments that are identified solely on the basis of their proximity to a Tn*5* insertion linked to the gene of interest, and in subsequent steps for localization of the particular gene within the cloned fragment by analysis of subcloned fragments, nested sets of deletions, or mutations introduced into the cloned DNA. As a representative example of this approach, Shimkets et

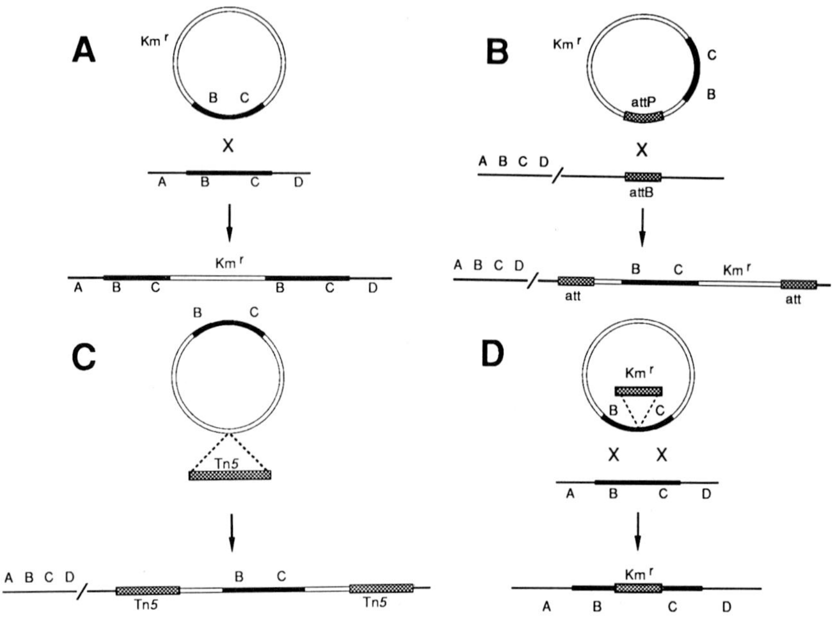

Figure 2. Pathways for integration of DNA into *M. xanthus* chromosome. (A) Integration of a circular plasmid by homologous recombination. The plasmid encodes Km^r and contains a segment of cloned *Myxococcus* DNA (black box) and cloned sequences designated B and C. A single recombination between the cloned DNA and its chromosomal counterpart forms a merodiploid strain having two directly repeated copies of the cloned DNA separated by vector sequences (stippled box) at the novel joint. (B) Integration of a circular plasmid by site-specific recombination. The plasmid contains the phage attachment site *(attP)* of the temperate virus Mx8. A single site-specific recombination between *attP* and the bacterial attachment site *(attB)* integrates the entire plasmid at the *attB* locus. (C) Transposition-mediated integration of circular plasmids. The nonreplicating plasmid contains the cloned BC locus of *M. xanthus* and Tn*5* (cross-hatched box), which has transposed into the vector portion (stippled box) of the plasmid. The plasmid integrates randomly into the chromosome, and the integrated DNA is flanked by Tn*5* or single IS*50* elements. (D) Gene replacement by homologous recombination. The nonreplicating plasmid contains the cloned BC locus, into which a mutation has been introduced. In this example, the mutation results from insertion of a kanamycin resistance gene into the cloned BC locus. Upon introduction into *M. xanthus*, the plasmid may integrate by a single recombination (A). Alternatively, a double recombination may occur (D), once on each side of the mutation, which will replace the intervening chromosomal sequences by the mutant sequences on the plasmid with loss of all other plasmid sequences from the cell.

al. (1983) cloned in *E. coli* a 21-kbp *Bam*HI restriction fragment containing sequences adjacent to a Tn*5* insertion previously shown to be genetically linked to the *csg* locus (formerly *spoC*). This fragment extended from within the Tn*5* element outward into the adjacent *M. xanthus* DNA and so contained the Tn*5* gene for kanamycin resistance at one end. To determine whether this plasmid actually contained the *csg* locus, the plasmid was transferred to a *csg* mutant of *M. xanthus* by P1 specialized transduction and selection for kanamycin resistance. These transductants contained an integrated form of the plasmid with the predicted sequence organization, and approximately 80% (67 to 90%) of them were phenotypically Csg^+. This confirmed that the csg^+ allele was in fact located on the cloned fragment and suggested that *csg* mutations are recessive to wild type. As would be predicted from their sequence organizations, these strains were unstable, losing the integrated plasmid sequences at a frequency of about 0.2% per generation. For each of the five Csg^+ merodiploids examined, both Csg^+ and Csg^- segregants were found, indicating that the duplication strains were genotypically csg^+/csg merodiploids. Although most of the transductants were Csg^+, a minor fraction of the strains (10 to 33%) retained the mutant phenotype. These merodiploid strains contained integrated plasmids with the predicted sequence organizations. However, only *csg* mutant haploid segregants could be recovered, suggesting that these were *csg*/*csg* homozygous merodiploids that arose by gene conversion. Gene conversion of this sort has been observed in other bacteria and is a likely consequence of the recombination event that integrated the plasmid into the *M. xanthus* chromosome rather than a property of the particular cloned sequence. Gene conversion may involve either the donor or recipient allele, although an apparent preference for retaining the chromosomal allele has been observed. Observing apparent complementation of a chromosomal mutation by integration of a cloned DNA fragment does not necessarily mean that the entire locus is present. This point is illustrated in Fig. 3, which shows how integration of a cloned truncated gene may give rise to a phenotypically wild-type cell by production of a wild-type genetic recombinant rather than by *trans* complementation.

A related strategy helps delineate the bounds of cloned genes and transcription units (Fig. 3). The different genetic consequences of integrating plasmids containing only portions of cloned genes or operons into a wild-type recipient by homologous recombination can be predicted. If the cloned DNA contains at least one end of each transcription unit on the cloned DNA, then integration into a wild-type recipient chromosome will result in no disruption of gene expression. In contrast, if the integrated plasmid contains only an internal fragment of a gene, the cell will have only two truncated copies of the gene and gene function will be lost.

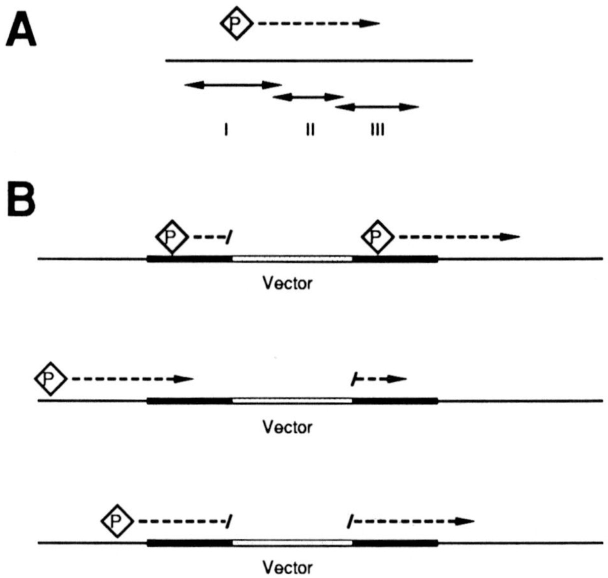

Figure 3. (A) Hypothetical gene or operon with its promoter (P) and transcript (dashed arrow). Three fragments are indicated by double-headed arrows. Fragment I contains the promoter region and the 5′ end of the transcriptional unit; fragment II is an internal fragment that does not contain either the 5′ or the 3′ end; fragment III minimally contains the 3′ end of the transcriptional unit. (B) Consequences of integrating plasmids containing fragment I or III (top and middle panels, respectively). In each case, a merodiploid is formed and the transcription unit is reconstituted. Integration of a plasmid containing only internal fragment II results in a strain in which the transcription unit is disrupted and is likely not to be normally expressed.

Likewise, if a plasmid contains a cloned fragment that is entirely from within a multigene operon but no intact promoter region or 3′ end of the operon, then integration of the plasmid will separate the downstream genes from their promoter and will likely result in a loss of expression of the downstream gene(s) of the operon.

Several studies have used merodiploids formed by homologous integration of recombinant plasmids to perform interallelic complementation tests between pairs of mutants. For this analysis, each of the mutant alleles to be tested is separately cloned and transferred to an *M. xanthus* cell having a different mutant allele. In the simplest interpretation, if no complementation occurs and the resulting strain retains its mutant phenotype, then the two mutations may lie in the same gene; conversely, if complementation occurs and the resulting strain has a wild-type phenotype, then the two mutations may be in different genes. In this way, it should be possible to determine the number of individual genes located within a particular locus. However, the interpretation of interallelic crosses has not been without complications. There are at least two ways in which an interallelic cross can give rise to a merodiploid having a wild-type phenotype in the absence of genetic complementation; both appear to be consequences of the recombination event itself. The first occurs when the recombination event that integrates the plasmid occurs within the sequences located between the two mutant alleles. Though rare, such an event gives a duplication strain in which both mutations lie on one arm of the duplication, while the other arm contains only wild-type sequences. This strain would have a wild-type phenotype not due to complementation. The second complication is a high frequency of apparent gene conversion in certain interallelic crosses.

The actual frequency of gene conversion in interallelic crosses varies from locus to locus and perhaps with the size of the particular DNA fragment used (Gill, unpublished data). Shimkets et al. (1983) reported crossing three different cloned *csg* mutations into four *csg* mutant recipients and finding that all of the nine different heteroallelic double heterozygotes had the mutant phenotype. From this, they concluded that all of the *csg* mutations belonged to the same complementation group. No phenotypically Csg^{+} transductants that would have resulted from gene conversion were recovered. Either gene conversion did not occur during these crosses, or it was too infrequent to be detected.

In dramatic contrast to the situation with the *csgA* locus, Stephens and Kaiser (1987) found that in interallelic crosses involving the cloned *mgl* locus, apparent gene conversion was a common occurrence. Interallelic crosses between several pairs of mutations which are now known by sequence analysis to lie in the same gene gave rise to merodiploids of which more than half had undergone apparent gene conversion and were Mgl^{+}. To discriminate between true complementation and artifacts arising from recombination is tedious at best and requires either segregation analysis, cloning, and determination of the genotype of each arm of the duplication or restriction enzyme analysis in those cases in which the mutations are physically distinguishable. These results clearly indicate that considering just the frequency of obtaining phenotypically wild-type merodiploids formed by homologous integration of plasmids may be misleading and that additional analysis is necessary to determine genetic complementation in intergenic crosses.

Homologous integration of in vitro-constructed *lacZ* fusions to cloned *Myxococcus* genes is widely used to study transcriptional regulation. Either operon fusions or protein fusions used in these studies were made

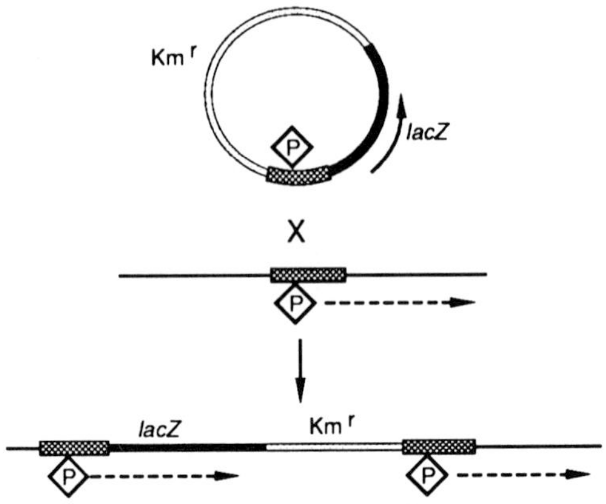

Figure 4. Homologous integration of a plasmid containing a *lacZ* fusion. The 5′ portion of a cloned *M. xanthus* gene (cross-hatched box) including the promoter region (P) is cloned upstream of a promoterless *lacZ* reporter gene (black box). Integration of the plasmid into the chromosome results in duplicates of the promoter-containing fragment separated by vector sequences (stippled box) that encode Km^r. The 5′ (left-end) copy of the promoter fragment is joined to sequences normally upstream of the promoter and transcribes *lacZ* sequences; the 3′ (right-end) copy of the promoter fragment transcribes an intact copy of the *M. xanthus* gene. Transcripts are indicated by the dashed arrows.

by cloning the 5′ portions of *M. xanthus* genes into vectors containing a *lacZ* reporter gene (for examples, see Downard and Zusman [1985], Gill and Bornemann [1988], and Downard et al. [1984]). Homologous integration of such plasmids, shown schematically in Fig. 4, forms a strain with a duplication of the cloned fragment. Integration will rejoin the promoter-containing fragment of the *lac* fusion to its normal upstream DNA sequences in the chromosome. Therefore, proper regulation of the fusion is expected. Furthermore, if the cloned fragment contains a fully functional promoter, then the downstream gene will also be properly expressed and the strain will be phenotypically wild type.

Gene replacements

In contrast to homologous integration of circular plasmids resulting from a single homologous recombination event, a double-recombination event is predicted to result in the replacement of the chromosomal sequences with those on the plasmid and loss of all plasmid vector sequences (gene replacement; Fig. 2D). This technique is routinely used for transferring mutations isolated in cloned DNA to the host chromosome for an assessment of phenotype. Gene replacements are facilitated if the mutant allele to be transferred to the chromosome is associated with a selectable marker such as an antibiotic resistance gene.

The gene replacement strategy is particularly valuable when *Myxococcus* genes are cloned on the basis of homology to other known genes. In these cases, isolation of corresponding mutations on the bacterial chromosome may be the most straightforward way to demonstrate a role for the particular cloned gene in developmental or cellular processes. A recent example of this strategy is the characterization of a gene *(pkn-1)* encoding a protein serine/threonine kinase (Pkn1) (Muñoz-Dorado et al., 1991). The protein kinase gene was initially detected and cloned by hybridization to a polymerase chain reaction product from *M. xanthus* by using primers representing highly conserved regions of eukaryotic protein serine/threonine kinases. Expression of *pkn-1* was developmentally regulated, and the Pkn1 was shown to be autophosphorylated when expressed in *E. coli*. To determine the function of the Pkn1 during development, a mutation in the cloned *pkn-1* gene in which the catalytic domain was removed and replaced by a gene encoding kanamycin resistance was constructed. This mutation was transferred to the chromosome of *M. xanthus* by gene replacement and was shown to result in premature differentiation and low spore yield. The behavior of this mutant demonstrated that Pkn1 is indeed required for normal development.

The isolation of gene replacements is occasionally hampered if the mutation to be transferred to the chromosome lies near one end of the cloned fragment or if the fragment itself is small (O'Connor and Zusman, 1986). Then, the length of the interval through which one of the recombination events must occur is quite short (less than 1 kbp), so that the corresponding recombination frequency is low. When the recombination frequency is low, the most frequent event is a single homologous crossover that integrates the entire plasmid. O'Connor and Zusman (1986) have reported that gene replacements can nevertheless be efficiently isolated in two steps. First, the merodiploid is isolated following a single crossover. Then, the merodiploid strain is used as a donor in generalized transduction by Mx4. A significant proportion of the transductants is made up of the desired gene replacements.

Plasmid integration by site-specific recombination

An alternative to integration of recombinant plasmids by homologous recombination makes use of the site-specific recombination system of the temperate myxophage Mx8. In a manner analogous to that of bacteriophage lambda, Mx8 integrates into the *M. xanthus* chromosome by site-specific recombination between a site on the phage DNA, *attP*, and a corresponding site on the bacterial chromosome, *attB* (Stellwag et al., 1985).

Plasmids containing the cloned Mx8 *attP* site and adjacent DNA have been shown to integrate into the chromosomal *attB* site (Stellwag et al., 1985) (Fig. 2B). Whether the site-specific recombination is mediated by recombination enzymes encoded on the *attP* fragment or by cryptic bacteriophage in the *M. xanthus* strains has not been determined. Nevertheless,

Mx8 *att*-mediated site-specific integration of plasmids is particularly efficient; Shimkets and Asher (1988) showed that a plasmid containing both an *attP* site and cloned *M. xanthus* DNA preferentially integrates through site-specific rather than homologous recombination. Unpublished results from our laboratory and others suggest that the high frequency of site-specific integration may be sufficient to allow efficient use of electroporation as a means of introducing recombinant DNA into these cells. Once formed, the integrated plasmids are stably maintained, segregating at only 0.0007% per generation, or only about 1/100 the frequency of merodiploids formed by homologous recombination (Shimkets and Asher, 1988).

Because of the properties of the process, *attP*-mediated integration of plasmids is particularly useful for analysis of cloned DNA and avoids many of the shortcomings of homologous integration. In sharp contrast to homologous recombination, gene conversion and marker rescue occur only rarely, if at all, upon integration via site-specific recombination (Shimkets and Asher, 1988; Stephens and Kaiser, 1987). This infrequency of occurrence certainly simplifies the interpretation of complementation studies with pairs of mutations. It will also be of particular value in current studies in which genes and regulatory sequences altered by site-directed mutagenesis must be introduced unaltered into *M. xanthus* for functional analysis. An additional attribute of site-specific integration is that, unlike homologous integration, it is not dependent on the size of the cloned *Myxococcus* DNA. This allows efficient genetic analysis of small DNA fragments such as those containing promoter sequences and their upstream regulatory regions.

Li and Shimkets (1988) have reported construction of a promoter cloning vector containing a promoterless *lacZ* reporter gene and the Mx8 *attP* site. To test whether such a vector would be suitable for studying regulation of cloned promoters, a developmental promoter region was cloned upstream of the reporter gene and the plasmid was integrated via Mx8 *attP*-mediated recombination (Li and Shimkets, 1988). They demonstrated that the promoter was properly regulated during development, although the maximum level of expression was only about half that of the same promoter at its normal chromosomal site.

One difficulty with using gene fusions, whether introduced by site-specific recombination or by other means, is assessing the accuracy with which the accumulation of β-galactosidase reflects the accumulation of the native gene product. Slight differences in mRNA stability, translation efficiency, or protein stability could result in a great disparity between the level of the native gene product and the level of β-galactosidase. A remarkable correlation was observed with the *csgA* gene, which suggests that β-galactosidase-specific activity is roughly proportional to the level of the CsgA gene product. A series of nested deletions were constructed across the upstream regulatory region of a *csgA-lacZ* operon fusion and integrated by site-specific recombination. As the length of the deletion increased, the maximum amount of β-galactosidase produced during development decreased (Li et al., 1992). The same set of nested deletions was placed next to a functional *csgA* gene, and the cells were tested for developmental proficiency. As the length of the deletion increased, the cells ceased development at successively earlier stages. When the levels of β-galactosidase in the fusion strains were compared with the terminal developmental phenotypes observed with cells containing the truncated *csgA* genes, it appeared that rippling, aggregation, and sporulation were induced at successively higher CsgA concentrations. These results support the suggestion that CsgA acts as a developmental timer whose concentration rises gradually during development and entrains the morphological stages of development in the proper temporal order at successively higher concentrations (Kim and Kaiser, 1991; Li et al., 1992).

The findings that promoter regions cloned in this manner retain their regulated expression and that plasmids integrated by site-specific recombination are stably maintained in single copy and are unaltered by gene conversion or marker rescue make this an valuable cloning method for studying promoter structure and function. However, if proper promoter function requires that upstream DNA be folded in a particular way, homologous integration may be a better method.

Tn*5*-mediated plasmid integration

An alternative to site-specific or homologous integration of transduced plasmids has been discovered for plasmids containing a Tn*5* transposon in the vector portion of the molecule (Downard, 1988) (Fig. 2C). This mechanism is observed when the plasmids transduced by P1 contain only a small *M. xanthus* DNA insert, so that the frequency of integration via homologous recombination would be low. Under these circumstances, Downard found that the integrated plasmids were arranged in a quite unique organization: (i) each integrant contained one or more intact copy of the plasmid; (ii) integration took place at several, perhaps random sites in the bacterial chromosome, and these sites had no apparent homology to the cloned DNA; and (iii) the integrated sequences were flanked by either a Tn*5* or an IS*50* element located at the junctions between plasmid and chromosomal sequences.

Prospects for an Autonomously Replicating Plasmid

O'Connor et al. (submitted) have recently reported construction of a plasmid, pMx4T, that is capable of autonomous replication in the myxobacterial species *M. xanthus* and *Stigmatella aurantiaca* and in *E. coli*. The plasmid utilizes an origin of replication derived from bacteriophage Mx4 and is stably maintained in *M. xanthus* DZF1 at approximately 100 copies per cell and in *E. coli* at 10 to 40 copies per cell depending on the strain. No evidence for integration of the plasmid into the host chromosome was found. The high copy number of pMx4T should provide a means of overexpressing the products of cloned genes in *M. xanthus* for the purpose of purification or for assessment of the effects of overexpression of regulatory proteins or multiple copies of regulatory sites.

Inducible Expression of Cloned Genes

The ability to manipulate the concentration of a gene product in vivo is a potentially powerful research tool that has a number of applications, including visualization of gene products and analysis of regulatory pathways. The gene whose product is to be regulated is generally placed under the control of a promoter whose expression is sensitive to well-defined environmental stimuli and that is carried on a plasmid referred to as an expression vector. One of the most frequently used expression systems in *E. coli* involves the use of the hybrid *tac* promoter, whose expression is inhibited by the *lacI* gene product and is induced with isopropyl-β-D-thiogalactopyranoside (IPTG). When this system was evaluated for use as an expression system in *M. xanthus,* IPTG induced expression only about eightfold (Letouvet-Pawlak et al., 1990). The light-inducible *carQRS* promoter induced expression about 47-fold in the presence of light, offering great promise as an expression system. However, the light intensities necessary to induce expression of the promoter also inhibit development and limit the usefulness of this expression system for the analysis of developmental genes (Li et al., 1992). An expression vector that can be used during fruiting body development is still lacking.

Acknowledgments. We thank Elizabeth Hager for helpful suggestions during the preparation of this manuscript.

This work is supported by National Institute of General Medical Sciences grant GM31900 to R.E.G. and National Science Foundation grant DCB-9001755 to L.J.S.

REFERENCES

Avery, L., and D. Kaiser. 1983. In situ transposon replacement and isolation of a spontaneous tandem duplication. *Mol. Gen. Genet.* **191:**99–109.

Campos, J. M., J. Geisselsoder, and D. R. Zusman. 1978. Isolation of bacteriophage Mx4, a generalized transducing phage for *Myxococcus xanthus*. *J. Mol. Biol.* **119:**167–178.

Chen, H.-W., I. M. Keseler, and L. J. Shimkets. 1990. The genome size of *Myxococcus xanthus* determined by pulsed-field gel electrophoresis. *J. Bacteriol.* **172:**4206–4213.

Chen, H.-W., A. Kuspa, I. M. Keseler, and L. J. Shimkets. 1991. Physical map of the *Myxococcus xanthus* chromosome. *J. Bacteriol.* **173:**2109–2115.

Downard, J. S. 1988. Tn*5*-mediated transposition of plasmid DNA after transduction to *Myxococcus xanthus*. *J. Bacteriol.* **170:**4939–4941.

Downard, J. S., D. Kupfer, and D. R. Zusman. 1984. Gene expression during development of *Myxococcus xanthus:* analysis of the genes for protein S. *J. Mol. Biol.* **175:**469–492.

Downard, J. S., and D. R. Zusman. 1985. Differential expression of protein S genes during *Myxococcus xanthus* development. *J. Bacteriol.* **161:**1146–1155.

Fink, J. M., M. Kalos, and J. F. Zissler. 1989. Isolation of cell surface antigen mutants of *Myxococcus xanthus* by use of monoclonal antibodies. *J. Bacteriol.* **171:**2033–2041.

Fink, J. M., and J. F. Zissler. 1989. Defects in motility and development of *Myxococcus xanthus* lipopolysaccharide mutants. *J. Bacteriol.* **171:**2042–2048.

Furuichi, T., M. Inouye, and S. Inouye. 1985. Novel one-step cloning vector with a transposable element: application to the *Myxococcus xanthus* genome. *J. Bacteriol.* **164:**270–275.

Geisselsoder, J., J. M. Campos, and D. R. Zusman. 1978. Physical characterization of bacteriophage Mx4, a generalized transducing phage for *Myxococcus xanthus*. *J. Mol. Biol.* **119:**179–189.

Gill, J. S., and M. Dworkin. 1986. Cell surface antigens during submerged development of *Myxococcus xanthus* examined with monoclonal antibodies. *J. Bacteriol.* **168:**505–511.

Gill, J. S., and M. Dworkin. 1988. Isolation of additional monoclonal antibodies directed against cell surface antigens of *Myxococcus xanthus* cells undergoing submerged development. *J. Bacteriol.* **170:**5953–5955.

Gill, J. S., B. W. Jarvis, and M. Dworkin. 1987. Inhibition of development in *Myxococcus xanthus* by monoclonal antibody 1604. *Proc. Natl. Acad. Sci. USA* **84:**4505–4508.

Gill, R. E. Unpublished data.

Gill, R. E., and M. C. Bornemann. 1988. Identification and characterization of the *Myxococcus xanthus bsgA* gene product. *J. Bacteriol.* **170:**5289–5297.

Gill, R. E., M. G. Cull, and S. Fly. 1988. Genetic identification and cloning of gene required for developmental cell interactions in *Myxococcus xanthus*. *J. Bacteriol.* **170:**5279–5288.

Hagen, D. C., A. P. Bretscher, and D. Kaiser. 1978. Synergism between morphogenetic mutants of *Myxococcus xanthus*. *Dev. Biol.* **64:**284–296.

Hereford, L. M., and L. H. Hartwell. 1974. Sequential gene function in the initiation of *S. cerevisiae* DNA synthesis. *J. Mol. Biol.* **84:**445.

Hodgkin, J., and D. Kaiser. 1977. Cell-to-cell stimulation of movement in nonmotile mutants of *Myxococcus*. *Proc. Natl. Acad. Sci. USA* **74:**2938–2942.

Kaiser, D. 1984. Genetics of Myxobacteria, p. 163–184. *In* E. Rosenberg (ed.), *Myxobacteria: Development and Cell Interactions*. Springer-Verlag, New York.

Kalos, M., and J. Zissler. 1990. Transposon tagging of genes for cell-to-cell signaling in *Myxococcus xanthus*. *Proc. Natl. Acad. Sci. USA* **87:**8316–8320.

Kaplan, H. B., A. Kuspa, and D. Kaiser. 1991. Suppressors that permit A-signal-independent developmental gene expression in *Myxococcus xanthus*. *J. Bacteriol.* **173:**1460–1470.

Kim, S. K., and D. Kaiser. 1991. C-factor has distinct aggregation and sporulation thresholds during *Myxococcus* development. *J. Bacteriol.* **173:**1722–1728.

Kohara, Y., K. Akiyama, and K. Isono. 1987. The physical map of the whole *E. coli* chromosome: application of a new strategy for rapid analysis and sorting of a large genomic library. *Cell* **50:**495–508.

Kroos, L., and D. Kaiser. 1984. Construction of Tn*5lac*, a transposon that fuses *lacZ* expression to endogenous promoters, and its introduction into *Myxococcus xanthus*. *Proc. Natl. Acad. Sci. USA* **81:**5816–5820.

Kroos, L., and D. Kaiser. 1987. Expression of many developmentally regulated genes in *Myxococcus xanthus* depends on a sequence of cell interactions. *Genes Dev.* **1:**840–854.

Kroos, L., A. Kuspa, and D. Kaiser. 1986. A global analysis of developmentally regulated genes in *Myxococcus xanthus*. *Dev. Biol.* **117**:252–266.

Kroos, L., A. Kuspa, and D. Kaiser. 1990. Defects in fruiting body development caused by Tn*5lac* insertions in *Myxococcus xanthus*. *J. Bacteriol.* **172**:484–487.

Kuner, J., and D. Kaiser. 1981. Introduction of transposon Tn*5* into *Myxococcus* for analysis of developmental and other non-selectable mutants. *Proc. Natl. Acad. Sci. USA* **78**: 425–429.

Kuspa, A., L. Kroos, and D. Kaiser. 1986. Intercellular signalling is required for developmental gene expression in *Myxococcus xanthus*. *Dev. Biol.* **117**:267–276.

Kuspa, A., D. Vollrath, Y. Cheng, and D. Kaiser. 1989. Physical mapping of the *Myxococcus xanthus* genome by random cloning in yeast artificial chromosomes. *Proc. Natl. Acad. Sci. USA* **86**:8917–8921.

LaRossa, R., J. Kuner, D. Hagen, C. Manoil, and D. Kaiser. 1983. Developmental cell interactions of *Myxococcus xanthus:* analysis of mutants. *J. Bacteriol.* **153**:1394–1404.

Letouvet-Pawlak, B., C. Monnier, S. Barray, D. A. Hodgson, and J. F. Guespin-Michel. 1990. Comparison of β-galactosidase production by two inducible promoters in *Myxococcus xanthus*. *Res. Microbiol.* **141**:425–435.

Li, S.-F., B.-U. Lee, and L. J. Shimkets. 1992. *csgA* expression entrains *Myxococcus xanthus* development. *Genes Dev.* **6**:401–410.

Li, S., and L. J. Shimkets. 1988. Site-specific integration and expression of a developmental promoter in *Myxococcus xanthus*. *J. Bacteriol.* **170**:5552–5556.

Manoil, C., and J. Beckwith. 1985. Tn *phoA:* a transposon probe for protein export sequences. *Proc. Natl. Acad. Sci. USA* **82**:8129–8133.

Martin, S., E. Sodergren, T. Masuda, and D. Kaiser. 1978. Systematic isolation of transducing phage for *Myxococcus xanthus*. *Virology* **88**:44–53.

Michaelis, S., H. Inouye, D. Oliver, and J. Beckwith. 1983. Mutations that alter the signal sequence of alkaline phosphatase in *Escherichia coli*. *J. Bacteriol.* **154**:366–374.

Muñoz-Dorado, J., S. Inouye, and M. Inouye. 1991. A gene encoding a protein serine/threonine kinase is required for normal development of M. xanthus, a gram-negative bacterium. *Cell* **67**:995–1006.

O'Connor, K. A., S. Isassi-Ramirez, and D. R. Zusman. Submitted for publication.

O'Connor, K. A., and D. R. Zusman. 1983. Coliphage P1-mediated transduction of cloned DNA from *Escherichia coli* to *Myxococcus xanthus:* use for complementation and recombinational analyses. *J. Bacteriol.* **155**:317–329.

O'Connor, K. A., and D. R. Zusman. 1986. Genetic analysis of *Myxococcus xanthus* and isolation of gene replacements after transduction under conditions of limited homology. *J. Bacteriol.* **167**:744–748.

Rhie, H. G., and L. J. Shimkets. 1989. Developmental bypass suppression of *Myxococcus xanthus csgA* mutations. *J. Bacteriol.* **171**:3268–3276.

Rothstein, S. J., R. A. Jorgensen, J. C.-P. Yin, Z. Yong-Di, R. C. Johnson, and W. S. Reznikoff. 1981. Genetic organization of Tn*5*. *Cold Spring Harbor Symp. Quant. Biol.* **45**: 99–105.

Rudd, K., and D. R. Zusman. 1979. Rifampin-resistant mutants of *Myxococcus xanthus* defective in development. *J. Bacteriol.* **137**:295–300.

Sasakawa, C., and M. Yoshikawa. 1987. A series of Tn*5* variants with various drug-resistance markers and suicide vector for transposon mutagenesis. *Gene* **56**:283–288.

Shimkets, L. J. Unpublished data.

Shimkets, L. J., and S. J. Asher. 1988. Use of recombination techniques to examine the structure of the *csg* locus of *Myxococcus xanthus*. *Mol. Gen. Genet.* **211**:63–71.

Shimkets, L. J., R. E. Gill, and D. Kaiser. 1983. Developmental cell interactions in *Myxococcus xanthus* and the *spoC* locus. *Proc. Natl. Acad. Sci. USA* **80**:1406–1410.

Silhavy, T. J., and J. R. Beckwith. 1985. Uses of *lac* fusions for the study of biological problems. *Microbiol. Rev.* **49:**398–418.

Sodergren, E., Y. Cheng, L. Avery, and D. Kaiser. 1983. Recombination in the vicinity of insertions of transposon Tn*5* in *Myxococcus xanthus. Genetics* **105:**281–291.

Sodergren, E., and D. Kaiser. 1983. Insertions of Tn*5* near genes that govern stimulatable cell motility in *Myxococcus. J. Mol. Biol.* **167:**295–310.

Stellwag, E., J. M. Fink, and J. Zissler. 1985. Physical characterization of the genome of the *Myxococcus xanthus* bacteriophage Mx8. *Mol. Gen. Genet.* **199:**123–132.

Stephens, K., and D. Kaiser. 1987. Genetics of gliding motility in *Myxococcus xanthus*. Molecular cloning of the *mgl* locus. *Mol. Gen. Genet.* **207:**256–266.

Sternberg, N., M. Powers, M. Yarmolinski, and S. Austin. 1981. Group Y incompatibility and copy control of P1 prophage. *Plasmid* **5:**138–149.

Wu, T. T. 1966. A model for three-point analysis of random generalized transduction. *Genetics* **54:**405–410.

Yarmolinsky, M. B. 1977. Genetic and physical structure of bacteriophage P1 DNA, p. 721–732. *In* A. I. Bukhari, J. A. Shapiro, and S. L. Adhya (ed.), *DNA Insertion Elements, Plasmids and Episomes*. Cold Spring Harbor Laboratory, Cold Spring Harbor, N.Y.

Zusman, D. R. 1982. Frizzy mutants: a new class of aggregation-defective developmental mutants of *Myxococcus xanthus. J. Bacteriol.* **150:**1430–1437.

Myxobacteria II
Edited by Martin Dworkin and Dale Kaiser
© 1993 American Society for Microbiology, Washington, DC 20005

Chapter 7

Genetics of Regulation and Pathway of Synthesis of Carotenoids

David A. Hodgson and Francisco J. Murillo

INTRODUCTION

One trait that contributes to the attractive appearance of the myxobacteria is the vivid color of their colonies and sporangioles. This characteristic is due in some cases to the presence of carotenoids. Other pigments that accumulate in significant amounts have been described, but their chemical structures have only rarely been identified. The focus of this chapter will be exclusively on carotenoids, with special emphasis on the recent molecular genetic analysis of carotenogenesis in myxobacteria. The reader should consult Reichenbach and Kleinig (1984) for a more general review on myxobacteria pigments.

The Carotenoids

Carotenoids are among the most ubiquitous pigments in nature. They accumulate in the chloroplasts of all green leaves and are also produced by algae, photosynthetic bacteria, and many fungi and nonphotosynthetic bacteria. Animals do not synthesize carotenoids but can transform those they ingest. A classic example is the breakdown of carotenes to produce retinoids. Transformed and untransformed dietary carotenoids are responsible for the bright coloration of many animals, from crustaceans to birds. In the past 2 decades, many books and reviews have dealt with various aspects of carotenoids: structure and chemophysical properties, distribution, biological roles and applications, and biosynthesis and regulation

David A. Hodgson • Department of Biological Sciences, University of Warwick, Coventry CV4 7AL, England. **Francisco J. Murillo** • Departamento de Genetica y Microbiologia, Facultad de Biologia, Universidad de Murcia, 30071 Murcia, Spain.

(Isler, 1971; Goodwin, 1980, 1988; Britton and Goodwin, 1982; Bramley and MacKenzie, 1988; Britton, 1991). A brief summary of some of these aspects follows.

Natural carotenoids are C_{40} terpenes made up of eight isoprenoid units, although C_{30}, C_{45}, and C_{50} carotenoids have also been described. Carotenoid hydrocarbons are known as carotenes, and their general structure is well represented by the acyclic $C_{40}H_{56}$ conjugated polyene lycopene (Fig. 1). Frequent variations of this structure include cyclization at one or both ends of the molecule, hydrogenation or dehydrogenation, and insertion of oxygen in various forms. Chemical functions observed in oxygenated carotenoids (collectively known as xanthophylls) are hydroxyl, aldehyde, keto, epoxy, methoxyl, and carboxyl. A modification that seems to occur only in some bacteria is the fusion of hydroxylated carotenoids to a sugar molecule. The sugar moiety might be esterified, as in the myxobacterial carotenoids (Fig. 1; see below).

Carotenoids and Photoprotection

Carotenoids protect cells against photodynamic sensitization. In photosynthetic organisms, they have an additional role as accessory light-harvesting pigments. Photodynamic sensitization results from the absorption of light energy by certain molecules (photosensitizers). The energized photosensitizers cause cell damage or pass the energy on to other molecules that cause damage.

In unexcited (ground-state) molecules, electrons are paired in discrete, low-energy (nucleus-proximal) orbitals. If paired, the electrons spin in opposite (antiparallel) directions (the singlet state). If the molecule is excited by light energy, an electron is promoted to a higher-energy orbital. If the electron retains the same direction of spin as it had in the ground state, the molecule is said to be in the "excited singlet state." Once in an excited singlet state, one of the previously paired electrons may then alter its direction of spin so that both electrons have the same (parallel) direction. The molecule is then said to be in the "excited triplet state."

An important group of photosensitizing molecules contains the porphyrins. These molecules are ubiquitous, as they are components of cytochromes and chlorophylls, and they absorb strongly around 410 nm (the Soret band). The excited triplet-state porphyrin is relatively long-lived and can react directly with larger molecules, e.g., proteins, lipids, and nucleic acids, leading to oxidation of the substrate or formation of porphyrin adducts. These type I reactions (so called because they involve oxidation of the substrate by the photosensitizer without mediation of oxygen [Foote, 1991]) cause direct damage of cellular components (Hader and Trevini, 1987; Malik et al., 1990).

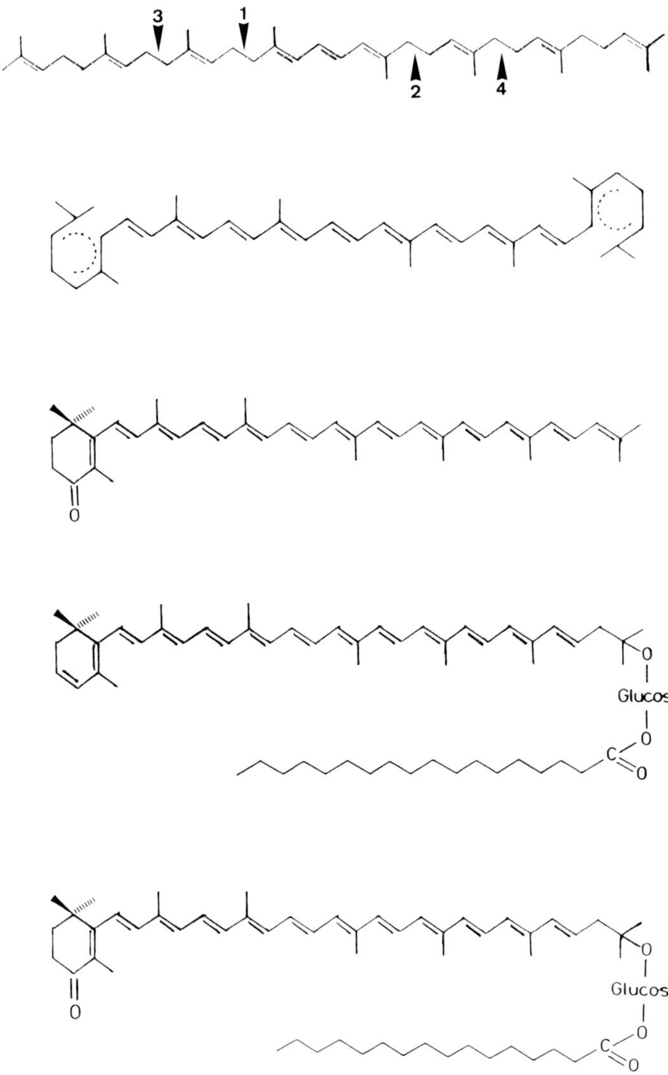

Phytoene

Lycopene

4-Keto-torulene

Myxobactin stearate

Myxobacton palmitate

Figure 1. Some representative carotenoid structures. Phytoene is the universal first C_{40} precursor of the biosynthesis of carotenoids. The numbers in this molecule indicate the order of the four successive desaturation reactions proposed by Porter and Lincoln (1950) for the conversion of phytoene into lycopene (see text). Keto-torulene, myxobactin stearate, and myxobacton palmitate are the major carotenoids found in myxobacteria of the suborder *Cystobacterineae*.

Alternatively, excited triplet-state porphyrin can interact with oxygen, forming a high-energy species that causes cell damage by oxidation—the type II reaction (Foote, 1991). An important type II event is the formation of singlet oxygen ($^1O_2$). Oxygen is unusual in that the ground state contains two unpaired electrons spinning in parallel; i.e., its ground state is triplet state, and when it gains energy, one of the unpaired electrons changes the direction of its spin. $^1O_2$ is relatively long-lived, lasting 2 to 4 μs in aqueous environments but 25 to 100 μs in hydrophobic environments. The species can cause peroxidation of lipids, cross-linking of proteins, and oxidation of any biomolecules sensitive to electrophilic attack (Knox and Dodge, 1985; Malik et al., 1990).

Carotenoids prevent photooxidative damage by quenching $^1O_2$ (Di Mascio et al., 1989) and triplet porphyrin, thus preventing formation of $^1O_2$ (Dirks et al., 1980; Cogdell, 1988).

Biosynthesis and Regulation

The universal C_5 building block for carotenoid synthesis is isopentenyl PP_i (IPP). IPP is isomerized to dimethylallyl PP_i before the two are condensed to form a C_{10} intermediate called geranyl PP_i. Sequential condensations with two further IPP molecules produce first farnesyl PP_i (C_{15}) and then geranylgeranyl PP_i (GGPP; C_{20}). Two GGPP molecules are joined to form phytoene, the first C_{40} carotenoid precursor (Fig. 1). Many years ago, Porter and Lincoln (1950) proposed that phytoene was converted to lycopene by a series of four desaturation reactions. These involved the removal of two hydrogen atoms at a time from alternate sides of the molecule, which sequentially generates phytofluene, ζ-carotene, neurosporene, and lycopene. The last compound might then be subject to one or more of the modifications discussed above. The Porter-Lincoln series has been corroborated in many different plant and fungal species, where a single enzyme, phytoene dehydrogenase, carries out the four desaturation reactions. In some bacteria, however, neurosporene, not lycopene, is subject to hydroxylation, cyclization, and/or other modifications. Such is the case in *Rhodobacter* and *Myxococcus* species, where the last desaturation reaction, carried out by a separate dehydrogenase enzyme to phytoene dehydrogenase, takes place on an already modified neurosporene molecule (Guiliano et al., 1988; Armstrong et al., 1990b) (Fig. 2).

Carotenoid synthesis is tightly regulated by exogenous and endogenous factors (reviewed in Bramley and MacKenzie [1988]), with light being the environmental agent having the most widespread influence. This is particularly so in plants, where carotenoid accumulation is closely associated with red-light-induced development of chloroplasts. Carotenoid pro-

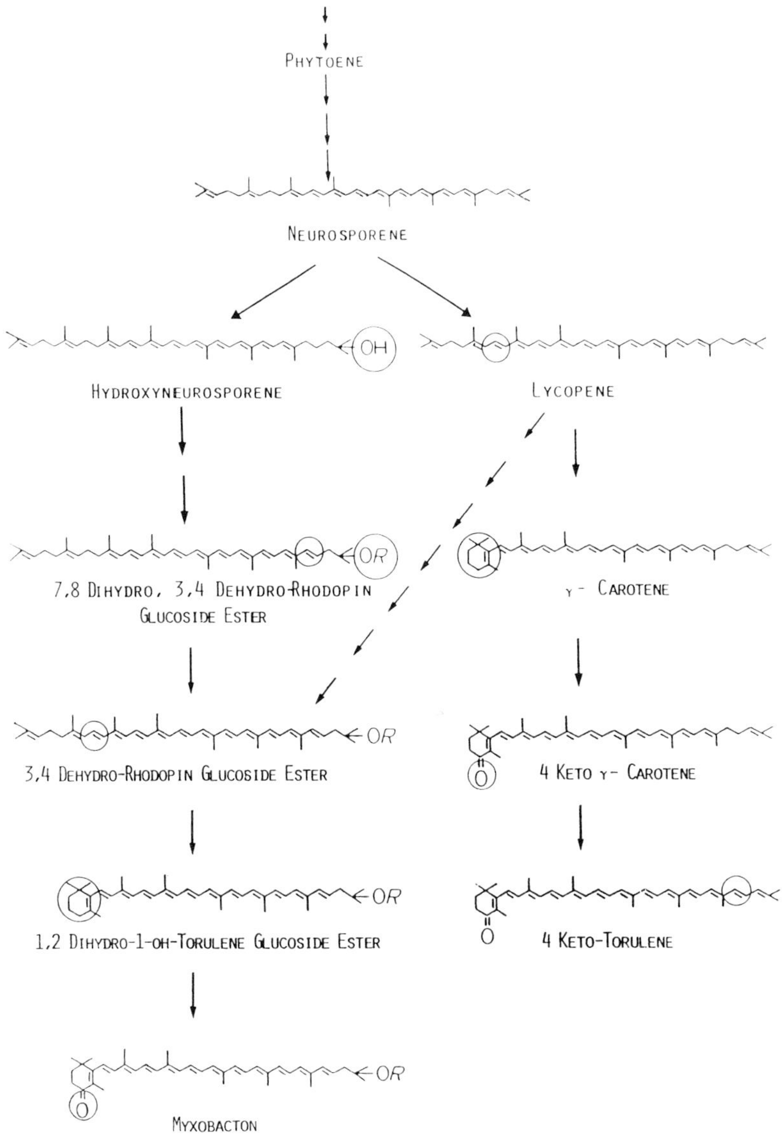

Figure 2. Pathway for the synthesis of carotenoids in *M. fulvus* (Kleinig, 1975). R, glucoside ester.

duction is also photoregulated in some algae, while it is constitutive in others. Carotenogenesis in many fungi and nonphotosynthetic bacteria is activated by blue light. In some species, the presence of light is absolutely required, whereas in others, light serves to increase the concentration from a significant, constitutive level.

A major endogenous factor that affects carotenoid production in plants is fruit ripening. The transformation of chloroplasts into chromoplasts associated with ripening has strong quantitative and qualitative influences on carotenoid accumulation. In fungi, carotenoids are often considered secondary metabolites, as they accumulate after nutrient limitation or growth arrest. We will see below that the greatest accumulation of carotenoids occurs after the transition from exponential to stationary phase of growth in myxobacterial cultures.

Carotenogenic Genes

Detailed genetic analysis of carotenoid synthesis in a few fungal and bacterial species has been reported. In the fungus *Neurospora crassa,* mutations at any of three genes (*al-1, al-2,* and *al-3*) block the production of carotenoids. Genes *al-1* and *al-3* have recently been cloned and shown to be transcriptionally activated by blue light. Gene *al-1* encodes the phytoene dehydrogenase, whereas *al-3* appears to encode the GGPP synthase (Nelson et al., 1989; Schmidhauser et al., 1990). In the fungus *Phycomyces blakesleeanus,* mutants affected in phytoene dehydrogenase and lycopene cyclase have been isolated, but the corresponding genes have not yet been cloned. In this fungus, the carotene pathway is feedback regulated by the end product of the pathway (β-carotene) and is stimulated by several environmental factors: blue light, proximity of cells of the other mating type, and certain chemicals. Many *Phycomyces* mutants affected in feedback or environmental regulation have been isolated. These mutants have been subject to genetic analysis, which has revealed the presence of several independent regulatory loci (Cerda-Olmedo, 1987). The molecular basis of the mode of action of these regulatory genes is not known.

Rhodobacter capsulatus, a facultative phototrophic bacterium, accumulates carotenoids in a light-independent manner. Nine carotenoid synthesis *(crt)* genes were identified within a big (46-kb) cluster of genes involved in construction of the photosynthetic apparatus. Eight of the *crt* genes are tightly linked in an 11-kb region that was separated by 12 kb from the ninth gene. The entire 11-kb region has been sequenced, and further analysis revealed the cluster to be transcribed in four mRNAs. Each of the eight *crt* genes was assigned to a particular biosynthetic step between GGPP and spheroidene and spheroidenone, the two acyclic, oxygenated carotenoids accumulated by this species. No regulatory genes have been reported (Yen and Marrs, 1976; Biel and Marrs, 1985; Armstrong et al., 1989, 1990b).

Finally, carotenogenic genes have been studied in two species of the phytopathogenic bacterial genus *Erwinia* (*Erwinia herbicola* and *Erwinia uredovora*). In both species, the genes were clustered. A more detailed sequence and mutant analysis carried out on the *E. uredovora* genes revealed six genes for the conversion of GGPP into a diglucoside derivative of zeanthin, a bicyclic, hydroxylated carotenoid (Perry et al., 1986; Misawa et al., 1990).

CAROTENOIDS OF MYXOBACTERIA

Dynamics of Photolysis and Carotenogenesis in *Myxococcus xanthus*

In the early 1960s, Robert Burchard, then a student of Martin Dworkin, attempted to isolate myxobacteriophage and noticed what appeared to be plaques in lawns of *M. xanthus* that had been left out on a bench. The lawns had also turned from the usual yellow-tan to orange. Further work revealed that the plaques had not been caused by bacteriophage but by sunlight-induced cell lysis-photolysis. The orange color was caused by photoinduction of carotenoids (Dworkin, personal communication).

Burchard and Dworkin (1966) reported that carotenoids accumulated in light-grown cells. However, these pigments appeared in quantity only after the cells had reached the stationary phase of growth. Photolysis could be induced in dark-grown cells when they were exposed to light. The amount of light necessary to induce cell lysis also depended on the growth phase of the cell culture. The light fluences needed to lyse stationary-phase cells were lower than those necessary for exponential phase cells. There appeared to be a dynamic equilibrium between photolysis and carotenogenesis. If the cells were in good physiological condition, carotenogenesis resulted. However, if the cells were in bad physiological condition, photolysis resulted.

Protoporphyrin IX appeared to be the photosensitizer for photolysis, and the concentration of this pyrrole in the cells increased 16-fold as the culture entered the stationary phase of growth. Photolysis was dependent on the photosensitizer, light intensity, and presence of oxygen and monovalent cations. It was independent of temperature and could be inhibited by the presence of any one of a number of divalent cations or by the presence of carotenoids in the cell.

The action spectra of photolysis and carotenogenesis were examined in greater detail. Burchard et al. (1966) confirmed the similarity of the action spectrum of photolysis and the absorption spectrum of protoporphyrin IX and reported the dose-response curves when monochromatic light of different wavelengths was used. The action spectrum of carotenogenesis was also similar to the absorption spectrum of protoporphyrin IX (Burchard and Hendricks, 1969). This implied that the same photosensitizer was acting in both processes.

Structures of Myxobacterial Carotenoids

The reader is recommended to the review of Reichenbach and Kleinig (1984) for a more detailed discussion of the structure and biosynthesis of myxobacterial carotenoids.

In an extensive analysis of light-grown *Myxococcus fulvus,* Reichenbach and Kleinig (1971) reported the presence of some 50 to 60 different carotenoids, most of which occurred at concentrations too low to allow identification. However, those authors did obtain enough material to identify 24 of the carotenoids. The most abundant of the 24 were the monocyclic carotenoid 4-keto-torulene and the fatty acid esters of two glycosylated, monocyclic carotenoids, myxobactin (1′,2′-dihydro-1′-hydroxy-3,4-dehydro-torulene glucoside) and myxobacton (1′,2′-dihydro-1′-hydroxy-4-keto-torulene glucoside) (Fig. 1). The fatty acids varied, but in the main, they were unbranched, straight-chain residues linked to the carotenoid via the glucose residue. Fatty acid esters of myxobactin and myxobacton were also the major carotenoids of *Stigmatella aurantiaca* (Kleinig et al., 1970). Carotenoid glycosides had previously been reported in bacteria, but the acetylation appeared to be a novel feature. However, Kleinig et al. (1970) did warn that this may have resulted from the method of preparation of carotenoids by other experimenters. The discovery of an acetylated carotenoid that did not contain a sugar moiety (2′-acyl-1′,2′-dihydro-1′,2′-dihydroxy-4-keto-torulene) in *M. fulvus* was the first example of such a compound in bacteria.

In *M. fulvus,* light led to a 10-fold increase in total carotenoids during the exponential phase of growth. By stationary phase, the total carotenoid content had increased 20-fold over that in dark-grown cells. The myxobacton esters made up 70% (by weight) of the final carotenoid content. The myxobactin esters made up 10%, and a further 10% was made up of either 4-keto-torulene or torulene. The relative proportion of the last two carotenoids depended on culture conditions. The remaining 10% was made up of the low-abundance carotenoids mentioned above (Reichenbach and Kleinig, 1971). Kleinig (1975) reported that myxobacton ester made up 80% and 4-keto-torulene made up 15% of the total carotenoids in *M. fulvis.* As might be expected, cell fractionation studies revealed that the carotenoid esters were membrane bound (Kleinig, 1972).

The carotenoids found in representatives of the genera of the *Cysto-*

bacterineae (*Myxococcus, Cystobacter,* and *Stigmatella*) appeared to be very similar. However, when carotenoids were isolated and characterized from representatives of the genera of the *Sorangineae* (*Chondromyces, Nannocystis,* and *Sorangium*), they proved to be different from those of the *Cystobacterineae* and to vary from genus to genus. Indeed, there were two novel carotenoid rhamnosides that were found only in one of the two species of *Sorangium* analyzed (Reichenbach and Kleinig, 1984).

Biosynthesis of Carotenoid Glucoside Esters in *M. fulvus*

The effects of various inhibitors and oxygen starvation on carotenogenesis were used to identify the pathways of synthesis of the carotenoids. The addition of low concentrations of nicotine inhibited cyclization and caused the accumulation of an acyclic carotenoid glucoside ester. Removal of the drug led to cyclization of the latter compound. This implies that cyclization was a late step in carotenogenesis. Removal of nicotine in anaerobic conditions led to the reversible accumulation of a monocyclic precursor of myxobacton ester that lacked the keto function, the implication being that inclusion of a keto group was a postcyclization step and required molecular oxygen. At higher nicotine concentrations, hydroxylation at the 1′ position was blocked, and hence, glucosidation of carotenoids was blocked. Aglycolytic carotenoids, principally (95%) lycopene, accumulated (Kleinig and Reichenbach, 1973).

The effects of a number of other carotenogenesis inhibitors were examined by Kleinig (1974, 1975). 2-(4-Chlorophenylthio)triethylamine (CPTA) had the same effects as L- and D-nicotines. Desaturation of carotenoids was blocked by the addition of diphenylamine or 4-chloro-5-(dimethylamino)-2-,α,α,α(trifluoro-*m*-tolyl)-3(2H)-pyridazinone (San 6706). The latter inhibitor had the advantage that it could be washed from the cell, and transient inhibition effects could thus be studied. From these studies, it became clear that the pathway of synthesis of myxobacton ester and 4-keto-torulene was not a simple series and that a number of steps could occur in a different order and yet the same product or closely related product would be formed. It appeared that only a minority of myxobacton ester was formed via lycopene, i.e., the final product of the Lincoln-Porter series. Evidence was presented that the 1′ carbon of neurosporene was hydroxylated and then glycosylated and esterified. It was this intermediate that was finally desaturated (probably by an enzyme other than phytoene dehydrogenase, as it was diphenylamine resistant) before cyclization and transfer of a keto group onto the ring structure (Fig. 2).

It was proposed that of the subpopulation of phytoene molecules that were converted to lycopene, only a minority were glycosylated, esterified, and shunted into the myxobacton ester pool. The majority of the lycopene was thought to be left unhydroxylated at the 1′ position, cyclized at the 1-to-6 position, and then oxygenated to generate a keto group on the ring. The resulting intermediate was then converted to 4-keto-torulene by desaturation of the 3′-4′ carbon-carbon bond. Isolation and characterization of the carotenoids generated by *M. xanthus* mutants blocked in the biosynthesis of carotenoids have confirmed this branched scheme (Ruiz-Vaquez and Murillo, unpublished data).

MOLECULAR GENETIC ANALYSIS OF CAROTENOGENESIS

Genetic analysis of carotenogenesis and its regulation has been applied almost exclusively to *M. xanthus*. This arises because of the development of classic genetic (Kaiser, 1984) and molecular genetic (see chapter 6) methods for this species. These methods have made possible the identification and characterization of the modes of action of both structural and regulatory genes for carotenogenesis.

As mentioned above, carotenogenesis in *M. xanthus* is absolutely dependent on illumination with blue light. No carotenoids or carotene precursors are found in dark-grown cultures of this organism (Martinez-Laborda et al., 1990). The yellow appearance of dark-grown cultures is due to an uncharacterized noncarotenoid pigment. When exposed to light, *M. xanthus* colonies or broth cultures turn orange-red because of the accumulation of carotenoids. Loss of the ability to produce carotenoids or to regulate carotenoid production is not detrimental to the cell under normal conditions, so mutants are easily isolated. It has not proven possible to devise a direct mutant selection procedure. However, the conspicuousness of the color change makes the screening for mutants easy, even in a dense background of wild-type colonies. The notation *car* is used to describe genes affecting carotenogenesis in *M. xanthus*.

Both spontaneous and induced *car* mutations have been isolated. The latter include chemical-, UV-, and transposon-induced mutations. The transposons used were Tn*5* (streptomycin, bleomycin, and kanamycin [Km^r] resistances) and its derivatives: Tn*5*-132 (tetracycline resistance) and the promoter probe Tn*5 lac* (Km^r). In addition, cloned *car* genes were disrupted in vitro and introduced into *M. xanthus* by homologous recombination. Tn*5* insertions that were next to *car* genes but did not affect carotenogenesis were isolated by using a technique pioneered by Kuner and Kaiser (1981). Such transposon-tagged *car* loci were used in transduction-mediated gene-mapping studies, epistasis studies, and cloning of *car* genes. The analysis of gene expression was made by fusion of a promoterless *lacZ* gene to *car* gene promoters, using either the Tn*5 lac* transposon mentioned above or in vitro insertion into any one of a number of plasmid vectors that integrated into the *M. xanthus* chromosome.

The *car* Mutants

M. xanthus car mutants belong to one of two distinct phenotypic classes: those that form orange-red colonies independent of light and those that form colonies that do not turn orange-red in the light. Members of the former class constitutively form carotenoids (the Car^c phenotype), while members of the latter class cannot form carotenoids under any circumstances (the Car^- phenotype).

The first spontaneous *car* mutant isolated was of the Car^c phenotype and was used by Burchard and Dworkin (1966) to provide evidence of the photoprotective role of carotenoids. Several other Car^c mutants have been isolated in different laboratories, usually while researchers were looking for other types of mutant. The stability and striking appearance of the phenotype have led to its use as a genetic marker, particularly in the elegant analysis of phenotypic complementation of morphogenesis mutants (Hagen et al., 1978; Mayo and Kaiser, 1989). Car^c mutations clearly affect regulation of carotenogenesis by light.

Car^- mutants have been isolated by screening large numbers of light-grown colonies derived from mutagenized cells or by looking for mutations that suppress the Car^c phenotype. The Car^- phenotype could result either from lack of an enzyme for a metabolic step before a colored carotene is formed or from loss of a positive regulator of carotenogenesis.

Regulatory Genes

carQRS genes

Martinez-Laborda et al. (1986) reported the mapping of six independent Car^c mutants. Five of the mutations mapped to a single locus near the two independent Tn*5* insertion sites ΩMR136 and ΩDK4611. A third Tn*5* insertion, ΩDK1910, mapped to this same locus, but unlike the others, this insertion caused a Car^c phenotype. ΩDK1910 and the five point mutations appeared to define a new gene named *carR* (Balsalobre et al., 1987). The sixth Car^c mutation was not linked by transduction to the *carR* locus. This mutation defined a new regulatory gene, *carA* (see below). Analysis of one of the *carR* point mutations confirmed that it accumulated the same carotenoids in the dark that the wild type accumulated in the light (Martinez-Laborda et al., 1990).

The Car^c mutant phenotype of the Tn*5* insertion ΩDK1911 (which was identical to ΩDK1910) was dominant when it was cloned. This fact

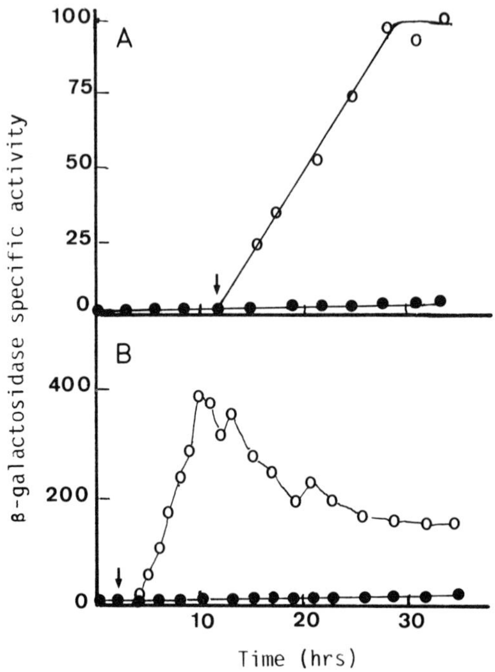

Figure 3. Light activation of *carB* (A) and *carQRS* (B). In each case, β-galactosidase activity was measured in strains carrying a *lacZ* transcriptional fusion. Cultures incubated in the dark (closed symbols) were illuminated (open circles) at an early stage of exponential growth phase (indicated by an arrow). Redrawn from Balsalobre et al., 1987 (A), and Hodgson, 1993 (B).

led to the predictions that a light-inducible promoter was responsible for transcription of a positive regulator of carotenogenesis and that the ΩDK1911 insertion had made the promoter constitutive. Fusion of the putative promoter region to a promoterless *Escherichia coli lacZ* gene revealed a promoter that was essentially inactive in the dark but very powerfully activated in the light (Fig. 3). The promoter region from a ΩDK1911 mutant was active in both the light and the dark because of an outward-reading promoter in the IS*50* of the Tn*5* that occluded the light-inducible promoter. An in vitro-generated deletion of the DNA downstream of the promoter generated a Car$^-$ phenotype when it was used to replace the region in an otherwise wild-type strain, thus confirming all three predictions. Expression of the light-inducible promoter was examined in different Car^c mutations. The promoter became constitutively expressed in the dark in all the *carR* point mutants tested. Complementation tests revealed that the *carR* mutations mapped downstream of the light-inducible promoter. Therefore, two genetic functions were controlled by the promoter: negative regulation of the promoter and positive regulation of carotenogenesis (Hodgson, 1987, 1993).

Sequencing of the *carR* region revealed three translationally coupled open reading frames (ORFs) downstream of the light-inducible promoter region (McGowan and Hodgson, unpublished data). When the *carR* region point mutations were sequenced, all of them affected the second ORF. Therefore, this ORF must encode the negative regulator of the promoter and was termed *carR*. When the third ORF (now called *carS*) was deleted, the strain bearing the deletion became Car$^-$ and the genetically unlinked *carB* structural gene (see below) could not be activated. However, the promoter still retained its light inducibility. When *carS* alone was coupled to a constitutive promoter, the strain became Car^c, and *carB* became constitutive. Therefore, *carS* is the positive regulator of carotenogenesis identified previously (McGowan and Hodgson, unpublished data).

The first ORF, named *carQ,* was disrupted by a sequence that did not disrupt the translational coupling of the other two genes. A strain bearing this mutation was Car$^-$, and the *carQRS* promoter was inactivated (Gorham et al., unpublished data). Therefore, *carQ* and *carR* were responsible for control of the *carQRS* promoter. The *carS* gene was a positive regulator of carotenogenesis but did not regulate the *carQRS* promoter. Martinez-Laborda and Murillo (1989) reported the isolation of three spontaneous mutations that generated a Car$^-$ phenotype, suppressed a *carR* mutation, blocked expression of the unlinked structural genes (see below), and were linked to the *carR* locus. Genetic analysis revealed that one of these mutations appeared to be a deletion. Subsequent Southern blot analysis has confirmed this prediction and revealed a complete deletion of the locus (Hodgson, unpublished data). Sequence analysis revealed that the two point mutations affect the *carQ* gene (McGowan, unpublished data).

Balsalobre (1989) isolated a mutation linked to the *carR* region that led to constitutive expression of one of the structural genes (*carB;* see below). However, in itself, this mutation did not cause a Car^c phenotype. Surprisingly, sequence analysis (McGowan, unpublished data) revealed that this mutation truncated the *carS* gene product. Therefore, deletion of the *carS* gene caused loss of structural gene expression, while a nonsense mutation caused constitutive expression of the same genes.

***carA* gene**

As mentioned above, one of the original Car^c mutations defined a new locus, *carA*. The original mutation affected a gene that was linked to a Tn*5* insertion (ΩMR134) that had no effect on carotenogenesis (Martinez-Laborda et al., 1986). Recently, two new Car^c mutations have been mapped to the *carA* locus (Hodgson, 1993). Colonies of cells carrying *carA* mutations and the ΩDK1910 mutation are less intensely pigmented (orange) than the *carR* point mutations (deep red).

Strains bearing the *carA1* allele accumulate large amounts (ca. 6 mg/g of protein) of all-*trans* phytoene (see above) in the dark. They also produce minor quantities (ca. 1 mg/g of protein) of phytoene, ζ-carotene, and neurosporene (see above). The last two are colored and probably account for the color of the *carA* strains. In the light, the strains accumulate almost normal amounts of the usual end products but also accumulate abnormal amounts of phytoene, i.e., ca. 4 mg/g of protein, or four times the amount found in similarly treated wild-type cells (Martinez-Laborda et al., 1990). Thus, the mutation results in constitutive expression of phytoene in the dark and simultaneous impairment of phytoene metabolism in the light. Loss of either *carQ* or *carS* gene function has no effect on the *carA1* phenotype, which implies that the action of the *carA* gene occurs later in the regulatory cascade than that of the *carQRS* gene cluster.

The *carD* gene

A Car$^-$ Tn*5* insertion strain that when starved failed to aggregate and form fruiting bodies (Fru$^-$) has recently been isolated. Transduction analysis revealed that the mutation was linked to no other *car* locus and that the Fru$^-$ and Car$^-$ phenotypes cosegregated with the Tn*5* Km^r marker. The new gene has been named *carD*. The mutation *(carD1)* blocks light activation of the *carQRS* gene cluster and the *car* structural genes (see below). However, it does not suppress the phenotype of the *carA1* mutation. Expression of several developmentally activated genes is suppressed by *carD1,* which seems to explain the Fru$^-$ phenotype of the mutant (Nicolas et al., unpublished data). Why the light-induced and starvation-induced responses of *M. xanthus* are integrated through the action of the *carD* gene is an intriguing question. One is tempted to speculate that *carD* coordinates responses to environmental stresses placed on the cell.

Structural Genes

carB gene cluster

Various *car* mutations have served to identify two loci that include genes that encode the enzymes of carotenoid biosynthesis (the structural genes). All structural gene mutations, as expected, are epistatic to the *carQRS* and *carA* Car^c mutations. Several independent transposon insertions that caused a Car$^-$ phenotype are tightly linked to the *carB* locus. The *carB* locus is linked to the *carA* locus, being separated by ca. 7 kb of DNA (Martinez-Laborda et al., 1986; Balsalobre, 1989; Martinez-Laborda and Murillo, 1989).

DNA from the *carB* locus has been sequenced, and an ORF with homology to the phytoene synthetase enzymes of *Erwinia* spp. (Armstrong et al., 1989) and *Rhodobacter* spp. (Misawa ct al., 1990) has been identified. Other *car* genes between the *carB* and *carA* genes (confirmed by site-directed mutagenesis) appear to form an operon and encode prephytoene PP_i synthase, neurosporene hydroxylase, and hydroxyneurosporene dehydrogenase (Ruiz-Vazquez et al., unpublished data).

carC gene

A single transposon mutation identified the second, unlinked structural-gene locus *carC* (Balsalobre et al., 1987). In the light, the *carC* mutant accumulated almost exclusively all-*trans* phytoene. In the dark, no carotenoids or precursors were produced (Martinez-Laborda et al., 1990). Prokaryotic and eukaryotic phytoene dehydrogenases have two conserved domains: a nucleotide (NAD or FAD)-binding site at the amino terminus and a carboxy terminus domain of unknown function (Armstrong et al., 1990a). Sequencing of the *carC* gene revealed an ORF with homology to both conserved domains (Fontes, 1992).

Table 1 summarizes our present knowledge of the *car* genes of *M. xanthus* and includes results of the gene expression studies discussed below.

Table 1. *car* genes of *M. xanthus*

Type of gene and name	Phenotype	Carotenoid contest[a]	Description
Regulatory			
carQ	Car^-	None in light or dark	First gene of light-inducible *carQRS* operon. CarQ is required for activation of P^{carQRS} and P^{carC}.
carR	Car^c	Wild type in light and dark	Second gene of light-inducible *carQRS* operon; translationally coupled to *carQ*. CarR is membrane bound and inactivates P^{carQRS} in dark. Light inactivates action of CarR. Mutations in *carQ* are epistatic to *carR* mutations.
carS	Car^- (*carS*$^{\Delta}$)	ND	Third gene of light-inducible *carQRS* operon; translationally coupled to *carR*; required for light activation of *carB*.

(continued)

Table 1. *(continued)*

Type of gene and name	Phenotype	Carotenoid contest[a]	Description
	Car$^+$ *carB*^{c} [*carS*(Um)]	ND	*carS* nonsense mutation (umber) causes constitutive expression of *carB* in light and dark but no Car^c phenotype.
carD	Car$^-$ Agg$^-$ Fru$^-$	ND	Distinct genetic locus. CarD is required for expression of *carQRS* operon and hence all structural *car* genes. CarD is also required for expression of several developmentally regulated genes.
carA	Car^c	Dark: phytoene; light: wild type plus excess phytoene	Linked to *carB* cluster of structural genes. Mutation *carAl* causes constitutive expression of *carB* in dark and light while reducing light activation of *carC*. *carA1*effect on *carB* is epistatic over effects of *carQ* or *carD* mutations.
Structural			
carB	Car$^-$	ND	ORF with homology to phytoene synthetases of other bacteria; grouped with other structural genes in single, light-inducible operon linked to *carA*. Mutations at *carQ, carS* (deletion), or *carD* block expression of operon. Mutations at *carR, carS*(Um), and *carA* cause light-independent expression.
carC	Car$^-$	Dark: no carotenoids; light: all-*trans* phytoene	Distinct genetic locus. ORF has homology to phytoene dehydrogenase. Activated by light only after cells enter stationary phase of growth. Mutations in *carQ* or *carD* block light activation. Mutations in *carR* cause light-independent expression. Mutations in *carS* have no effect on *carC*. Mutation *carA1* reduces stimulatory effect of light on *carC* expression.

[a] ND, not determined.

Expression of the *car* Genes

Structural genes

Expression of the *carB* and *carC* genes is tightly regulated by light. This was revealed by individual fusion of each gene with a promoterless *E. coli lacZ* gene and assay of the specific activity of β-galactosidase in cells grown in the light and the dark. Maximum light stimulation of the *carB* promoter was about 20-fold (Fig. 3), while that of the *carC* promoter was 400-fold (Balsalobre et al., 1987; Martinez-Laborda and Murillo, 1989; Fontes, 1992). A remarkable difference between the *carB* and *carC* promoters was that the former was stimulated by light at any time during the growth cycle while the latter was not activated by light unless the cells had entered the stationary phase of growth. The fact that the product of the *carC* gene (phytoene dehydrogenase) converts colorless phytoene into colored carotenoids explains the observation of Burchard and Dworkin (1966) that maximum accumulation of colored carotenoids in *M. xanthus* occurred after the cells had entered stationary phase.

Fontes (1992) demonstrated that the growth phase dependence of *carC* light responsiveness was due to exhaustion of the carbon source but not of the nitrogen source. It is interesting that in *E. coli,* carbon source starvation stimulates the cell's response to oxidative stress (Farr and Kogoma, 1991), and as discussed above, carotenoids quench photochemically generated oxidants.

Regulatory genes

Mutation of *carR* removes the light requirement for expression of both *carB* and *carC* (Balsalobre et al., 1987; Fontes, 1992), whereas mutation in either *carQ* or *carD* blocks all expression of the structural genes (Martinez-Laborda and Murillo, 1989; Nicolas, unpublished data). Deletion of *carS* blocks expression of *carB* but not *carC* (McGowan, unpublished data; Fontes, 1992), while a nonsense mutation in *carS* causes expression of the *carB* gene in the dark but has no effect on *carC* (Balsalobre, 1989; Fontes, 1992; McGowan, unpublished data). Thus, *carD, carQ, carR,* and *carS* integrate the separate structural gene loci in a single, light-regulated *car* regulon. Mutation of the *carA* gene also affects both structural gene loci but in different ways. It causes expression of *carB* in the dark, while greatly reducing light stimulation of *carC* expression. This suggests that *carA* plays a dual function in the *car* regulon, acting as a negative regulator of *carB* in the dark and a positive regulator of *carC* in the light (Fontes, 1992).

Light regulation of the *car* regulon is a consequence of the effect of light on the expression of the *carQRS* operon (Hodgson, 1987, 1993; McGowan et al., unpublished data). Light inducibility of the *carQRS* promoter is controlled by interactions between the products of the *carD, carQ,* and *carR* genes. As discussed above, loss of *carR* function leads to high-level constitutive expression of the *carQRS* promoter. Sequence analysis of *carR* revealed that this gene encoded an integral membrane protein. Fusion of the *carR* ORF with assayable proteins showed that the fusions were transported to the membrane fraction of *M. xanthus* and the inner membrane fraction in *E. coli*. We have yet to resolve the membrane fractions of *M. xanthus* (Gorham, unpublished data). Thus, it appears that regulation of the light responsiveness of the *carQRS* promoter and hence the *car* regulon involves an integral membrane protein.

Loss of *carQ* leads to loss of all expression of the *carQRS* promoter. Sequence analysis of the *carQ* ORF revealed no homology with any protein sequence in the EMBL data base. Mutation of the *carD* gene blocked all expression of the *carQRS* promoter and was epistatic over the *carR* mutations but not the ΩDK1911 mutation. This implies that the *carD* gene product is necessary to activate the *carQRS* promoter, but if the promoter is made redundant, as is the case of ΩDK1911 (where the outward-reading IS*50* promoter causes occlusion), it is no longer necessary.

LIGHT REGULATION OF THE *car* REGULON OF *M. XANTHUS*

Reception of Light by the Cell

Carotenogenesis in *M. xanthus* appears to be dependent on protoporphyrin IX as photosensitizer (Burchard and Dworkin, 1966; Burchard and Hendricks, 1969). Robson (1992) repeated the isolation of a porphyrin from *M. xanthus* and confirmed its identity as protoporphyrin IX. He also demonstrated by using laser light that the expression of the *carQRS* promoter was dependent on 410-nm (blue) light; 410 nm is the main absorption wavelength (Soret band) of protoporphyrin IX. Much work on the photochemistry of porphyrins and their role in the oxygen-dependent light inhibition of both prokaryotic (Malik et al., 1990) and eukaryotic cells, especially the targeted photolysis of cancer cells (Kessel, 1984), has been reported.

As discussed above, triplet-state porphyrins are efficient generators of $^1O_2$. It is probably $^1O_2$ that causes most damage to the cell, as it is long-lived (especially in membranes) and highly diffusible. Protoporphyrin IX could cause activation of the *car* regulon by either of two mechanisms: direct involvement of the triplet porphyrin (type I reaction) or involvement via $^1O_2$ (type II reaction). Carotenoids are capable of quenching both $^1O_2$ and triplet porphyrin.

Light and Gene Action

As discussed above, the *carQRS* locus is central to light regulation of the *car* regulon. The gene products of this locus mediate light activation of the *car* structural genes and the *carQRS* genes themselves.

The *carR* product (the integral membrane protein CarR) blocks expression of the *carQRS* promoter in the dark. The *carQ* product (CarQ) is required for *carQRS* promoter expression in the light. It is tempting to suggest that CarQ is an intermediary between the membrane-bound CarR and the promoter. This suggestion is supported by the observation that when there were more copies of the intact *carQ* gene than of the intact *carR* gene present in a strain, the *carQRS* promoter became constitutive in the dark. When there were more copies of the intact *carR* gene than of the intact *carQ* gene, then the promoter remained light inducible (McGowan et al., unpublished data). As noted above, the *carQ* and *carR* genes and the *carR* and *carS* genes are translationally coupled; i.e., the AUG start codon of the downstream gene overlaps the UGA stop codon of the upstream gene to form the DNA sequence ATGA (McGowan and Hodgson, unpublished data). Translational coupling has been shown to result in the production of stoichiometric amounts of the coupled gene products (Oppenheim and Yanofsky, 1980). It appears, therefore, that the balance of production of CarR and CarQ is very precise.

If membrane-bound CarR is responsible for monitoring the interaction of blue light with protoporphyrin IX, CarQ could be the intermediate that transfers information about that interaction to the *carQRS* promoter. One mechanism of transfer might be the sequestration of CarQ to the membrane by CarR in the dark. Interaction of blue light with protoporphyrin IX might lead to release of CarQ, which could then activate the promoter. Activation of the *carQRS* promoter would lead to increased expression of the *carQ, carR,* and *carS* genes. CarQ and the *carS* product, CarS, would then activate the rest of the *car* regulon, possibly through the action of other genes, e.g., *carA*. Hence, carotenogenesis would be initiated. The membrane sequestration of CarQ by CarR is favored, because this process would explain the requirement for precise one-to-one stoichiometry of the proteins. Chemical modification of one protein by another would not require such stoichiometry (Hodgson, unpublished data).

The effect of the *carD1* mutation on expression of the *carQRS* promoter is the same as the effects of the *carQ* mutations. The *carD* gene product, CarD, might interact with CarQ to activate the *carQRS* promoter, or CarD might be involved in the release of CarQ from CarR in the membrane.

Light is not the only effector of *carQRS* expression. When the *carQRS* promoter was fused to *lacZ* and the cell was exposed to continuous illumination, β-galactosidase specific activity increased within the cell. However, just after the onset of carotenoid production, the β-galactosidase activity decreased until it leveled off at 40-fold greater than the activity in uninduced cells (Fig. 3). This might indicate that the carotenoids are causing feedback inhibition of *carQRS* promoter activity. This hypothesis is supported by the observation that in a *carA* mutant, which constitutively produces carotenoids, the *carQRS* promoter was induced by light to a lesser extent than in a *carA*$^+$ strain (Hodgson, 1993). The hypothesis was further supported by the observation that the *carQRS* promoter was not inactivated during continuous illumination in Car$^-$ mutants affected in *carB* or *carS* (Robson, 1992; McGowan, unpublished data).

The amount of phytoene produced in the light by the *carC* mutant was two- to threefold greater than the total amount of carotenoid produced by the wild type exposed to the same amount of light. This observation also led to the proposal that the pathway was subject to feedback repression by the end products (Martinez-Laborda et al., 1990). Presumably, phytoene cannot repress feedback, while the myxobacton and myxobactin esters can.

Role for Singlet Oxygen in Activation of the *carQRS* Operon?

Robson (1992) investigated the role of oxygen in the induction of carotenogenesis and its interaction with light-activated protoporphyrin IX. Using cells growing in a chemostat, he demonstrated that the *carQRS* promoter could be activated, albeit to only a low level (twofold), by hyperoxygenation in the dark. This implied that excess oxygen could short-circuit the need for light by the promoter, which implied that oxygen might have a role in control of its expression.

Direct evidence for the role of $^1O_2$ was obtained by using toluidine blue. This dye generates $^1O_2$ when it is exposed to oxygen and red light. Red light had no effect on the expression of the *carQRS* promoter unless 100 pM dye was present. The addition of 100 pM toluidine blue in the dark had no effect on expression of the promoter (Robson, 1992).

Indirect evidence for the role of $^1O_2$ comes from the observation that expression of the *carQRS* promoter ceased upon production of carotenoids. As discussed above, carotenoids are very efficient quenchers of $^1O_2$. If $^1O_2$ is important in promoter induction, the mechanism of feedback repression could be explained by the quenching of this intermediate. The effect of addition of another $^1O_2$ quencher on light induction of the promoter was assessed. 1,4-Diazobicyclo-[2,2,2]-octane (DABCO) is water soluble and structurally unrelated to the carotenoids. Addition of increasing amounts of DABCO (9 to 12.5 mM) greatly decreased the light inducibility of the promoter. At these concentrations, DABCO had no effect on the growth rate of the cell. DABCO also decreased induction of the promoter by red light in the presence of 100 pM toluidine blue (Robson, 1992).

Further circumstantial evidence for a role of $^1O_2$ is the observation that this species is more long-lived in hydrophobic environments (25 to 100 μs; see above) such as membranes. It is highly probable that free protoporphyrin IX is membrane bound, as it is a very hydrophobic compound. It should also be remembered that the proposed receptor of the light signal, the *carR* receptor protein, is also membrane bound. Finally, there is direct, physical evidence that the carotenoids are membrane bound in *M. fulvus* (Kleinig, 1972) and so they are in the appropriate location to quench the system.

The Present Model

Our proposed model is presented in Fig. 4. In it, blue light interacts with protoporphyrin IX to form the excited triplet-state species. $^1O_2$ is

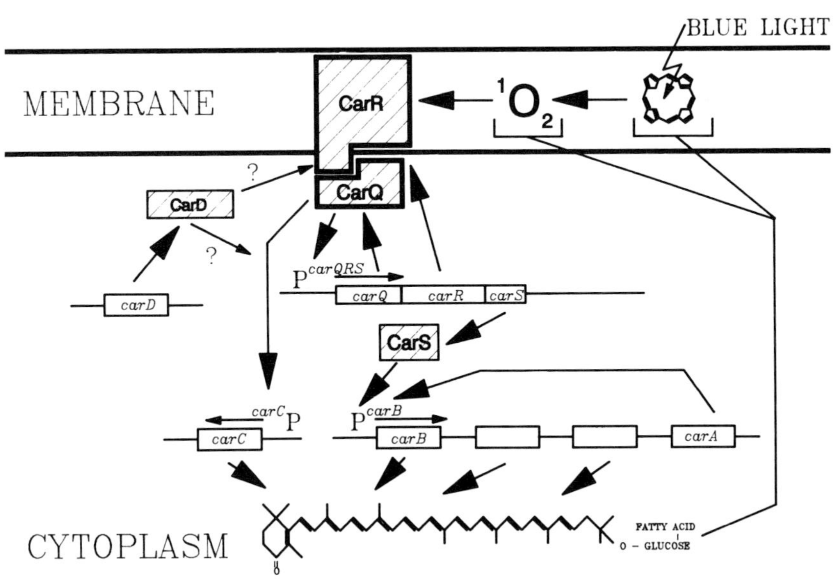

Figure 4. *car* regulon of *M. xanthus* (see text for explanation).

created by intersystem crossover. Interaction of the membrane-bound *carR* gene product (CarR) with $^1O_2$ leads to release of the *carQ* gene product (CarQ) from the membrane. CarQ then causes increased expression of P^{carQRS} (the *carQRS* promoter). The product of the *carD* gene (CarD) is involved in the release of CarQ or its stimulation of P^{carQRS}. CarQ also activates the promoter of the *carC* gene (P^{carC}). Increased expression of the *carQRS* regulon increases expression of the *carS* gene product (CarS). This protein is necessary in conjunction with the *carA* gene for expression of the *carB* gene cluster. Expression of the *carB* and *carC* gene clusters leads to production of the carotenoids myxobacton and myxobactin esters, etc. These carotenoids quench the triplet-state protoporphyrin IX and $^1O_2$, so the expression of the *car* regulon is down-regulated.

Certainly, many loose ends remain. For example, we do not know the mechanism of gene activation by CarQ or CarD. The sequences of the proteins do not look like any others in the protein sequence data bases, so if they are DNA-binding proteins, they are novel. We compared the two CarQ-activated promoters P^{carQRS} and P^{carC} but found no obvious common sequences (unpublished results from our laboratories). It is possible that CarQ does not activate either promoter or that both promoters directly interact with another gene product. It is also apparent that other genes must be involved in the growth phase regulation of promoter activity and the differential effects that some mutations have on different promoters within the *car* regulon. These and other unexplained aspects of the *M. xanthus* response to blue light should keep us busy for some time to come.

REFERENCES

Armstrong, G. A., M. Alberti, and J. E. Hearst. 1990a. Conserved enzymes mediate the early reactions of carotenoid biosynthesis in nonphotosynthetic and photosynthetic prokaryotes. *Proc. Natl. Acad. Sci. USA* **87:**9975–9979.

Armstrong, G. A., M. Alberti, F. Leach, and J. E. Hearst. 1989. Nucleotide sequence, organization and nature of the protein products of the carotenoid biosynthesis gene cluster of *Rhodobacter capsulatus. Mol. Gen. Genet.* **216:**254–268.

Armstrong, G. A., A. Schmidt, G. Sandmann, and J. E. Hearst. 1990b. Genetic and biochemical characterization of carotenoid biosynthesis mutants of *Rhodobacter capsulatus. J. Biol. Chem.* **265:**8329–8338.

Balsalobre, J. M. 1989. Induccion por la luz de la expression genica y la carotenogenesis en *Myxococcus xanthus*. Ph.D. thesis. University of Murcia, Murcia, Spain.

Balsalobre, J. M., R. M. Ruiz-Vazquez, and F. J. Murillo. 1987. Light induction of gene expression in *Myxococcus xanthus. Proc. Natl. Acad. Sci. USA* **84:**2359–2362.

Biel, A. J., and B. L. Marrs. 1985. Oxygen does not directly regulate carotenoid biosynthesis in *Rhodobacter capsulata. J. Bacteriol.* **162:**1320–1321.

Bramley, P. M., and A. MacKenzie. 1988. Regulation of carotenoid biosynthesis. *Curr. Top. Cell. Regul.* **29:**291–343.

Britton, G. 1991. The biosynthesis of carotenoids: a progress report. *Pure Appl. Chem.* **63:** 101–108.

Britton, G., and T. W. Goodwin (ed.). 1982. *Carotenoid Chemistry and Biochemistry.* Pergamon Press, Oxford.

Burchard, R. P., and M. Dworkin. 1966. Light-induced lysis and carotenogenesis in *Myxococcus xanthus. J. Bacteriol.* **91:**535–545.

Burchard, R. P., S. A. Gordon, and M. Dworkin. 1966. Action spectrum for the photolysis of *Myxococcus xanthus. J. Bacteriol.* **91:**896–897.

Burchard, R. P., and S. B. Hendricks. 1969. Action spectrum for carotenogenesis in *Myxococcus xanthus. J. Bacteriol.* **97:**1165–1168.

Cerda-Olmedo, E. 1987. Carotene, p. 199–222. *In* E. Cerda-Olmedo and E. D. Lipson (ed.), *Phycomyces.* Cold Spring Harbor Laboratory, Cold Spring Harbor, N.Y.

Cogdell, R. 1988. The function of pigments in chloroplasts, p. 183–230. *In* T. Goodwin (ed.), *Plant Pigments.* Academic Press, London.

Di Mascio, P., S. Kaiser, and H. Sies. 1989. Lycopene as the most efficient biological carotenoid singlet oxygen quencher. *Arch. Biochem. Biophys.* **274:**532–538.

Dirks, G., A. L. Moore, T. A. Moore, and D. Gust. 1980. Light absorption and energy transfer in polyene-porphyrin esters. *Photochem. Photobiol.* **32:**277–280.

Dworkin, M. Personal communication.

Farr, S. B., and T. Kogoma. 1991. Oxidative stress responses in *Escherichia coli* and *Salmonella typhimurium. Microbiol. Rev.* **55:**561–585.

Fontes, M. 1992. Estructura y regulacion por la luz del gen de la deshidrogensa del fitoeno de la bacteria *Myxococcus xanthus.* Ph.D. thesis. University of Murcia, Murcia, Spain.

Foote, C. S. 1991. Definition of type I and type II photosensitized oxidation. *Photochem. Photobiol.* **54:**659. (Guest editorial.)

Giuliano, G., D. Pollock, H. Stapp, and P. A. Scolnik. 1988. A genetic-physical map of the *Rhodobacter capsulatus* carotenoid biosynthesis gene cluster. *Mol. Gen. Genet.* **213:** 78–83.

Goodwin, T. W. 1980. *The Biochemistry of Carotenoids,* vol. 1. *Plants.* Chapman and Hall, Ltd., London.

Goodwin, T. W. (ed.). 1988. *Plant Pigments.* Academic Press, London.

Gorham, H. C. Unpublished data.

Gorham, H. C., P. R. H. Robson, S. J. McGowan, and D. A. Hodgson. Unpublished data.

Hadar, D.-P., and M. Trevini. 1987. *General Photobiology.* Pergamon Press, Oxford.

Hagen, D. C., A. P. Bretscher, and D. Kaiser. 1978. Synergism between morphogenetic mutants of *Myxococcus xanthus. Dev. Biol.* **64:**284–296.

Hodgson, D. A. 1987. Light inducible promoter. International patent application no. PCT/GB87/0040. New British patent application (1986) no. 8615263.

Hodgson, D. A. 1993. Light-induced carotenogenesis in *Myxococcus xanthus:* genetic analysis of the *carR* region. *Mol. Microbiol.* **7:**471–488.

Hodgson, D. A. Unpublished data.

Isler, O. 1971. *Carotenoids.* Birkhauser Verlag, Basel.

Kaiser, D. 1984. Genetics of myxobacteria, p. 163–184. *In* E. Rosenberg (ed.), *Myxobacteria: Development and Cell Interactions.* Springer-Verlag, New York.

Kessel, D. 1984. Hematoporphyrin and Hpd: photolysis, photochemistry and phototherapy. *Photochem. Photobiol.* **39:**851–859.

Kleinig, H. 1972. Membranes from *Myxococcus fulvus (Myxobacterales)* containing carotenoid glucosides. I. Isolation and composition. *Biochim. Biophys. Acta* **274:**489–498.

Kleinig, H. 1974. Inhibition of carotenoid synthesis in *Myxococcus fulvus (Myxobacterales). Arch. Microbiol.* **97:**217–226.

Kleinig, H. 1975. On the utilization *in vivo* of lycopene and phytoene as precursors for the formation of carotenoid glucoside ester and on the regulation of carotenoid biosynthesis in *Myxococcus fulvus. Eur. J. Biochem.* **57:**301–308.

Kleinig, H., and H. Reichenbach. 1973. Biosynthesis of carotenoid glucóside esters in *Myxococcus fulvus (Myxobacterales):* inhibition by nicotine and carotenoid turnover. *Biochim. Biophys. Acta* **306:**249–256.

Kleinig, H., H. Reichenbach, and H. Achenbach. 1970. Carotenoid pigments of *Stigmatella aurantiaca (Myxobacterales).* II. Acylated carotenoid glucosides. *Arch. Mikrobiol.* **74:** 223–234.

Knox, J. P., and A. D. Dodge. 1985. Singlet oxygen and plants. *Phytochemistry* **24:**889–896.

Kuner, J. M., and D. Kaiser. 1981. Introduction of transposon Tn*5* into *Myxococcus* for analysis of development and other selectable mutants. *Proc. Natl. Acad. Sci. USA* **78:** 425–429.

Malik, Z., J. Hanania, and Y. Nitzan. 1990. New trends in photobiology (invited review). Bactericidal effects of photoactivated porphyrins—an alternative approach to antimicrobial drugs. *J. Photochem. Photobiol. B* **5:**281–293.

Martinez-Laborda, A., J. M. Balsalobre, M. Fontes, and F. J. Murillo. 1990. Accumulation of carotenoids in structural and regulatory mutants of the bacterium *Myxococcus xanthus. Mol. Gen. Genet.* **223:**205–210.

Martinez-Laborda, A., M. Elias, R. Ruiz-Vazquez, and F. J. Murillo. 1986. Insertions of Tn*5* linked to mutations affecting carotenoid synthesis in *Myxococcus xanthus. Mol. Gen. Genet.* **205:**107–114.

Martinez-Laborda, A., and F. J. Murillo. 1989. Genic and allelic interactions in carotenogenic response of *Myxococcus xanthus* to blue light. *Genetics* **122:**481–490.

Mayo, K. A., and D. Kaiser. 1989. *asgB,* a gene required early for developmental signaling, aggregation and sporulation of *Myxococcus xanthus. Mol. Gen. Genet.* **218:**409–418.

McGowan, S. J. Unpublished data.

McGowan, S. J., H. C. Gorham, and D. A. Hodgson. Unpublished data.

McGowan, S. J., H. C. Gorham, P. R. H. Robson, and D. A. Hodgson. Unpublished data.

McGowan, S. J., and D. A. Hodgson. Unpublished data.

Misawa, N., M. Nakagawa, K. Kobayashi, S. Yamano, Y. Izawa, K. Nakamura, and K. Harashima. 1990. Elucidation of the *Erwinia uredovora* carotenoid biosynthetic pathway by functional analysis of gene products expressed in *Escherichia coli. J. Bacteriol.* **172:** 6704–6712.

Nelson, M. A., G. Morelli, A. Carattoli, N. Romano, and G. Macino. 1989. Molecular cloning of *Neurospora crassa* carotenoid biosynthetic gene (albino-3) regulated by blue light and the products of the white collar genes. *Mol. Cell. Biol.* **9:**1271–1276.

Nicholas, F. J. Unpublished data.

Nicholas, F. J., R. Ruiz-Vazquez, and F. J. Murillo. Unpublished data.

Oppenheim, D. S., and C. Yanofsky. 1980. Translational coupling during expression of the tryptophan operon of *Escherichia coli. Genetics* **95:**785–795.

Perry, K. L., T. A. Simonitch, K. J. Harrison-Lavoie, and S.-T. Liu. 1986. Cloning and regulation of *Erwinia herbicola* pigment genes. *J. Bacteriol.* **168:**607–612.

Porter, J. W., and R. E. Lincoln. 1950. Lycopersicon selections containing a high content of carotenes and colorless polyenes. II. The mechanism of carotene biosynthesis. *Arch. Biochem. Biophys.* **27:**390–403.

Reichenbach, H., and H. Kleinig. 1971. The carotenoids of *Myxococcus fulvus* (Myxobacterales). *Arch. Mikrobiol.* **76:**364–380.

Reichenbach, H., and H. Kleinig. 1984. Pigments of myxobacteria, p. 128–137. *In* E. Rosenberg (ed.), *Myxobacteria: Development and Cell Interactions.* Springer-Verlag, New York.

Robson, P. R. H. 1992. Towards a mechanism of carotenogenesis in *Myxococcus xanthus.* Ph.D. thesis. University of Warwick, Coventry, England.

Ruiz-Vazquez, R., J. A. Botella, and F. J. Murillo. Unpublished data.

Ruiz-Vazquez, R., and F. J. Murillo. Unpublished data.

Schmidhauser, T. J., F. R. Lauter, V. E. A. Russo, and C. Yanofsky. 1990. Cloning, sequence and photoregulation of *al-1,* a carotenoid biosynthetic gene of *Neurospora crassa. Mol. Cell. Biol.* **10:**5064–5070.

Yen, H.-C., and B. L. Marrs. 1976. Map of genes for carotenoid and bacteriochlorophyl biosynthesis in *Rhodobacter capsulata. J. Bacteriol.* **126:**619–629.

Myxobacteria II
Edited by Martin Dworkin and Dale Kaiser
© 1993 American Society for Microbiology, Washington, DC 20005

Chapter 8

Transcriptional Regulation of Developmental Gene Expression in *Myxococcus xanthus*

John Downard and Lee Kroos

Myxococcus xanthus is a bacterium with a distinctive life cycle that involves large populations of cells and group-dependent behaviors. Perhaps the most unusual aspect of this multicellular life cycle is found during development, when many thousands of cells aggregate into mounds called fruiting bodies. Within these multicellular fruiting structures, individual cells differentiate to produce heat- and desiccation-resistant spores. Development (fruiting body formation) is triggered by starvation and requires a sufficient population of cells and a solid surface. Much of the interest in *M. xanthus* as an experimental system has focused on analysis of the role of cell-cell signaling during development. Several intercellular signaling systems have been identified, and the characteristics of these systems are under investigation (see chapter 4).

Driving development is a complex program of gene expression. Developmentally regulated genes are subject to control by nutrient availability; to temporal regulation; and also, in certain cases, to a form of spatial control depending on the positional relationship of a cell to its neighbors. Virtually all developmentally regulated genes appear to require cell-cell signaling for normal expression, and it seems likely that the coordination of developmental gene expression between cells is one of the most basic roles for cell-cell signaling.

From the information that is currently available about the *M. xanthus*

John Downard • Department of Botany and Microbiology, University of Oklahoma, Norman, Oklahoma 73019. **Lee Kroos** • Department of Biochemistry and Department of Microbiology, Michigan State University, East Lansing, Michigan 48824.

system and from what is generally known about the responses of cells to extracellular stimuli, a model can be proposed to explain how developmental gene expression might be controlled by cell-cell signaling. According to this model, a cell responds to signals released extracellularly by the population of cells through specific receptors, and this information is transmitted into the cell (signal transduction) to control the production or activity of transcriptional regulatory proteins. In certain cases, the information provided from cell-cell signaling may have to be integrated with information on the nutritional state of the cell or on other aspects of the environment. A hierarchical series of these intercellular signaling events, each involving a particular signal transduction pathway and culminating in the activation of a particular set of genes, may serve the primary purpose of ensuring that cells within the population pass through the various developmental stages in unison and that particular genes are expressed only when the products are required.

In order to understand the molecular nature of the particular intercellular signaling pathways involved in *M. xanthus* development and how these systems direct the transcription of developmental genes, information about the regulation of specific developmental genes will be required. The regulatory sites within these genes will need to be located, and the regulatory proteins that recognize these sites will have to be identified. To understand the mechanisms involved in the transcriptional regulation of these genes, it will be necessary to develop in vitro transcription systems that faithfully reproduce the regulatory responses observed during development.

The identification and characterization of genes encoding the transcriptional regulatory proteins will be essential in these studies and may allow the association of *M. xanthus* gene products with families of regulatory proteins involved in signal transduction phenomena in other bacteria and higher organisms. For example, a two-component gene family is widely utilized to respond to a variety of environmental cues like nitrogen and phosphate limitation (Albright et al., 1989), and a related system(s) could be involved in *M. xanthus* development. In addition, the role of multiple σ factors in the sporulation process of *Bacillus subtilis* has been established, and it is clear that multiple σ factors are present in *M. xanthus* (Apelian and Inouye, 1990; Inouye, 1990); *M. xanthus* also contains a family of genes encoding eukaryotic-like protein kinases, some of which participate in developmental signal transduction (Munoz-Dorado et al., 1991). Eventually, it will be necessary to associate specific environmental and intercellular signaling systems with the transcriptional regulatory systems that they directly control.

Relatively little is known about the mechanisms for transcriptional regulation involved in *M. xanthus* development. However, many developmentally regulated genes have been identified in recent years, and basic information about the timing of expression and regulation of these genes is accumulating. The regulatory regions of a few genes have been defined in some detail, and this information will surely be utilized to identify the corresponding regulatory proteins. The cloning and sequencing of developmentally regulated genes will also allow comparisons of regulatory DNA sequences and will help identify genes under common regulatory control. Since the regulation of developmental gene expression may be quite complex, it is important that a variety of biochemical and genetic approaches be available for studies of *M. xanthus* transcription. In this chapter, we summarize what is known about *M. xanthus* developmental gene transcription and describe some of the approaches that may be useful in advancing these studies.

DEVELOPMENTALLY REGULATED GENES

The expression of many genes is altered during *M. xanthus* development. An initial glimpse of the complexity of this regulation was provided by pulse-labeling cells with radioactive amino acids at different times during development and examining the pattern of protein synthesis (Inouye et al., 1979a). The rates of synthesis of at least nine soluble proteins and six membrane proteins increased at various stages of development. The most abundant soluble protein, protein S, has been studied intensively; it is found on the surfaces of mature spores (Inouye et al., 1979b, 1981; Nelson and Zusman, 1983). Gene cloning and hybridization studies revealed the presence of two very similar genes separated by 1.4 kb that could potentially encode protein S (Inouye et al., 1983a, 1983b; Downard et al., 1984). Mutational and expression studies showed that the downstream gene, *tps*, is expressed early in development and encodes protein S, whereas the upstream gene, *ops*, is expressed much later, when spores form (Downard et al., 1984; Inouye, 1984; Komano et al., 1984; Downard and Zusman, 1985; Teintze et al., 1985a, 1985b). The regulatory regions of *tps* and *ops* have been characterized in considerable detail (see below).

Another well-characterized, developmentally regulated protein whose corresponding gene has been cloned is myxobacterial hemagglutinin (Cumsky and Zusman, 1979, 1981; Romeo et al., 1986). This lectin-like protein appears to be associated with cell poles and plays a subtle role in aggregation (Nelson et al., 1981; Romeo and Zusman, 1987). Induction of the gene encoding myxobacterial hemagglutinin *(mbhA)* during development involves an increase in both the rate of transcription and the stability of the mRNA (Romeo and Zusman, 1991). The unusual stability of *mbhA* mRNA in developing cells requires sequences within the 3′ untranslated region and appears to involve an interaction between these sequences and the translational machinery (Romeo and Zusman, 1992). Like *mbhA* mRNA, *tps* mRNA has an unusually long half-life in developing cells (Downard et al., 1984), but the determinants of this unusual stability have not yet been elucidated.

Several developmentally regulated genes have been identified by mutations that interfere with development. The techniques for cloning genes identified in this way are described in chapter 9. *csgA* encodes a 17-kDa protein that serves as an essential extracellular signal during development (see chapter 12). *csgA* is expressed weakly in growing cells, and its expression increases throughout development, reaching a peak during sporulation (Kim and Kaiser, 1991; Li et al., 1992). The product of *csgA,* called CsgA or C factor, appears to stimulate *csgA* transcription. This positive feedback loop may order the sequence of developmental events, because successively higher concentrations of CsgA are required for rippling, then aggregation, and finally, sporulation. The loop also appears to require that cells move into alignment for efficient C signaling (Kim and Kaiser, 1990a, 1990b); hence, expression of *csgA* and expression of genes that depend on C signaling appear to be subject to a form of spatial control.

Another genetic locus identified by mutations is "frizzy," which contains six genes (*frzA, -B, -CD, -E, -G,* and *-F*), at least four of which appear to be cotranscribed *(frzABCDE)* (see chapter 13). Mutations in the *frz* genes alter the pattern of developmental aggregation, but spores are produced normally. The *frz* gene products are similar to proteins of the enteric chemotactic signal transduction pathway and control the frequency with which *M. xanthus* cells reverse the directions of their gliding movements. The *frzABCDE* operon is expressed in growing cells and is induced at the onset of development, reaching a peak during the aggregation stage (Weinberg and Zusman, 1989).

A general approach to the identification of developmentally regulated genes in *M. xanthus* has employed a transposable promoter probe, Tn*5lac,* that generates transcriptional fusions to the *Escherichia coli lacZ* gene (Kroos and Kaiser, 1984). Of 2,374 insertion-containing strains, 36 increased β-galactosidase expression more than threefold during development, and 29 of these appeared to have Tn*5lac* inserted in different transcription units (Kroos et al., 1986). This type of analysis led to an estimate that several hundred genes increase their expression during *M. xanthus* development. Each of the 29 distinct *lacZ* fusion-containing strains begins to express β-galactosidase at a particular time during development, and these expression times range from the onset of starvation to the onset of sporulation. Interestingly, most of the Tn*5lac* insertions in developmentally regulated regions did not cause a discernible developmental defect.

Cloning of *M. xanthus* genes via sequence similarity with genes from other organisms has so far resulted in the characterization of two developmentally regulated genes. *sigB* exhibits considerable sequence similarity to prokaryotic σ factors, is expressed at the onset of sporulation, and is required for spore viability upon prolonged incubation (Apelian and Inouye, 1990). *pkn1* exhibits sequence similarity with the catalytic domain of eukaryotic protein serine/threonine kinases, and its product is autophosphorylated at both serine and threonine residues when expressed in *E. coli* (Munoz-Dorado et al., 1991). Like *sigB* expression, *pkn1* expression is induced at the onset of sporulation. Deletion of *pkn1* results in premature and incomplete aggregation and a reduced spore yield.

The approximate times of expression of most of the known developmentally regulated *M. xanthus* genes as well as of other developmental markers can be deduced from Fig. 2 of chapter 12. This diagram also indicates the dependence of these developmental markers on cell-cell interactions. Obviously, many different dependence patterns are observed, even for markers expressed with similar timing. Because many of the markers are transcriptional fusions to *lacZ,* the different dependence patterns probably reflect different mechanisms of transcriptional control. Conversely, *lacZ* fusions that exhibit similar temporal and dependence patterns may be coordinately regulated. It should be noted that this diagram may overestimate the complexity of transcriptional control mechanisms for at least two reasons. First, the expression times are approximate (± 3 h in many cases), so some *lacZ* fusions that exhibit the same dependence pattern and slightly different expression times (according to the diagram) may in fact be coordinately regulated. Second, distinctions between partial and absolute dependence on each cell-cell interaction are made. Fusions expressed with similar timing but exhibiting dependence on the same cell-cell interactions and differing only in the degree of dependence (i.e., partial versus absolute) could well employ similar mechanisms of transcriptional control.

Other criteria have also been used to classify developmentally regulated *M. xanthus* genes. For example, *tps* is expressed when cells are starved in a shaking liquid culture, whereas *ops, mbhA,* and *frzABCDE* are not expressed (Downard and Zusman, 1985; Weinberg and Zusman, 1989; Romeo and Zusman, 1991). A survey of 11 developmentally regulated Tn*5lac* insertions showed that in general, early-expressed fusions that do not depend on C signaling for their expression are expressed upon starvation in liquid, whereas later-expressed fusions that depend on C signaling are not expressed (Kroos, 1986). The results suggest that A signaling can occur in shaking liquid culture (provided the cell density is sufficiently high) but that C signaling cannot take place. This is consistent with the finding that A factor is a diffusible signal, whereas C factor is membrane associated and C signaling requires close contact between cells (see chapter 4). *tps, ops, pkn1,* and 13 developmentally regulated Tn*5lac* insertions have also been tested for expression upon induction of sporulation by glycerol (Downard and Zusman, 1985; Kroos, 1986; Munoz-Dorado et al., 1991). Only *ops* and Tn*5lac* Ω4530 were expressed, indicating that glycerol sporulation bypasses much of the developmental program expressed during fruiting body formation.

REGULATORY SITES IN DEVELOPMENTAL GENES

The approaches being employed to identify the regulatory regions of developmental genes in *M. xanthus* are typical of the modern approaches that are being used successfully to study genes from a variety of organisms. Generally, *M. xanthus* studies involve cloning of the gene of interest, in vitro alteration of the upstream DNA sequences of the cloned gene, and introduction of the altered gene into *M. xanthus* to determine the effect of the change on expression of the gene. The expression of the altered gene may be monitored by detection of the native RNA or protein products of the gene or, more frequently, by the assay of β-galactosidase produced by a *lacZ* reporter gene fused with the gene of interest. Cloned DNA containing precisely defined alterations may be introduced into *M. xanthus* by utilizing DNA packaged by the generalized transducing phage P1 (O'Connor and Zusman, 1983) or by electrotransformation (Kuspa and Kaiser, 1989). Since an *M. xanthus* plasmid shuttle vector has not been available, studies of gene expression have required the transferred DNA to be integrated into the *M. xanthus* chromosome for stable maintenance. DNA has been integrated into the genome by homologous recombination between common sequences in the transferred DNA and the *M. xanthus* chromosome (Downard and Zusman, 1985), by a site-specific recombination event based on the integration system of the temperate myxophage Mx8 (see chapter 9) or by transposition into the genome promoted by transposon Tn*5* (Downard, 1988).

By using approaches of this type, the regulatory regions of the *tps* and *ops* genes have been analyzed in some detail. These genes are highly similar in sequence and encode abundantly produced spore-associated proteins. As discussed above, these closely linked genes are differentially expressed, with the activation of *tps* expression occurring early in development and *ops* activation beginning much later. The transcription initiation sites for these genes (+1) appear to be at corresponding positions 50 bp upstream from the respective translational start sites (Downard, 1987).

Studies have indicated that the regulatory regions of the *tps* and *ops* genes consist of at least two types of sites: promoters for RNA polymerase binding, which are located immediately upstream of the transcriptional start sites, and upstream activation sites (UAS) for binding of transcription factors that may stimulate transcription initiation. The existence of UAS regions was initially suggested by the finding that a segment of DNA located relatively far upstream from the *ops* transcriptional start (−131 to −311) is required for a normal level of *ops* gene transcription (Downard et al., 1988). To study this *ops* DNA segment in more detail, it was placed upstream from the *tps* promoter (Kil et al., 1990). This promoter was missing *tps* DNA sequences upstream from −66 and failed to direct developmentally regulated transcription. Surprisingly, the *ops* UAS segment, a regulatory DNA segment from a gene expressed only late in development, restored the ability of the *tps* promoter to direct a *tps* temporal pattern of early developmental expression (Table 1). Furthermore, the *ops* UAS

Table 1. UAS-directed β-Galactosidase expression during development

UAS (location)[a]	No. of copies	Promoter[b]	Relative β-galactosidase sp act during: Early development[c]	Late development[d]
		tps	2.5	
ops (−311 to −131)	1	*tps*	20	
ops (−311 to −131)	2	*tps*	46	
ops (−290 to −269)	1	*tps*	8.4	
tps (−375 to −173)	1	*tps*	13	
tps (−375 to −173)	3	*tps*	41	
tps (−110 to −82)	1	*tps*	13	
		ops		1.9
tps (−375 to −173)	1	*ops*	1.9	32
ops (−311 to −131)	1	*ops*	1.5	39

[a] The sequence indicated (relative to the transcription initiation site at +1) was placed upstream from the promoter region indicated.

[b] The *tps* promoter region included DNA from -66 to +104, and the *ops* promoter region included DNA from −130 to +104. Each promoter region was fused with the *lacZ* reporter gene to produce a translational fusion.

[c] β-Galactosidase specific activity produced during early development (18 h) was determined and divided by the specific activity of vegetative cells. Values represent the averages determined from at least five transductants.

[d] β-Galatosidase specific activity produced during late development (48 h) was determined and divided by the specific activity of vegetative cells. Values represent the averages determined from at least five transductants.

segment functioned to activate transcription in either orientation with respect to the promoter, and the presence of an additional UAS segment resulted in incrementally higher gene expression from the promoter. Thus, this *ops* DNA segment shares properties with UAS or enhancer elements found in other bacteria and in eukaryotes (Atchison, 1988; Struhl, 1989).

A gel retardation assay was utilized to show that a DNA-binding activity specific for the *ops* UAS segment also appears early in development. A binding site for this activity was mapped to an 18-bp sequence (−273 to −290) by an in situ DNA footprinting analysis of a protein-*ops* UAS segment complex (Kuwabara and Sigman, 1987; Brown, 1990). A synthetic fragment containing the −273 to −290 sequence was also able to direct a modest three- to fourfold increase in the level of developmental transcription when it was placed alone upstream from the *tps* promoter (Table 1). This is a relatively low level of expression compared with that of the native *tps* regulatory region, which can direct a >100-fold induction, but the expression directed by the synthetic fragment was readily detectable and shared the temporal pattern of expression exhibited by the intact *tps* gene. Moreover, at least part of the explanation for high-level expression of the *tps* gene seems to be the presence of multiple sites for the binding of upstream activators and the cumulative effect of these sites on the level of gene expression. While these results suggest the importance of the −273 to −290 region in upstream activation, three other related sequences exist in the upstream region of the *ops* gene (Table 2), and they may also function in developmental gene regulation.

Two regions of the *tps* gene also appear to have UAS activity. These regions are the −82 to −110 and the −173 to −375 DNA segments. Both segments of DNA were observed to activate early developmental

Table 2. Related upstream regions of *ops* gene

Region[a]	Sequence[b]	Identity[c]
−290 to −273	CGACCCGCTCCTGGCGGC	
−198 to −181	CATCCCCCTGCTCGCGGC	13/18
−156 to −139	CGGCCCCCTGCACTGGGC	11/18
−54 to −37	CCGCGCGGTGCTGTAGGC	11/18

[a] Related sequences of DNA located upstream from the *ops* gene transcriptional start site (+1). The −290 to −273 segment was identified in a footprinting analysis of an *ops*-protein complex formed in a crude extract of developing cells (Brown, 1990).

[b] Related DNA sequences. Boldface letters in the −290 to −273 sequence indicate nucleotides shared by all four sequences at the same relative positions. Boldface letters in the other three sequences indicate nucleotides that each of these sequences shared with the −290 to −273 sequence.

[c] Degree of identity with the −290 to −273 sequence (number of nucleotides in boldface type in column 2).

transcription (a *tps*-like temporal pattern) from the *tps* promoter, and DNA-binding activity for the larger segment also appeared early in development (Table 1; Brown, 1990). In addition, the −173 to −375 segment was able to replace the −131 to −311 *ops* segment to direct late developmental transcription (an *ops*-like temporal pattern) from the *ops* promoter (Table 1; Kil, 1991). This result and those described above indicate that the −131 to −311 *ops* UAS segment and the −173 to −375 *tps* UAS segment are interchangeable with respect to *tps* and *ops* promoter activation. Since the *ops* and *tps* genes are located close together on the *M. xanthus* chromosome, the interesting possibility is raised that the *ops* UAS region may participate not only in *ops* regulation but also in regulation of the *tps* gene located about 2 kbp away. Consistent with this idea, the separation of the *tps* regulatory region from the *ops* upstream sequences by the insertion of DNA between them resulted in a 60 to 70% decrease in the level of *tps* RNA in developmental cells (Kil et al., 1990). The possibility of a reciprocal effect of the *tps* UAS region on *ops* expression has not yet been investigated.

From the information at hand, it is possible to begin to construct molecular models for developmental *tps* and *ops* regulation. For the *tps* gene, the appearance of binding activity for UAS segments may be the critical event for transcriptional activation. Since a low level of *tps* transcription has been detected in vegetative cells (Downard, 1987), a vegetative form of RNA polymerase may be used for *tps* RNA synthesis. In the case of the *ops* gene, the appearance of binding activity for the UAS region seems to be a requirement for a normal level of gene expression, but this event is not sufficient by itself to activate transcription. Another event involving *ops* sequences close to the transcriptional start appears to be required to trigger transcription initiation. This event might be the appearance of a unique form of RNA polymerase containing a development-specific σ factor such as SigB. The production of the *ops* protein product has been shown to be reduced in a *sigB* mutant strain (Apelian and Inouye, 1990), but a direct role for SigB in *ops* expression must be demonstrated to confirm this idea. Since the *tps* and *ops* UAS segments appear to be interchangeable, it is possible that a single transcriptional activator is involved in regulation of both genes.

While these relatively simple models serve to explain the basic characteristics of *tps* and *ops* gene regulation, there are indications that regulation may actually be more complicated. First, there are no obvious DNA sequence similarities between the different *tps* and *ops* segments with UAS activity. This observation raises the possibility that different (but apparently functionally interchangeable) transcriptional activators may be involved. Second, conversion of the *tps* promoter sequence to that of the corresponding region of the *ops* gene (+49 to −27) by introducing five nucleotide substitutions and a nucleotide deletion in *tps* DNA did not result in a gene that was expressed in an *ops* temporal pattern as would be expected if the appearance of an altered form of RNA polymerase were the only requirement (besides *ops* UAS binding) for late expression (Kim and Downard, unpublished data). Finally, site-specific alterations in the *tps* and *ops* promoter regions have suggested that both genes may be subject to negative as well as positive regulation (Downard et al., 1988).

Analysis of other developmental genes has indicated that large regulatory regions may be a general feature of developmentally regulated genes. Of particular interest in this respect is the *csgA* gene that encodes C factor (see above). Analysis of *csgA* has shown that a segment of DNA located between 729 and 930 bp upstream from the transcription initiation site is required for full developmental expression (Li et al., 1992). This analysis of *csgA* deletions has also suggested that important regulatory sites are found in the −676 to −729, −400 to −676, −336 to −400, and −266 to −336 upstream DNA segments of the gene. The large size of the *csgA* regulatory region is intriguing in light of the fact that careful control of the amount of the CsgA signal produced in cells seems to be an important feature in coordination of the timing of developmental aggregation and sporulation. The expression of the myxobacterial hemagglutinin gene, *mbhA*, has also been investigated. This gene is expressed from a σ^{54}-like promoter, and an upstream DNA sequence located between −150 and −307 is required for expression as well (Romeo and Zusman, 1991).

The regulation of two developmentally regulated Tn*5lac* insertions has been examined by using the site-specific integration system of myxophage Mx8 (see chapter 9). In each case, various segments of *M. xanthus* DNA upstream of the Tn*5lac* insertion site were tested for the ability to direct developmental β-galactosidase expression when fused to *lacZ* in a plasmid integrated in single copy at the Mx8 phage attachment site in the *M. xanthus* chromosome. In the case of Ω4435, 11- and 1.3-kbp segments of upstream *M. xanthus* DNA directed *lacZ* expression with similar timing but at half the level as observed for the strain containing Tn*5lac* Ω4435 (Li and Shimkets, 1988). One possible explanation for these results is that the 1.3-kbp segment contains all the regulatory elements required for proper developmental expression, but one of the regulatory elements requires a positive-acting factor that is present in a limiting amount. Because the strain contains two copies of the regulatory region (one at the phage attachment site and one at the native site) and only one of these is fused to *lacZ*, the putative positive-acting factor would partition equally between the two copies and the strain would exhibit half as much *lacZ* expression as the strain containing Tn*5lac* Ω4435. In the case of Ω4403,

8.5-, 4.0-, 2.0-, and 1.0-kbp segments of upstream *M. xanthus* DNA directed *lacZ* expression with timing similar to that observed for the strain containing Tn*5lac* Ω4403 but at only 20% the level (Semancik, 1991). Titration of a limiting positive-acting factor does not provide an adequate explanation for these results, especially since a strain containing the 8.5-kbp segment fused to *lacZ* and integrated by homologous recombination directed developmental *lacZ* expression at 80% the level observed for the strain containing Tn*5lac* Ω4403. Instead, these results suggest that DNA beyond 8.5 kbp upstream of the Tn*5lac* Ω4403 insertion site is required for full developmental expression from a transcriptional start site that, according to the results of S1 nuclease mapping, lies about 380 bp upstream of the insertion site. Alternatively, since both Tn*5lac* Ω4435 and Tn*5lac* Ω4403 are expressed late in development and depend absolutely on C signaling for their expression (Kroos and Kaiser, 1987), perhaps genes in this class cannot be fully expressed during development when they are present at the phage attachment site in the chromosome. The upstream region of at least one Tn*5lac* insertion, Ω4499, which is expressed earlier and depends only partially on C signaling for its expression, is capable of directing full developmental expression from the phage attachment site (Fisseha and Kroos, unpublished data). Thus, further studies will be needed to determine whether the genes identified by these Tn*5 lac* insertions have large regulatory regions, as do the other developmentally regulated *M. xanthus* genes examined so far.

The most obvious explanation for the existence of large regulatory regions for developmental genes is that they are controlled, at least in part, by UAS-binding proteins that can act at a distance from the regulated promoters. This mechanism for developmental gene regulation has certain advantages. First, by having UAS elements that are recognized by different transcriptional regulators, a gene may become responsive to multiple signal transduction systems. In this way, the input from various sensory systems may be integrated at the level of expression of particular genes to achieve an optimal level of the protein product. The *csgA* regulatory region appears to be an excellent example of this strategy, since it responds to the levels of nutrient, peptidoglycan components, B signal, and C signal (Li et al., 1992). Second, since upstream activation places fewer constraints on the location of regulatory sites than other mechanisms of gene regulation, it is easier to envision the evolution of a gene subject to complex control from a primordial gene found in an ancestral organism. Third, the level of gene expression may be controlled by the number of regulatory sites available for the binding of positive regulators. This appears to be the explanation for the increased level of transcription from the *tps* promoter when the number of *tps* or *ops* UAS segments is increased from one to two or three (Table 1; Kil et al., 1990). In any event, it will be interesting to determine whether the large regulatory regions of these developmental genes are truly a reflection of the complexity of developmental gene regulation.

IDENTIFICATION OF TRANSCRIPTIONAL REGULATORY PROTEINS AND GENES

While relatively little is known about the transcriptional regulatory proteins involved in the expression of specific developmental genes, it appears that more-rapid progress in this area will be possible in the near future. Several approaches that should prove fruitful in these studies are available. First, it may be possible to utilize similarity between previously identified transcriptional regulatory genes from other bacteria and their *M. xanthus* homologs to clone the *M. xanthus* genes. Null alleles of these genes can then be created to examine the role of the genes in development. This approach has been employed successfully to identify *sigB,* a gene that encodes a σ factor that appears to be involved in developmental transcription (Apelian and Inouye, 1990). However, the target genes whose expression is directly controlled by this σ factor have not yet been identified.

Another approach that should be useful is examination of genes that are part of developmentally regulated operons for roles in transcriptional regulation. In some cases, regulatory genes have been found within the operons they control. For example, it appears that the *frzCD* and *frzE* proteins act as transcriptional activators to control the expression of the *frzABCDE* operon, which encodes these proteins (Weinberg and Zusman, 1989). The mechanisms of action of these gene products have not been established, although potential roles have been suggested from DNA sequence analysis. Interestingly, the *frzCD* gene product may be a σ factor, since a portion of the protein shows 32% identity to the *B. subtilis* σ^{43}.

Finally, with the demonstration of DNA-binding activities specific for the regulatory regions of developmental genes in crude cell extracts (Kil et al., 1990), it should be possible to purify the proteins responsible for these activities and test them for regulatory functions. An important part of these studies will be the use of purified proteins to clone the cognate genes (reverse genetics), which can then be mutated and introduced into *M. xanthus* as gene replacements (Romeo and Zusman, 1987). This type of genetic analysis will lead to a better understanding of the roles of specific regulatory proteins in development. The value of this approach is that the regulatory protein(s) that directly controls the expression of a particular gene may be identified. Other approaches, such as the isolation of mutants defective in the expression of a developmental gene, will frequently identify genes that are only indirectly involved in control of the gene of interest.

IN VITRO STUDIES OF TRANSCRIPTION

Studies of transcription in vitro have so far focused on characterizing RNA polymerase isolated from growing *M. xanthus* cells. Rudd and Zusman (1982) purified *M. xanthus* RNA polymerase more than 300-fold and found that the enzyme could be fractionated into core and holoenzyme fractions by using single-stranded DNA agarose chromatography. The core enzyme showed a subunit structure similar to that of *E. coli* core RNA polymerase, and the holoenzyme showed, in addition, two prominent polypeptides that could be σ factors. These two polypeptides displayed similar peptide maps. Subsequently, it was shown that antibodies raised against *B. subtilis* σ^{43} (the major σ factor in growing cells) recognize two polypeptides with similar sizes as the putative σ factors in a whole-cell extract of growing *M. xanthus* (Inouye, 1990). It was suggested that the two polypeptides are products of the *M. xanthus sigA* gene, with the smaller polypeptide possibly being a breakdown product of the larger polypeptide or resulting from translation initiating at a downstream, in-phase initiation codon. The high degree of similarity between *E. coli rpoD* (encoding σ^{70}), *B. subtilis sigA* (encoding σ^{43}), and *M. xanthus sigA* in the regions thought to be involved in promoter recognition (i.e., regions 2.4 and 4.2; Helman and Chamberlin, 1988) predicts that holoenzymes containing these σ factors should transcribe similar promoters. Indeed, the holoenzyme fraction isolated by Rudd and Zusman (1982) produced transcripts similar to that of *E. coli* σ^{70} RNA polymerase when *E. coli, B. subtilis,* or *M. xanthus* phage DNA served as the template for in vitro transcription.

By following the procedure developed by Rudd and Zusman (1982) through the double-stranded DNA cellulose chromatography step, Semancik (1991) identified fractions eluting at slightly different salt concentrations that had different transcriptional properties in vitro. The fractions in relatively low salt concentrations transcribed a nonspecific template [poly(dA–dT)] strongly but transcribed the *B. subtilis veg* (Moran et al., 1982) or *E. coli lacUV5* (Silverstone et al., 1970) promoter (both of which conform strongly in the -10 and -35 regions to the *E. coli* σ^{70} consensus; Rosenberg and Court, 1979) weakly when a run-off transcription assay was used. Fractions eluting at slightly higher salt concentrations transcribed the nonspecific template weakly and the specific promoters strongly. These results suggested that the low-salt fractions contain core RNA polymerase and the high-salt fractions contain holoenzyme. The putative holoenzyme failed to transcribe the *M. xanthus vegA* promoter (Komano et al., 1987), which conforms strongly in the −35 region but only weakly in the −10 region to the *E. coli* σ^{70} consensus. Western blotting (immunoblotting) with antibodies raised against *B. subtilis* σ^{43} revealed two bands in the putative holoenzyme fraction that were not detected in the putative core fraction, one relatively faint that comigrated with *E. coli* σ^{70} and the other relatively intense that migrated at about 50 kDa. Inouye (1990) had observed bands at these positions plus one that would correspond to a polypeptide slightly larger than *E. coli* σ^{70} in a Western blot of a whole-cell extract of growing *M. xanthus* probed with anti-α^{43} antibodies. After elution from a sodium dodecyl sulfate-polyacrylamide gel and renaturation (Hager and Burgess, 1980), a 50-kDa polypeptide from the high-salt fraction could stimulate the putative core fraction to transcribe the *B. subtilis veg* promoter. No polypeptides from the 70-kDa region of the gel exhibited this activity. These results suggest that the 50-kDa protein is a σ factor. The 50-kDa protein could be a breakdown product of the *sigA* gene product(s) or the product of another gene.

Characterization of RNA polymerase from growing *M. xanthus* cells has laid the foundation for attempts at reconstructing transcription of developmentally regulated genes in vitro. This task will probably not be easy, especially since the developmentally regulated genes analyzed so far appear to have large regulatory regions. It may be necessary to first purify proteins that bind to upstream elements and then add these DNA-binding proteins to partially purified fractions of RNA polymerase from developing cells in order to reconstruct developmental gene transcription in vitro.

SUMMARY AND PROSPECTS

To date, much of the attention on *M. xanthus* development has focused on the identification of developmentally regulated genes and the nature of cell-cell interactions. From these studies, it has become clear that regulation of this "simple" multicellular form of development is a complicated matter involving a number of intracellular and extracellular signaling systems. It is likely that the targets of many and perhaps most of the signaling systems involved are the regulation of many developmentally expressed genes. In order to understand more fully the dynamic aspects of this multicellular process, it will be necessary to identify the specific transcriptional regulatory proteins that are involved and their mechanisms of action. The association of specific signaling events with particular gene activation patterns will be a serious challenge, but considering the experimental advantages of the *M. xanthus* system, the prospects are exciting.

REFERENCES

Albright, L. M., E. Huala, and F. M. Ausubel. 1989. Prokaryotic signal transduction mediated by sensor and regulator protein pairs. *Annu. Rev. Genet.* **23:**311–336.

Apelian, D., and S. Inouye. 1990. Development-specific σ-factor essential for late-stage differentiation of *Myxococcus xanthus. Genes Dev.* **4:**1396–1403.

Atchison, M. L. 1988. Enhancers: mechanisms of action and cell specificity. *Annu. Rev. Cell Biol.* **4:**127–153.

Brown, G. L. 1990. Ph.D. dissertation. University of Oklahoma, Norman.

Cumsky, M. G., and D. R. Zusman. 1979. Myxobacterial hemagglutinin: a development-specific lectin of *Myxococcus xanthus. Proc. Natl. Acad. Sci. USA* **76:**5505–5509.

Cumsky, M. G., and D. R. Zusman. 1981. Purification and characterization of myxobacterial hemagglutinin, a development-specific lectin of *Myxococcus xanthus. J. Biol. Chem.* **256:** 12581–12588.

Downard, J. S. 1987. Identification of the RNA products of the *ops* gene of *Myxococcus xanthus* and mapping of the *ops* and *tps* RNAs. *J. Bacteriol.* **169:**1522–1528.

Downard, J. S. 1988. Tn*5*-mediated transposition of plasmid DNA after transduction to *Myxococcus xanthus. J. Bacteriol.* **170:**4939–4941.

Downard, J. S., S.-H. Kim, and K.-S. Kil. 1988. Localization of the *cis*-acting regulatory DNA sequences for the *Myxococcus xanthus tps* and *ops* genes. *J. Bacteriol.* **170:** 4931–4938.

Downard, J. S., D. Kupfer, and D. R. Zusman. 1984. Gene expression during development of *Myxococcus xanthus:* analysis of the genes for protein S. *J. Mol. Biol.* **175:**469–492.

Downard, J. S., and D. R. Zusman. 1985. Differential expression of protein S genes during *Myxococcus xanthus* development. *J. Bacteriol.* **161:**1146–1155.

Fisseha, M., and L. Kroos. Unpublished data.

Hager, D. A., and R. R. Burgess. 1980. Elution of proteins from sodium dodecyl sulfate-polyacrylamide gels, removal of sodium dodecyl sulfate, and renaturation of enzymatic activity: results with sigma subunit of *Escherichia coli* RNA polymerase, wheat germ DNA topoisomerase, and other enzymes. *Anal. Biochem.* **109:**76–86.

Helman, J. D., and M. J. Chamberlin. 1988. Structure and function of bacterial sigma factors. *Annu. Rev. Biochem.* **57:**839–872.

Inouye, M., S. Inouye, and D. R. Zusman. 1979a. Gene expression during development of *Myxococcus xanthus:* pattern of protein synthesis. *Dev. Biol.* **68:**579–591.

Inouye, M., S. Inouye, and D. R. Zusman. 1979b. Biosynthesis and self-assembly of protein S, a development-specific protein of *Myxococcus xanthus. Proc. Natl. Acad. Sci. USA* **76:**209–213.

Inouye, S. 1984. Identification of a development-specific promoter of *Myxococcus xanthus. J. Mol. Biol.* **174:**113–120.

Inouye, S. 1990. Cloning and DNA sequence of the gene coding for the major sigma factor from *Myxococcus xanthus. J. Bacteriol.* **172:**80–85.

Inouye, S., T. Franceschini, and M. Inouye. 1983a. Structural similarities between the development-specific protein S from a Gram-negative bacterium, *Myxococcus xanthus,* and calmodulin. *Proc. Natl. Acad. Sci. USA* **80:**6829–6833.

Inouye, S., W. Harada, D. Zusman, and M. Inouye. 1981. Development-specific protein S of *Myxococcus xanthus:* purification and characterization. *J. Bacteriol.* **148:**678–683.

Inouye, S., Y. Ike, and M. Inouye. 1983b. Tandem repeat of the genes for protein S, a development-specific protein of *Myxococcus xanthus. J. Biol. Chem.* **258:**38–40.

Kil, K.-S. 1991. Ph.D. dissertation. University of Oklahoma, Norman.

Kil, K.-S., G. L. Brown, and J. S. Downard. 1990. A segment of *Myxococcus xanthus ops*

DNA functions as an upstream activation site for *tps* gene transcription. *J. Bacteriol.* **172:** 3081–3088.

Kim, S.-H., and J. Downard. Unpublished data.

Kim, S. K., and D. Kaiser. 1990a. Cell motility is required for the transmission of C-factor, an intercellular signal that coordinates fruiting body morphogenesis of *Myxococcus xanthus*. *Genes Dev.* **4:**896–905.

Kim, S. K., and D. Kaiser. 1990b. Cell alignment required in differentiation of *Myxococcus xanthus*. *Science* **249:**926–928.

Kim, S. K., and D. Kaiser. 1991. C-factor has distinct aggregation and sporulation thresholds during *Myxococcus* development. *J. Bacteriol.* **173:**1722–1728.

Komano, T., T. Franceschini, and S. Inouye. 1987. Identification of a vegetative promoter in *Myxococcus xanthus:* a protein that has homology to histones. *J. Mol. Biol.* **196:**517–524.

Komano, T., T. Furuichi, M. Teintze, M. Inouye, and M. Inouye. 1984. Effect of deletion of the gene for development-specific protein S on differentiation in *Myxococcus xanthus*. *J. Bacteriol.* **158:**1195–1197.

Kroos, L. 1986. Ph.D. thesis. Stanford University, Stanford, Calif.

Kroos, L., and D. Kaiser. 1984. Construction of Tn*5lac*, a transposon that fuses *lacZ* expression to exogenous promoters, and its introduction into *Myxococcus xanthus*. *Proc. Natl. Acad. Sci. USA* **81:**5816–5820.

Kroos, L., and D. Kaiser. 1987. Expression of many developmentally regulated genes in *Myxococcus* depends on a sequence of cell interactions. *Genes Dev.* **1:**840–854.

Kroos, L., A. Kuspa, and D. Kaiser. 1986. A global analysis of developmentally regulated genes in *Myxococcus xanthus*. *Dev. Biol.* **117:**252–266.

Kuspa, A., and D. Kaiser. 1989. Genes required for developmental signaling in *Myxococcus xanthus:* three *asg* loci. *J. Bacteriol.* **171:**2762–2772.

Kuwabara, M. D., and D. S. Sigman. 1987. Footprinting DNA-protein complexes *in situ* following gel retardation assays using 1,10-phenanthroline-copper ion: *Escherichia coli* RNA polymerase-*lac* promoter complexes. *Biochemistry* **26:**7234–7238.

Li, S., B. Lee, and L. Shimkets. 1992. *csgA* expression entrains *Myxococcus xanthus* development. *Genes Dev.* **6:**401–410.

Li, S., and L. J. Shimkets. 1988. Site-specific integration and expression of a developmental promoter in *Myxococcus xanthus*. *J. Bacteriol.* **170:**5552–5556.

Moran, C. P., Jr., N. Lang, S. F. J. LeGrice, G. Lee, M. Stephens, A. L. Sonenshein, J. Pero, and R. Losick. 1982. Nucleotide sequences that signal the initiation of transcription and translation in *Bacillus subtilis*. *Mol. Gen. Genet.* **186:**339–346.

Munoz-Dorado, J., S. Inouye, and M. Inouye. 1991. A gene encoding a protein serine/threonine kinase is required for normal development of *M. xanthus*, a Gram-negative bacterium. *Cell* **67:**995–1006.

Nelson, D. R., M. G. Cumsky, and D. R. Zusman. 1981. Localization of myxobacterial hemagglutinin in the periplasmic space and on the cell surface of *Myxococcus xanthus* during developmental aggregation. *J. Biol. Chem.* **256:**12589–12595.

Nelson, D. R., and D. R. Zusman. 1983. Transport and localization of protein S, a spore coat protein, during fruiting body formation by *Myxococcus xanthus*. *J. Bacteriol.* **154:** 547–553.

O'Connor, K., and D. R. Zusman. 1983. Coliphage P1-mediated transduction of cloned DNA from *Escherichia coli* to *Myxococcus xanthus:* use for complementation and recombinational analyses. *J. Bacteriol.* **155:**317–329.

Romeo, J. M., B. Esmon, and D. R. Zusman. 1986. Nucleotide sequence of the myxobacterial hemagglutinin gene contains four homologous domains. *Proc. Natl. Acad. Sci. USA* **83:** 6332–6336.

Romeo, J. M., and D. R. Zusman. 1987. Cloning of the gene for myxobacterial hemagglutinin and isolation and analysis of structural gene mutations. *J. Bacteriol.* **169:**3801–3808.

Romeo, J. M., and D. R. Zusman. 1991. Transcription of the myxobacterial hemagglutinin gene is mediated by a σ^{54}-like promoter and a *cis*-acting upstream regulatory region of DNA. *J. Bacteriol.* **173:**2969–2976.

Romeo, J. M., and D. R. Zusman. 1992. Determinants of an unusually stable mRNA in the bacterium *Myxococcus xanthus. Mol. Microbiol.* **6:**2975–2988.

Rosenberg, M., and D. Court. 1979. Regulatory sequences involved in the promotion and termination of transcription. *Annu. Rev. Genet.* **13:**319–353.

Rudd, K. E., and D. R. Zusman. 1982. RNA polymerase of *Myxococcus xanthus:* purification and selective transcription in vitro with bacteriophage templates. *J. Bacteriol.* **151:**89–105.

Semancik, M. 1991. M.S. thesis. Michigan State University, East Lansing.

Silverstone, A., R. Arditi, and B. Magasanik. 1970. Catabolite-insensitive repression of *lac* promoter mutants. *Proc. Natl. Acad. Sci. USA* **66:**773–779.

Struhl, K. 1989. Molecular mechanisms of transcriptional regulation in yeast. *Annu. Rev. Biochem.* **58:**1051–1077.

Teintze, M., T. Furuichi, R. Thomas, M. Inouye, and S. Inouye. 1985a. Differential expression of two homologous genes coding for spore-specific proteins in *Myxococcus xanthus,* p. 253–260. *In* J. Hoch and P. Setlow (ed.), *Molecular Biology of Microbial Differentiation.* American Society for Microbiology, Washington, D.C.

Teintze, M., R. Thomas, T. Furuichi, M. Inouye, and S. Inouye. 1985b. Two homologous genes coding for spore-specific proteins are expressed at different times during development of *Myxococcus xanthus. J. Bacteriol.* **163:**121–125.

Weinberg, R. A., and D. R. Zusman. 1989. Evidence that the *Myxococcus xanthus frz* genes are developmentally regulated. *J. Bacteriol.* **171:**6174–6186.

Myxobacteria II
Edited by Martin Dworkin and Dale Kaiser
© 1993 American Society for Microbiology, Washington, DC 20005

Chapter 9

Development-Specific Gene Expression: Protein Serine/Threonine Kinases and Sigma Factors

Sumiko Inouye and Masayori Inouye

Many genes are specifically expressed during differentiation of *Myxococcus xanthus*. The first such genes to be investigated were the genes for protein S and protein S1. The expression of the former gene, designated *tps*, is dramatically induced immediately after cells start to differentiate. The latter gene, designated *ops*, is expressed in a much later stage of development (Teintze et al., 1985; Downard and Zusman, 1985). Many hundreds of such development-specific genes are probably associated with differentiation and morphogenesis of *M. xanthus*. Indeed, one of the most intriguing questions in developmental biology is how the stage-specific expression of developmental genes is regulated and what signals are responsible for switching these genes on and off.

To answer these questions, we have taken two approaches. One approach is to explore the possibility that *M. xanthus* may contain eukaryotic-like protein serine/threonine or tyrosine kinases known to play essential roles in development in eukaryotic cells. The other is to search for development-specific sigma factors, which are required for transcription initiation by RNA polymerase. To our surprise, *M. xanthus* was found to contain a large family of serine/threonine kinases (Munoz-Dorado et al., 1991; Zhang et al., 1992). In addition, *M. xanthus* contains a family of sigma factors, two of which have been shown to be development specific (Apelian and Inouye, 1990, 1993).

Sumiko Inouye and Masayori Inouye • Department of Biochemistry, Robert Wood Johnson Medical School, University of Medicine and Dentistry of New Jersey, Piscataway, New Jersey 08854.

In this chapter, we briefly describe our approach to demonstrating the existence of development-specific protein kinases and sigma factors and discuss their roles in the development of *M. xanthus*.

EUKARYOTIC-LIKE PROTEIN KINASES

The developmental cycle and multicellular morphogenesis of *M. xanthus* resemble those of eukaryotic slime molds such as *Dictyostelium discoideum*. *D. discoideum* is known to possess a signal transduction system consisting of a receptor, a G protein, an effector, and a multigene family of protein serine/threonine kinases that are developmentally regulated and are considered to play important roles in development (Kaiser, 1986; Kessin, 1988; Devreotes, 1989). These protein kinases belong to a large family of protein serine/threonine kinases widely found in eukaryotic cells (Edelman et al., 1987).

Prokaryotes, on the other hand, possess various signal transduction systems required for adaptation to numerous environmental signals that are known to consist of protein histidine kinases (Stock et al., 1989). Signal transduction through protein serine/threonine or tyrosine kinases has been thought to be specific to eukaryotic cells (Edelman et al., 1987; Hunter and Cooper, 1985) and has not been shown to exist in the prokaryotes until recently (Munoz-Dorado et al., 1991). Because of the resemblance of the *M. xanthus* life cycle to that of *D. discoideum,* we speculated that signal transduction systems similar to those in eukaryotic cells might play an important role during *M. xanthus* development (Munoz-Dorado et al., 1991). The catalytic domain of each eukaryotic protein kinase consists of 11 subdomains, some of which contain highly conserved consensus amino acid sequences (Hanks and Quinn, 1991). Oligonucleotides designed on the basis of consensus sequences in subdomains VI and VIII of protein serine/threonine kinases were used as primers for polymerase chain reactions (PCRs), with the *M. xanthus* chromosomal DNA as a template (Munoz-Dorado et al., 1991). Gel analysis of the PCR products revealed a major band migrating at approximately 175 bp. After the band had been cloned into a plasmid, its DNA sequence was determined. Three different putative amino acid sequences were identified on the basis of the amino acid sequences deduced from DNA sequences of cloned DNA fragments. In all of the sequences, subdomain VII containing the highly conserved amino acid sequence Asp-Phe-Gly was flanked by the amino acid sequences of subdomains VI and VIII in the same reading frame.

This result suggested that *M. xanthus* contains at least three eukaryotic-like protein kinases. The first *M. xanthus* protein kinase gene to be determined was *pkn1* (Munoz-Dorado et al., 1991). The amino acid sequence deduced from the DNA sequence consists of 693 residues, and its amino-terminal half contains all the consensus sequences found in the catalytic domains of eukaryotic protein serine/threonine kinases. Ca^{2+}-calmodulin-dependent protein kinase II (CAM-kinase II) is known to have a catalytic domain at the amino-terminal region and a regulatory domain at the carboxy-terminal region. When the amino acid sequence from residues 57 to 264 in Pkn1 is compared with the catalytic domain of rat CAM-kinase II (Tobimatsu and Fujisawa, 1989) as shown in Fig. 1, the identity between them is 31%. Identity in the same region is 27% between Pkn1 and rabbit protein kinase Cα (Ohno et al., 1987), 28% between Pkn1 and yeast cyclic AMP-dependent protein kinase (Toda et al., 1987), and 30% between Pkn1 and human cyclic GMP-dependent protein kinase (Sandberg et al., 1989). The carboxy-terminal domain of CAM-kinase II is involved in regulation, subunit assembly, and subcellular localization (Soderling, 1990). In addition, this kinase can recognize the sequence RXXS/T and phosphorylate the serine or threonine residue within the kinase sequence (Colbran et al., 1988).

Phosphorylated amino acid residues in Pkn1 were determined after *pkn1* was expressed by using the T7 RNA polymerase system (Munoz-Dorado et al., 1991). Serine and threonine residues but not tyrosine residues were equally phosphorylated. This autophosphorylation reaction was blocked when the lysine residue at position 88 in Pkn1 was replaced with an asparagine residue. The sequence recognized as the phosphorylation site by CAM-kinase II (RXXS/T) exists at three locations in Pkn1: RQRS (residues 517 to 520), RLRT (residues 540 to 543), and RARS (residues 689 to 692) (Munoz-Dorado et al., 1991). These similarities in sequence and domain structure between Pkn1 and CAM-kinase II are of particular interest in light of the facts that the major Ca^{2+}-binding proteins (protein S and S1; Inouye et al., 1983) are induced during myxospore formation and that these proteins resemble calmodulin (Inouye et al., 1983).

β-Galactosidase activity of *M. xanthus* cells harboring a *pkn1–lacZ* fusion gene showed that the *pkn1* gene was expressed during development but not during vegetative growth. When these *M. xanthus* cells carrying the *pkn1–lacZ* gene were placed on a low-nutrient agar (clone fruiting agar) plate with 5-bromo-4-chloro-3-indolyl-β-D-galactopyranoside (X-Gal) as a *lacZ* indicator at 30°C, a blue color was evident after approximately 30 h of incubation, a time at which fruiting bodies were well formed. Moreover, *pkn1* expression began before sporulation and continued inside myxospores (Munoz-Dorado et al., 1991).

Δ*pkn1* deletion mutants were constructed to investigate the function of *pkn1* during development. These mutant cells grew normally in rich

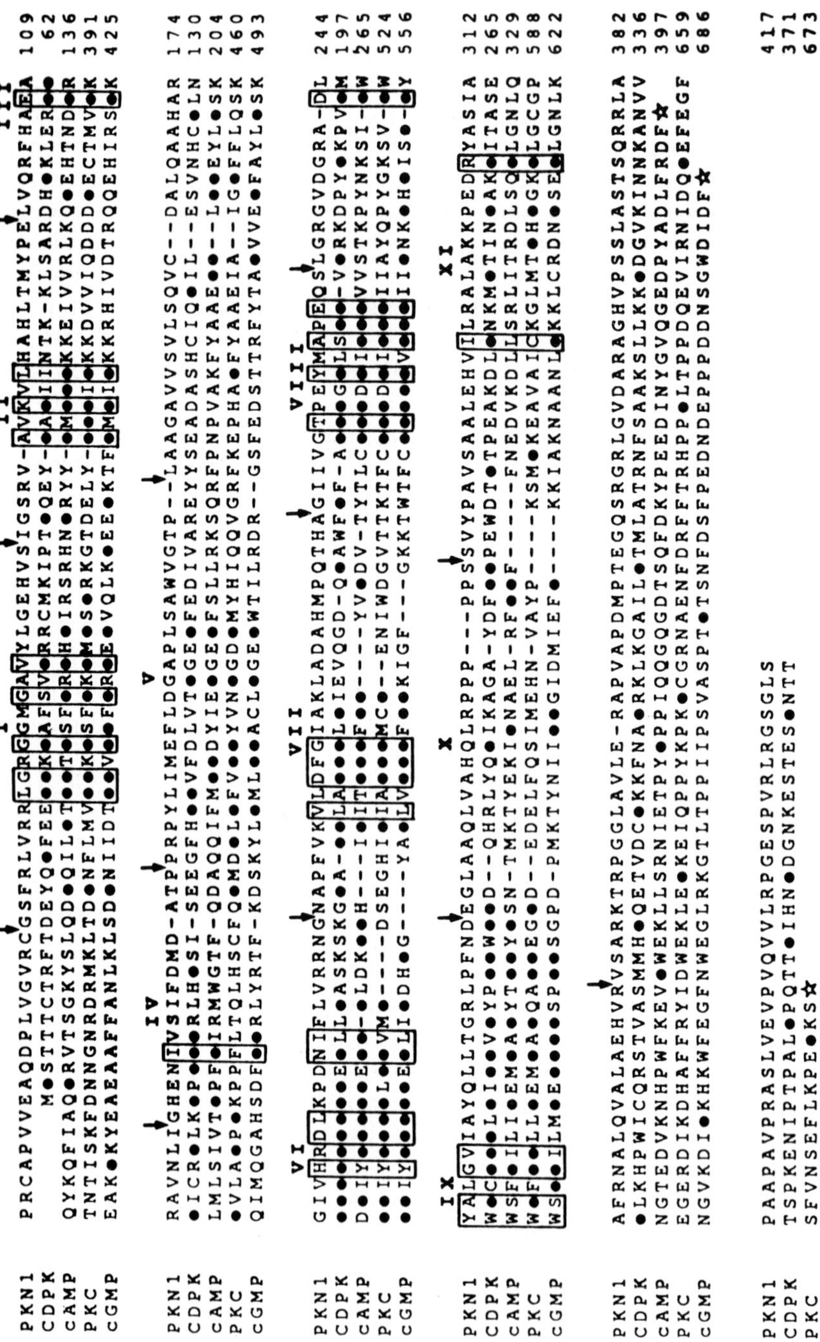

media but were defective in fruiting body formation (Munoz-Dorado et al., 1991). Their fruiting bodies were smaller than and not as compact as those of the wild-type strain. Spore formation in the Δ*pkn1* strains started earlier and reached a plateau much sooner than that of the wild-type strain. As a result, the spore yields of the Δ*pkn1* strains were approximately 35% of that of the wild-type strain. These results indicate that *pkn1* is required for the normal development of *M. xanthus* and that it is probably involved in sensing a signal that initiates fruiting body formation.

During the investigation of the *pkn1* gene with a PCR product (PCRPK1) to probe Southern blots of a *Xho*I digest of the chromosomal DNA, another band in addition to the *pkn1* band hybridized with the probe (Munoz-Dorado et al., 1991). Hybridization under less stringent conditions with the same probe (Zhang et al., 1992) revealed 12 additional bands with different intensities. During the course of the experiments identifying PCRPK1 from the PCR products described above, two other DNA fragments consisting of 173 and 176 bp were obtained. These sequences included well-conserved domains VI, VII, and VIII and thus represented other *pkn* genes. These three PCR products (PCRPK1, PCRPK2, and PCRPK3) shared the 16-base sequence $^{5'}$TGGACTTCGGCATCGC$^{3'}$ within subdomain VII. With this sequence as a probe, at least 23 bands were detected in the *Xho*I digest of the chromosomal DNA, and 12 of these hybridized to PCRPK1 (Zhang et al., 1992). DNA fragments corresponding to these bands were identified with an *M. xanthus* genomic λ library by using a PCRPK1 probe. Partial DNA sequences that include the catalytic domains of some of these putative protein kinases have been obtained. Amino acid sequences of the region encompassing domains VI, VII, and VIII were deduced from the DNA sequences of seven independent clones and compared with that of Pkn1 in Fig. 2 Eight independent and different protein serine/threonine kinases are represented in Fig. 2.

Application of DNA fragments of *pkn2, -3, -5, -6, -11, -13,* and *-15* as probes in Southern blot hybridization implies a total of at least 26 *pkn* genes in *M. xanthus* (Zhang et al., 1992).

Figure 1. Sequence alignment in the region between domains I and XI of Pkn1 and other eukaryotic protein kinases. CDPK, rat CAM-kinase II (Tobimatsu and Fujisawa, 1989); cAMP, yeast cyclic AMP-dependent protein kinase (Toda et al., 1987); PKC, rabbit protein kinase Cα (Ohno et al., 1987); cGMP, human cyclic GMP-dependent protein kinase (Sandberg et al., 1989). Amino acid residues identical to amino acids in Pkn1 are represented by solid circles. Bars represent gaps. The assignment of subdomains was defined by Hanks and Quinn (1991).

```
Consensus in
 eukaryotes    D K  N                               D G                                   PE
Pkn 1       IVHRDLKPDNIFLVRRNG-----NAPFVKVLDFGIAKLADAHMPQTH------AGIIVGTPEYMAPEQS
Pkn 2       ●I●●●●●●E●V●ISKSARG------EQARL●●●●●●●R●VEPDAASSVSQ----I●VVL●●●●●LS●●●A
Pkn 3       VI●●●●●●E●●MVEP●RNEP-----D●●●●●●●●●●●●IT●STDDGPALTR---●●FVC●●●●●●S●●●A
Pkn 5       L●●●●●●●S●●MVDDDR---------Q●RLM●●●L●●FLADDAAI●E------●●KL●●●YR●●●●●●I
Pkn 6       LI●●●VS●●●●LVS●QG---------A●●●V●●●●●●●V●GQGHRTL-------T●VVK●KVA●●P●●●L
Pkn 11      L●●●●VT●H●VL●SFDG---------A●●LT●●●●●●AGNKLTQ---------P●VLK●KFA●●S●●●A
Pkn 13      VI●●●VAKE●●MVTYEG---------VT●L●●●●●●●●SLARADRTA-------V●MVK●●SG●●S●●●I
Pkn 15      VI●●●●●●E●VMVREDG---------VL●LM●●●●●R●M●IEERM●V------T●TL●●S●AH●●●●II

Subdomain          VIb                           VII                             VIII
```

Figure 2. Sequence comparisons among *M. xanthus* protein kinases. The regions for subdomains VIb, VII, and VIII are shown for Pkn1 (Munoz-Dorado et al., 1991), -2, -3, -5, -6, -11, -13, and -15. Residues identical to those in Pkn1 are represented by solid circles. Subdomain sequences are separated by dashed lines. Highly conserved residues in eukaryotic protein serine/threonine kinases (Hanks and Quinn, 1991) are shown at the top as consensus residues.

SIGMA FACTORS

As pointed out above, many development-specific genes are expressed during differentiation of *M. xanthus*. Different stage-specific patterns are found: some genes are expressed in early stages, while others are expressed late. We believe that most of these genes are regulated at the level of transcription. In prokaryotes, core RNA polymerase consists of β, β′, and 2α subunits. Associated with the core polymerase is a factor called sigma, which is required for promoter recognition. In *Escherichia coli,* there is one major primary sigma factor, sigma-70, and several other sigma factors associated with, for example, heat shock promoters (Lonetto et al., 1992).

Bacillus subtilis, a spore-forming gram-positive bacterium, has several development-specific sigma factors, each of which is responsible for the expression of particular developmental-stage-specific genes (Stragier and Losick, 1990). Induction of the genes for these developmental sigma factors is regulated by a cascade of sigma factors in such a way that the first sigma factor induces the early developmental genes including the second sigma factor, which then induces the genes of the next developmental stage including the third developmental sigma factor, and so on.

In light of this information, one might ask whether *M. xanthus* has a developmental sigma cascade. To start, we isolated the *M. xanthus* gene for the vegetative sigma factor, using the *E. coli* sigma-70 gene as a hybridization probe (Inouye, 1990). The sigma factor thus cloned from *M. xanthus* was designated sigma factor A (coded for by the *sigA* gene). The amino acid sequence of SigA is similar to those of the primary sigma factors of *E. coli* (sigma-70) and *B. subtilis* (sigma-43). SigA has 708 amino acid residues. The carboxy-terminal 242 residues, which are required for promoter recognition and core recognition, show 78 and 72% amino acid identity with sigma-70 and sigma-43, respectively.

To identify other sigma factors in *M. xanthus,* we used a 324-bp fragment from a highly conserved region of *sigA* as a probe to identify genes for other sigma factors. As shown in Fig. 3, we were able to identify several DNA fragments in Southern blot hybridizations of *M. xanthus* chromosomal DNA. For example, the *Xho*I digest of chromosomal DNA yielded four distinct bands at 5.5, 6.3, 7.4, and 8.4 kb in addition to the 2.7-kb *sigA* fragment. We have cloned and characterized two of these *Xho*I fragments, one at 7.4 kb and the other at 8.4 kb (marked with a and b in Fig. 3).

When these fragments were sequenced, they contained open reading frames that could encode proteins having sequences similar to those of the *sigA* sigma factor. As shown in Fig. 4, the 7.4-kb *Xho*I fragment con-

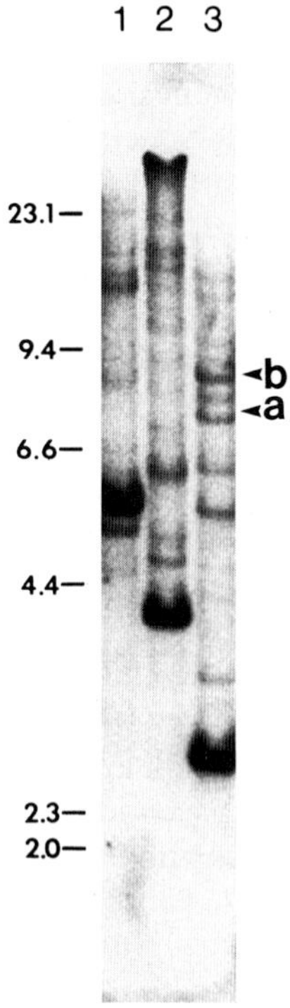

Figure 3. Southern blot analysis of chromosomal DNA from *M. xanthus* DZF1 with a nick-translated *M. xanthus sigA* fragment as a probe. The DNA was digested with *Pst*I (lane 1), *Sal*I (lane 2), and *Xho*I (lane 3). Numbers at left represent molecular sizes in kilobases. a and b indicate the 7.4- and 8.4-kb *Xho*I fragments, respectively.

tains an open reading frame for a polypeptide of 295 amino acid residues (Apelian and Inouye, 1990), and the 8.4-kb *Xho*I fragment also contains an open reading frame encoding a polypeptide of 295 residues (Apelian and Inouye, 1993). These two genes were designated *sigB* and *sigC*, respectively, and there are extensive (78%) identities throughout the entire sequence between the two gene products.

The amino acid sequences of both SigB and SigC are similar to that of the C-terminal region of SigA from residues 475 to 693. Certain structural and functional features of sigma factors (four regions in particular) are conserved (Lonetto et al., 1992; Helmann and Chamberlin, 1988). Region 1 includes subregions 1.1 and 1.2; subregion 1.1 is thought to exist only in primary sigma factors such as SigA. The absence of region 1.1 in both SigB and SigC suggests that B and C may be secondary, regulated sigmas. Within the 26-residue sequence of subregion 1.2 are 22 identities between *E. coli* sigma-70 and SigA, while there are only 11 and 10 identities between SigA and SigB and between SigA and SigC, respectively. In contrast, there are 16 identities between SigB and SigC in subregion 1.2.

```
Mx SigA  DPVRLYLRKMGSVSLLTREGEVEIAKRIEDGEKEVL    232
Mx SigB  GSLAMYLSEINHYSLLKVEEEQELARRFLKGD        43
Mx SigC  DSLSTYLSEINQYPLLTQPQEQELSKRFRAGD        42
                        1.2

Mx SigA  RAERAKSELVEANLRLVVSIAKKYTNRGLQFLDLIQEGNIGLMKAVDKFEYKRGYKFSTYATWWIRQAITRAIADQARTIRIPVHM   551
Mx SigB     LAAGHRMVTANLRFVVKVAYEYRSYGIKMSDLIQEGNIGLMKAVQKFDPDKGIRLISYAVWWIRAYIQNYILKSWSLVKLGTT   126
Mx SigC     LAAGHQLVTANLRFVVKVAYEYRSYGLKMSDLIQEANIGLMKAVQKFDPDKGIRLISYAVWWIRAYIQNCILKNWSLVKLGTT   125
               2.1              2.2              2.3              2.4

Mx SigA  IETINKLIRTSRYLVQE---IGREPT----PEEIAEKMELPLDKVRKV-LKIAKEPISLETPIGEEE------DSHLGDFIEDKSLVSPADAV   630
Mx SigB  QAHGKLFFSLARTRRELEKFGSGDAAV-NVDDIARRLHVKPGEVREMEQRMGGRDLSLDAPMGEDGGNSHVDFVVSAAAPQDDEFADKEEA   216
Mx SigC  QAQRRLFFSLARTRRELEKMGAGDANVVNAEEIARKLNVKASEVREMEQRMGGRDLSLDAPMGEDGDATHLDFVESESVSAVDEVADRQQA   216
                                  3.1                                         3.2

Mx SigA  INMNLAEQTRKVLATLTPREEKVLRMRFGIGEKSDHTLEEVGQDFEVTRERIRQIEAKALRKLRHPSRSKRLRSFVES        708
Mx SigB     GLINARVRTALMRLDPRERFIIEQRVM-NE-RPMTLKELGEHFGFSRERARQLEIRAKDKLKSELAALMAEVDPEAVAAQQ   295
Mx SigC     NLTRELVQRALRRLDPRERFIIEQRVM-GD-AEMTLSELGEHFGFSRERARQLEIRAKDKLKLALVTLMAEAGVDESTLNA   295
                       4.1                      4.2
```

Figure 4. Amino acid sequence alignment of SigA, SigB, and SigC. Subregions are denoted by lines under the aligned sequences. Vertical lines between sequences indicate identical residues. The lines are drawn between SigA and SigB only when all three proteins share an identical residue. Numbers at the right represent residue numbers from the N-terminal ends.

Region 2 consists of four subregions (2.1, 2.2, 2.3, and 2.4), as indicated in Fig. 4, and is the region most conserved in all known sigma factors (Lonetto et al., 1992; Helmann and Chamberlin, 1988). The highest identities are, in fact, in this region in SigA and the two sigma factors: 47% identity between SigA and SigB and 46% identity between SigA and SigC. The identity between SigB and SigC is 93% in this region. Within region 2, subregion 2.1 is thought to be involved in the binding of sigma factors to core RNA polymerase (Lonetto et al., 1992). Subregion 2.2 is the most conserved domain among sigma factors, but its function has not been determined. Subregion 2.3 together with subregion 2.1 has been proposed to participate in DNA strand opening. Subregion 2.4 is thought to be involved in recognition of the −10 region of the promoter.

Region 3 (Fig. 4) is divided into two subdomains. This region is highly conserved among primary sigma factors but not among other sigma factors. The sequence identity between *E. coli* sigma-70 and SigA is 72%, while identities between *E. coli* sigma-70 and SigB and between *E. coli* sigma-70 and SigC are 15 and 14%, respectively. This region is thought to have the helix-turn-helix DNA-binding structure, but its function has not been defined. Region 4 (Fig. 4) is known to participate in recognition of the −35 promoter sequence. Again, this region is highly homologous between SigB and SigC but not between SigA and the two other sigma factors.

The comparisons in Fig. 4 suggest that both SigB and SigC are sigma factors associated with specific regulatable genes. They are highly homologous, particularly in regions 2, 3.1, 4.1, and 4.2 (93, 84, 94, and 93% identities, respectively). In order to examine whether these sigma factors are produced during specific stages of the developmental cycle, we constructed fusions with the *lacZ* gene. From an analysis of β-galactosidase expression, it was found that *sigB* is expressed from the middle to late stages of development, corresponding to the period from the onset of sporulation to late development. *sigC* expression, on the other hand, is induced in a very early stage of development and persists in nonspore cells during development (no *sigC* expression occurs in myxospores).

A deletion mutant of *sigB* displayed normal fruiting body formation and sporulation, but production of protein S1, a late developmental protein produced only in myxospores, was blocked. Furthermore, myxospores from the *sigB* deletion strain were severely defective in stability and viability during late development. These data indicate that SigB is a development-specific sigma factor essential for the maturation of myxospores at a late stage of *M. xanthus* differentiation. A *sigC* deletion mutant displayed abnormal fruiting bodies, indicating that SigC is also a development-specific sigma factor.

PROTEIN KINASE AND SIGMA CASCADES

The discovery of a large family of eukaryotic-like protein scrinc/threonine kinases in *M. xanthus* strongly suggests that this bacterium has signal transduction cascades that resemble those in eukaryotic cells. These cascades probably play important roles in the *M. xanthus* life cycle during development and also during growth. Many intriguing questions remain to be answered. Does *M. xanthus* also contain protein tyrosine kinases that may function as initial sensors for external signals? Does it contain eukaryotic-like protein phosphatases or G proteins? Because of the power of its genetics, *M. xanthus* may be an ideal system for studying signal transduction cascades that use protein kinases.

M. xanthus also may use cascades of sigma factors to regulate expression of stage-specific genes in development. Since *M. xanthus* may have sigma cascades similar to those for spore formation in *B. subtilis* (Stragier and Losick, 1990) and also protein kinase cascades that control cell growth and differentiation in eukaryotic cells, the two cascades may be interdependent (Karin and Smeal, 1992). Some kinase cascades may be targeted to specific sigma factors in either a negative or a positive fashion. At the same time, some developmental sigma factors may be responsible for the expression of some protein kinase genes.

Acknowledgments. We thank Susan Eagle for critical reading of this chapter. This work was supported by grant GM 26843 from the U.S. Public Health Service.

REFERENCES

Apelian, D., and S. Inouye. 1990. Development-specific sigma factor essential for late stage differentiation of *Myxococcus xanthus*. *Genes Dev.* **4:**1396–1403.

Apelian, D., and S. Inouye. 1993. A new putative sigma factor of *Myxococcus xanthus*. *J. Bacteriol.* **175:**3335–3342.

Colbran, R. J., C. M. Schworer, Y. Hashimoto, Y. L. Fong, D. P. Rich, M. K. Smith, and T. R. Soderling. 1988. Calcium/calmodulin-dependent protein kinase II. *Biochem. J.* **258:** 313–325.

Devreotes, P. N. 1989. *Dictyostelium discoideum:* a model system for cell-cell interactions in development. *Science* **245:**1054–1058.

Downard, J. S., and D. R. Zusman. 1985. Differential expression of protein S genes during *Myxococcus xanthus* development. *J. Bacteriol.* **161:**1146–1155.

Edelman, A. M., D. K. Blumenthal, and E. G. Krebs. 1987. Protein serine/threonine kinases. *Annu. Rev. Biochem.* **56:**567–613.

Hanks, S. K., and A. M. Quinn. 1991. Protein kinase catalytic domain sequence database: identification of conserved features of primary structure and classification of family members. *Methods Enzymol.* **200:**38–61.

Helmann, J. D., and M. J. Chamberlin. 1988. Structure and function of bacterial sigma factors. *Annu. Rev. Biochem.* **57:**839–872.

Hunter, T., and J. A. Cooper. 1985. Protein-tyrosine kinases. *Annu. Rev. Biochem.* **54:** 897–930.

Inouye, S. 1990. Cloning and DNA sequence of the gene coding for the major sigma factor of *Myxococcus xanthus*. *J. Bacteriol.* **172:**80–85.

Inouye, S., T. Franceshini, and M. Inouye. 1983. Structural similarities between the development-specific protein S from a Gram-negative bacterium, *Myxococcus xanthus,* and calmodulin. *Proc. Natl. Acad. Sci. USA* **80:**6829–6833.

Kaiser, D. 1986. Control of multicellular development: Dictyostelium and Myxococcus. *Annu. Rev. Genet.* **20:**539–566.

Karin, M., and T. Smeal. 1992. Control of transcription factors by signal transduction pathways: the beginning of the end. *Trends Biochem. Sci.* **17:**418–422.

Kessin, R. H. 1988. Genetics of early *Dictyostelium discoideum* development. *Microbiol. Rev.* **52:**29–49.

Lonetto, M., M. Gribskov, and C. A. Gross. 1992. The σ^{70} family: sequence conservation and evolutionary relationships. *J. Bacteriol.* **174:**3843–3849.

Munoz-Dorado, J., S. Inouye, and M. Inouye. 1991. A gene encoding a protein serine/threonine kinase is required for normal development of *M. xanthus,* a gram-negative bacterium. *Cell* **67:**995–1006.

Ohno, S., H. Kawasaki, S. Imajoh, and H. Suzuki. 1987. Tissue-specific expression of three different types of rabbit protein kinase C. *Nature* (London) **325:**161–166.

Sandberg, M., V. Natarajan, I. Ronander, D. Kalderon, U. Walter, S. M. Lohmann, and T. Jahnsen. 1989. Molecular cloning and predicted full-length amino acid sequence of the type I beta isozyme of cGMP-dependent protein kinase from human placenta. *FEBS Lett.* **255:**321–329.

Soderling, T. R. 1990. Protein kinases, regulation by autoinhibitory domains. *J. Biol. Chem.* **265:**1823–1826.

Stock, J. B., A. J. Ninta, and A. M. Stock. 1989. Protein phosphorylation and regulation of adaptive responses in bacteria. *Microbiol. Rev.* **53:**450–490.

Stragier, P., and R. Losick. 1990. Cascades of sigma factors revisited. *Mol. Microbiol.* **4:** 1801–1806.

Teintze, M., R. Thomas, T. Furuichi, M. Inouye, and S. Inouye. 1985. Two homologous genes coding for spore-specific proteins are expressed at different times during development of *Myxococcus xanthus*. *J. Bacteriol.* **163:**121–125.

Tobimatsu, T., and H. Fujisawa. 1989. Tissue-specific expression of four types of rat calmodulin-dependent protein kinase II mRNAs. *J. Biol. Chem.* **264:**17907–17912.

Toda, T., S. Cameron, P. Sass, M. Zoller, and M. Wigler. 1987. Three different genes in *S. cerevisiae* encode the catalytic subunits of the cAMP-dependent protein kinase. *Cell* **50:** 277–287.

Zhang, W., M. Munoz-Dorado, M. Inouye, and S. Inouye. 1992. Identification of a putative eukaryotic-like protein kinase family in the developmental bacterium *Myxococcus xanthus*. *J. Bacteriol.* **174:**5450–5453.

Myxobacteria II
Edited by Martin Dworkin and Dale Kaiser
© 1993 American Society for Microbiology, Washington, DC 20005

Chapter 10

Developmental Lysis and Autocides

Amy Rosenbluh and Eugene Rosenberg

LYSIS AND AUTOLYSIS

Biologically mediated cell lysis has been observed and studied in a wide variety of organisms and tissues. The phenomenon of cellular lysis can be divided into two broad categories: (i) lysis induced by the members of one species in members of another species and (ii) lysis carried out by a member of one species on itself or on other members of the same species, i.e., autolysis (Saunders, 1966; Kerr et al., 1972; Wyllie et al., 1980; Ellis et al., 1991). It has been known from virtually the beginning of myxobacterial research that the myxobacteria engage in the former type of lysis as a major means of obtaining nutrients from the environment. Myxobacteria produce a plethora of extracellular lytic agents, including lipases, proteases, antibiotics, nucleases, cell wall-degrading enzymes, and bacteriocins. Some of these substances have been shown to be active lysing agents, but the function(s) of the large number of antibiotics produced by myxobacteria (see chapter 16) is not clear. The lysis of a wide variety of microorganisms by myxobacteria has been previously reviewed (Rosenberg and Varon, 1984) and will not here be dealt with further. The major focus of this chapter is the autolytic phenomenon observed in myxobacteria and specifically *Myxococcus xanthus*.

Lysis within a population of a single species relatively independent of external influences constitutes an integral part of the life cycles of multicellular organisms. This phenomenon has been observed in numerous biological systems, from mammalian to amoeboid, and has been termed programmed cell death (Ellis et al., 1991). This term indicates that only specific cells are committed to die as part of a general program of

Amy Rosenbluh • Department of Membrane Research and Biophysics, The Weizmann Institute of Science, 76100 Rehovot, Israel. **Eugene Rosenberg** • Department of Microbiology, Tel Aviv University, 69978 Ramat Aviv, Israel.

development and differentiation. This commitment to die remains reversible to a specific stage in the developmental program of the organism, after which it becomes irreversible. There are several biological systems in which it has been shown that perturbation of the process of cell death results in a major upset in the development and differentiation of the organism. In the cellular slime mold, cell death may be essential to subsequent development. *Dictyostelium discoideum* has a complex life cycle that includes several multicellular stages, the end product of which is the formation of a spore-filling fruiting body (for a review, see Bonner [1967]). It has been shown that under optimal conditions, 20% of the 10^5 cells constituting the fruiting body are committed early in the developmental program to become stalk cells. Cells committed to becoming part of the stalk (prestalk cells) subsequently die, and their membranes harden to become the physical scaffolding upon which the spore-containing sporangiole is supported. The proportion of prestalk cells can be altered in proportioning mutants, and as in more-complicated systems, cell history can affect cell fate (Kessin, 1988; Devreotes, 1989).

The developing forespore of *Bacillus subtilis,* a gram-positive bacterium, matures within the surrounding mother cell. The differentiation into forespore and mother cell is the result of an unequal septation event, which is characterized by subsequent differential gene expression in the two compartments (Losick et al., 1988; Errington, 1991). Essential to the culmination of the sporulation process, i.e., the production of a mature endospore, is release of the forespore from the mother cell by lysis, the mechanism of which is not understood.

In contrast to the examples given above, where disruption of lysis can halt development, mutants of the nematode *Caenorhabditis elegans* in which the programmed cell death process is defective may nevertheless develop normally and display normal behavior, although some more-subtle effects have been observed (Ellis et al., 1991). Understanding of the molecular basis of programmed cell death has increased in recent years with the identification of genes in *C. elegans* necessary for either the execution or the prevention of programmed cell death (Ellis and Horvitz, 1986; Hengartner et al., 1992). The nonessential nature of cell death in this system may reflect the greater complexity of the organism, with the possibility for a variety of redundant systems.

The elucidation of the role and mechanism(s) of programmed cell death has provided an intriguing line of research involving the issues of cell fate and cell commitment. A simple model system displaying differentiation and/or development in the presence of lysis would greatly assist us in approaching the questions posed by the existence of this widespread phenomenon.

DEVELOPMENTAL LYSIS IN *M. XANTHUS*

Autolysis has been observed in growing and stationary-phase cultures of *M. xanthus* under various conditions (Mason and Powelson, 1958; Burchard and Dworkin, 1966; Jones and Barr, 1983; Gonzalez et al., 1992), but this lysis appears to be nonselective. The question of whether selected cells in prokaryotic systems die at the expense of surviving cells first became relevant when the multicellular nature of the myxobacterial life cycle came under investigation. Indeed, developmental lysis in *M. xanthus* was first described in 1975 by Wireman and Dworkin and subsequently became accepted as an integral part of the developmental pathway in this organism, serving as a useful marker for normal development. Initial reports of cell lysis (Wireman and Dworkin, 1975, 1977) were based on the results of two types of assays: (i) determination of the loss of acid-precipitable labeled DNA in developing cultures of cells that had been prelabeled with [^{3}H]thymidine and (ii) determination of cell numbers observed by direct counting of intact cells in a Petroff-Hausser chamber as a function of time of development. This second assay was coupled with the quantitation of mature, viable spores by destroying all vegetative cells and immature spores through heating and sonication and subsequently plating the cells on rich solid medium. Each of these assays yielded results indicating that 80 to 90% of the initial cell population lysed during the first 72 h of development. This result complemented the data obtained for spore numbers; i.e., the yield of mature spores from a developing culture was shown to be 10 to 20%. In addition, it was observed that after a defined period of development on starvation agar had elapsed, a proportion of the developing cells was committed to lyse, even if they were removed from the developmental milieu and transferred to liquid culture in rich growth medium. These cells were shown to constitute a subset of the total cell population in which DNA synthesis had ceased (Wireman and Dworkin, 1975, 1977). What determines whether a given cell will lyse or sporulate remained an open question.

Developmental cell lysis provided a means of explaining various observations made during the development of *M. xanthus*. These observations included the finding that protein S, the protein most abundantly synthesized during sporulation, appeared on the outside of mature spores, although no signal sequence that would target it to the membrane for secretion had been identified (Teintze et al., 1985a). Lysis of a subpopulation of developing cells would provide a means by which this abundant protein could find its way out of the cell, allowing it to assemble onto spores. This hypothesis was tested in several different experiments. Teintze et al. (1985a) assayed wild-type cells for the ability to supply protein

S through a membrane to a sporulating mutant that itself did not produce protein S. They demonstrated that wild-type cells produce diffusible protein S in sufficient concentrations to assemble onto both wild-type sporulating cells and sporulating cells of the protein S-deficient mutant. Lysis would provide a mechanism by which such concentrations could accumulate in the extracellular space in the absence of secretion. In addition to protein S, which is synthesized early in development (12 to 16 h), developing *M. xanthus* cells produce a protein highly homologous to protein S that is synthesized late in development (48 to 52 h) (Inouye et al., 1983; Downard et al., 1984). This protein, S-1, does not assemble onto spores, ostensibly because it is synthesized after developmental lysis has ceased and therefore is trapped within cells that have already completed lysis. In order to test this hypothesis, expression of the structural gene for protein S-1 was placed under control of the promoter of the protein S gene (Teintze et al., 1985b). In such a construct, protein S-1, like protein S, was produced early in development and, also like protein S, assembled onto spores. This suggested that protein S-1 per se is as likely as protein S to assemble onto spores and that only its synthesis late in development prevents this from happening. Developmental cell lysis, occurring as it does within this time window, could again provide a mechanism to explain the observations.

Although an effort to explore alternative modes of active transport of protein S through the cell membrane has been made (Nelson and Zusman, 1983), lysis of a subset of the developing population still appears the most likely avenue of export. In contrast to protein S, protein U, a spore coat protein synthesized late in development, does appear to be actively secreted from the cell. Determination of the amino-terminal amino acid sequence of protein U revealed features characteristic of the prokaryotic signal peptide (Gollop et al., 1991), indicating that targeting of protein S and protein U to the spore outer coat is accomplished by different mechanisms.

Several developmental mutants that had previously been identified as incapable of normal sporulation were subsequently observed to be defective in lysis under developmental conditions (Janssen and Dworkin, 1985; Shimkets and Kaiser, 1982b). Although this correlation was not rigorously tested for all developmental mutants, observation of it was widespread enough for developmental lysis to become a hallmark and an easily assayable feature of normal development.

Janssen and Dworkin (1985) studied the phenomenon of cell-cell interactions and developmental lysis in four groups of signal-defective developmental mutants (Hagen et al., 1978). Each of these groups of mutant strains (groups A, B, C, and D, now termed *asg*, *bsg*, *csg*, and *dsg*, respectively) is defective in sporulation and fruiting body formation. However, when mixtures of two different signaling mutant strains were tested under developmental conditions, they were capable of complementing each other (Hagen et al., 1978). Janssen and Dworkin (1985) found that under developmental conditions, each of these mutant strains lysed at a level at or below 30% compared with >90% in the wild-type control. They showed that mixtures of a C-group strain with a strain from any of the other groups (A, B, or D) resulted in sporulation and developmental lysis of the mixed strains. In addition, the extent of rescue of sporulation was tightly correlated with the extent of rescue of lysis in the strain mixture. The authors did not determine which of the complementation partners lysed and sporulated in each of these successful mixtures. They showed, however, that a cell extract (DS) of the B group could rescue sporulation and lysis in the C-group mutant but not vice versa, suggesting that rescue in the original cell mixture was unidirectional. A carbohydrate component identified in chemical studies of substance DS led Janssen and Dworkin (1985) to test various sugars for rescue of development of the C-group mutant strains. Rescue was effected only by glucosamine (GlcN) and mannosamine (ManN), the former complementing at a lower concentration and more fully than the latter. The possible role of GlcN in the developmental lysis of *M. xanthus* is discussed below.

In a series of experiments studying the sporulation defect of the C-group mutant (Shimkets and Kaiser, 1982a, 1982b), it was shown that peptidoglycan and each of its four components could induce rippling, a coordinated movement observed in developing *M. xanthus* cultures that is absent in starving cultures of the C-group mutants. In addition, peptidoglycan components were shown to rescue sporulation and lysis in this group of developmental mutant strains as well. Lysis of a significant proportion of the cells would provide a means for a developing population to release sufficient concentrations of peptidoglycan as part of the developmental process.

QUANTITATION OF DEVELOPMENTAL LYSIS

In order to quantitate cell number by direct cell counting during development, it was necessary to remove cells from the developmental milieu, fix them in 2 to 5% glutaraldehyde, and sonicate them briefly to disperse aggregates of cells and fruiting bodies (Wireman and Dworkin, 1975, 1977) before enumeration. This procedure became the standard method for determining the number of cells at any given stage of development. However, under certain conditions, fixation of cells may not be sufficient to prevent some level of cell disruption during sonication, and this observation was one of several that led O'Connor and Zusman (1988) to reexamine the extent of autolysis during development of *M. xanthus*.

In the O'Connor and Zusman study (1988), the Petroff-Hausser and labeling assays of Wireman and Dworkin (1975, 1977) were repeated with similar results; i.e., only 10 to 20% of the cells entering development could be accounted for as spores 5 days later. O'Connor and Zusman were uncomfortable with these results, however, because of their observations that (i) glutaraldehyde-fixed cells were damaged during sonication, and the total cell number actually increased as development proceeded beyond 48 h; (ii) the number of spores that represented CFU was always smaller than the number of refractile spores observed in the direct cell count method; and (iii) [^{3}H]thymidine label was quenched by spores that had not been solubilized by the various tissue solubilization cocktails employed in the original study. In an attempt to ameliorate these discrepancies, O'Connor and Zusman attempted to improve upon the methodology employed and thereby came to the conclusion that autolysis occurs at a much lower level than previously reported. This conclusion was based on the following observations.

(i) In an effort to determine the stability of developing cells in harvesting buffers, cells were removed from the developing culture, suspended in several test buffers, and subsequently stored on ice. O'Connor and Zusman found that cell number decreased and the concentration of protein recovered in the test buffer increased over time. They attributed an underestimation of cell populations during development to the existence of a class of highly fragile cells that were destroyed by the classical methods employed for their enumeration. Including 1 mM KH_2PO_4 and/or 10% Ficoll 400 in the harvesting buffer decreased the amount of released protein and increased the final yield of cells obtained at various times during development. When using a harvesting buffer containing both 1 mM KH_2PO_4 and 10% Ficoll 400, O'Connor and Zusman recovered 34 to 36% of the maximum number of cells (as measured at 17 h) after 96 to 144 h of development and 58% after 160 h. They explained this surprising result by the observation that fruiting bodies began to deliquesce by 7 days and could then be more completely dissociated and by the fact that more spores were mature and not disrupted by the means employed to disperse them. Results of this assay suggested that the proportion of lysing cells in a developing culture was somewhat lower than previously estimated.

(ii) To circumvent the problem introduced by quenching of [^{3}H]thymidine detection by spores, O'Connor and Zusman (1988) assayed for the accumulation of acid-soluble label outside of cells instead of assaying for the loss of radioactive label in developing cells. To this end, they employed the submerged-culture technique, whereby cells develop beneath a layer of liquid buffer (Kuner and Kaiser, 1982). This technique is particularly conducive to the addition to, and removal of, the supernatant fluid in developing cultures. In this study, overlying medium was removed from the developing cultures, and the level of soluble [^{3}H]thymidine was determined. A slow release of radioactive label was detected over the course of 7 days of development in a developmentally proficient strain, and a similar release was observed for a *csg* developmental mutant strain, a hallmark of which had been the absence of developmental lysis. O'Connor and Zusman concluded from the results obtained with this assay that there is no lysis at a discrete stage of development, although low-level, gradual lysis, not unlike the lysis observed in old cultures of many microorganisms, does occur.

A major problem with this assay is that the experimental design allowed for detection of soluble [^{3}H]thymidine in the bulk liquid only. Since a slime layer encases *M. xanthus* cells under the developmental conditions described, it is probable that a large proportion of the label released from lysing cells was sequestered in intercellular spaces and within the slime layer. Since no attempt was made to recover solubilized label from the slime or from between intact cells, admittedly a difficult task, it is questionable whether the released label detected adequately reflected the level of developmental autolysis under these conditions. In fact, it has been observed (Rosenbluh and Kaiser, unpublished data) that upwards of 70% of the initial amount of protein in a submerged-culture developmental sample is lost within the first 20 h of development, which suggests that a large proportion of cells undergo sufficient damage to the cell envelope to allow leakage of cellular protein into the overlying supernatant fluid, where it is purportedly broken down.

Results of the O'Connor and Zusman study suggest both that developmental conditions can significantly alter the absolute level of developmental autolysis and that investigator-induced manipulations can alter the level of lysis observed. These findings, along with a recent report of the existence of a population of cells distinct from vegetative and spore cells during development, i.e., peripheral rod cells (O'Connor and Zusman, 1991), must be taken into account when conclusions are drawn from studies of developmental lysis.

SPORULATION AND LYSIS IN MICROBEADS AND STARVATION BUFFER

Additional methods of studying the question of lysis and sporulation in starving cultures of *M. xanthus* have been developed in recent years. Sporulation of cells suspended in a liquid starvation buffer produced nor-

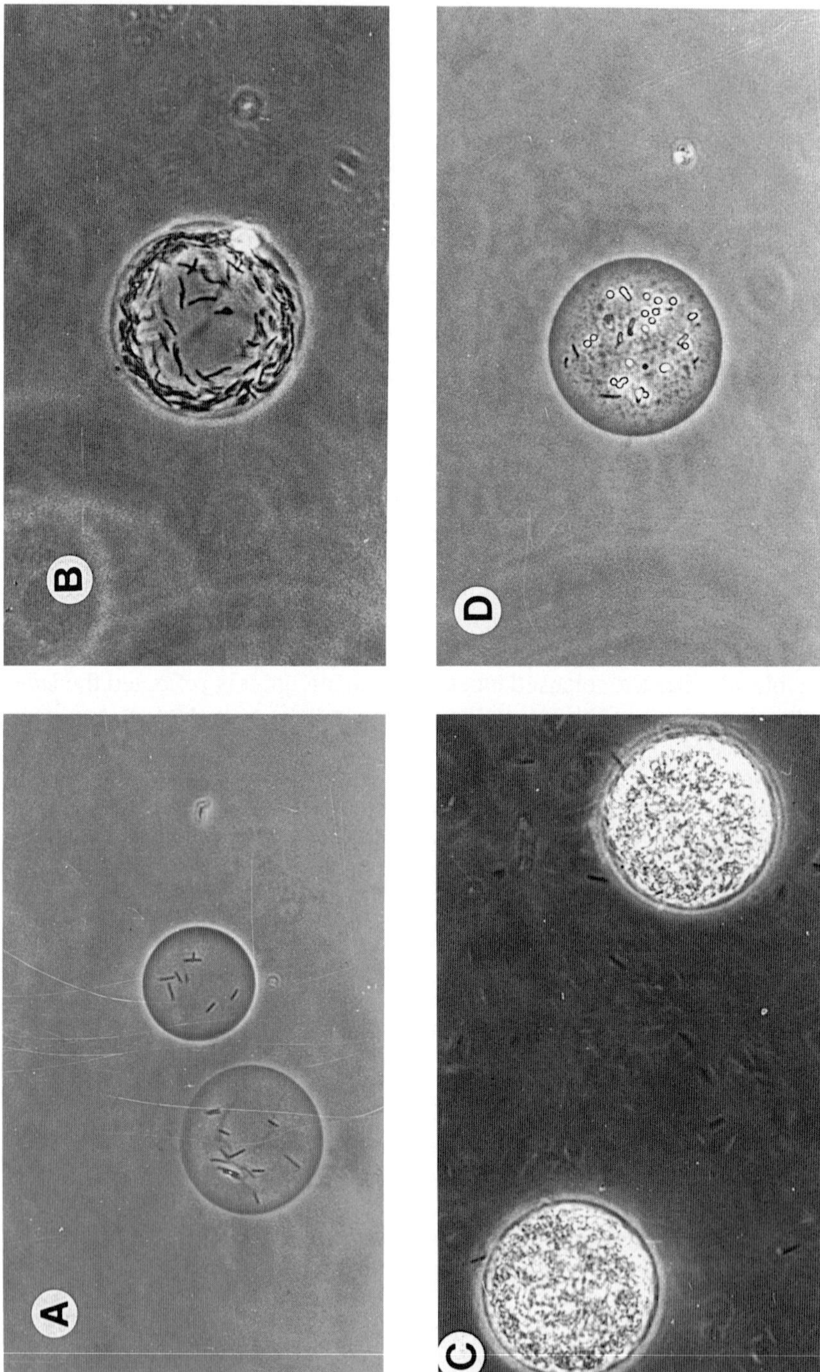
A
B
C
D

mal spores, resulting in the sporulation of no more than 15% of the initial number of cells (Rosenbluh and Rosenberg, 1989b). This sporulation process was similar in several respects to fruiting body sporulation: (i) it proceeded with kinetics similar to those seen in fruiting body sporulation, (ii) only a small proportion of the cells sporulated, (iii) the spores were normal for all resistance and biochemical characteristics tested, and (iv) sporulation was accompanied by a dramatic decrease in the turbidity of the cell culture. Most of the *spo* mutants tested did not sporulate under these conditions, and this inability to sporulate was correlated with a significantly smaller drop in turbidity. In light of these results, we suggested that at least some of the drop in turbidity was due to lysis of cells that did not sporulate. Mueller and Dworkin subsequently determined cell numbers during sporulation in this shake flask system and demonstrated that by 48 h, >90% of the cells had lysed, indicating that the drop of turbidity was indeed due to developmental cell lysis (Mueller and Dworkin, 1991).

It was desirable to develop a method by which starving cells could be assayed for sporulation and lysis with a minimum of manipulation by the investigator and ideally by which individual cells could be observed. Rosenbluh et al. (1989) found that by entrapping small numbers of cells in agarose microbeads and incubating them in liquid starvation buffer, the course of lysis and sporulation could be closely monitored. Cells of *M. xanthus* strains entrapped in 0.6% agarose microbeads were capable of growing and dividing when incubated in rich medium and of sporulating when incubated in starvation buffer (Fig. 1). We demonstrated that sporulation of cells in the microbeads incubated in starvation buffer was always accompanied by lysis of a proportion of the nonsporulating cells and vice versa. The percentage of lysed cells was tightly correlated with the efficiency of sporulation; i.e., in an experiment in which 46% of the cells lysed, 3% of the survivors sporulated, whereas when 82% of the cells lysed, 44% of the survivors sporulated. In that study, it was also demonstrated that lysis and sporulation were both cell density and strain dependent. In the series of experiments presented, cells at an initial density of 0.37 cells per bead neither lysed nor sporulated, whereas cells at an initial density of 4.18 cells per bead began lysing within 42 h and sporulating

---

Figure 1. Entrapment, growth, and sporulation of *M. xanthus* cells in agarose microbeads. (A) Exponential-phase cells of *M. xanthus* ER1500 entrapped in 25-μm-diameter agarose microbeads appear healthy and intact. The microbeads were incubated in either 0.5 CT growth medium for 24 h (B) or 40 h (C) or in MCM sporulation medium (10 mM MOPS, 2 mM $CaCl_2$, 4 mM $MgSO_4$, pH 7.2) for 48 h (D). Magnification, ×800.

Table 1. Distribution of cells and spores of ER1500 in microbeads under sporulation conditions

Expt	Time (h)	Total no. of:			No./bead	
		Beads	Cells	Spores	Cells	Spores
1	0	801	300		0.37	
	42	389	207		0.53	
	68	418	262		0.63	
	192	392	251		0.64	
2	0	618	1,027		1.66	
	42	331	752		2.27	
	68	242	458		1.89	
	192	535	652	21	1.22	0.04
3	0	474	1,982		4.18	
	42	369	604		1.64	
	68	234	307	52	1.31	0.22
	192	695	591	259	0.85	0.37

within 68 h (Table 1). Results of experiments studying cell density dependence in this system suggested the existence of two separate cell density signals, one based on local cell density (i.e., that within a single microbead) and one based on a more general cell density (i.e., that within the overall culture).

A strain incapable of sporulating under fruiting body conditions failed to sporulate or lyse in the microbead system, even when the microbeads were filled to capacity with the mutant cells. Although development in this system was outside of fruiting bodies, the technique allowed for observation of sporulation and lysis in the absence of investigator-induced manipulations. These data strongly reinforced the original observation that lysis and sporulation are tightly interrelated processes.

Close to 30 years ago, Dworkin and Gibson (1964) demonstrated that vegetative cells are capable of carrying out a large portion of the sporulation process when they are suspended in liquid cultures containing glycerol or phenethyl alcohol. In contrast to fruiting body sporulation and sporulation in starvation buffer, however, glycerol-induced sporulation in liquid growth medium resulted in the production of abnormal spores (Dworkin and Gibson, 1964; Bacon and Eiserling, 1968; Bacon et al., 1975; Inouye et al., 1979) that were sensitive to a stage-specific autolytic process under extreme conditions (Kottel and White, 1974). In addition to the increased sensitivity of these spores under certain conditions, glycerol-induced sporulation proceeded in the absence of lysis, with sporulation (i.e., appearance of round or ovoid refractile spores) at >90% of the initial cell number (Komano et al., 1980). Glycerol spores lacked the thick, layered spore coat seen in electron micrographs of fruiting body spores (Inouye et al., 1979) and showed reduced resistance to extreme conditions compared with fruiting body spores. It is possible that the missing spore coat(s) of these spores is composed of protein S, since it has been shown that protein S is not synthesized in glycerol spores (Komano et al., 1980), whereas protein U is synthesized. The relationship between the reduced resistance of glycerol-induced spores and the fact that this type of sporulation proceeds in the absence of lysis has not been investigated.

AUTOCIDES

Autocides is the term applied to a group of at least five substances produced by *M. xanthus* that were observed to induce lysis in growing or resting cells of the producing and related strains (Varon et al., 1984). Autocidal activity was originally detected as a contaminant during the purification process of antibiotic TA (Nathan, unpublished data). Autocides were purified by extraction of the cells with ethanol and of the extracellular fluid of growing cells with chloroform (Varon et al., 1984). These autocides, the majority of which were shown to be cell associated, were purified and characterized by thin-layer and column chromatography. Autocidal-activity assays were developed for testing autocides both in liquid cultures and on agar plates, and in each case, the activity could be quantified and a unit of activity could be precisely defined.

The terms given to the two major autocides (by percentage of total autocidal activity), AMI and AMV, reflect their relative positions of increasing polarity in thin-layer chromatography (Varon et al., 1984). AMI is composed of a mixture of free fatty acids (Varon et al., 1986). This mixture was fractionated to yield an almost equal concentration of saturated and monounsaturated fatty acids, with a small proportion of diunsaturates. As can be seen in Table 2, the lytic activities of these fractions on a tester strain of *M. xanthus* were determined, with the diunsaturated fatty acids having the highest specific activity, the monounsaturated fraction having the next highest, and the mixture of saturated fatty acids displaying the lowest specific activity. There was no synergistic effect observed for lytic activities in mixtures of the various AMI component fatty acids, and it is of interest that no single model fatty acid displayed an activity as high as that of the natural AMI mixture (1,000 U/mg). Fatty acids have been shown to induce lysis in other prokaryotic systems (e.g., see Tsuchido et al., 1985) although at considerably higher concentrations (500 μM, compared with 10 μM AMI in *M. xanthus*). The mechanism of action for these other systems has been postulated to involve activation of autolysins.

Table 2. Fractionation of AMI-C by $AgNO_3$-silicic acid chromatography

Fatty acid fraction	Major fatty acids (% of fraction)	Abundance (mg/sample)[a]	Activity (U/sample)[a]	Sp act (U/mg)
Saturated	n-14:0 (9.4) i-15:0 (57–75) i-17:0 (9) n-16:0 (12)	17.1 (46.4)	1,500 (7.5)	100
Monosaturated	n-16:1 (2.2) n-16:1 (79)	18.2 (49.3)	15,000 (75.0)	800
Diunsaturated	n-16:2 (65) i-17:2 (32)	1.6 (4.3)	3,500 (17.5)	2,220

[a] Numbers in parentheses indicate percent abundance or activity in each fraction.

The autocide story acquired further interest when Gelvan et al. (1987) characterized the lytic activity of AMV. Whereas AMI gave a clear plaque with well-defined borders on agar plate assays, the typical area of lysis produced by AMV was complex, consisting of several concentric rings of clear and turbid areas (Fig. 2). The reason for this observation was not obvious. Upon purification of AMV, it was observed that the peak of active material eluted together with that of organic phosphorus, which suggests that AMV is a phospholipid. Strengthening this assumption further was the observation that one of the model phospholipids tested, phos-

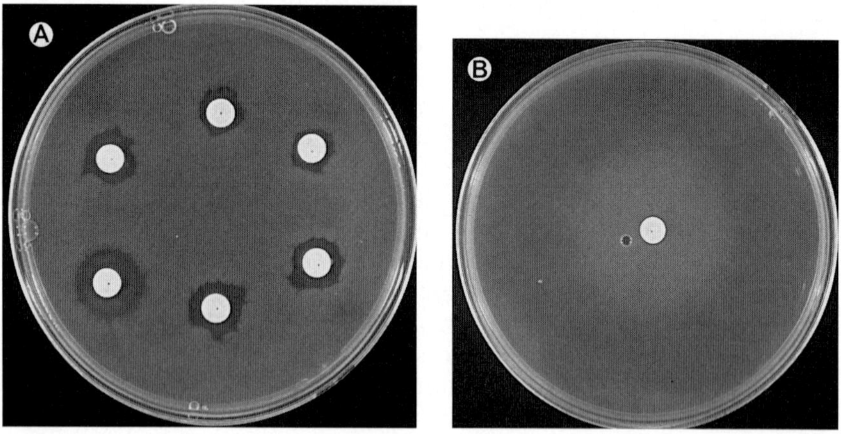

Figure 2. Agar plate assay for autocidal lytic activity. The clear plaques produced by AMI and model fatty acids (A) and the complex plaque produced by AMV (B) are observed on a lawn on *M. xanthus* ER-1 growing on 1 CT agar (1% Casitone, 0.2% $MgSO_4 \cdot H_2O$, 1.8% agar).

phatidylethanolamine (PE), ran on thin-layer chromatographic plates at a position similar to that of AMV. In addition, both synthetic PE and PE derived from *Escherichia coli* gave complex, AMV-like plaques in the agar plate assay.

One crucial observation allowed the unraveling of the complex plaque puzzle and helped elucidate the relationship between AMI and AMV. During the course of studying AMV, it was observed that on assay plates that had AMV incorporated into the agar, lysis of the tester cells was observed only if a minimum of 2×10^6 cells per ml was plated (Gelvan et al., 1987). A lower density of cells allowed unimpaired growth without AMV-induced lysis. Cell density dependence of AMV-induced lysis was relieved by the addition of cell-free supernatant fluid from exponentially growing cultures, although this activity was shown to be heat labile. In addition, it was observed that snake venom phospholipase A_2 allowed AMV-induced lysis of low-density cultures. Taken together, these observations led Gelvan et al. (1987) to postulate that AMV is PE and that its lytic activity is a result of its being hydrolyzed to free fatty acids (i.e., AMI) by the action of a critical concentration of extracellular phospholipase(s). *M. xanthus* cell membranes had previously been shown to contain both phospholipase A and lysophospholipase activity (Orndorff and Dworkin, 1980) capable of breaking down PE into its constituent fatty acids. The necessity for a critical concentration of phospholipase(s) would explain the observed cell density dependence of the AMV lytic activity. It appeared that AMI was produced in sufficient concentrations to lyse *M. xanthus* only when local cell densities reached a specific threshold, as might occur during aggregation of cells during development.

Arias et al. (1983), in working with a *Myxococcus corraloides* strain, studied autolysis of stationary-phase cultures and identified and partially characterized a substance from supernatant fluids of old cultures that induced lysis of exponential-phase cultures. This substance was shown to be stable at 100°C for 20 min. The relationship of this activating substance to autocides of *M. xanthus* has not been determined.

AUTOCIDES, DEVELOPMENT, AND DEVELOPMENTAL LYSIS

Previous observations of the existence of developmental lysis coupled with the identification of autocides that induced lysis specifically of *M. xanthus* and closely related strains led Varon et al. (1986) to search for a possible link between autocides and development. Spontaneous mutants as well as UV- and *N*-methyl-*N'*-nitro-*N*-nitrosoguanidine-induced mutants were obtained by plating *M. xanthus* on AMI-containing agar. These mutants were resistant to lysis by AMI and by model fatty acids. As

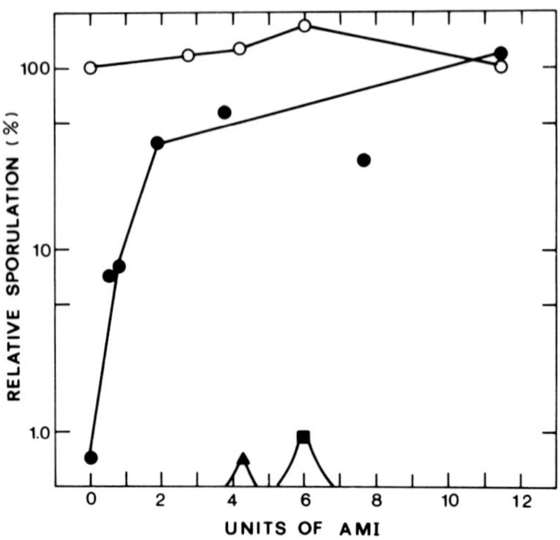

Figure 3. Effect of AMI on the formation of myxospores in submerged cultures of wild type and of sporulation-deficient mutants. The ordinate is the number of heat-resistant spores of each strain after 144 h of development relative to the number of spores of the wild-type strain (DK1622) in the absence of AMI. Wild type (○) and the *dsg* mutant (DK3960) (●) were exposed to concentrations of AMI of 0 to 11.5 U ml^{-1}; *asg* (DK5057) (■), *bsg* (M380), and *csg* (DK2653) (▲) mutants were exposed to 0, 2.8, 4.2, 6.0, and 7.7 U of AMI ml^{-1}. Values below 0.5% sporulation are not shown on this figure.

expected, these mutants were also resistant to lysis by AMV and by PE. All such mutants were defective in fruiting body formation to various degrees, whereas sporulation was generally not affected. These mutants could be maintained only under selective pressure and were highly unstable.

Because of the instability of these AMI-resistant mutants, an alternative approach was sought to study the possible relationship between autocides and developmental lysis. Several classes of developmental mutants defective in signaling had previously been isolated and characterized (Hagen et al., 1978). In studying the effect of autocide AMI on aggregation and sporulation of these mutant strains, we (Rosenbluh and Rosenberg, 1989a) found that one class of signaling mutants, the *dsg* class, could be rescued for both aggregation and sporulation in developmental submerged cultures in the presence of low concentrations (0.5 to 2.0 μg/ml) of AMI. This rescue was specific for *dsg* and was not observed for the three other classes of signaling mutants (Fig. 3). Cheng and Kaiser (1989) showed that the *dsg* mutation was leaky in that under certain conditions, *dsg* mutants could sporulate in fruiting bodies after many days of developmental conditions. The rescue by AMI, however, allowed for apparently normal development within the time frame observed for wild-type cultures. AMI was also shown to substantially accelerate the rate of aggregation and sporulation in wild-type cultures and to increase the final spore yield by as much as 70% (Rosenbluh and Rosenberg, 1989a).

A new transposon insertion mutant that was incapable of any aggregation or significant sporulation in submerged cultures even after prolonged incubations was isolated (Rosenbluh and Rosenberg, 1990). This mutant, ER304, the development of which could be rescued by AMI, was subsequently employed to characterize the AMI-induced rescue phenomenon.

Rescue of ER304 development was effected by low concentrations (3 to 6 μg/ml) of AMI. The three fatty acid fractions that constitute the natural AMI mixture (saturated, monounsaturated, and diunsaturated fatty acids) each displayed, in addition to the lytic activities reported by Varon et al. (1986), developmental rescue activity (Rosenbluh and Rosenberg, 1990). It is of interest that the most active lytic fraction, the diunsaturates, also had the highest development rescue specific activity (700 U/mg compared with 115 U/mg for monounsaturated and 68 U/mg for saturated fatty acids). Despite this relationship between lytic and rescue activities, the ratios of rescue activity to lytic activity for different natural components found in the autocide mixtures proved to be more complex and suggested an uncoupling of the two activities. Uncoupling of lysis and rescue of development was further observed in studies of the activities of model fatty acids and phospholipids. In this case, it was observed that a particular lipid that displayed a high lytic activity did not necessarily display a high development rescue activity and vice versa (Table 3). The rescue of development was shown to be an early event. ER304 could be

Table 3. Development rescue of ER304 by model fatty acids and phospholipids[a]

Lipid	Maximum no. of spores/ml	Lytic activity (U/mg)	Aggregation
Saturated fatty acids			
n-10:0	1×10^6	25	+
n-12:0	1×10^6	50	+
n-14:0	3×10^6	100	+
i-15:0	3×10^5	100–200	+
Ante-i-16:0	3×10^6	200–250	+
n-16:0	1×10^5	~250	−
i-18:0	8×10^4	50	−
Monounsaturated fatty acid (n-16:1, *cis* 9)	1×10^6	250–500	+
Polyunsaturated fatty acid (n-18:3, *cis* 6,9,12)	3×10^5	850	−
Phospholipids			
Synthetic PE	1×10^6	17	+
E. coli PE	5×10^2	25	−

[a] To test development rescue, up to 40 μg of lipid was added to 1 ml of submerged cultures of ER304. Aggregation was scored by visual examination of the wells; sporulation was determined after 6 days of submerged culture as described in the text.

rescued by addition of AMI for up to 10 h after the initiation of starvation conditions in submerged cultures. After this time, addition of AMI caused increased cell death in the absence of developmental rescue.

Taken together, the above-mentioned observations led us (Rosenbluh and Rosenberg, 1990) to propose a model for the action of autocides during development. We suggested that development rescue activity is the result of increased permeabilization of the cell membrane rather than the result of lysis. This conclusion was based on three observations: (i) lower concentrations of autocide AMI were necessary for rescue activity than for lysis, (ii) lytic and development rescue activities could be uncoupled, and (iii) development rescue could be effected by AMI only at times earlier than those observed for initiation of developmental lysis. Increased permeabilization of the membrane of the developing cell would allow developmental signals to be exchanged. In addition to rescuing mutants with altered membranes, a permeabilizing agent might allow earlier and more efficient signaling in wild-type cultures, thus accelerating development and allowing more-efficient sporulation, as observed in the study. Since cell death is a consequence of addition of autocide AMI later in development (after 13 h), lysis may be a secondary effect of autocides later in development.

GLUCOSAMINE AND DEVELOPMENTAL LYSIS

The work on autocides and development was complemented and extended by Mueller and Dworkin (1991), who set out to study the effect of GlcN on development in *M. xanthus* (Janssen and Dworkin, 1985). As described above, GlcN and, to a lesser extent, ManN at millimolar concentrations were capable of specifically rescuing the *csg* class of developmental signaling mutant strains. Mueller and Dworkin (1991) showed that GlcN induced lysis of exponentially growing cells as well as of cells incubating in liquid sporulation medium. The extent of this lysis was highly dependent on the proportions of tan and yellow phase variants in the original cell population, an effect observed in previous studies (Janssen and Dworkin, 1985). Mutants resistant to this GlcN-induced lysis were isolated following mutagenesis with ethyl methanesulfonate, and of the two mutant strains chosen for further study, one was developmentally defective, while the other was developmentally proficient. The developmental defect of the former was rescued by addition of the amino sugar GlcN. GlcN-induced lysis of vegetatively growing cells was strongly correlated with sporulation of a significant proportion of the survivors. Mutant strains resistant to GlcN-induced lysis did not sporulate at any concentration of GlcN tested, a result consistent with results of previous studies that showed a close correlation between sporulation and developmental lysis.

Interestingly, addition of GlcN to wild-type developing cultures caused inhibition of fruiting body formation. This observation led Mueller and Dworkin (1991) to measure levels of autocide production in vegetative and developing cells in order to investigate the possibility that inhibition of wild-type development by GlcN might be due to an early or augmented lysis event in the developing cells. In addition to measuring levels of AMI, they determined the levels of phospholipase A and lysophospholipase, reasoning, as did Gelvan et al. (1987), that AMI activity reflected phospholipase activities that are responsible for hydrolysis of PE (i.e., AMV) in the cell. They found that AMI levels and phospholipase activity in vegetative cells increased three- and eightfold, respectively, in the presence of GlcN. Mutants resistant to GlcN-induced lysis did not show this increase in AMI production or in phospholipase activity. Whereas developing wild-type cells normally showed a threefold increase in AMI production within the first 24 h of development, the presence of GlcN accelerated this increased production of AMI by 6 to 12 h, with a concomitant slight increase in the amount of AMI produced. Similar results were obtained for phospholipase activity. The time scale reported for this acceleration of GlcN-induced AMI production is reminiscent of that described above for AMI-induced acceleration of development (Rosenbluh and Rosenberg, 1989a). It is interesting that the GlcN-resistant mutant strain that did not sporulate did not show any increase in AMI production or phospholipase activity under developmental conditions and that GlcN rescue of its development correlated with resultant wild-type levels in AMI and in phospholipase activity.

An unexpected outcome of the study by Mueller and Dworkin (1991) was the observation that GlcN-induced spores from vegetatively growing cultures resembled glycerol-induced spores in electron micrographs. Indeed, GlcN was found to induce a transient rise in the intracellular levels of glycerol. The model proposed by Mueller and Dworkin (1991) coincides in many aspects with that of Gelvan et al. (1987) and with ours (Rosenbluh and Rosenberg, 1990) and focuses on the generation of two signals, one in the sporulation pathway and the other in the developmental lysis pathway (Fig. 4). Mueller and Dworkin (1991) proposed that GlcN serves to activate cellular phospholipases, which in turn hydrolyze cell envelope phospholipids (i.e., PE), yielding glycerol and autocide AMI. Glycerol then activates the sporulation pathway, while AMI initiates permeabilization of the cell membrane and ultimately results in lysis of a proportion of the nonsporulating cell population.

Extraneously added AMI accelerated development in wild-type cul-

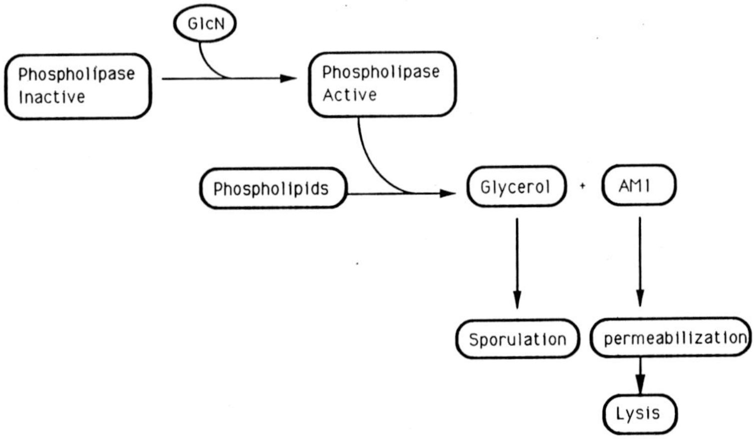

Figure 4. Proposed model for GlcN regulation of lysis and sporulation. Reprinted with permission (Mueller and Dworkin, 1991).

tures, but addition of amino sugars inhibited development in wild-type cells. Mueller and Dworkin (1991) showed that when concentrations of exogenously added GlcN were lowered to ca. 30% of the concentration necessary to inhibit development, sporulation in the wild-type strain was significantly enhanced. Apparently, the effects of GlcN and AMI are highly concentration dependent, suggesting a sensitive balance between lysis and sporulation. In addition, GlcN specifically rescues *csg* mutants, while AMI rescues development of *dsg* mutants specifically. Again, this may be due to the different concentrations of the effector substances or to the fact that GlcN induced the release of glycerol and perhaps other unidentified substances in addition to AMI. The complex interplay of several phenomena can result in a cellular response different from the response to a single event.

The question of cell fate was also addressed in the Mueller and Dworkin study (1991). Yellow-tan phase variation has previously been described for *M. xanthus* cell cultures (Burchard and Dworkin, 1966; Burchard et al., 1977), and differential developmental fates and a difference in several additional traits have been observed for the two phases (Wireman and Dworkin, 1975, 1977; Janssen and Dworkin, 1985; Laue and Gill, 1989; Shimkets, 1990). Mueller and Dworkin observed that yellow cells were more likely to lyse than tan cells, and they suggested that GlcN preferentially lysed yellow cells via AMI release, while tan cells preferentially sporulated, possibly in response to the induced glycerol release. These observations have opened the way for a better understanding of the possible role(s) of tan-yellow phase variation of *M. xanthus* as well as of the larger question of cell fate in the multicellular development of this organism.

CONCLUSIONS

The observation and description of developmental cell lysis over 15 years ago have opened up a new area of research in myxobacteriology as studies have been initiated to quantify the extent of lysis, characterize its mechanism, and understand the role it plays in the multicellular developmental program of *M. xanthus*. In light of the facts that programmed cell death plays an important part in the development and differentiation of many if not all of the eukaryotic multicellular systems studied and that it involves active participation of and specific gene expression in the lysing cell (Hengartner et al., 1992), it was of great interest to observe that this effect exists in a prokaryote as well. There is still much that is not understood about this phenomenon, and continued research into the subject is necessary to further elucidate the mechanism of developmental lysis on a molecular level.

REFERENCES

Arias, J. M., A. Fernandez-Vivas, and E. Montoya. 1983. Evidence for an activating substance related to autolysis in *M. coralloides D. Arch. Microbiol.* **134:**164–166.

Bacon, K., D. Clutter, R. H. Kottel, M. Orlowski, and D. White. 1975. Carbohydrate accumulation during myxospore formation in *Myxococcus xanthus. J. Bacteriol.* **124:**1635–1636.

Bacon, K., and F. A. Eiserling. 1968. A unique structure in microcysts of *Myxococcus xanthus. J. Ultrastruct. Res.* **21:**378–382.

Bonner, J. T. 1967. *The Cellular Slime Moulds,* 2nd ed. Princeton University Press, Princeton, N.J.

Burchard, R. P., A. C. Burchard, and J. H. Parish. 1977. Pigmentation phenotype instability in *Myxococcus xanthus. Can. J. Microbiol.* **23:**1657–1662.

Burchard, R. P., and M. Dworkin. 1966. Light-induced lysis and carotenogenesis in *Myxococcus xanthus. J. Bacteriol.* **91:**535–545.

Cheng, Y., and D. Kaiser. 1989. *dsg,* a gene required for cell-cell interaction early in *Myxococcus* development. *J. Bacteriol.* **171:**3719–3726.

Devreotes, P. 1989. *Dictyostelium discoideum:* a model system for cell-cell interactions in development. *Science* **245:**1054–1058.

Downard, J. S., D. Kupfer, and D. R. Zusman. 1984. Gene expression during development of *Myxococcus xanthus:* analysis of the genes for protein S. *J. Mol. Biol.* **175:**469–492.

Dworkin, M., and S. M. Gibson. 1964. A system for studying microbial morphogenesis: rapid formation of microcysts in *Myxococcus xanthus. Science* **146:**243–244.

Ellis, R. E., and H. R. Horvitz. 1986. Genetic control of programmed cell death in the nematode *C. elegans. Cell* **44:**817–829.

Ellis, R. E., J. Yuan, and H. R. Horvitz. 1991. Mechanisms and functions of cell death. *Annu. Rev. Cell Biol.* **7:**663–698.

Errington, J. 1991. A model for asymmetric septum formation during sporulation in *Bacillus subtilis*. *Mol. Microbiol.* **5:**785–789.

Gelvan, I., M. Varon, and E. Rosenberg. 1987. Cell-density-dependent killing of *Myxococcus xanthus* by autocide AMV. *J. Bacteriol.* **169:**844–848.

Gollop, R., M. Inouye, and S. Inouye. 1991. Protein U, a late-developmental spore coat protein of *Myxococcus xanthus,* is a secretory protein. *J. Bacteriol.* **173:**3597–3600.

Gonzalez, F., M. A. Goicoechea, A. Fernandez-Vivas, J. M. Arias, and E. Montoya 1992. Phosphate mediation in the autolysis of *Myxococcus xanthus*. *Microbios* **70:**7–12.

Hagen, D. C., A. P. Bretscher, and D. Kaiser. 1978. Synergism between morphogenetic mutants of *Myxococcus xanthus*. *Dev. Biol.* **64:**284–296.

Hengartner, M. O., R. E. Ellis, and H. R. Horvitz. 1992. *Caenorhabditis elegans* gene *ced-9* protects cells from programmed cell death. *Nature* (London) **356:**494–499.

Inouye, M., S. Inouye, and D. R. Zusman. 1979. Biosynthesis and self assembly of protein S, a development specific protein of *Myxococcus xanthus*. *Proc. Natl. Acad. Sci. USA* **76:**209–213.

Inouye, S., Y. Ike, and M. Inouye. 1983. Tandem repeat of the genes for protein S, a development-specific protein of *Myxococcus xanthus*. *J. Biol. Chem.* **258:**38–40.

Janssen, G. R., and M. Dworkin. 1985. Cell-cell interactions in developmental lysis of *Myxococcus xanthus*. *Dev. Biol.* **112:**194–202.

Jones, M. V., and V. E. Barr. 1983. Phosphate-stimulated autolysis and cellular morphogenesis of *Myxococcus xanthus*. *FEMS Microbiol. Lett.* **16:**359–362.

Kerr, J. F. R., A. H. Wyllie, and A. R. Currie. 1972. Apoptosis: a basic biological phenomenon with wide ranging implications in tissue kinetics. *Br. J. Cancer* **26:**239–257.

Kessin, R. H. 1988. Genetics of early *Dictyostelium discoideum* development. *Microbiol. Rev.* **52:**29–49.

Komano, T., S. Inouye, and M. Inouye. 1980. Patterns of protein production in *Myxococcus xanthus* during spore formation induced by glycerol, dimethyl sulfoxide, and phenethyl alcohol. *J. Bacteriol.* **144:**1076–1082.

Kottel, R., and D. White. 1974. Autolytic activity associated with myxospore formation in *M. xanthus*. *Arch. Microbiol.* **95:**91–95.

Kuner, J., and D. Kaiser. 1982. Fruiting body morphogenesis in submerged cultures of *Myxococcus xanthus*. *J. Bacteriol.* **151:**458–461.

Laue, B. E., and R. E. Gill. 1989. Genetic analysis of phase variation in *Myxococcus xanthus*. *In Proceedings of the Sixteenth International Conference on the Biology of the Myxobacteria.*

Losick, R., L. Kroos, J. Errington, and P. Youngman. 1988. Pathways of developmentally regulated gene expression in *Bacillus subtilis,* p. 221–242. *In* D. Hopwood and K. E. Chater (ed.), *Genetics of Bacterial Diversity*. Academic Press, New York.

Mason, O. J., and D. Powelson. 1958. Lysis of *Myxococcus xanthus*. *J. Gen. Microbiol.* **19:**65–70.

Mueller, C., and M. Dworkin. 1991. Effects of glucosamine on lysis, glycerol formation, and sporulation in *Myxococcus xanthus*. *J. Bacteriol.* **173:**7164–7175.

Nathan, P. Unpublished data.

Nelson, D. R., and D. R. Zusman. 1983. Transport and localization of protein S, a spore coat protein, during fruiting body formation by *Myxococcus xanthus*. *J. Bacteriol.* **154:**547–553.

O'Connor, K., and D. R. Zusman. 1988. Reexamination of the role of autolysis in the development of *Myxococcus xanthus*. *J. Bacteriol.* **170:**4103–4112.

O'Connor, K., and D. R. Zusman. 1991. Development in *Myxococcus xanthus* involves differentiation into two cell types, peripheral rods and spores. *J. Bacteriol.* **173:**3318–3333.

Orndorff, P. E., and M. Dworkin. 1980. Separation and properties of the cytoplasmic and outer membranes of vegetative cells of *Myxococcus xanthus*. *J. Bacteriol.* **141:**914–927.

Rosenberg E., and M. Varon. 1984. Antibiotics and lytic enzymes, p. 109–125. *In* E. Rosenberg (ed.), *Myxobacteria: Development and Cell Interactions*. Springer-Verlag, New York.

Rosenbluh, A., and D. Kaiser. Unpublished data.

Rosenbluh, A., R. Nir, E. Sahar, and E. Rosenberg. 1989. Cell-density-dependent lysis and sporulation of *Myxococcus xanthus* in agarose beads. *J. Bacteriol.* **171:**4923–4929.

Rosenbluh, A., and E. Rosenberg. 1989a. Autocide AMI rescues development in *dsg* mutants of *Myxococcus xanthus*. *J. Bacteriol.* **171:**1513–1518.

Rosenbluh, A., and E. Rosenberg. 1989b. Sporulation of *Myxococcus xanthus* in liquid shake flask cultures. *J. Bacteriol.* **171:**4521–4524.

Rosenbluh, A., and E. Rosenberg. 1990. Role of autocide AMI in development of *Myxococcus xanthus*. *J. Bacteriol.* **172:**4307–4314.

Saunders, J. W., Jr. 1966. Death in embryonic systems. *Science* **154:**604–612.

Shimkets, L. J. 1990. Social and developmental biology of the myxobacteria. *Microbiol. Rev.* **54:**473–501.

Shimkets, L. J., and D. Kaiser. 1982a. Induction of coordinated cell movement in *Myxococcus xanthus*. *J. Bacteriol.* **152:**451–461.

Shimkets, L. J., and D. Kaiser. 1982b. Murein components rescue developmental sporulation of *Myxococcus xanthus*. *J. Bacteriol.* **152:**462–470.

Teintze, M., T. Furuichi, R. Thomas, M. Inouye, and S. Inouye. 1985a. Differential expression of two homologous genes coding for spore-specific proteins in *Myxococcus xanthus*, p. 253–260. *In* J. Hoch and P. Setlow (ed.), *Spores IX: the Molecular Biology of Microbial Differentiation*. American Society for Microbiology, Washington, D.C.

Teintze, M., R. Thomas, T. Furuichi, M. Inouye, and S. Inouye. 1985b. Two homologous genes coding for spore-specific proteins are expressed at different times during development of *Myxococcus xanthus*. *J. Bacteriol.* **163:**121–125.

Tsuchido, T., T. Hiraoka, M. Takano, and I. Shibasaki. 1985. Involvement of autolysin in cellular lysis of *Bacillus subtilis* induced by short- and medium-chain fatty acids. *J. Bacteriol.* **162:**42–46.

Varon, M., S. Cohen, and E. Rosenberg. 1984. Autocides produced by *Myxococcus xanthus*. *J. Bacteriol.* **160:**1146–1150.

Varon, M., A. Tietz, and E. Rosenberg. 1986. *Myxococcus xanthus* autocide AMI. *J. Bacteriol.* **167:**356–361.

Wireman, J. W., and M. Dworkin. 1975. Morphogenesis and developmental interactions in myxobacteria. *Science* **189:**516–523.

Wireman, J. W., and M. Dworkin. 1977. Developmentally induced autolysis during fruiting body formation by *Myxococcus xanthus*. *J. Bacteriol.* **129:**796–802.

Wyllie, A. H., J. F. R. Kerr, and A. R. Currie. 1980. Cell death: the significance of apoptosis. *Int. Rev. Cytol.* **68:**251–306.

Myxobacteria II
Edited by Martin Dworkin and Dale Kaiser
© 1993 American Society for Microbiology, Washington, DC 20005

Chapter 11

Protein Secretion in Myxobacteria

J. F. Guespin-Michel, B. Letouvet-Pawlak, and F. Petit

Many gram-negative bacteria secrete extracellular proteins in spite of the presence of a secretion barrier in the outer membrane. In most cases, this is bona fide secretion in that cell lysis is not responsible for the appearance of the proteins in the medium. In a few species, their close relationship to *Escherichia coli* and/or their amenability to genetic studies have already made it possible to obtain a great deal of information concerning the mechanisms whereby a protein crosses their three-layered envelopes. These mechanisms most often fall into one of four different classes (Pugsley, 1988; Letouvet-Pawlak, 1991b).

Myxobacteria, however, display particular characteristics with regard to the production of extracellular proteins that justify the study of protein secretion in this family. On the one hand, as scavenging bacteria, myxobacteria produce a large variety of extracellular proteins that are absolutely required for their growth in a variety of natural environments. Furthermore, the optimal utilization of extracellular enzymes often depends on the social behavior of the organisms. For instance, the predation of a cyanobacterium by *Myxococcus xanthus* occurs in large interspecies clumps in which the cells of the cyanobacterium are lysed (Burnham et al., 1981). Rosenberg et al. (1977) have demonstrated that for a dispersed growing mutant of *M. xanthus*, the use of casein as a substrate is cell density dependent.

On the other hand, other extracellular proteins must be important as signals for social behavior during both vegetative growth (swarming behavior) and the developmental cycle. Indeed, protein secretion also

---

J. F. Guespin-Michel, B. Letouvet-Pawlak, and F. Petit • Laboratoire de Microbiologie, URA 203 CNRS, Faculté des Sciences et Techniques, Université Rouen, 76134 Mont Saint Aignan, France.

occurs at least during the early phases of development. Though the role of secreted proteins is not always understood, the involvement of some of them as developmental signals has been demonstrated (see chapter 12). Identification of the extracellular proteins produced by myxobacteria during growth or development and elucidation of the regulation of their synthesis and mechanism(s) of secretion may thus provide new and important clues for understanding the biology of these bacteria.

Finally, as a result of increased progress in these areas and the broad range of biological activities of these organisms, the myxobacteria have become increasingly interesting to the industrial world. For example, antibiotics and other secondary metabolites produced by the myxobacteria (Reichenbach et al., 1988; see chapter 16) have proven to be industrially promising. In addition, the diversity of their secreted proteins together with the wide range of foreign proteins that they could be engineered to secrete (Breton et al., 1986; Breton and Guespin-Michel, 1987) point to another aspect of their biotechnological potential.

PROTEIN SECRETION DURING VEGETATIVE GROWTH OF MYXOBACTERIA

Is Extracellular Protein Production a Real Protein Secretion Process? Methodological Considerations

In the absence of international agreement on the terminology concerning the production of extracytoplasmic proteins, we shall define secretion as the overall production of extracellular proteins provided it is not caused by cell lysis, while export will specifically refer to protein translocation throughout the inner membrane.

During their vegetative growth, myxobacteria produce numerous extracellular proteins, the number and diversity of which are at least 1 degree of magnitude higher than in gram-negative bacteria in which protein secretion has been studied. It is therefore of prime importance to establish that the production of extracellular proteins observed with myxobacteria corresponds to protein secretion per se rather than to cell lysis. This, however, is not easy to do because of the presence of several proteases that are themselves secreted in the growth medium (Sudo and Dworkin, 1973; Rosenberg and Varon, 1984; Coletta and Miller, 1986) and of a slime that enhances protein denaturation and thus protease accessibility (Gnosspelius, 1978). Thus, the activity of the intracellular malate dehydrogenase from *M. xanthus* cannot be used as an assay of the degree of cell lysis, since the activity disappears extremely rapidly when it is incubated with the supernatant of a suspension of growing cells of *M. xanthus* (unpublished results). Even the β-galactosidase expressed in an *M. xanthus* recombinant containing the *E. coli lacZ* gene is hydrolyzed in the medium of a wild-type culture and can be used as a lysis marker only in a strain that produces less external protease (Letouvet-Pawlak, 1991a). In such a strain, however, the level of cell lysis contributes very little if at all to extracellular protein production.

Only indirect arguments are available for the other strains of *M. xanthus:* most extracellular activities are not cell bound, which means either that they are secreted as soon as they are produced or that the process of secretion activates an inactive precursor. Regardless of what causes this possible activation, it is unlikely to be achieved through cell lysis. In addition, the amount of total extracellular protein produced per unit dry mass (or per cell) is constant throughout exponential growth when measured by the method of Sedmak and Grossberg (1977) (with Coomassie blue) that assays only peptides larger than 5 kDa (Nicaud et al., 1984). The absence of increase in total amount of protein secreted per cell during stationary phase compared with exponential phase supports the hypothesis that cell lysis is not likely to account significantly for extracellular protein production; in parallel, the absence of a drop in the amount of total extracellular protein suggests that proteolysis of secreted proteins might be limited. However, bacteriolytic and proteolytic activities (Nicaud et al., 1984) and an acid, milk-clotting endoprotease (Mcp) (Carias et al., 1990; Petit and Guespin-Michel, 1992) also produced throughout the exponential phase of growth decrease per unit dry weight during this period, possibly reflecting partial proteolysis. These discrepancies further complicate the studies. However, the arguments presented above support the position that at least in *M. xanthus* during vegetative growth, the appearance of extracellular proteins can be ascribed to protein secretion proper.

Production of extracellular proteins in gram-negative bacteria pertains to two main processes, i.e., export, which is translocation of proteins through the inner membrane that results in the presence of periplasmic proteins, and secretion per se. Periplasmic proteins in myxobacteria have been described, although with some ambiguities that may reflect the mechanism(s) of secretion. Periplasmic restriction enzymes have been described; they seem to be tightly bound to the inner membrane (Mayer and Reichenbach, 1978). Phosphatase activities, both acidic and alkaline, have been shown to be mainly periplasmic in *M. xanthus* and in four other species of myxobacteria, their remaining activities (10 and 26%, respectively) being present in the growth medium (Nicaud et al., 1984; Gonzalez et al., 1989). A DNase activity, although detected as clear lysis halos surrounding colonies on plates containing DNA methyl green, has been found exclusively cell bound in liquid cultures of cells of five species of myxobacteria (Munoz et al., 1989), but it has not been precisely localized.

Additional difficulties relate to localizing those exported proteins found in the extracellular matrix, or slime. Depending on the criteria used, the slime, which can make up 5 to 10% of the dry weight of the cell, has been described as a high-molecular-weight complex of proteins, lipid, and saccharides (Gnosspelius, 1978), as extracellular fibrils capable of fixing Congo red and induced under conditions that lead to cell aggregation (Arnold and Shimkets, 1988), or as extracellular polysaccharides (Sutherland and Thomson, 1975). There has also been disagreement regarding its association with the cells. For instance, Kim and Kaiser (1990a) have purified the C factor, the product of gene *csgA*, and described it as a protein exclusively associated with the membrane fraction of developing cells, while Shimkets and Rafiee (1990), using immunodetection, have shown it to be present in the extracellular matrix. Clemans et al. (1991) and Behmlander and Dworkin (1991) have isolated fibril antigens and fibril-associated antigens from cells previously grown or starved on agar. Sutherland and Thomson (1975) and Gnosspelius (1978), however, extracted the extracellular polysaccharides or the slime directly from the supernatants of growing cultures. It would be useful to clarify this matter. Could these proteins be loosely attached to the cells when the cells are growing in liquid medium but more tightly when the cells are growing on a solid surface or undergoing development? Could two different extracellular polysaccharides be produced even though little differences in composition have been evidenced in the glycan fractions extracted from the two types of cells (Sutherland and Thomson, 1975)?

Here, we consider the proteins that are linked to the extracellular matrix as secreted proteins, regardless of whether they are extractable from the growth medium.

Diversity of Proteins Secreted by Myxobacteria

It has been known for a long time that myxobacteria produce numerous extracellular enzymes, and several studies concerning cell lysis activities were undertaken in the 1970s or even earlier (for a review, see Rosenberg and Varon [1984]).

An increasing number of proteins have been discovered lately. Systematic surveys of all proteins secreted or exported have recently been undertaken for *Myxococcus coralloides* (Farez-Vidal et al., 1990; Gonzalez et al., 1989, 1991) and for *Stigmatella aurantiaca* and *M. xanthus* (Julien, personal communication).

It becomes more and more evident that the myxobacteria may be an outstanding source of diverse extracellular degradative enzymes as well as of secondary metabolites. Thus, the chymosin-like milk-clotting protease (Mcp) produced by *M. xanthus* has a potential biotechnological application based on its site specificity (Carias et al., 1990). This enzyme is also interesting, however, because of its inhibition by aspartyl-protease inhibitors (Petit and Guespin-Michel, 1992), which is quite an unusual property among prokaryotic proteases and has prompted discoveries of analogous enzymes in other myxobacteria, e.g., *S. aurantiaca* (Julien, personal communication). The unexpected presence of a blood anticoagulant glycopeptide in the supernatant of *M. xanthus* (Akoum et al., 1987; Masson et al., 1991) also will certainly open interesting prospects.

Constancy in Rate of Protein Secretion in *M. xanthus*

As stated above, the amount of protein secreted per unit of dry mass is constant when *M. xanthus* is grown in the standard liquid Castione medium. Moreover, modifications of the culture conditions that significantly change the rate of accumulation of certain extracellular enzymes have no effect on the total secretion rates of proteins.

Thus, Nicaud et al. (1984) showed that the concentration of Casitone in the growth medium regulates (though by different processes) the rate of liberation of proteolytic and bacteriolytic activities, while total protein secretion remains constant. Gonzalez et al. (1989) have shown that varying the carbon source or the phosphate concentration in the medium results in modifications in the production of phosphatases without alterations in the overall rate of protein secretion by *M. coralloides*. Immobilization of *M. xanthus* cells in carrageenan (Younes et al., 1987) or in alginate beads (Fortin and Vuillemard, 1990) enhances the extracellular proteolytic activity but does not modify the total protein secretion rate. Younes (1986) tested the effect of adding ions, organic acids, or detergents to the growth medium. Only slight variations of the secretion rate were observed. The most drastic effect was an increase (40%) of the secretion rate when the cells were cultivated at 34°C in a fermentor with the pH held constant at 7.6.

These results indicate that protein secretion is a tightly regulated process but do not give any hint as to which step of the production pathway is controlled. Could it be directly at the level of export or secretion, or could there be a global regulation of synthesis of degradative enzymes such as takes place, for instance, in *Bacillus subtilis* (Msadek et al., 1990)?

Mutations That Modify Rate of Protein Secretion in *M. xanthus*

There have been repeated attempts to look for Tn*5* insertion mutants that might be impaired in either the overall process of secretion or the production of one or several extracellular enzymes. Different screening media have been used; clones were toothpicked onto a layer of soft agar containing a substrate allowing the formation of halos of activity surrounding the colonies (Nicaud et al., 1984). The size of the halo was used as the selection criterion. The putative mutant clones were assayed in liquid cultures for extracellular protein secretion and the production of particular enzyme activities.

No mutant totally impaired in protein secretion was ever found, but two classes of mutants, Exc^{++} and Exc$^{\pm}$, were defined as secreting, respectively, more or less total proteins. Electrofocusing profiles of the supernatants from late-exponential-phase cultures of four of these mutants have revealed striking similarities to and very few differences from the parent (Nicaud et al., 1984). The mutants produced either more or less of most extracellular enzyme activities assayed: general proteolytic and bacteriolytic activities, amylase (Nicaud et al., 1984), acid, milk-clotting endoprotease (Mcp), endoglucanase, and even extracellular polysaccharide production (Masson and Guespin-Michel, 1988) (Table 1). These activities were detected only in the growth medium for all strains, which means that no accumulation of the enzymes in the periplasms of Exc$^{\pm}$ mutants was ever detected. Degrees of activity varied from one activity to another and differed in each mutant. However, since the assays were performed on crude supernatants in which proteases were present in various amounts depending on the mutants, a precise analysis of the data might not be meaningful.

Some activities, however, varied in a manner opposite to that of the bulk of extracellular proteins; DNase and RNase activities (although they were detected only as halos surrounding the colonies on appropriate media

Table 1. Phenotypes of some Exc$^{\pm}$ and Exc^{++} mutants[a]

Mutant	Phenotype	Total mass secreted protein	Activity			
			Proteolytic	Bacteriolytic	Endoglucanase	Mcp
CM019	Exc$^{\pm}$	60	28	70		
CM031	Exc$^{\pm}$	60	53	45	40	70
CM040	Exc$^{\pm}$	70	50		57	70
CM033	Exc^{++}	150	161	150		
CM130	Exc^{++}	150	145	140		
CM057	Exc^{++}	170	140		130	130

[a] All activities are expressed as percentages of the parental strain measured under the same conditions. Total protein, proteolytic activity, endoglucanase, and Mcp were measured in the supernatants of late exponential cultures, and bacteriolytic activities were measured in the supernatants of mid-log-phase cultures. The activities are expressed per unit of cell dry weight.

[Nicaud et al., 1984]) varied in this way. It must be recalled, however, that Munoz et al. (1989) demonstrated that DNase could not be detected in the growth medium and thus may be periplasmic. Another periplasmic activity, an acid phosphatase, was also produced as an inverse function of extracellular protease production, a result that has not yet been explained (Breton et al., 1984). Finally, the production of only one extracellular protein, the blood anticoagulant glycopeptide designated myxaline and synthesized during growth (Akoum et al., 1987), was not modified in any of the $Exc^{\pm}$ or Exc^{++} mutants (Masson and Guespin-Michel, 1988).

All these mutants seemed to affect genes that regulate the rate of production of most extracellular proteins. Several genes are therefore involved in this control. The lack of true secretory mutants (i.e., mutants totally unable to secrete at least some of the proteins) might reflect the complex nature of the secretory mechanism or indicate that mutations that prevent extracellular protein production are lethal in *M. xanthus*.

Secreted Proteins Involved in Motility

The gliding motility of *M. xanthus* is regulated by two distinct systems, termed A (adventurous or individual motility) and S (social or group motility) (see chapter 13).

Concerning A motility, the most detailed information available on extracellular proteins comes from studies by Kalos and Zissler (1990), who used the Tn*phoA* fusion system of Manoil and Beckwith (1985). At least three extracellular proteins seem to be required for A motility. Since $Exc^{\pm}$ mutants display normal A motility, these proteins are likely not regulated concomitantly with extracellular degradative enzymes.

Concerning S motility, the molecular composition of the fimbriae (or pili) and fibrils associated with S motility is not precisely known (Arnold and Shimkets, 1988; see chapter 3). Proteins involved in the formation of these structures may not be modified by the $Exc^{\pm}$ mutations, which do not seem to impair S motility either.

Protein Secretion and Phase Variation in *M. xanthus*

Protein secretion should be added to the list of numerous differences between the tan and yellow variants of *M. xanthus* (Yee and Inouye, 1984; Reichenbach and Kleinig, 1984). On the one hand, tan variants secrete less of most proteins than their yellow counterparts do, although they are much less affected than the $Exc^{\pm}$ mutants (unpublished results). On the other hand, all $Exc^{\pm}$ mutants isolated so far are tan, whereas the Exc^{++} mutants are yellow. This fact should be taken into account when secretory mutants are studied.

PROTEIN SECRETION DURING DEVELOPMENT IN *M. XANTHUS*

Extracellular Proteins Produced during Development

The problem of cell lysis during development, regardless of its precise extent, obviously complicates the study of protein secretion.

In contrast to sporulating bacteria such as *Bacillus* sp., *M. xanthus* does not respond to starvation by a massive production of degradative enzymes. Some extracellular proteins are, however, produced during development, and it is likely that their production is related to this process.

Table 2 summarizes our present knowledge of extracellular proteins released during development, but the status of these proteins with regard to secretion may differ substantially. Three proteins (S, U, and C) are assembled on the spore surface, from which they can be more or less easily released (for example, protein S is released by a 1 M NaCl wash), and are considered spore coat proteins. They are referred to as secretory proteins but are never found in the medium.

The major early spore coat protein, protein S, has no signal peptide. Two hypotheses have been proposed to explain how protein S is released. According to the exospore hypothesis (Teintze et al., 1985), protein S is released by cells that undergo developmental autolysis and is then assembled on the surface of cells that sporulate. According to the second hypothesis, protein S can be translocated across the membrane without a cleavable signal peptide (Nelson and Zusman, 1983).

In contrast, protein U is produced at a late stage of development, with a typical signal peptide of 25 amino acid residues, and may be secreted and assembled on the spore surface (Gollop et al., 1991).

Table 2. Known extracellular proteins produced during *M. xanthus* development

Protein	Localization	Period of secretion	Reference(s)
S	Spore coat	Late	Nelson and Zusman, 1983
C	Spore coat	Late	McCleary et al., 1991
U	Spore coat	Late	Gollop et al., 1991
VGP	Extracellular medium	Early	Glufka and Maeba, 1991
CSA 1604	50% in extracellular medium with agitation	Throughout	Jarvis and Dworkin, 1989
C factor	Extracellular matrix	Early	Shimkets and Rafiee, 1990; Kim and Kaiser, 1990a
FA-1	Extracellular matrix		Behmlander and Dworkin, 1991
A factor	Extracellular medium	Very early	Plamann et al., 1992
Mcp	Extracellular medium	Early and late	Petit and Guespin-Michel, 1992

The VGP glycoprotein is a surface protein that represents 1% of the cell protein of vegetative cells. During development, it is released from the cell surface by an unknown mechanism. Does this mean that it is required as a secreted protein for the development cycle, or that its binding to the cell surface would prevent some step of development?

The cell surface antigen CSA 1604 is exclusively cell bound when cells are grown under nonagitated conditions but becomes 50% extracellular when agitation is applied. Should it then be considered a secreted protein? It is present during growth and development but may be involved only in development (see chapter 3).

At least two developmental proteins are associated with the fibrils of the extracellular matrix. C factor (the product of gene *csgA*) is a cell-cell signaling protein of paramount importance throughout development. It is not found free in the medium but is only fibril associated. Its range of action is narrow, as attested to by the fact that cell alignment in close position is required for its activity (Kim and Kaiser, 1990b; see chapter 12). The role of the second protein, the FA-1 antigen, which appears only during development, is unknown.

Mcp is a truly secreted protein during both growth and development, when it displays two phases of production. Two of the Tn*5lacZ* fusions in developmentally regulated genes (Kroos et al., 1986) prevent Mcp production during early development but allow normal fruiting. However, the fact that Mcp production is impaired during development in *asg, bsg,* and *dsg* mutants as well as in the two insertion fusions shows that it is a developmentally regulated secretory activity.

The cell-to-cell signaling A factor contains a heat-stable part (Kuspa et al., 1992) that seems to be generated by an extracellular protease(s) produced during early aggregation (see chapter 12).

It is quite likely that not all the proteins secreted during development have been discovered. The diversity of the biological problems raised by these proteins may yet surprise us.

Developmental Mutants Impaired in Protein Secretion during Growth

Four Exc$^{\pm}$ mutants analyzed so far (Nicaud, 1983; Merah, 1992) were unable to fruit on CF agar and belonged to at least two unlinked loci tentatively called *excA* and *excB*.

In addition, Merah (1992) has compared the production of extracellular proteins and enzymatic activities in six mutants belonging to all the known loci in which phenotypically complementable mutations that impair sporulation have been found (Hagen et al., 1978; see chapter 12). All *asg* mutants (Kuspa and Kaiser, 1989) behaved as Exc$^{\pm}$ mutants, which agrees with Kuspa and Kaiser's suggestion (1989) that the *asg* mutants might be secretory mutants. (One *bsg* mutant [Gill et al., 1988] was impaired in the production of only some extracellular activities.)

Thus, at least five different loci (*asgA, asgB, asgC, excA,* and *excB*) involved both in the production of total extracellular proteins during growth and the developmental cycle have been identified.

The effect of mutations in these genes on protein secretion during development has been only partly documented. Early Mcp secretion is drastically reduced during submerged development of all Exc$^{\pm}$ mutants (Petit and Guespin-Michel, 1992). In contrast, C factor is produced by *asg* mutants, while *excA* and *excB* mutants are able to rescue development and A-factor-dependent expression of a *lacZ* fusion in an *asgB* mutant (unpublished data) and thus presumably are able to secrete this factor.

The relationship between extracellular protein production and the development cycle is still largely speculative. The simple hypothesis that some of the extracellular proteins required for development would be produced at too low a level in all these mutants is contradicted, but not altogether precluded, by the mere fact that *asg* mutants are rescued by *exc* mutants. Another hypothesis can be proposed by analogy with the *B. subtilis* sporulation pathway, during which the regulation of degradative enzymes on the one hand and the onset of sporulation on the other are controlled at the transcriptional level by two different but interrelated signal transduction systems (Klier et al., 1992; Grossman, 1991). It is also possible that the *exc* genes products belong to the secretory machinery and that the integrity of this structure is required for complete development. In consequence, the study of the secretory mechanism(s) in *M. xanthus* will certainly help us understand the regulation of development.

MECHANISM(S) OF PROTEIN SECRETION IN *M. XANTHUS*

Mechanisms of Protein Secretion in Gram-Negative Bacteria

Protein secretion in gram-negative bacteria can only be a complex process, considering the presence of the two membranes that all secreted proteins must cross. It is only recently, therefore, that intensive studies have been carried out with several gram-negative bacteria. Two main routes have been described, depending on whether the protein is first exported into the periplasm or crosses both membranes in a single step (for reviews, see Pugsley [1988] and Letouvet-Pawlak [1991b]).

The export mechanism involved in the first route has been thoroughly studied only in *E. coli,* but similar processes have been demonstrated in other bacteria. The protein is translated into a precursor with an NH_2-terminal signal peptide and processed during translocation across the inner membrane. Once in the periplasm, it is translocated across the outer membrane by mechanisms specific to each or common to a few proteins. Some proteases (e.g., the immunoglobulin A protease from *Neisseria gonorrhoeae*) are autosecretory proteins in which the COOH extremity is required for the outer membrane translocation that also involves autoproteolysis of this fragment (Halter et al., 1984; Pohlner et al., 1987). Conversely, at least 13 different secretion proteins are necessary to ensure other membrane translocation of a single protein, the pullulanase produced by *Klebsiella oxytoca* (Pugsley and Reyss, 1990; Reyss and Pugsley, 1990). Finally, in other bacteria *(Pseudomonas aeruginosa, Erwinia chrysanthemi, Aeromonas hydrophila),* several secretion proteins that display definite homology with some of those in *K. pneumoniae* constitute a secretory apparatus common to several secreted proteins. These mechanisms, however, are very specific with regard to the protein(s) that is secreted. Mutations in the corresponding genes prevent secretion and lead to accumulation of the secreted proteins in the periplasms of the mutated cells (Wretlind and Pvlovskis, 1984; Andro et al., 1984; Thurn and Chatterjee, 1985; Howard and Buckley, 1983).

In the second route, the secreted protein is translocated in one step through both membranes and is not dependent on a signal peptide export machinery. In the case of the *E. coli* α-hemolysin, three proteins constitute a channel allowing specific translocation through both membranes, the corresponding secreted protein possessing a C-terminal topogenic recognition sequence (Holland et al., 1989). Several other secretion mechanisms belong to this type, as is shown by the homologies between secretion proteins.

As for the colicins (A and EII), they are synthesized in the cytoplasm of the producing strain but are released into the medium through the activity of a so-called lysis protein, the action of which results in transient permeabilization of the membranes (allowing a nonspecific release of other proteins in addition to the colicin [Baty et al., 1987]). Mutations in the genes coding for the lysis proteins result in the accumulation of colicin in the cytoplasm.

When no secretion mutants are easily available in the endogenous strain, the genes corresponding to the secreted proteins have been cloned and expressed in *E. coli,* the localization of the peptide in this strain being generally indicative of the general route of secretion used in the original host (Table 3).

Sequence comparisons of the proteins involved in the secretory systems of different proteins have shown that most of them belong to these four families, regardless of the bacterial species in which they are found, but different mechanisms can coexist in the same bacterium.

Table 3. Comparison of the four main secretory mechanisms of gram-negative bacteria with the secretory mechanism of *M. xanthus*

Class of mechanism	Signal peptide	Protein expressed from gene cloned in *E. coli*		Secretory mechanism	Mutants
		Localization	Secretion		
Hemolysin family	None	Cytoplasmic	Yes if 3 secretory genes are also provided	Limited to protein displaying C-terminal topogenic sequence	
Colicin family	None	Cytoplasmic	Yes if gene encoding lysis protein is provided	Nonspecific; other proteins also released	
Immunoglobulin A protease family	Yes	Secreted (C-terminal part of protein is required)		Conformation of C-terminal part of proprotein allows outer membrane translocation.	
Pullulanase family	Yes	Periplasmic	Yes if numerous secretory genes are provided	Very specific	Secretion mutants accumulate the proteins in their periplasms.
M. xanthus	Yes, at least in some cases[a]			Very nonspecific	$Exc^{\pm}$ and Exc^{++} mutants display global increase or decrease of all secreted proteins.

[a] Signal peptide-processed foreign proteins can be secreted, chimeric proteins obtained after Tn*phoA* fusions may be secreted, and some secreted proteins have a signal peptide precursor.

Signal Peptide Export Pathway in *M. xanthus*

The presence of a signal peptide export pathway involved in either exported or secreted proteins in *M. xanthus* is poorly documented. The arguments that suggest a signal peptide export are largely indirect. (i) Tn*phoA* random insertions yield chimeric proteins that are either cell bound or secreted (Kalos and Zissler, 1990; Breton et al., 1990). However, the very low frequency of fusions expressing PhoA activity obtained by both groups indicates a difference from the situation observed when the same fusion approach is used for members of the family *Enterobacteriaceae*. (ii) Foreign proteins expressed from genes coding for a signal peptide-processing protein are secreted totally or partly by recombinant *M. xanthus* strains (Breton et al., 1986; Breton and Guespin-Michel, 1987). (iii) The presence of a signal peptide has been deduced from the sequences of a few extracellular *M. xanthus* proteins that are released during development, e.g., protein U (Gollop et al., 1991) and possibly C factor (Hagen and Shimkets, 1990).

In contrast, protein S (Nelson and Zusman, 1983; Teintze et al., 1985) and developmental protein MBHA (myxobacterial hemagglutinin) located in the periplasm and on the cell surface (Romeo et al., 1986) do not have signal peptide-containing precursors. In addition, an autocide factor, AMI, is produced during development and has been proposed to increase the permeability of the cells in a manner reminiscent of the lysis protein involved in colicin secretion (Rosenbluh and Rosenberg, 1990).

It is thus very likely that different secretory mechanisms coexist in *M. xanthus,* as is the case in many other secretory bacteria. However, the degradative enzymes secreted during growth might be first exported into the periplasm via a signal peptide route.

Mechanism Involved in Secretion of Foreign Proteins by *M. xanthus*

Low specificity of *M. xanthus* secretory mechanism(s)

One of the most striking differences between *M. xanthus* and the other secretory systems studied so far is the absence of specificity towards the proteins secreted by the former. The five pectate lyases from *E. chrysanthemi* are readily secreted by *M. xanthus* recombinants expressing the five cloned genes (Breton et al., 1986). In their host, the proteins are secreted through the "pullulanase-like" mechanism. A homologous mechanism is also present in *P. aeruginosa*. Homologies have been detected by DNA hybridization between the secretion genes *(out)* of *P. aeruginosa* and those of other species such as *Xanthomonas campestris*. Though the GC content of these species is similar to that of *M. xanthus*, attempts to detect any significant hybridization between these *out* genes and the chromosome of *M. xanthus* have been unsuccessful (Filloux, personal communication).

Even more unexpected was the fact that both the TEM-1 β-lactamase coded for by RP4 and located in the periplasm of all tested gram-negative bacteria and the *E. coli* periplasmic pH 2.5 acid phosphatase expressed from the *appA* gene are at least partly secreted when synthesized in *M. xanthus*. $Exc^{\pm}$ and Exc^{++} mutations generated by the random insertion of the vector used to insert the *appA* gene into the *M. xanthus* chromosome respectively decreased or increased both the cell-bound and the secreted AppA activities of the recombinants (Breton and Guespin-Michel, 1987). This demonstrates that $Exc^{\pm}$ and Exc^{++} mutations, which modify the extracellular production of native proteins, also impair the total production of the foreign protein AppA. Since AppA was expressed under the double control of its own promoter and of the constitutive *tet* promoter of plasmid pBR322, it can be speculated that $Exc^{\pm}$ and Exc^{++} mutations release a control that is active at a posttranscriptional level. To understand this control, we have taken advantage of the lack of specificity of the secretory mechanism and analyzed protein AppA secretion. Its advantage over the native proteins is that 70% of its activity remains cell bound, thus allowing an easy distinction between export and secretion.

Protein diffusion and crossing of outer envelopes by protein AppA

The principle of these experiments on protein diffusion and crossing of the outer envelopes by protein AppA (Letouvet-Pawlak, 1991a) is that the extracellular production of AppA reflects translocation of the protein through all external envelopes, while the cell-bound activity reflects export of the protein. The observed secretion is not due to cell lysis, and the cell-bound activity is periplasmic in *M. xanthus,* as it is in *E. coli.*

The *appA* construct (a transferable cointegrate consisting of a conjugative IncP plasmid inserted at random in the *M. xanthus* chromosome and a pTTQ18 derivative in which gene *appA* is under the control of the isopropyl-β-D-thiogalactopyranoside [IPTG]-inducible *tac* promoter) was first inserted into strain DZ1. This strain is a pleiotropic mutant, nonmotile, fruitless, and $Exc^{\pm}$. Extracellular AppA is not degraded in the growth medium of this strain, and the kinetics of production and secretion of this protein after induction with IPTG can be monitored over several generations until steady state is reached.

Not only was the AppA protein not totally secreted, but the kinetics of secretion were slow with regard to the initial accumulation of AppA inside the cells. This suggests that although the cell envelope is rather permeable to many foreign proteins, this permeability is not due to a mere leakage of any polypeptide that accumulates in the periplasm. The secretion rate and the amount of pH 2.5 acid phosphatase released directly depended on the intracellular amount of the enzyme, and the release started only 5 h after initial accumulation of the enzyme in the periplasmic space (i.e., after IPTG induction). These results could be explained if protein diffusion through the periplasmic gel and/or the outer membrane was the secretory mechanism or the limiting step in the process. In this hypothesis, the initial stage of enzyme retention in the cells that precedes secretion would therefore correspond to the time required for diffusion of the protein through one and/or the other of these structures, and the percentage of enzyme released in the extracellular medium would depend on its diffusion coefficient. However, the fact that these measures reflect the permeability of a foreign protein might introduce an artifact due to the mere presence of the protein. It was therefore necessary to check the reliability of the method. AppA secretion was thus compared in different *M. xanthus* strains.

The strategy described above that was effective with an Exc$^{\pm}$ strain could not be applied to an Exc^{+} strain because of proteolysis of the secreted AppA protein by this strain. A simpler approach consisting of resuspending the cells in fresh medium was thus devised. Under these conditions, accumulation of extracellular acid phosphatase activity (expressed per milligram of total cell protein) was linear, the slope of the curve was proportional to the concentration of the cell-bound AppA, and the cellular retention time was unchanged (unpublished results).

This method was used to monitor protein secretion in a DK1622 wild-type recombinant harboring the *appA* construct and in an Exc$^{\pm}$ mutant obtained by transduction of the Tn*5* insertion Ω031 (Table 1) into this recombinant. The two strains were isogenic except for a small DNA segment inducing the Ω031 Tn*5* insertion responsible for the Exc$^{\pm}$ phenotype. In both cases, the secretion curves were linear, indicating that the mechanism of exit was not altered by the Exc$^{\pm}$-Exc^{+} state of the host. However, protein permeation in the Exc$^{\pm}$ mutated cells was drastically impaired, as evidenced by the large increase in retention time (5 h compared with 1 h) and the lowering of the secretion rate. Therefore, the Exc$^{\pm}$ mutation must result in a modification(s) of the structure of the envelopes that affects the permeation characteristics. In addition, the Ω031 insertion decreased cell-bound phosphatase activity in the Exc^{+} parental strain, as did the Exc$^{\pm}$ mutations obtained by Breton and Guespin-Michel (1987).

Thus, the single Exc$^{\pm}$ mutation decreased both the total production (export and secretion) of the foreign protein and the permeability of the outer envelope. This suggests either that outer-envelope permeability and rate of export are somehow correlated or that the amount of periplasmic activity would decrease when the retention time was increased (e.g., due to protein turnover). In any case, both hypotheses imply that Exc$^{\pm}$ mutants are impaired in the secretory apparatus.

Secretory Mechanism for Native Proteins in Myxobacteria

Study of the secretion of the AppA protein has proved a fruitful model for the understanding of protein secretion in *M. xanthus,* but it has not answered all questions. For instance, why are native active proteins never found in the periplasm? The hypothesis that the periplasmic form might be an inactive precursor that is activated upon secretion has been ruled out at least for the glucanase, since a cell-bound activity can be assayed in *E. coli* recombinants, expressing the corresponding *M. xanthus* genes (unpublished results). In addition, other foreign proteins such as the pectate lyases from *E. chrysanthemi* are totally secreted by *M. xanthus* recombinants. In contrast, several endogenous enzymes remain mainly periplasmic in growing myxobacteria as well as in developing cells (e.g., MBHA, although it is also partly exposed to the outer surface of the outer membrane [Nelson et al., 1981]).

Several hypotheses attempting to reconcile these observations can be proposed. Totally secreted proteins might be secreted through a specific yet undetected apparatus that would not recognize AppA. This would imply either that *exc* genes also regulate the functioning of this apparatus or that diffusion of the proteins through the outer envelope depends on the integrity of the *exc* gene products and is the limiting step in the secretion of all proteins. Conversely, the genuinely periplasmic proteins might actually be anchored on the outer face of the inner membrane, and the free periplasmic proteins might be partially or totally secreted at a rate that would depend on their diffusion constant through the outer envelope. In addition, native secreted proteins might have common topogenic sequences that would either enhance their diffusion through the envelope or allow their subsequent active secretion. Comparison of the genes coding for extracellular proteins, many of which have been sequenced (and more are being sequenced), might shed some light on that question. One must, however, keep in mind the likelihood of the existence of several secretory mechanisms, several of which might function simultaneously, especially during development.

CONCLUSION

Protein secretion is an extremely important process at all stages of the growth and development of *M. xanthus*. Most of the known extracellular proteins secreted during growth are degradative enzymes implicated in cell nutrition. The systematic survey of these proteins is only beginning, and the available results seem promising both from fundamental and applied viewpoints.

A set of genes defined by their mutated phenotypes, Exc$^{\pm}$ or Exc^{++}, regulates the production of most proteins secreted during growth. This suggests a common secretory pathway for these proteins, the mechanism of which would be altogether different from those known in other gram-negative bacteria. Table 3 summarizes the main known functional differences. The secretion of most proteins in *M. xanthus* seems to require previous export into the periplasm, and no autosecretory mechanism has been yet described. However, the process also differs form the pullulanase-like mechanism in that it lacks specificity and that no secretory mutations (i.e., mutations that prevent translocation of most proteins through the outer membrane) have been isolated. This supports our proposal that *M. xanthus* possesses at least one secretory mechanism that is still unknown.

It is equally important that all known Exc$^{\pm}$ mutations (scattered among five genes, including the three *asg* genes) also prevent normal development, though they do not seem to hinder the secretion of all the proteins secreted during development. Could this indicate the existence in *M. xanthus* of an adaptive regulation common to the synthesis of degradative enzymes and to the developmental cycle, similar to that demonstrated in *B. subtilis* (Msadek et al., 1990; Klier et al., 1992)?

One Exc$^{\pm}$ mutation, however, simultaneously decreases the total production of the foreign protein AppA expressed under the control of the inducible promoter p^{tac} and the permeability of the outer envelope to this protein (the other Exc$^{\pm}$ mutations were not tested in this regard). It seems difficult to reconcile these results with a model of transcriptional regulation of the production of secreted proteins unless proteins involved in envelope permeability are coregulated with the degradative enzymes.

The fact that the Exc$^{\pm}$ mutants are affected at different steps of the developmental cycle (since *exc* mutants can rescue *asg* mutants) suggests that it would not be the global deficiency in protein secretion that causes the defect in development but that the genes in which these mutations occur play a direct role in both processes.

A modification of the biogenesis of the outer envelope that would modulate its permeability constants and interfere with the developmental cycle would be more likely at the present stage of our studies to explain the Exc$^{\pm}$ phenotypes and the effect of at least one *exc* mutation on the kinetics of secretion of a foreign protein. One hypothesis could be that this modification concerns the export and amount of some components of the envelope (periplasm or outer membrane); alternatively, a decrease in envelope permeability due to structural changes could retroinhibit export.

This hypothesis highlights the importance of protein secretion in relation to *M. xanthus* social behavior, both during its vegetative scavenging life and during its multicellular response to nutrient starvation.

REFERENCES

Akoum, E. I., M. Vidjalakshmi, P. Cardon, B. Fournet, M. Sigot, and J. F. Guespin-Michel. 1987. *Myxococcus xanthus* produces an extracellular glycopeptide that displays blood anticoagulant properties. *Enzyme Microb. Technol.* **9:**426–446.

Andro, T., J. P. Chambost, A. Kotoujansky, J. Cattaneo, Y. Bertheau, F. Barras, F. Van Gijsegem, and A. Coleno. 1984. Mutants of *Erwinia chrysanthemi* defective in secretion of pectinase. *J. Bacteriol.* **160:**1199–1203.

Arnold, J. W., and L. J. Shimkets. 1988. Cell surface properties correlated with cohesion in *Myxococcus xanthus*. *J. Bacteriol.* **170:**5771–5777.

Baty, D., R. Lloubès, V. Geli, C. Lazdunski, and S. P. Howard. 1987. Extracellular release of colicin A is non-specific. *EMBO J.* **6:**2463–2468.

Behmlander, R. M., and M. Dworkin. 1991. Extracellular fibrils and contact-mediated cell interactions in *Myxococcus xanthus*. *J. Bacteriol.* **173:**7810–7821.

Breton, A., and J. F. Guespin-Michel. 1987. *Escherichia coli* pH 2.5 acid phosphatase and β-lactamase TEM-2 are secreted into the medium by *Myxococcus xanthus*. *FEMS Microbiol. Lett.* **40:**183–188.

Breton, A., J. M. Nicaud, and J. F. Guespin-Michel. 1984. *Myxococcus xanthus,* a Gram-negative nonpathogenic bacterium, that secretes proteins into the extracellular growth medium, is a potential cloning host for excreted proteins production. *Verlag Chem.* **3:** 441–446.

Breton, A. M., I. Buon, and J. F. Guespin-Michel. 1990. Use of Tn*phoA* to tag exported proteins in *Myxococcus xanthus*. *FEMS Microbiol. Lett.* **67:**179–186.

Breton, A. M., G. Younes, F. VanGijsegem, and J. F. Guespin-Michel. 1986. Expression in *Myxococcus xanthus* of foreign genes coding for secreted pectate lyases of *Erwinia chrysanthemi*. *J. Biotechnol.* **4:**303–311.

Burnham, J. C., S. A. Collart, and B. W. Highison. 1981. Entrapment and lysis of the cyanobacterium *Phormidium luridum* by aqueous colonies of *Myxococcus xanthus* PCO_2. *Arch. Microbiol.* **129:**285–294.

Carias, J. R., J. Raingeaud, C. Mazaud, G. Vachon, N. Lucas, Y. Cenatiempo, and R. Julien. 1990. A chymosin like extracellular acidic endoprotease from *Myxococcus xanthus* DK101: a potential new tool for protein engineering. *FEBS Lett.* **262:**97–100.

Clemans, D. L., C. M. Chance, and M. Dworkin. 1991. A development-specific protein in *Myxococcus xanthus* is associated with the extracellular fibrils. *J. Bacteriol.* **173:** 6749–6759.

Coletta, P. L., and P. G. G. Miller. 1986. The extracellular proteases of Myxococcus xanthus. *FEBS Microbiol. Lett.* **37:**203–207.

Farez-Vidal, M. E., A. Fernandez-Vivas, and J. M. Arias. 1990. Amylase programming during the life cycle of *Myxococcus coralloides*. *J. Appl. Bacteriol.* **69:**119–124.

Filloux, A. Personal communication.

Fortin, C., and J. C. Vuillemard. 1990. Elucidation of the mechanism involved in the regulation of protease production by immobilized *Myxococcus xanthus* cells. *Biotechnol. Lett.* **12:**913–918.

Gill, R. E., M. G. Gull, and S. Fly. 1988. Genetic identification and cloning of a gene required for developmental cell interaction in *Myxococcus xanthus. J. Bacteriol.* **170:**5279–5288.

Glufka, R., and M. Maeba. 1991. Release of a cell-surface protein during development of *Myxococcus xanthus. J. Bacteriol.* **173:**7988–7991.

Gnosspelius, G. 1978. Myxobacterial slime and proteolytic activity. *Arch. Microbiol.* **116:** 51–59.

Gollop, R., M. Inouye, and S. Inouye. 1991. Protein U, a late developmental spore coat protein of *Myxococcus xanthus,* is a secretory protein. *J. Bacteriol.* **173:**3597–3600.

Gonzalez, F., J. Munoz, J. M. Arias, and E. Montoya. 1989. Production of acid and alkaline phosphatases by *Myxococcus coralloides. Folia Microbiol.* **34:**185–194.

Gonzalez, F., A. Vargas, J. M. Arias, and E. Montoya. 1991. Phosphatase activity during development cycle of *Myxococcus xanthus. Can. J. Microbiol.* **37:**74–77.

Grossman, A. D. 1991. Integration of developmental signals and the initiation of sporulation in *Bacillus subtilis. Cell* **65:**5–8.

Hagen, D. C., and L. J. Shimkets. 1990. Nucleotide sequence and transcriptional products of the *csg* locus of *Myxococcus xanthus. J. Bacteriol.* **172:**15–23.

Hagen, T. J., A. P. Bretscher, and D. Kaiser. 1978. Synergism between morphogenetic mutants of *Myxococcus xanthus. Dev. Biol.* **64:**284–296.

Halter, R., J. Pohlner, and T. F. Meyer. 1984. IgA protease of *Neisseria gonorrhoeae:* isolation and characterization of the gene and its extracellular product. *EMBO J.* **3:** 1595–1601.

Holland, I. B., R. Wang, S. J. Séror, and M. Blight. 1989. Haemolysin secretion and other protein translocation mechanisms in gram-negative bacteria, p. 219–254. *In* S. Banmberg, I. Hunter, and M. Rhodes (ed.), *Microbial Products: New Approaches.* Cambridge University Press, Cambridge.

Howard, S. P., and J. T. Buckley. 1983. Intracellular accumulation of extracellular proteins by pleiotropic export mutants of *Aeromonas hydrophila. J. Bacteriol.* **154:**413–418.

Jarvis, B. W., and M. Dworkin. 1989. Purification and properties of *Myxococcus xanthus* cell surface antigen 1604. *J. Bacteriol.* **171:**4655–4666.

Julien, R. Personal communication.

Kalos, M., and J. Zissler. 1990. Transposon tagging of genes for cell-cell interactions in *Myxococcus xanthus. Proc. Natl. Acad. Sci. USA* **87:**8316–8320.

Kim, S. K., and D. Kaiser. 1990a. Purification and properties of *Myxococcus xanthus* C-factor, an intercellular signaling protein. *Proc. Natl. Acad. Sci. USA* **87:**3635–3639.

Kim, S. K., and D. Kaiser. 1990b. Cell alignment required in differentiation of *Myxococcus xanthus. Science* **249:**926–928.

Klier, A., T. Msadek, and G. Rapoport. 1992. Positive regulation in the Gram-positive bacterium: *Bacillus subtilis. Annu. Rev. Microbiol.,* **46:**429–459.

Kroos, L., A. Kuspa, and D. Kaiser. 1986. Intercellular signalling is required for developmental gene expression in *Myxococcus xanthus. Dev. Biol.* **117:**267–276.

Kuspa, A., and D. Kaiser. 1989. Genes required for developmental signalling in *Myxococcus xanthus:* three *asg* loci. *J. Bacteriol.* **171:**2762–2772.

Kuspa, A., L. Plamann, and D. Kaiser. 1992. Identification of heat-stable A-factor from *Myxococcus xanthus. J. Bacteriol.* **174:**3319–3326.

Letouvet-Pawlak, B. 1991a. Mécanisme de sécrétion des protéines chez *Myxococcus xanthus:* étude de la sécrétion d'une protéine étrangère après clonage du gène en aval d'un promoteur inductible. Ph.D. thesis. Université de Rouen, Rouen, France.

Letouvet-Pawlak, B. 1991b. La sécrétion des protéines chez les bactéries à Gram négatif. *Ann. Inst. Pasteur/Actualités* **4:**290–309.

Manoil, C., and J. Beckwith. 1985. Tn*pho*A: a transposon probe for protein export signals. *Proc. Natl. Acad. Sci. USA* **82:**8129–8133.

Masson, P., D. Gilbert, and J. F. Guespin-Michel. 1991. Immunological analysis of a blood-anticoagulant activity prepared from *Myxococcus xanthus* culture broth. *Enzyme Microb. Technol.* **13:**327–331.

Masson, P. J., and J. F. Guespin-Michel. 1988. An extracellular blood-anticoagulant glycopeptide produced exclusively during vegetative growth by *Myxococcus xanthus* and other myxobacteria is not co-regulated with other extracellular macromolecules. *J. Gen. Microbiol.* **134:**801–806.

Mayer, H., and H. Reichenbach. 1978. Restriction endonucleases: general survey procedure and survey of gliding bacteria. *J. Bacteriol.* **136:**708–713.

McCleary, W. R., B. Esmon, and D. R. Zusman. 1991. *Myxococcus xanthus* protein C is a major spore surface protein. *J. Bacteriol.* **173:**2141–2145.

Merah, M. 1992. Obtention et étude de mutants de *Myxococcus xanthus* altérés dans la production d'activités enzymatiques exocellulaires. Ph.D. thesis. Université de Rouen, Rouen, France.

Msadek, T., F. Kunit, D. Henner, A. Klier, G. Rapoport, and R. Dedonder. 1990. Signal transduction pathway controlling synthesis of a class of degradative enzymes in *Bacillus subtilis:* expression of the regulatory genes and analysis of mutation in *degS* and *degU*. *J. Bacteriol.* **172:**824–834.

Munoz, J., F. Gonzalez, M. M. Martinez-Canamero, A. M. Goiocoechea, A. L. Extremèra, and J. M. Arias. 1989. Deoxyribonuclease and phosphatase activity in myxobacteria. *Microbios* **58:**43–47.

Nelson, D. R., M. G. Cumsky, and D. R. Zusman. 1981. Localization of myxobacterial hemagglutinin in the periplasmic space and on the cell surface of *Myxococcus xanthus* during developmental aggregation. *J. Biol. Chem.* **256:**12589–12595.

Nelson, D. R., and D. R. Zusman. 1983. Transport and localization of protein S, a spore coat protein, during fruiting body formation by *Myxococcus xanthus*. *J. Bacteriol.* **150:** 547–553.

Nicaud, J. M. 1983. Etude de mutants affectés dans l'excrétion d'enzymes extracellulaires chez *Myxococcus xanthus*. Thèse de troisième cycle. Spécialité: microbiologie génie enzymatique et bioconversion. Université de Technologie de Compiègne, Compiègne, France.

Nicaud, J. M., A. Breton, G. Younes, and J. F. Guespin-Michel. 1984. Mutants of *Myxococcus xanthus* impaired in protein secretion: an approach to study of a secretory mechanism. *Appl. Microbiol. Biotechnol.* **20:**344–350.

Petit, F., and J. F. Guespin-Michel. 1992. Production of an extracellular milk-clotting activity during development in *Myxococcus xanthus*. *J. Bacteriol.* **174:**5136–5140.

Plamann, L., A. Kuspa, and D. Kaiser. 1992. Proteins that rescue A-signal defective mutants of *Myxococcus xanthus*. *J. Bacteriol.* **174:**3311–3318.

Pohlner, J., R. Halter, K. Beyreuther, and T. F. Meyer. 1987. Gene structure and extracellular secretion of *Neisseria gonorrhoeae* IgA protease. *Nature* (London) **325:**458–462.

Pugsley, A. P. 1988. Protein secretion across the outer membrane of gram-negative bacteria, p. 607–652. *In* R. A. Das and P. N. Robbins (ed.), *Protein Transfer and Organelle Biogenesis*. Academic Press, Inc., New York.

Pugsley, A. P., and I. Reyss. 1990. Five genes at the 3′ end of the *Klebsiella pneumoniae pulC* operon are required for pullulanase secretion. *Mol. Microbiol.* **4:**365–379.

Reichenbach, H., K. Gerth, H. Irschik, B. Kunze, and C. Höfle. 1988. Myxobacteria: a source of new antibiotics. *Trends Biotechnol.* **6:**115–121.

Reichenbach, H., and H. Kleinig. 1984. Pigments of myxobacteria, p. 128–137. *In* E. Rosenberg (ed.), *Myxobacteria: Development and Cell Interactions*. Springer Verlag, New York.

Reyss, I., and A. P. Pugsley. 1990. Five additional genes in the pulC-O operon of the gram-negative bacterium Klebsiella oxytoca UNF5023 which are required for pullulanase secretion. *Mol. Gen. Genet.* **222:**176–184.

Romeo, J. M., B. Esmon, and D. R. Zusman. 1986. Nucleotide sequence of the myxobacterial hemagglutinin gene contains four homologous domains. *Proc. Natl Acad. Sci. USA* **83:** 6332–6336.

Rosenberg, E., and M. Varon. 1984. Antibiotics and lytic enzymes, p. 109–125. *In* E. Rosenberg (ed.), *Myxobacteria: Development and Cell Interactions.* Springer-Verlag, New York.

Rosenberg, E. K., H. Keller, and M. Dworkin. 1977. Cell density-dependent growth of *Myxococcus xanthus* on casein. *J. Bacteriol.* **129:**770–777.

Rosenbluh, A., and E. Rosenberg. 1990. Role of autocide AMI in development of *Myxococcus xanthus. J. Bacteriol.* **172:**4307–4314.

Sedmak, J., and S. Grossberg. 1977. A rapid sensitive and versatile assay for protein using Coomassie brilliant blue G250. *Anal. Biochem.* **79:**544–552.

Shimkets, L. J., and H. Rafiee. 1990. CsgA, an extracellular protein essential for *Myxococcus xanthus* development. *J. Bacteriol.* **172:**5299–5306.

Sudo, S. Z., and M. Dworkin. 1973. Bacteriolytic enzymes produced by *Myxococcus xanthus. J. Bacteriol.* **110:**236–245.

Sutherland, I. W., and S. Thomson. 1975. Comparison of polysaccharides produced by *Myxococcus* strains. *J. Gen. Microbiol.* **89:**124–132.

Teintze, M., R. Thomas, T. Furuichi, M. Inouye, and S. Inouye. 1985. Two homologous genes coding for spore-specific proteins are expressed at different times during development of *Myxococcus xanthus. J. Bacteriol.* **163:**121–125.

Thurn, K. K., and A. K. Chatterjee. 1985. Single-site chromosomal Tn*5* insertions affect the export of pectolytic and cellulolytic enzymes in *Erwinia chrysanthemi* EC16. *Appl. Environ. Microbiol.* **6:**894–898.

Wretlind, B., and O. R. Pavlovskis. 1984. Genetic mapping and characterization of *Pseudomonas aeruginosa* mutants defective in the formation of extracellular proteins. *J. Bacteriol.* **158:**801–808.

Yee, T., and M. Inouye. 1984. Genome of Myxococcus xanthus, p. 69–90. *In* E. Rosenberg (ed.), *Myxobacteria: Development and Cell Interactions.* Springer-Verlag, New York.

Younes, G. 1986. Effet des conditions de croissance sur la production de protéines endogènes ou étrangères par *Myxococcus xanthus.* Ph.D. thesis. Université de Technologie de Compiègne, Compiègne, France.

Younes, G., A. M. Breton, and J. F. Guespin-Michel. 1987. Production of extracellular native and foreign proteins by immobilized growing cells of *Myxococcus xanthus. Appl. Microbiol. Biotechnol.* **25:**507–512.

Myxobacteria II
Edited by Martin Dworkin and Dale Kaiser
© 1993 American Society for Microbiology, Washington, DC 20005

Chapter 12

Intercellular Signaling

Dale Kaiser and Lee Kroos

Myxobacterial cells feed, move, and develop cooperatively. Genetic, biochemical, and cell mosaic studies show that cells coordinate their multicellular behaviors by transmission of intercellular signals. Some of the signal molecules have been identified chemically. Other signals can be inferred from the behavior of cells, in some cases by a dependence on cell density. The genetic simplicity (see chapter 10) and tractability (see chapter 9) of *Myxococcus xanthus* make it an excellent experimental system for exploring how intercellular signals control multicellular behavior. Other chapters in this volume that contain information relevant to intercellular signaling describe the cell surface (chapter 3), the control of gene expression (chapters 8 and 9), cell motility (chapter 13), fruiting body morphogenesis (chapter 14), protein export (chapter 11), and the autocides (chapter 10).

DENSITY-DEPENDENT GROWTH

Myxobacteria feed on particulate organic matter in the soil by means of extracellular bacteriolytic, proteolytic, cellulolytic, and other digestive enzymes (Reichenbach and Dworkin, 1992). On the basis of their secretion of lytic enzymes, Dworkin (1973) proposed that myxobacteria feed like "packs of microbial wolves." Rosenberg et al. (1977) measured the growth rate when the only source of carbon and nitrogen for *M. xanthus* cells in liquid culture was the polymeric substrate casein, so that proteolysis was required for growth. A twofold increase in growth rate was observed as the cell density was raised above 10^4 cells per ml. When intact

---

Dale Kaiser • Departments of Biochemistry and Developmental Biology, Stanford University School of Medicine, Stanford, California 94305. **Lee Kroos** • Departments of Biochemistry and Microbiology, Michigan State University, East Lansing, Michigan 48824.

casein was replaced with enzymatically hydrolyzed casein, the cells grew at the more rapid rate independent of cell density. Furthermore, cultures at densities below 10^3 cells per ml formed no colonies when plated on agar containing undigested casein as nutrient source. Evidently, extracellular digestion of protein (casein) is enhanced by cooperation between cells. The dynamics of this cooperation are not yet clearly understood, but it is thought that the higher local concentration of enzymes might increase the overall efficiency of release of amino acids and small peptides from casein. It also seems likely that the capture of hydrolytic products is more efficient at a higher cell density, so that conversion of organic matter to myxobacterial cell substance is greater at the higher densities.

The growth advantage of cooperative feeding may have been a major selective force behind the evolution and extensive use of extracellular signals by myxobacteria. A swarm may be the unit of efficient cooperative feeding. Signals may have evolved to keep swarm cells together while they "hunt" new sources of nutrient in soil. A fruiting body may be large enough to be transported by wind, water, or small animals in soil, thus ensuring efficient dispersal (Shimoura and Bonner, 1982). Dispersal may be efficient because rapid growth can follow germination at a new site where nutrient is available. Kühlwein and Reichenbach (1968) showed that a single germinating sporangiole of a *Chondromyces apiculatus* fruiting body forms an active swarm that behaves much like a swarm of bees.

CELL INTERACTIONS THAT CONTROL CELL MOVEMENT

Myxobacteria move by gliding (see chapter 11), which permits them to translocate efficiently on surfaces, including the surfaces of other cells (Doetsch and Hageage, 1968). One cannot watch the behavior of cells in swarms, illustrated in the time-lapse films of Reichenbach (Kühlwein and Reichenbach, 1968; Reichenbach, 1966), without perceiving interactions between the moving cells. Several hundred mutants of *M. xanthus* that change the arrangement of cells in swarms have been isolated (Burchard, 1970; Hodgkin and Kaiser, 1979; Zusman, 1982). Among these mutants, six *frz* genes that control the frequency of reversal of gliding direction have been identified (Blackhart and Zusman, 1985, 1986; McCleary and Zusman, 1990). Changes in the methylation pattern of the *frz* C/D protein accompany the starvation and aggregation phases of fruiting body development (see chapter 11) and suggest that reversal frequency may in some way be altered by neighboring cells.

About 30 genes that control the swarming pattern belong to two other sets of genes, called motility system A and motility system S. From wild-type cells, which have wild-type alleles of all the genes of systems A and

S and so are designated A^+S^+, two classes of single mutants have been derived: A^+S^- and A^-S^+. These two mutant classes swarm in patterns that are complementary. Many single cells are evident at the edge of a swarm of A^+S^- mutants, where the genes belonging to system A (A for adventurous) all have their wild-type alleles, but any one of the S genes is mutant. The A pattern arises because single A^+S^- cells can move even when they are well separated from all other cells. Twenty-three genes belong to system A (Fink and Zissler, 1989; Hodgkin and Kaiser, 1979; Sodergren and Kaiser, 1983). Some A-motility genes participate in the synthesis of cell surface lipopolysaccharides (Fink and Zissler, 1989). Insertion of the transposon Tn*phoA* oriented properly within certain A-system genes yields active alkaline phosphatase, indicating that the products of these genes are exported to the cell membrane or the periplasmic space (Kalos and Zissler, 1990a; Manoil and Beckwith, 1985).

A second set of pattern mutants, A^-S^+ mutants, express what is called social, or S, motility because A^-S^+ cells tend to cluster in groups of 10 to 20 cells with few isolated cells (Burchard, 1970; Hodgkin and Kaiser, 1979). These A^-S^+ cells move only when they are within a cell's length of one another, hence the term social motility (Kaiser and Crosby, 1983). Nine S-motility genes are known. One of the genes of system S, called *dsp,* is needed for production of an extracellular meshwork of fibrils mainly composed of polysaccharide (Arnold and Shimkets, 1988) (see also chapter 5; *dsp* was isolated by D. Morandi). Three other S-motility genes, *sglB, sglG,* and *tgl,* are needed for the production of pili, which are located at one pole of *M. xanthus* cells (Kaiser, 1979). The *tgl* gene specifies a cell surface lipoprotein (Rodriguez, unpublished data).

From mutations in different A and S genes, more than 50 A^-S^- mutants have been built, and all are nonmotile (Hodgkin and Kaiser, 1979). In addition, they have lost the abilities to build fruiting bodies and to form myxospores in response to starvation. The nonmotility of all A^-S^- mutants implies that the A- and S-gene systems perform parallel motility functions that apparently cannot be provided any other way.

Wild-type cells are A^+S^+; their swarm patterns combine adventurous and social motilities. Both A and S motilities are modulated by interactions between cells. This is evident from the fact that the swarm expansion rate increases at least sixfold with increasing initial cell density (Fig. 1). A-motile (A^+S^-) cells have the ability to move autonomously, as reflected in a baseline rate of about 0.2 μm/min at the lowest cell density and the presence of many isolated cells at the edge of a swarm. A motility increases rapidly as initial cell density is increased, while S motility (in an A^-S^+ mutant) increases more gradually (Fig. 1). S motility starts from a zero baseline, reflecting the complete dependence of motility on neigh-

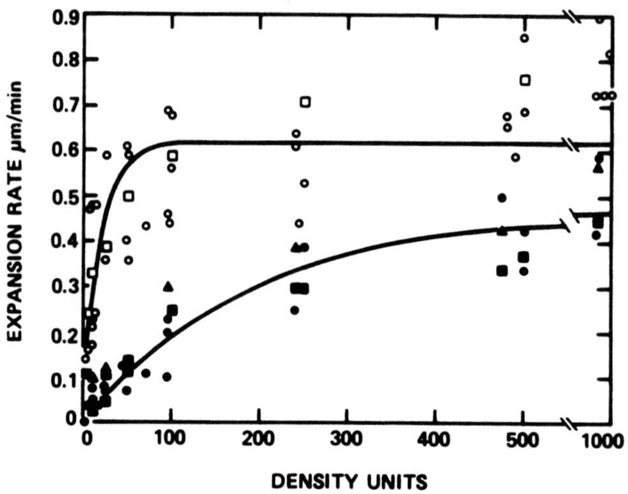

Figure 1. Rate of swarm expansion versus initial cell density for three A^-S^+ strains (closed symbols) and two A^+S^- strains (open symbols). Microdroplets of cell suspensions at the densities shown (100 density units = 4×10^8 cells per ml) were placed on nutrient agar, and the plates were incubated at 33°C. Circular swarms that expand linearly with time are produced from the droplets. Expansion rates were measured by taking the slopes of diameter versus time plots. Movement contributed 88% to the expansion of wild-type swarms; the contribution of growth was measured with nonmotile mutants. These data are from Kaiser and Crosby (1983).

boring cells and explaining the absence of isolated cells at the edge of an S-motile swarm. S motility continues to increase at intermediate cell densities at which A motility has already been saturated (Fig. 1). Wild-type (A^+S^+) cells show the 0.2-µm/min baseline characteristic of A motility and its rapidly increasing response at low cell density (not shown in Fig. 1, but see Kaiser and Crosby [1983]). Wild-type cells slowly increase their responses in the mid-density range, which is like S motility. In these respects, A^+S^+ cells combine the A and S patterns. However, either A or S motility alone generates a spreading movement of about 0.5 µm/min (Fig. 1). Wild-type swarms expand three times faster, reaching a maximum of 1.5 µm/min, or about 50% more than the sum of A and S. Moreover, the A^+S^+ pattern includes a wider variety of cell clusters than either A or S motility alone. Both properties suggest synergism between the two motility systems. Cell density dependence and the patchy distribution of cells in swarms suggest cell interactions, but neither the changes in single cell behavior nor the cell interactions that underlie them have yet been defined.

Part of those interactions may be specifically revealed in the "stimulatable" members of the A and S systems. When certain pairs of nonmotile strains are mixed on the surface of agar, one or both become temporarily motile and, in the case of *tgl*, temporarily piliated (Hodgkin and Kaiser, 1977; Kaiser, 1979). This effect is termed stimulation to emphasize that it is transitory. Stimulation has been observed only when cells of the two interacting types are able to touch each other (Hodgkin and Kaiser, 1977). The recipient strain must carry a *cgl* mutation (in motility system A) or a *tgl* mutation (in motility system S), and the donor strain must carry the corresponding cgl^+ or tgl^+ allele (Hodgkin and Kaiser, 1977). Stimulation is thus genetically constrained, but gene transfer does not occur, because the stimulated recipient cells retain their original genotypes (Hodgkin and Kaiser, 1977; Kaiser, 1979). Although the molecular basis of motility stimulation is not understood, stimulation is efficient. A mixed pair of nonmotile stimulatable strains can swarm at rates approaching those of wild type. Moreover, stimulation can generate sufficient cell movement to support aggregation in fruiting body development (Kalos and Zissler, 1990b; Kroos et al., 1988). The requirement for cell contact suggests that the relevant molecules are normally cell bound.

INTERCELLULAR SIGNALING IN FRUITING BODY DEVELOPMENT

The best-studied cases of intercellular signaling serve essential roles in the program that regulates fruiting body development. These signals are necessary for the progressive and reliable creation of order from disorder that is evident in fruiting body morphogenesis. Starvation for amino acids, cells at sufficiently high density, and location on a solid surface lead to aggregation of about 100,000 cells in a fruiting body, which ultimately becomes filled with dormant, environmentally resistant myxospores. A solid surface is needed for gliding cells to aggregate on (Kroos et al., 1988; Shimkets and Dworkin, 1981).

While the cells are aggregating and sporulating, the pattern of protein synthesis changes. Synthesis of at least 15 proteins, identified as bands separated by gel electrophoresis, is increased during development (Inouye et al., 1979). Isolation of some of the more abundant proteins has allowed the generation of antibodies and cloning of the corresponding genes, permitting studies of developmental gene expression with antibody and nucleic acid probes as well as with gene fusions. These studies provided some of the first biochemical "markers" of *Myxococcus* development, such as protein S, a spore coat protein expressed from the *tps* gene beginning at about 5 h poststarvation, and the very similar protein S-1, which is synthesized inside spores from the *ops* gene (located 1.4 kb upstream from *tps*) beginning at about 24 h poststarvation (Downard and Zusman, 1985; Downard et al., 1984; Inouye, 1984; Inouye et al., 1983a, 1983b; Teintze et al., 1985). MBHA (myxobacterial hemagglutinin), a protein with lectin activity (Cumsky and Zusman, 1979), is expressed from *mbhA* at about 10 h (Romeo and Zusman, 1991), and protein C, a spore coat protein, is expressed early in development (McCleary and Zusman, 1990). Three development-specific phosphatase activities with different pH optima also serve as useful biochemical markers (Weinberg and Zusman, 1990), as do changes in cell surface hydrophobicity (i.e., cells undergo a "hydrophilic shift" during development [Kupfer and Zusman, 1984]) and antigenicity (Gill and Dworkin, 1986, 1988). Monoclonal antibodies to some of the cell surface antigens block development (Gill et al., 1987; Jarvis and Dworkin, 1989), suggesting that these antigens may be involved in contact-mediated cell interactions (see chapter 3). One antigen, dFA-1, is associated with extracellular fibrils (Clemans et al., 1991). Another approach that has yielded many useful developmental markers employs a transposable promoter probe, Tn*5 lac,* that forms transcriptional fusions (Kroos and Kaiser, 1984). In one study, random insertions of Tn*5 lac* identified 29 genes that increase their activities according to a regular time schedule during development (Kroos et al., 1986). The program of fruiting body development combines differential gene expression with morphological change.

Cell-to-cell signals govern progress through this developmental program. McVittie et al. (1962) isolated a pair of mutants that complemented each other for fruiting body development. However, the significance was obscured by the possibility of cross-feeding, since one of the mutants was a pyrimidine auxotroph. Different signals were implicated by four complementary classes of nonautonomous developmental mutants isolated by Hagen et al. (1978). Cell density dependence was employed to find nonautonomous mutants, including those deficient in signal production. Mutants of the *asg, bsg, csg,* and *dsg* classes cannot sporulate when they are alone, but their sporulation can be rescued by codevelopment with wild-type cells. The classes were distinguished by mixing mutants with each other. Mutants of the same class fail to complement each other; mutants of different classes do complement each other. The *asg, bsg, csg,* and *dsg* mutants behave as if they are defective in the production of different extracellular signals, which we will refer to as A, B, C, and D factors, respectively. A fifth class, *esg,* has recently been discovered and is described below (Downard, 1992).

Each class of mutants affects the developmental program in a different way. Figure 2 summarizes the results of measuring the expression of different developmental markers in *asg, bsg, csg,* and *dsg* mutants. If expression of the marker was reduced or abolished in a mutant, the marker was said to exhibit partial or absolute dependence, respectively, on the corresponding extracellular factor. The dependence pattern of any marker in Fig. 2 can easily be read by noting the dependences (or lack of dependences) encountered as the horizontal and vertical lines are followed to the lower left corner of the diagram. Thus, horizontal lines connect markers with identical dependence patterns. Vertical lines roughly mark the times at which the A, B, C, and D factors are first required for normal developmental-gene expression, and they indicate the cumulative nature of the dependences. A and B factors appear to function very early in development; D and C factors are first required slightly later, at about 4 and 6 h poststarvation, respectively. Different markers expressed with similar timing exhibit different dependence patterns, necessitating many branches in the diagram. This is very different from the dependence patterns of developmental genes in *Bacillus subtilis,* which can be accommodated by a simple linear pathway of cumulative dependences early in sporulation and two branches, corresponding to forespore- and mother cell-expressed genes, late in sporulation (Losick and Kroos, 1989). On the other hand, the regulatory circuits governing initiation of *B. subtilis* sporulation appear to be very complex (Smith, 1989; Sonenshein, 1989). The different dependence patterns of *Myxococcus* genes that are expressed with similar timing suggest that these genes may possess complex regulatory regions capable of responding to multiple signals. To the extent that they have been examined, the regulatory regions of developmentally regulated *Myxococcus* genes do appear to contain multiple upstream elements (see chapter 8).

The different effects on the developmental program imply a different signal substance for each class of mutants. Identification of *asg* and *csg* signal molecules from wild-type cells has proven that these two classes are deficient in production of different signals. The following sections review progress toward understanding the chemical nature of intercellular signals in *Myxococcus* spp., the mechanisms by which signals are transmitted and recipient cells respond, and the function of these cell-cell interactions in the developmental program.

asg MUTANTS

Approximately half of the signaling mutants recovered by Hagen et al. (1978) were of the A class. Their relatively high frequency is explained by the fact that all *asg* mutants are secretion defective, and apparently

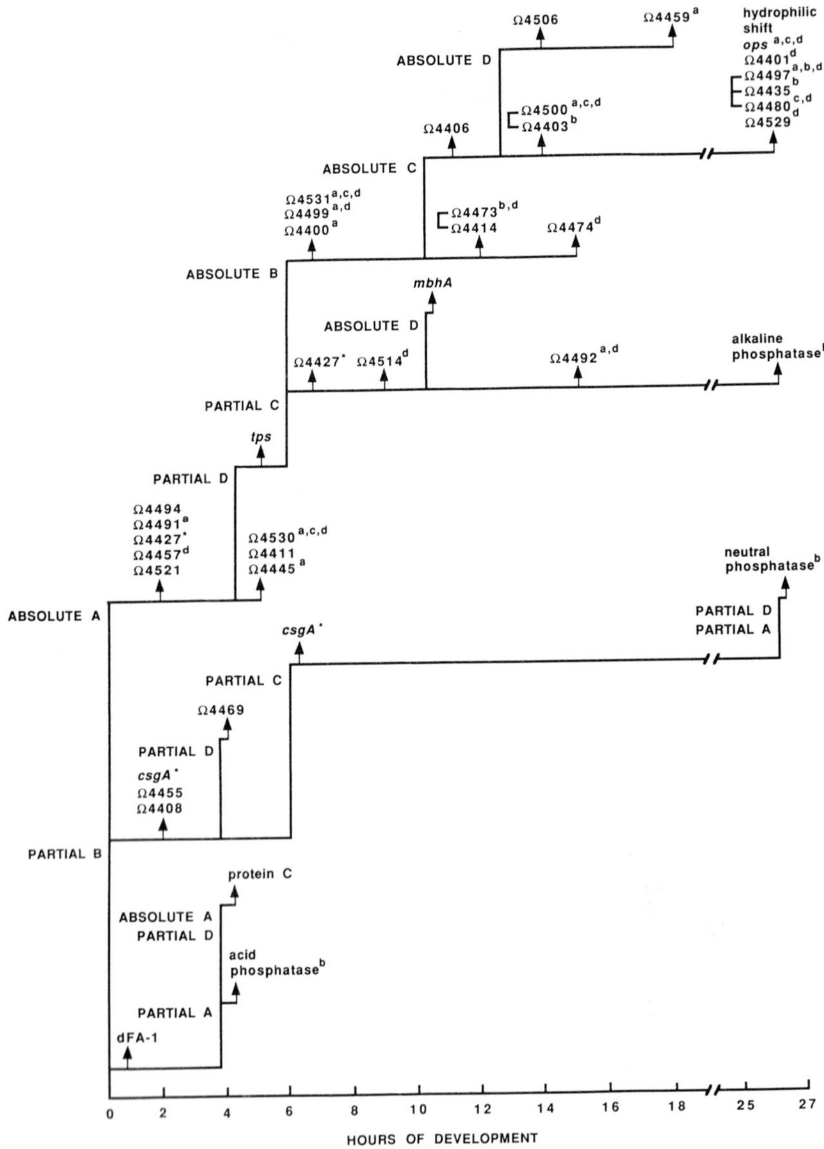

Figure 2. Dependence of developmental markers on A, B, C, and D factors. Each marker is placed above an arrow positioned along the *x* axis according to the time at which its expression normally begins during development. Position along the *y* axis indicates the marker's dependence on cell-cell interactions, as determined by measuring its expression in *asg, bsg, csg,* and *dsg* mutants. For example, a marker whose expression is reduced in an *asg* mutant exhibits partial dependence on A factor (abbreviated "partial A"). Similarly, a marker whose expression is abolished in a *bsg* mutant exhibits absolute dependence on

there are many ways to block secretion (Plamann et al., 1992). With linked insertions of the transposon Tn*5*, 13 *asg* mutants were mapped to three genetic loci: *asgA*, *asgB*, and *asgC* (Kuspa and Kaiser, 1989; Mayo and Kaiser, 1989). Development in submerged culture produces a severer starvation than is produced on agar because commercial agar contains protein (Bretscher and Kaiser, 1978; Kuner and Kaiser, 1982). Mutants with mutations in the various loci differ in their residual developmental capacities. In submerged culture, *asgA* and *asgB* mutants produce 10^5-fold fewer spores than wild type, and the *asgC* mutant produces 50-fold fewer spores than wild type (Kuspa and Kaiser, 1989). In submerged culture, *asgA* and *asgB* cells do not aggregate and *asgC* cells form small, irregular clusters of cells that do not mature into fruiting bodies. Simple mixing of wild-type cells with *asgA*, *asgB*, or *asgC* cells rescues both aggregation and sporulation of the mutants to near wild-type levels.

Since the *asgA* and *asgB* mutants are defective at different loci, their common stage of developmental arrest indicates when A signaling is required (Kuspa et al., 1986). They arrest early in development before aggregation, and they fail to express many of the developmentally regulated *lacZ* fusions (Fig. 2). Several fusions are expressed in the *asg* mutants, indicating that the mutants begin to develop before they arrest about 2 h into the program. One of the earliest A-factor-dependent genes, identified by Tn*5* *lac* fusion Ω4521, is normally expressed at 1 to 2 h by developing wild-type cells. Expression from the Ω4521 promoter rises 20-fold above vegetative levels in wild-type cells but increases less than 2-fold in *asg* mutants (Kaplan et al., 1991; Kuspa et al., 1986). Correctly timed expression of Ω4521 and other *asg*-dependent genes is restored in *asg* mutants by mixing them with wild-type cells. (For these experiments, the Ω4521 fusion is present in the *asg* mutant cells but not in the wild-type cells.)

B factor (abbreviated "absolute B"). Dependences are cumulative, so a higher position along the *y* axis indicates greater dependence on cell-cell interactions. Superscript a, b, c, or d means that the marker has not been tested in *asg*, *bsg*, *csg*, or *dsg* mutant cells, respectively. Tn*5* *lac* insertions are indicated by Ω followed by a four-digit number, and brackets connect insertions that probably lie in the same transcription unit (Kroos et al., 1986). Ω4427 and *csgA* are starred and appear twice in the diagram because they exhibit no dependence on C factor before 6 h of development but a partial dependence on C factor after 6 h. References for dependences are as follows: dFA-1, Clemans et al. (1991); acid, neutral, and alkaline phosphatases, Weinberg and Zusman (1990); protein C, McCleary et al. (1991); Tn*5* *lac* insertions, Cheng and Kaiser (1989a), Kroos and Kaiser (1987), and Kuspa et al. (1986); *csgA*, Kim and Kaiser (1991) and Li et al. (1992); *tps*, Cheng and Kaiser (1989a), Gill and Cull (1986), Kroos and Kaiser (1987), and Kuspa et al. (1986); *mbhA*, La Rossa et al. (1983); *ops*, Gill and Cull (1986); and hydrophilic shift, Kupfer and Zusman (1984).

Different *asg* mutants mixed with each other fail to show this level of rescue of β-galactosidase or of sporulation, confirming the integrity of the *asg* class (Kuspa and Kaiser, 1989; Mayo and Kaiser, 1989). When wild-type cells rescue expression of the earliest *asg*-dependent gene and later genes as well, the time of β-galactosidase expression becomes that of the wild type. Under these conditions, *asg* mutant development continues smoothly through the stage that would otherwise be blocked by the *asg* mutation.

Rescue of β-galactosidase expression from an Ω4521 fusion provides a convenient assay for A factor, the putative A-signal molecule (Kuspa et al., 1986). A factor assayed this way is released into conditioned medium by wild-type cells at the same time that *asg* mutants arrest their development, but *asg* mutants release less than 5% of wild-type levels of A factor (Kuspa et al., 1986). A factor from wild-type cells rescues the aggregation and sporulation defects of the *asgB* and *asgC* mutants and restores the aggregation of *asgA* mutants (Kuspa et al., 1986, 1992a, 1992b).

A FACTOR

A-factor rescue of β-galactosidase expression from the Ω4521 fusion in an *asgB* mutant is at the level of mRNA accumulation (Kaplan et al., 1991). RNA that hybridizes to *M. xanthus* DNA just upstream of the *lacZ* fusion Ω4521 was absent in growing cells but present at 2 and 4 h of development; the accumulation of this mRNA parallels β-galactosidase expression from the *lacZ* part of the Ω4521 fusion. The message was absent in an *asg* mutant. Wild-type cells or partially purified A factor from medium conditioned by developing wild-type cells restores mRNA accumulation; thus, A factor governs expression of A-factor-dependent genes by controlling their mRNA.

Half the A-factor activity in conditioned medium (measured in β-galactosidase units) is heat labile and nondialyzable, and the other half is heat stable and passes through dialysis membranes (Plamann et al., 1992). Heat-labile A factor has been purified from conditioned medium and found to consist of a mixture of proteases (Plamann et al., 1992). A-factor proteins of 27 and 10 kDa have been identified, and each protein has proteolytic activity with a different substrate specificity. These two proteins represent neither all of the heat-labile A factor in conditioned medium nor all of its proteolytic activity, suggesting that other proteases should have A-factor activity as well. Indeed, pronase, chymotrypsin, trypsin, and proteinase K have A-factor activity. Bovine trypsin has about one-half the A-factor specific activity of the highly purified 27-kDa myxobacterial protein. Titration of trypsin's proteolytic activity by the specific trypsin inhibitor protein decreased trypsin's A-factor activity in parallel with its protease activity. This demonstrated that the proteolytic activity of trypsin is required for its A-factor activity. Pure proteases restore normal aggregation to all *asg* mutants and restore near-wild-type levels of sporulation to *asgB* and *asgC* mutants but not to *asgA* mutants (Plamann et al., 1992; Kuspa et al., 1992b).

Heat-stable A factor has also been purified. It is a set of amino acids and the small peptides that contain those amino acids (Kuspa et al., 1992a). Wild-type cells release a reproducible mix of amino acids when they are releasing A-factor activity. Moreover, particular amino acids have high A-factor specific activity. Proline, tyrosine, phenylalanine, tryptophan, leucine, isoleucine, and alanine have the highest specific A-factor activities. Each of these amino acids added alone at concentrations ranging from 0.1 to 0.5 mM rescues Ω4521 gene expression to near-normal levels. Many small peptides have heat-stable A-factor activity, and their activities are close to those expected from a sum of their constituent amino acids. Heat-stable A factor can be accounted for by a mixture of free amino acids and many peptides.

Single amino acids rescue normal aggregation in all of the *asg* mutants and restore near-wild-type levels of sporulation to the *asgB and asgC* mutants but not to *asgA* mutants (Plamann et al., 1992; Kuspa et al., 1992b). It appears that amino acids are the primary A-signal molecules, while the extracellular release of proteases and proteins by wild-type cells generates first peptides and then amino acids.

RESPONSE TO A FACTOR

Rescue of Ω4521 gene expression by amino acids does not extrapolate to zero β-galactosidase at zero concentration of amino acids (Kuspa et al., 1992b). Rather, it exhibits a threshold concentration. Amino acids with high A-factor specific activity have a threshold of about 10 μM. Because of their threshold concentrations and the fact that amino acids and peptides in conditioned medium are released by cells in proportion to the cell density, extracellular amino acids specify a minimum cell density required for wild-type development. This minimum has been demonstrated by diluting wild-type cells to the point that the calculated concentration of released extracellular amino acids falls short of threshold levels. At these cell densities (between 3×10^8 and 5×10^8 cells per ml), Ω4521 gene expression is decreased 80 to 90% and aggregation and sporulation are reduced (Kuspa et al., 1992b). As expected from their failure to use A signal, the dilute wild-type cells behave like *asg* mutants. Expression of the Ω4521 gene can be restored to dilute wild-type cells by amino acids or by proteases. The same concentrations of amino acids or proteases needed to rescue the *asg* mutants also rescue dilute wild-type cells (Kuspa et al., 1992b).

Use of amino acids as signals during development raises a question of signal specificity: how can amino acids that must be abundant in the extracellular environment for growth serve as specific signals for continuing development at the 2-h stage? A part of the answer is that the lowest concentration of an equimolar mixture of the 20 common amino acids required for growth is approximately 10 mM, while the threshold concentration for A-factor activity is 20-fold lower, at 0.05 mM (Kuspa et al., 1992a, 1992b). But what is to prevent amino acids in the growth medium from prematurely signaling development? In fact, the higher levels of amino acids needed to support growth actually prevent development. When the amino acid levels are high enough to support growth, β-galactosidase is not expressed from Ω4521. When amino acid levels are below the 10 mM growth threshold, Ω4521 is expressed (Kuspa et al., 1992b). Signal specificity is also provided temporally, because Ω4521 is not expressed prior to 2 h of starvation. Perhaps the actions of the products of *asg*-independent genes enhance cellular response to low levels of extracellular amino acids; some of the *asg*-independent genes might encode parts of an amino acid receptor system, for instance.

At least one gene whose product is likely to mediate the cellular response to A factor has been identified. Suppressor mutations isolated by Kaplan et al. (1991) in an *asgB* strain carrying the Ω4521 *lacZ* fusion restore the ability to express β-galactosidase from Ω4521 in the absence of A factor. Nine of the mutations mapped to the same locus, *sasA* (*sas* for suppressor of *asg*). A-factor production is not restored, nor is *sasA* in the Ω4521 gene itself. Rather *sasA* mutants are second-site suppressors of the *asgB* defect, and they restore Ω4521 expression to *asgA* and *asgC* mutants as well as to the *asgB* mutant in which they were selected. However, *sasA* mutations do not restore complete development or sporulation to an *asgB* strain. They are specific suppressors of Ω4521 and Ω4445, another A-factor-dependent fusion; other A-factor-dependent *lac* fusions were not suppressed. The *sasA* mutations are recessive, and their frequency of occurrence is that expected for loss-of-function mutations. The Ω4521 gene expresses β-galactosidase at the normal time in development in a *sasA asgB* double mutant. Since *sasA* mutants render Ω4521 expression independent of A factor, the *sasA*$^{+}$ gene product seems to repress Ω4521 gene expression. In accord with this idea, *sasA* mutations cause a developmental defect on their own (i.e., in an *asg*$^{+}$ genetic background), reducing sporulation 10- to 100-fold and changing the expression of at least four developmental genes in addition to Ω4521 (Kaplan et al., 1991).

csg MUTANTS

Like the *asg* mutants, the *csg* mutants were isolated as nonautonomous developmental mutants that failed to develop alone but could aggregate and sporulate when mixed with wild-type cells (Hagen et al., 1978). Unlike the *asg* mutants, all existing members of the *csg* class mapped to the same gene, called *csgA* (Shimkets and Asher, 1988; Shimkets et al., 1983). The developmental block in *csg* mutants is at a later developmental stage than in *asg* mutants. *csg* mutants (but not *asg* mutants) synthesize developmental markers that are expressed before 6 h of development normally, but expression is reduced or abolished for markers that begin to be expressed after 6 h (Fig. 2). Also, *csg* mutants produce A factor (Kuspa, 1989).

Unlike wild-type cells, which aggregate into compact mounds by 12 h on a defined starvation agar medium, *csgA* mutants form ridges of cells and diffuse mounds, but only after 18 h (Shimkets and Kaiser, 1982b). In the more stringent starvation of submerged culture, *csgA* cells fail to aggregate at all; they also fail to ripple or sporulate (Shimkets and Kaiser, 1982b). Mixing them with wild-type cells restores expression of developmental markers and sporulation to *csgA* mutants (Kroos and Kaiser, 1987).

C FACTOR

The DNA sequence of *csgA* implies that its product should be a 17-kDa protein (Hagen and Shimkets, 1990). The amino-terminal sequence of the predicted *csgA* protein resembles the signal sequence used by the enteric bacteria, but processing has not been investigated (Hagen and Shimkets, 1990). Polyclonal antibodies raised against a *lacZ-csgA* translational fusion protein produced in *Escherichia coli* identify a 17-kDa polypeptide in extracts made from $csgA^+$ cells but not from vegetative or *csgA* cells (Kim and Kaiser, 1990d; Shimkets and Rafiee, 1990). Antibodies specific for *csgA* gene product also block development aggregation and sporulation of wild-type cells (Shimkets and Rafiee, 1990).

Like the identification of A factor, rescue of the developmental defects of *csgA* mutants was the basis of a specific bioassay for C factor. The molecule responsible for rescue was purified by column chromatography from sonic extracts of nascent fruiting bodies of a wild-type strain (Kim and Kaiser, 1990d). A membrane-associated protein of 17,000 M_r

was obtained (Kim and Kaiser, 1990a). When added at 1 nM, this C-factor protein rescues *csgA* mutant aggregation, sporulation, and gene expression to wild-type levels. Its size, the amino acid sequence of an undecapeptide fragment, the recognition of C factor by purified anti-*csgA* antibodies, and its absence from extracts of a *csgA* mutant demonstrate that C factor is the product of the *csgA* gene (Kim and Kaiser, 1990a, 1990d; Shimkets and Rafiee, 1990).

C-SIGNAL TRANSMISSION AND RESPONSE

Requirements for C-signal transmission from cell to cell are different from those for A signal: while A-factor activity is recovered from medium conditioned by starving wild-type cells, C factor is found in the pellet after high-speed centrifugation (Kim and Kaiser, 1990a). Gold-labeled anti-C-factor antibody bound to material associated with the cell surface (Shimkets and Rafiee, 1990). Separation of wild-type and *asg* strains by a porous membrane impermeable to cells still permits rescue of *asg* mutant development; *csg* mutants do not develop, however, when separated from wild-type cells by a membrane (Kim and Kaiser, 1990a). While A factor is a diffusible signal that can act over distances greater than a cell's length, C factor acts short range and requires close contact between producing and responding cells.

A requirement for cell motility also distinguishes transmission of the A and C signals. Expression of an A-signal-dependent *lacZ* fusion like Ω4521 is normal in a cell rendered nonmotile by mutation in the *mglA* gliding gene (Kroos et al., 1988; Stephens et al., 1989). Thus, A-factor transmission is independent of cell motility. However, expression of C-dependent *lacZ* fusions in nonmotile cells is reduced or abolished to the same degree as expression in *csgA* cells. Nonmotile cells, like *csgA* cells, fail to aggregate, ripple, or sporulate. These observations suggest that motility is necessary for proper C signaling and that nonmotile cells fail to develop because they are unable to complete C signaling (Kroos et al., 1988).

Following on this suggestion of a link between motility and C signaling, C factor purified from wild-type cells was found capable of restoring sporulation and gene expression to wild-type levels in *mglA* strains, showing that the nonmotile mutants can respond to C factor. (Nonmotile-cell aggregation was not restored because C factor did not rescue the basic motility defect.) However, mixing intact wild-type cells (that would present C factor) with *mglA* cells did not rescue the developmental defects of *mglA* cells despite opportunities for close cell contact in the mixtures. In addition, wild-type concentrations of active C factor could be purified from *mglA* cells, showing that nonmotile cells make C factor (Kim and

Kaiser, 1990c). Despite their abilities to make C factor, intact *mglA* cells were unable to rescue development of adjacent *csgA* cells. This set of observations implied that motility is required in both donor and responder cells for C-signal transmission between those cells.

One consequence of organized cell movement during fruiting body morphogenesis is a local increase in cell density of perhaps 50-fold. In a nascent fruiting body (roughly a 100-μm-diameter sphere), about 100,000 rod-shaped cells are organized in parallel arrays at a density of about 2×10^{11} cells per ml. At this density, cells are packed like bricks in a wall (Kuner and Kaiser, 1982; O'Connor and Zusman, 1989). The regularity of the cell arrangement suggested that cell movement might be necessary to achieve the dense packing of cells found in a fruiting body and that C signaling might depend on dense packing. In fact, an increase in the density of nonmotile cells by 10-fold resulted in a 1,000-fold increase in sporulation, to 1% of wild-type levels (Kroos et al., 1988). However, cells packed to a density of about 5×10^{10}/ml, the limits of this centrifuge method, were poorly organized. C-signal-dependent gene expression in these sedimented cells was not increased detectably (Kim and Kaiser, 1990b).

To align nonmotile cells mechanically, they were allowed to settle into the narrow grooves produced by gently scratching an agar surface with 10-μm-diameter aluminum oxide particles (emery paper). The aim was to test the idea that a critical feature of the organization of cells within a nascent fruiting body is their alignment. Normally, this alignment would result from directed cell movement, and alignment was crucial for C-factor transmission. Microscopic examination revealed that the long axes of nonmotile cells were indeed oriented within grooves parallel to the edges of the groove. Both C-signal-dependent gene expression and sporulation were evident among cells in the grooves but not in control cells on the smooth agar surface between grooves. No increase in sporulation was detected when isogenic *csgA* mutant cells were aligned (Kim and Kaiser, 1990b).

C factor appears to modulate the activities of more than 50 gene products whose functions include cell motility (Shimkets and Kaiser, 1982a), aggregation, and sporulation (Shimkets et al., 1983). Genes that may be involved in the response to C factor have been identified by genetic selection for cells with suppressors of a *csgA* insertion mutant (Rhie and Shimkets, 1989). These second-site suppressor mutations mapped to six different loci and exhibit a variety of phenotypes. One mutant bypassed two intermediate steps in the developmental pathway reported by *lacZ* gene fusions yet formed spore-filled fruiting bodies. Another suppressor mutant restored eight of nine expression reporters tested to wild-type or higher levels and partially restored sporulation yet failed to form fruiting bodies. The suppressor allele *(soc-500)* in this mutant has been cloned. It is dominant over the wild-type allele in merodiploids and permits low-temperature (15°C) induction of sporulation in nutrient-rich medium (Rhie and Shimkets, 1991). The suppressors corroborate other work that suggests that C signaling temporally and spatially coordinates several late developmental functions.

MORPHOGENETIC ROLE OF C FACTOR

Following successful transmission of A factor early in development, C factor stimulates aggregation of cells. This is clear from the failure of *csgA* mutant cells to aggregate and the rescue of aggregation by purified C factor (Kim and Kaiser, 1990a). Transmission of C signal may be inefficient in preaggregation stages because cells are in low-density arrangements that are poorly aligned. Following aggregation, cells achieve a geometry that favors maximum cell packing and thus efficient C-signal transmission.

Close packing of aligned cells would be expected to have several consequences. First, the cells would make many new end-to-end contacts. Second, side-to-side contacts would become more extensive. Third, the cell density would increase at least fourfold, from 5×10^{10} cells per ml (the concentration of nonmotile cells packed by centrifugation) to 2×10^{11} cells per ml (the concentration of cells aligned by their own movements into a closely packed array). An increase in cell density would be expected to amplify the C-factor signal concentration by positive autoregulation of the *csgA* promoter (Kim and Kaiser, 1991). A *csgA* mutant strain with a *csgA-lacZ* transcriptional fusion increased β-galactosidase expression twofold when development was induced by starvation (Kim and Kaiser, 1991). When the same strain was mixed with an equal number of wild-type cells or when 1 nM purified C factor was added, expression of *csgA* increased fourfold prior to the onset of sporulation (Kim and Kaiser, 1991). Accordingly, the net increase in local C-factor concentration might be as much as 16-fold, the product of a 4-fold increase in *csgA* transcription and a 4-fold increase in factor density (Kim and Kaiser, 1991).

Local C-factor concentration is very important, because depending on its concentration, C factor elicits distinct morphogenetic and transcriptional responses (Kim and Kaiser, 1991). Addition of 0.6 nM C factor had no detectable effect on cells. However, adding C factor to achieve a concentration of 0.8 nM rescued *csgA* cell aggregation and allowed full expression of the partially C-dependent *lacZ* fusion Ω4499, which is normally expressed in aggregating cells. This amount of C factor, however, fails to rescue *csgA* sporulation or expression of the absolutely C-factor-dependent fusion Ω4435, which is primarily expressed in sporulating cells (Kroos and Kaiser, 1987). Addition of 1 nM purified C factor led to aggregation, expression of both C-factor-dependent gene fusions, and sporulation. Parallel changes in C-factor concentration have been produced in vivo in a series of deletions of the DNA region upstream of the *csgA* mRNA start site (Li et al., 1992). Larger deletions terminated development earlier, with the *csg*-controlled behaviors of rippling, aggregation, and sporulation requiring about 20, 30, and 82%, respectively, of maximum *csgA* expression. Together, these results indicate that a time gradient of increasing C-factor concentration is responsible for the normal developmental time order: rippling, then aggregation, and finally, sporulation.

COMMON FEATURES OF A AND C SIGNALING

After starvation initiates development, *Myxococcus* spp. uses A factor to test cell density. A factor is a freely diffusible cell signal that does not require movement of the transmitting or receiving cells. If the A signal is not transmitted because the cell density is too low, the cells may use residual nutrient for (slow) growth until the appropriate density is reached. Once the proper density is reached, the cells express A-factor-dependent genes and proceed through the aggregation stages of the developmental program.

C factor provokes cells to aggregate into an organized mound and then triggers differentiation of spores. C factor is well suited for monitoring the quality of cell arrangements, because it appears to remain tightly associated with the cell surface. Efficient C-type signaling strictly requires that cells achieve the proper densely packed geometry found within the nascent fruiting body. Cells cannot passively accumulate but must move into the proper signaling position. C-type signaling allows cells to verify their positions and delays sporulation until the appropriate time.

The requirements for transmission of C signal differ from those for A signal in terms of cell density and arrangement. Nevertheless, A- and C-type signaling have common properties. Both apply tests of cell density. Both act to restrict further developmental choices: A signal commits to early development as opposed to growth, while C signal commits to aggregation and, at a higher concentration, sporulation. Density-dependent signaling may ensure that the *Myxococcus* fruiting body contains a sufficiently high density of spores for efficient feeding upon germination.

bsg MUTANTS AND THE *bsgA* GENE

A subset of the *bsg* mutants isolated by Hagen et al. (1978) map to the *bsgA* locus (Gill and Cull, 1986). New mutations, including insertions, have been isolated in this locus, and they all have the *bsgA* phenotype. Some *bsg* mutants are not linked to *bsgA*, but these mutant loci remain unmapped. *bsgA* mutants are defective in aggregation and sporulation (Gill and Cull, 1986) and either reduce or abolish expression of nearly all developmental markers examined, including many that are normally expressed during the first few hours of development (Fig. 2). Thus, *bsgA* appears to affect development at a very early stage. Nevertheless, the sporulation, aggregation, and gene expression defects of *bsgA* can be rescued by admixed wild-type cells (Gill and Cull, 1986). Aggregation and sporulation are delayed about 48 h (relative to those of wild type) in the rescue mixtures.

Growing colonies of *bsgA* mutants have a distinctive interaction with wild-type colonies. When the two approach one another, the leading edge of the mutant colony stops advancing and a prominent ridge of cells builds up at the colony edge. This behavior is specific for the *bsgA*–wild-type pair; it does not occur when two wild-type or two mutant colonies converge. The *bsgA* mutation appears to alter the motility response of growing cells when they encounter wild-type cells.

Gill et al. (1988) have cloned the $bsgA^+$ locus and demonstrated that *bsgA* mutants are recessive in partial diploid strains. Gill et al. (in press) expressed the $bsgA^+$ gene and showed that the 90-kDa protein product has ATP-dependent protease activity. The *bsgA* gene product shows 42% amino acid sequence identity and 57% similarity to the serine-protease La of *E. coli*. This suggests that the basis for *bsgA* mutants is a defect in intracellular proteolysis. There is no report as to the nature of the B signal itself.

dsg MUTANTS

dsg mutants are 10^3- to 10^5-fold deficient in sporulation compared with wild type and form abnormal aggregates but only after a delay (Cheng and Kaiser, 1989a). These developmental defects can be rescued by mixing a *dsg* mutant with wild-type cells (Hagen et al., 1978). The mutants either fail to express or express at a reduced level most developmental markers whose expression normally occurs after 4 h of development (Fig. 2).

The *dsg* gene has been located by a linked Tn*5* insertion, Ω1867. The gene has been cloned on the basis of its capacity to rescue a *dsg* mutant in partial diploid strains. The gene defined by rescue turns out to be a vital gene necessary for growth of *M. xanthus*. Insertions within the *dsg* gene are lethal. Partial diploids (*dsg*::Tn*5*/*dsg*$^+$) are viable, but no haploid *dsg*::Tn*5* segregants survive (Cheng and Kaiser, 1989b). A mixture of fatty acids and autocide AMI has been found to efficiently rescue the aggregation and sporulation defects of the *dsg-439* mutant (Rosenbluh and Rosenberg, 1989).

esg MUTANTS

The Tn*5* insertion Ω258 *(esg)* causes defects in sporulation and aggregation (Downard, 1992). Sporulation is decreased 10^4- to 10^5-fold, and mutant cells have only a slight tendency to aggregate, with mats of cells taking on a slightly roughened appearance. Developmental expression of the *tps* gene is decreased 10-fold or more in the mutant. Additional insertions in the same region of the genome also generate the *esg* phenotype. The sporulation defect of the mutants is corrected in mixing experiments, with approximately equal numbers of *esg* and wild-type spores formed in 1:1 mixtures. Representatives of the other extracellular complementation groups (*asg, bsg, csg,* and *dsg*) are also able to rescue sporulation and *tps* expression in the *esg* mutant, though less than is found with wild type. The *esg* mutants thus define a fifth extracellular complementation group, the E group.

ADENOSINE

Besides starvation for amino acids or other nutrients, the addition of various adenine compounds also induces aggregation into fruiting body-like structures (Campos and Zusman, 1975; Shimkets and Dworkin, 1981). Since adenine treatment of *E. coli* causes a decrease in 5-phosphoribosyl-1-pyrophosphate levels (Bagnara and Finch, 1974), the exposure of *M. xanthus* to adenine-related compounds may limit the synthesis of substances requiring phosphoribosyl PP_i for their synthesis, including thiamine, histidine, tryptophan, and uridine. Indeed, adenosine or cyclic AMP increases intracellular levels of guanosine pentaphosphate, indicating that adenine compounds may be inducing amino acid limitation (Manoil and Kaiser, 1980). To test whether it is essential for development that cells respond to adenine compounds in their environment, Manoil and Kaiser (1980) isolated a mutant that could grow in the presence of the toxic adenine analog 2,6-diaminopurine. A diaminopurine-resistant mutant, DK590, was not induced to aggregate by adenine, adenosine, cyclic AMP, or hypoxanthine, although any of these substances induced the diaminopurine-sensitive parent, DK501. The diaminopurine-resistant mutant aggregated and sporulated normally when starved for phenylalanine or for lysine, showing that DK590 was not developmentally deficient. Because the mutant can develop but no longer responds to exogenous adenine or adenosine, these compounds appear not to be necessary for aggregation.

Adenosine and other nucleic acid derivatives are released from cells during fruiting body development (Shimkets and Dworkin, 1981). At a cell density of $8 \times 10^5/\text{cm}^2$, which is below a critical density, general starvation fails to induce aggregation. However, the addition of adenine, adenosine, 5′-AMP, 3′-AMP, or cyclic AMP rescues aggregation of the low-density culture (Shimkets and Dworkin, 1981). No sporulation rescue was reported for these cultures. Guanine or pyrimidine nucleosides failed to rescue aggregation at the low cell density. Moreover, the level of adenosine secreted by wild-type cells into conditioned medium after 24 h at a high cell density was sufficient to induce fruiting body-like structures from low-density wild-type cells. Thus, adenosine, like A factor, enhances aggregation of wild-type cells at a low initial cell density. However, in the *asg* mutants, adenosine does not rescue aggregation, sporulation, or Ω4521 gene expression (Kuspa et al., 1992b), nor does adenosine rescue Ω4521 gene expression in low-density asg^+ cells (Kuspa et al., 1992b). Results with the *asg* mutants and the diaminopurine-resistant mutant imply that if adenosine serves to measure cell density, as proposed by Shimkets and Dworkin (1981), then unlike A factor, it is not obligatory and does not activate all developmentally regulated genes. Shimkets (1990) suggested that adenine and its derivatives may enhance starvation. Perhaps these compounds entering from the extracellular fluid facilitate the decisions of cells to respond to starvation as a homogeneous population. Relationships between starvation, adenosine, and A factor merit further investigation.

EXTRACELLULAR EFFECTORS

Several small molecules that have striking developmental effects when added to *M. xanthus* cells are described in this section.

Glycerol and Other Sporulation Inducers

Ordinarily, myxosporulation takes place within a maturing fruiting body. However, it is possible to bypass the early steps of cooperative multicellular morphogenesis and induce cellular morphogenesis (myxosporulation) directly by the addition of certain substances to a growing culture of vegetative cells (Dworkin and Gibson, 1964). Glycerol, dimethyl sulfoxide, ethylene glycol, phenylethanol, and indole are effective as sporulation inducers for *M. xanthus*. Glycerol, e.g., will induce more than 90% of cells to become spores. In *Stigmatella aurantiaca,* glycerol, *tert*-butanol, and a series of monovalent cations can induce sporulation (Gerth and Reichenbach, 1978; Reichenbach and Dworkin, 1970).

A high cell density, starvation, and a solid surface are all required for the formation of spores within fruiting bodies (Wireman and Dworkin, 1975), yet glycerol will induce sporulation in a culture at low density and in a nutritionally rich liquid medium. This suggests that each cell has within it a program for changes in cell shape and acquisition of the resistance properties characteristic of spores. This suggestion would also account for the fact that four-fifths of the mutants with temperature-sensitive defects in fruiting body sporulation are also temperature sensitive for glycerol-induced sporulation (Morrison and Zusman, 1979). This unicellular program would normally be activated late in fruiting body development by earlier developmental events. Glycerol, however, would activate this program directly, bypassing the normal requirement for starvation, high cell density, and a solid surface. The only markers listed in Fig. 2 that are known to be induced by glycerol are dFA-1 (Clemans et al., 1991), Tn*5 lac* Ω4530 (Kroos, 1986), *ops* (Downard and Zusman, 1985), and the hydrophilic shift (Kupfer and Zusman, 1984).

Glucosamine

Janssen and Dworkin (1985) discovered that glucosamine, mannosamine, or a polysaccharide-containing extract (called DS) of *M. xanthus* cells could rescue fruiting body formation and sporulation in certain developmentally defective mutants. Mueller and Dworkin (1991) showed that glucosamine induced intracellular glycerol formation. They observed that in response to glucosamine addition, the intracellular glycerol concentration transiently rose to 140 nmol/10^8 cells, or about 0.3 M. This glycerol concentration, if added extracellularly, would suffice to induce sporulation. The structures of glucosamine-induced spores resembled those of glycerol-induced spores. Glucosamine also induces lysis of *M. xanthus* cells (Janssen and Dworkin, 1985). Mueller and Dworkin (1991) found that glucosamine stimulates autocide AMI production, and since AMI induces lysis, they suggested that glucosamine works through the autocide.

Peptidoglycan Components

Rhythmically advancing waves of cells, called ripples, arise spontaneously and transiently during the aggregation of *M. xanthus* and many other myxobacteria into fruiting bodies (Reichenbach, 1965). Rippling can be induced and maintained without fruiting body aggregation by addition of *N*-acetylglucosamine, *N*-acetylmuramic acid, diaminopimelate, and D-alanine to *M. xanthus* cultures that are growing slowly because of reduction of the nutrient level (0.5% Casitone) (Shimkets and Kaiser, 1982a). The four compounds are all constituents of *Myxococcus* peptidoglycan, and murein purified from a variety of different bacteria is capable of inducing rippling. There is considerable turnover of peptidoglycan as fruiting bodies form (Dawson and Jones, 1979), and it is possible that rippling is a consequence of the release and uptake of these peptidoglycan constituents.

In addition to peptidoglycan components, an intact C-signaling system is required; *csgA* mutants fail to ripple. Moreover, a mutation in any A- or any S-motility gene prevents rippling (Shimkets and Kaiser, 1982a). In view of the requirement of C signaling for rippling, it is surprising that the same peptidoglycan components at the same concentrations rescued sporulation in *csg* mutants. These same constituents also rescue sporulation in about one-third of the *bsg* mutants, but they do not rescue *asg* or *dsg* mutants or the remaining two-thirds of the *bsg* mutants (Shimkets and Kaiser, 1982b). *N*-Acetylglucosamine, *N*-acetylmuramic acid, diaminopimelate, and D-alanine each increase the number of viable spores produced by a *csg* mutant. When all four components are included at 2.5 mM, they have a synergistic effect, raising the number of spores produced by the mutant to the same level as that of wild type. These effects are stereospecific: L-alanine will not substitute for D-alanine, and neither lysine nor ornithine will substitute for diaminopimelate. Peptidoglycan-rescued spores are resistant to heat and sonic vibration, and they germinate when plated on a nutrient-rich medium. They appear to be identical to fruiting body myxospores in their ultrastructure, protein composition, and resistance to boiling sodium dodecyl sulfate. Rescue of sporulation by peptidoglycan, like normal fruiting body sporulation, requires high cell density, low nutrient level, and a solid surface. Because the peptidoglycan components rescue the sporulation but not the rippling effects of *csg* mutants, they may bypass C signaling rather than replace it.

Autocides and Fatty Acids

Autocides are extracellular effectors as the term is understood here, and the reader is referred to chapter 10 in this volume.

PERSPECTIVE

The experimental opportunity offered by the myxobacteria to probe multicellularity was foreseen by Roland Thaxter 100 years ago (Thaxter,

1892). Today, some important goals of *Myxococcus* research are to identify, characterize, and link up the cellular response elements necessary for transduction of information contained in the intercellular signals. These goals appear within the reach of currently available experimental techniques. Intercellular signaling is a general regulatory strategy used in the development of organisms ranging from bacteria to vertebrates. *Myxococcus* signals illustrate several different mechanisms that regulate development in both time and space. The signals may trap regulatory states that are appropriate for progress through the developmental program. Intercellular signaling thereby creates the cellular order that is so strikingly apparent in fruiting body development.

Acknowledgments. Special thanks are due to Adam Kuspa, Lynda Plamann, and Seung Kim for their contributions to formulating the arguments on A and C signaling.

Research was supported by grant GM23441 from the National Institutes of Health (NIH) on signaling and grant DCB 8903705 from the National Science Foundation on motility. L.K. was supported by the Michigan Agricultural Experiment Station and by grant GM47293 from NIH.

REFERENCES

Arnold, J. W., and L. Shimkets. 1988. Inhibition of cell-cell interactions in *Myxococcus xanthus* by Congo red. *J. Bacteriol.* **170:**5765–5770.

Bagnara, A., and L. Finch. 1974. The effects of bases and nucleosides on the intracellular contents of nucleotides and 5-phosphoribosyl 1-pyrophosphate in *Escherichia coli. Eur. J. Biochem.* **41:**421–430.

Blackhart, B. D., and D. Zusman. 1985. The frizzy genes of *Myxococcus xanthus* control directional movement of gliding motility. *Proc. Natl. Acad. Sci. USA* **82:**8767–8770.

Blackhart, B. D., and D. Zusman. 1986. Analysis of the products of the *Myxococcus xanthus frz* genes. *J. Bacteriol.* **166:**673–678.

Bretscher, A. P., and D. Kaiser. 1978. Nutrition of *Myxococcus xanthus,* a fruiting myxobacterium. *J. Bacteriol.* **133:**763–768.

Burchard, R. P. 1970. Gliding motility mutants of *Myxococcus xanthus. J. Bacteriol.* **104:** 940–947.

Campos, J., and D. Zusman. 1975. Regulation of development in *Myxococcus xanthus:* effect of cAMP, AMP, ADP, and nutrition. *Proc. Natl. Acad. Sci. USA* **72:**518–522.

Cheng, Y., and D. Kaiser. 1989a. *dsg,* a gene required for a cell-cell interaction early in *Myxococcus* development. *J. Bacteriol.* **171:**3719–3726.

Cheng, Y., and D. Kaiser. 1989b. *dsg,* a gene required for *Myxococcus* development, is necessary for cell viability. *J. Bacteriol.* **171:**3727–3731.

Clemans, D. L., C. M. Chance, and M. Dworkin. 1991. A development-specific protein in *Myxococcus xanthus* is associated with extracellular fibrils. *J. Bacteriol.* **173:**6749–6759.

Cumsky, M., and D. Zusman. 1979. Myxobacterial hemagglutinin: a development-specific lectin of *Myxococcus xanthus. Proc. Natl. Acad. Sci. USA* **76:**5505–5509.

Dawson, H., and M. V. Jones. 1979. Cell wall turnover during myxospore formation in *Myxococcus. J. Gen. Microbiol.* **112:**143–148.

Doetsch, R. N., and G. J. Hageage. 1968. Motility in procaryotic organisms: problems, points of view, and perspectives. *Biol. Rev.* **43:**317–362.

Downard, J. (University of Oklahoma). 1992. Personal communication.

Downard, J., and D. Zusman. 1985. Differential expression of protein S genes during *Myxococcus xanthus* development. *J. Bacteriol.* **161:**1146–1155.

Downard, J. S., D. Kupfer, and D. Zusman. 1984. Gene expression during development of *Myxococcus xanthus*. Analysis of the genes for protein S. *J. Mol. Biol.* **175:**469–492.

Dworkin, M. 1973. Cell-cell interactions in the Myxobacteria. *Symp. Soc. Gen. Microbiol.* **23:**125–147.

Dworkin, M., and S. M. Gibson. 1964. Rapid formation of microcysts in *Myxococcus xanthus*. *Science* **146:**243–244.

Fink, J. M., and J. F. Zissler. 1989. Defects in motility and development of *Myxococcus xanthus* lipopolysaccharide mutants. *J. Bacteriol.* **171:**2042–2048.

Gerth, K., and H. Reichenbach. 1978. Induction of myxospore formation in *Stigmatella aurantiaca:* general characterization of the system. *Arch. Microbiol.* **117:**173–182.

Gill, J., and M. Dworkin. 1986. Cell surface antigens during submerged development of *Myxococcus xanthus* examined with monoclonal antibodies. *J. Bacteriol.* **168:**505–511.

Gill, J. S., and M. Dworkin. 1988. Isolation of additional monoclonal antibodies directed against cell surface antigens of *Myxococcus xanthus* cells undergoing submerged development. *J. Bacteriol.* **170:**5953–5955.

Gill, J. S., B. W. Jarvis, and M. Dworkin. 1987. Inhibition of development in *Myxococcus xanthus* by monoclonal antibody 1604. *Proc. Natl. Acad. Sci. USA* **84:**4505–4508.

Gill, R., and M. G. Cull. 1986. Control of developmental gene expression by cell-to-cell interactions of *Myxococcus xanthus*. *J. Bacteriol.* **168:**341–347.

Gill, R., M. G. Cull, and S. Fly. 1988. Genetic identification and cloning of a gene required for developmental cell interactions in *Myxococcus xanthus*. *J. Bacteriol.* **170:**5279–5288.

Gill, R. E., M. Karlok, and D. Benton. *Myxococcus xanthus* encodes an ATP-dependent protease which is required for developmental gene transcription and intercellular signalling. In press.

Hagen, D. C., A. P. Bretscher, and D. Kaiser. 1978. Synergism between morphogenetic mutants of *Myxococcus xanthus*. *Dev. Biol.* **64:**284–296.

Hagen, T. J., and L. J. Shimkets. 1990. Nucleotide sequence and transcriptional products of the *csg* locus of *Myxococcus xanthus*. *J. Bacteriol.* **172:**15–23.

Hodgkin, J., and D. Kaiser. 1977. Cell-to-cell stimulation of movement in nonmotile mutants of *Myxococcus*. *Proc. Natl. Acad. Sci. USA* **74:**2938–2942.

Hodgkin, J., and D. Kaiser. 1979. Genetics of gliding motility in *M. xanthus* (Myxobacterales): two gene systems control movement. *Mol. Gen. Genet.* **171:**177–191.

Inouye, M., S. Inouye, and D. Zusman. 1979. Gene expression during development of *Myxococcus xanthus:* pattern of protein synthesis. *Dev. Biol.* **68:**579–591.

Inouye, S. 1984. Identification of a development-specific promoter of *Myxococcus xanthus*. *J. Mol. Biol.* **174:**113–120.

Inouye, S., T. Franceschini, and M. Inouye. 1983a. Structural similarities between the development-specific proteins from a gram-negative bacterium, *Myxococcus xanthus,* and calmodulin. *Proc. Natl. Acad. Sci. USA* **80:**6829–6833.

Inouye, S., Y. Ike, and M. Inouye. 1983b. Tandem repeat of the genes for protein S. *J. Biol. Chem.* **258:**38–40.

Janssen, G., and M. Dworkin. 1985. Cell-cell interactions in developmental lysis of *Myxococcus xanthus*. *Dev. Biol.* **112:**194–202.

Jarvis, B. W., and M. Dworkin. 1989. Role of *Myxococcus xanthus* cell surface antigen 1604 in development. *J. Bacteriol.* **171:**4667–4673.

Kaiser, D. 1979. Social gliding is correlated with the presence of pili in *Myxococcus xanthus*. *Proc. Natl. Acad. Sci. USA* **76:**5952–5956.

Kaiser, D., and C. Crosby. 1983. Cell movement and its coordination in swarms of *Myxococcus xanthus*. *Cell Motil.* **3:**227–245.

Kalos, M., and J. F. Zissler. 1990a. A transposon tagging of genes for cell-cell interactions in *Myxococcus xanthus*. *Proc. Natl. Acad. Sci. USA* **87:**8316–8320.

Kalos, M., and J. F. Zissler. 1990b. Defects in contact-stimulated gliding during aggregation by *Myxococcus xanthus*. *J. Bacteriol.* **172:**6476–6493.

Kaplan, H. B., A. Kuspa, and D. Kaiser. 1991. Suppressors that permit A-signal-independent developmental gene expression in *Myxococcus xanthus*. *J. Bacteriol.* **173:**1460–1470.

Kim, S. K., and D. Kaiser. 1990a. C-factor: a cell-cell signaling protein required for fruiting body morphogenesis of *M. xanthus*. *Cell* **61:**19–26.

Kim, S. K., and D. Kaiser. 1990b. Cell alignment required in differentiation of *Myxococcus xanthus*. *Science* **249:**926–928.

Kim, S. K., and D. Kaiser. 1990c. Cell motility is required for the transmission of C-factor, an intercellular signal that coordinates fruiting body morphogenesis of *Myxococcus xanthus*. *Genes Dev.* **4:**896–905.

Kim, S. K., and D. Kaiser. 1990d. Purification and properties of *Myxococcus xanthus* C-factor, an intercellular signaling protein. *Proc. Natl. Acad. Sci. USA* **87:**3635–3639.

Kim, S. K., and D. Kaiser. 1991. C-factor has distinct aggregation and sporulation thresholds during *Myxococcus* development. *J. Bacteriol.* **173:**1722–1728.

Kroos, L. 1986. Cell interactions govern gene expression during *Myxococcus* development. Ph.D. thesis. Stanford University, Stanford, Calif.

Kroos, L., P. Hartzell, K. Stephens, and D. Kaiser. 1988. A link between cell movement and gene expression argues that motility is required for cell-cell signalling during fruiting body development. *Genes Dev.* **2:**1677–1685.

Kroos, L,. and D. Kaiser. 1984. Construction of Tn*5* lac, a transposon that fuses lacZ expression to exogenous promoters, and its introduction into *Myxococcus xanthus*. *Proc. Natl. Acad. Sci. USA* **81:**5816–5820.

Kroos, L., and D. Kaiser. 1987. Expression of many developmentally regulated genes in *Myxococcus* depends on a sequence of cell interactions. *Genes Dev.* **1:**840–854.

Kroos, L., A. Kuspa, and D. Kaiser. 1986. A global analysis of developmentally regulated genes in *Myxococcus xanthus*. *Dev. Biol.* **117:**252–266.

Kühlwein, H., and H. Reichenbach. 1968. Schwarmentwicklung und Morphogeneses bei *Myxobakterien*. Film C893/1965. Institut Wissenschaftlichen Film, Göttingen, Germany.

Kuner, J., and D. Kaiser. 1982. Fruiting body morphogenesis in submerged cultures of *Myxococcus xanthus*. *J. Bacteriol.* **151:**458–461.

Kupfer, D., and D. R. Zusman. 1984. Changes in cell surface hydrophobicity of *Myxococcus xanthus* are correlated with sporulation-related events in the developmental program. *J. Bacteriol.* **159:**776–779.

Kuspa, A. 1989. Intercellular signalling in the regulation of early development in *Myxococcus xanthus*. Ph.D. thesis. Stanford University, Stanford, Calif.

Kuspa, A., and D. Kaiser. 1989. Genes required for developmental signalling in *Myxococcus xanthus:* three *asg* loci. *J. Bacteriol.* **171:**2762–2772.

Kuspa, A., L. Kroos, and D. Kaiser. 1986. Intercellular signaling is required for developmental gene expression in *Myxococcus xanthus*. *Dev. Biol.* **117:**267–276.

Kuspa, A., L. Plamann, and D. Kaiser. 1992a. Identification of heat-stable A-factor from *Myxococcus xanthus*. *J. Bacteriol.* **174:**3319–3326.

Kuspa, A., L. Plamann, and D. Kaiser. 1992b. A-signalling and the cell density requirement for *Myxococcus xanthus* development. *J. Bacteriol.* **174:**7360–7369.

LaRossa, R., J. Kuner, D. Hagen, C. Manoil, and D. Kaiser. 1983. Developmental cell interactions in *Myxococcus:* analysis of mutants. *J. Bacteriol.* **153:**1394–1404.

Li, S., B. U. Lee, and L. Shimkets. 1992. *csgA* expression entrains *Myxococcus xanthus* development. *Genes Dev.* **6:**401–410.

Losick, R., and L. Kroos. 1989. Dependence pathways for the expression of genes involved in endospore formation in *Bacillus subtilis,* p. 223–241. *In* I. Smith, R. Slepecky, and P. Setlow (ed.), *Regulation of Procaryotic Development.* American Society for Microbiology, Washington, D.C.

Manoil, C. C., and J. Beckwith. 1985. TnphoA: a probe for protein export signals. *Proc. Natl. Acad. Sci. USA* **82:**8129–8133.

Manoil, C., and D. Kaiser. 1980. Purine-containing compounds, including cyclic adenosine 3′,5′-monophosphate, induce fruiting of *Myxococcus xanthus* by nutritional imbalance. *J. Bacteriol.* **141:**374–377.

Mayo, K., and D. Kaiser. 1989. asgB, a gene required early for developmental signalling, aggregation, and sporulation of *Myxococcus xanthus. Mol. Gen. Genet.* **218:**409–418.

McCleary, W. R., B. Esmon, and D. Zusman. 1991. *Myxococcus xanthus* protein C is a major spore surface protein. *J. Bacteriol.* **173:**2141–2145.

McCleary, W. R., and D. R. Zusman. 1990. Purification and characterization of the *Myxococcus xanthus* FrzE protein shows that it has autophosphorylation activity. *J. Bacteriol.* **172:**6661–6668.

McVittie, A., F. Messik, and S. A. Zahler. 1962. Developmental biology of *Myxococcus. J. Bacteriol.* **84:**546–551.

Morrison, C., and D. Zusman. 1979. Mutants with temperature-sensitive stage-specific defects: evidence for independent pathways in development. *J. Bacteriol.* **140:**1036–1042.

Mueller, C., and M. Dworkin. 1991. Effects of glucosamine on lysis, glycerol formation, and sporulation in *Myxococcus xanthus. J. Bacteriol.* **173:**7164–7175.

O'Connor, K. A., and D. R. Zusman. 1989. Patterns of cellular interactions during fruiting body formation in *Myxococcus xanthus. J. Bacteriol.* **171:**6013–6024.

Plamann, L,. and A. Kuspa, and D. Kaiser. 1992. Proteins that rescue A-signal-defective mutants of *Myxococcus xanthus. J. Bacteriol.* **174:**3311–3318.

Reichenbach, H. 1965. Rhythmische Vorgänge bei der Schwarmentfaltung von Myxobakterien. *Ber. Dtsch. Bot. Ges.* **78:**102–105.

Reichenbach, H. 1966. *Myxococcus* spp. (Myxobacteriales) Schwarmentwicklung und Bildung von Protocysten. Film E778/1965. Institut Wissenschaftlichen Film, Göttingen, Germany.

Reichenbach, H., and M. Dworkin. 1970. Induction of myxospore formation in *Stigmatella aurantiaca* by monovalent cations. *J. Bacteriol.* **101:**325–326.

Reichenbach, H., and M. Dworkin. 1992. The myxobacteria, p. 3416–3487. *In* A. Balows, H. G. Trüper, M. Dworkin, W. Harder, and K.-H. Schleifer (ed.), *The Prokaryotes,* Springer-Verlag, New York.

Rhie, H. G., and L. J. Shimkets. 1989. Developmental bypass suppression of *Myxococcus xanthus csgA* mutations. *J. Bacteriol.* **171:**3268–3276.

Rhie, H.-G., and L. J. Shimkets. 1991. Low-temperature induction of *Myxococcus xanthus* developmental gene expression in wild-type and *csgA* suppressor cells. *J. Bacteriol.* **173:** 2206–2211.

Rodriguez, J. Unpublished data.

Romeo, J. M., and D. R. Zusman. 1991. Transcription of the myxobacterial hemagglutinin gene is mediated by a σ^{54}-like promoter and a *cis*-acting upstream regulatory region of DNA. *J. Bacteriol.* **173:**2969–2976.

Rosenberg, E., K. Keller, and M. Dworkin. 1977. Cell density-dependent growth of *Myxococcus xanthus* on casein. *J. Bacteriol.* **129:**770–777.

Rosenbluh, A., and E. Rosenberg. 1989. Autocide AMI rescues development in *dsg* mutants of *Myxococcus xanthus. J. Bacteriol.* **171:**1513–1518.

Shimkets, L., and D. Kaiser. 1982a. Induction of coordinated movement of *Myxococcus xanthus* cells. *J. Bacteriol.* **152:**451–461.

Shimkets, L., and D. Kaiser. 1982b. Murein components rescue developmental sporulation of *Myxococcus xanthus. J. Bacteriol.* **152:**462–470.

Shimkets, L. J. 1990. Social and developmental biology of the myxobacteria. *Microbiol. Rev.* **54:**473–501.

Shimkets, L. J., and S. J. Asher. 1988. Use of recombination techniques to examine the structure of the csg locus of *Myxococcus xanthus. Mol. Gen. Genet.* **211:**63–71.

Shimkets, L. J., and M. Dworkin. 1981. Excreted adenosine is a cell density signal for the initiation of fruiting body formation in *Myxococcus xanthus. Dev. Biol.* **84:**51–60.

Shimkets, L. J., R. E. Gill, and D. Kaiser. 1983. Developmental cell interactions in *Myxococcus xanthus* and the *spoC* locus. *Proc. Natl. Acad. Sci. USA* **80:**1406–1410.

Shimkets, L. J., and H. Rafiee. 1990. CsgA, an extracellular protein essential for *Myxococcus xanthus* development. *J. Bacteriol.* **172:**5299–5306.

Shimoura, O., H. Suthers, and J. T. Bonner. 1982. Chemical identity of the acrasin of the cellular slime mold *Polysphondylium violaceum. Proc. Natl. Acad. Sci. USA* **79:** 7376–7379.

Smith, I. 1989. The initiation of sporulation, p. 185–210. *In* R. Slepecky and I. Smith (ed.), *Regulation of Procaryotic Development*. American Society for Microbiology, Washington, D.C.

Sodergren, E., and D. Kaiser. 1983. Insertions of Tn5 near genes that govern stimulatable cell motility in Myxococcus. *J. Mol. Biol.* **167:**295–310.

Sonenshein, A. L. 1989. Metabolic regulation of sporulation and other stationary-phase phenomena, p. 109–130. *In* R. Slepecky and I. Smith (ed.), *Regulation of Procaryotic Development*. American Society for Microbiology, Washington, D.C.

Stephens, K., P. Hartzell, and D. Kaiser. 1989. Gliding motility in *Myxococcus xanthus:* the *mgl* locus, its RNA, and predicted protein products. *J. Bacteriol.* **171:**819–830.

Teintze, M., R. Thomas, T. Furuichi, M. Inouye, and S. Inouye. 1985. Two homologous genes coding for spore-specific proteins are expressed at different times during development of *Myxococcus xanthus. J. Bacteriol.* **163:**121–125.

Thaxter, R. 1892. On the Myxobacteriacae, a new order of Schizomycetes. *Bot. Gaz.* **17:** 389–406.

Weinberg, R. A., and D. R. Zusman. 1990. Alkaline, acid, and neutral phosphatase activities are induced during development in *Myxococcus xanthus. J. Bacteriol.* **172:**2294–2302.

Wireman, J., and M. Dworkin. 1975. Morphogenesis and developmental interactions in the Myxobacteria. *Science* **189:**516–523.

Zusman, D. 1982. Frizzy mutants: a new class of aggregation-defective developmental mutants of *Myxococcus xanthus. J. Bacteriol.* **150:**1430–1437.

Myxobacteria II
Edited by Martin Dworkin and Dale Kaiser
© 1993 American Society for Microbiology, Washington, DC 20005

Chapter 13

Motility and Tactic Behavior of *Myxococcus xanthus*

Mark J. McBride, Patricia Hartzell, and David R. Zusman

The myxobacteria are unusual prokaryotes in many respects, including their motility and tactic properties. They exhibit a form of motility called gliding motility, whose basis has been an unsolved biological mystery for 100 years (Thaxter, 1892). Myxobacteria do not have flagella, nor have any organelles of motility been convincingly identified (Pate, 1988; Burchard, 1984). In liquid medium, they are unable to swim. On a moist surface, they move slowly (500- to 1,000-fold slower than *Escherichia coli*) in the direction of their long axes and occasionally reverse their direction of movement (the leading end becomes the lagging end). Cells are able to move individually or in groups. Motility is essential for the social interactions of these bacteria, the most studied of which are aggregation and fruiting body formation.

Gliding motility, while considered unusual, is certainly not unique to the myxobacteria. Gliding motility occurs in at least six of the nine eubacterial divisions (Reichenbach et al., 1986). The mechanism of motility is not known for any of these bacteria, and it is not known whether these bacteria all use a similar means of movement. However, certain features are common to the motility systems of virtually all gliding bacteria (Pate, 1988; Burchard, 1981). (i) Gliding bacteria produce extracellular slime. No slime from the gliding bacteria has been completely characterized, but myxobacterial slime consists primarily of polysaccharide (Sutherland and

Mark J. McBride • Department of Biological Sciences, University of Wisconsin–Milwaukee, Milwaukee, Wisconsin 53211. **Patricia Hartzell** • Department of Microbiology and Molecular Genetics, University of California–Los Angeles, Los Angeles, California 90024. **David R. Zusman** • Department of Molecular and Cell Biology, University of California–Berkeley, Berkeley, California 94720.

Thomson, 1975). Lack of slime production has been correlated with loss of gliding ability of several bacteria (Grimm and Kuhlwein, 1973; Niklitschek, 1934). Slime may act as a reversible adhesive that allows cells to attach themselves to a surface yet move over that surface (Burchard, 1984). (ii) In order to exhibit motility, gliding cells require contact with a surface or interface. Most gliding bacteria require a film of water or a relatively well-hydrated surface in order to move. They cannot swim through liquid media. (iii) As with flagellar motility, the movement of most gliding bacteria appears to be powered by a membrane potential rather than by ATP directly (Ridgway, 1977; Duxbury et al., 1980; Pate and Chang, 1979; Halfen and Castenholz, 1971; Glagoleva et al., 1980). It is not clear that a membrane potential is the driving force for the motility of myxobacteria (Burchard, 1981, 1984). Gliding of *Myxococcus xanthus* was inhibited by proton-conducting uncouplers, but this effect was reversed by the addition of pyruvate. Movement was also inhibited by arsenate, which should not have dissipated the membrane potential.

On the basis of the properties of gliding cells, a variety of theories to explain gliding motility have been proposed. These include (i) directed extrusion of slime (Burchard, 1984), (ii) directed release of surfactants (Keller et al., 1983; Dworkin et al., 1983), (iii) microscopic contractions (Burchard, 1984), (iv) movement of outer membrane absorption sites along fixed tracks (Lapidus and Berg, 1982), and (v) interaction of rotary motors in the cell envelope with the substratum or the surface slime layer (Pate and Chang, 1979; Pate, 1988). Conclusive evidence for any one theory is presently lacking.

GLIDING MOTILITY OF MYXOBACTERIA

Recent advances in our understanding of gliding motility have come from studies of myxobacteria and of *M. xanthus* in particular. *M. xanthus* has distinct advantages as an experimental system for addressing questions regarding gliding motility. This organism is easily grown in the laboratory, and a wide variety of techniques are available for its genetic manipulation (Rosenberg, 1984).

Several investigators have conducted electron microscopic investigations to determine the nature of myxobacterial gliding motility. In general, these studies have found little to indicate the presence of obvious structures involved in motility. However, intracellular filaments that may be parts of the motility apparatus have been observed in some specimens (Burchard et al., 1977; Lunsdorf and Reichenbach, 1989). These could function as contractile elements involved in gliding.

GENETIC STUDIES OF GLIDING

Many mutants of *M. xanthus* with defects in motility have been isolated (Burchard, 1970; MacRae and McCurdy, 1976; Hodgkin and Kaiser, 1977, 1979a, 1979b). Genetic analysis of motility mutants revealed two systems of motility in *M. xanthus:* S (social) motility and A (adventurous) motility (Fig. 1) (Hodgkin and Kaiser, 1979b). The S motility system is responsible for movement of bacteria in groups. Cells must be in close proximity (within one cell length) for the S motility system to function properly (Kaiser and Crosby, 1983). The A motility system allows movement of isolated cells and does not require cell-cell contact. The two motility systems apparently operate independently. A^+S^- or A^-S^+ cells are still motile by gliding and form colonies that spread, although at a reduced rate. However, A^-S^- cells form colonies that do not spread. In addition to mutations in the A and S systems, mutations at one other locus, *mgl* (mutual function for gliding), have been observed to affect motility (Hodg-

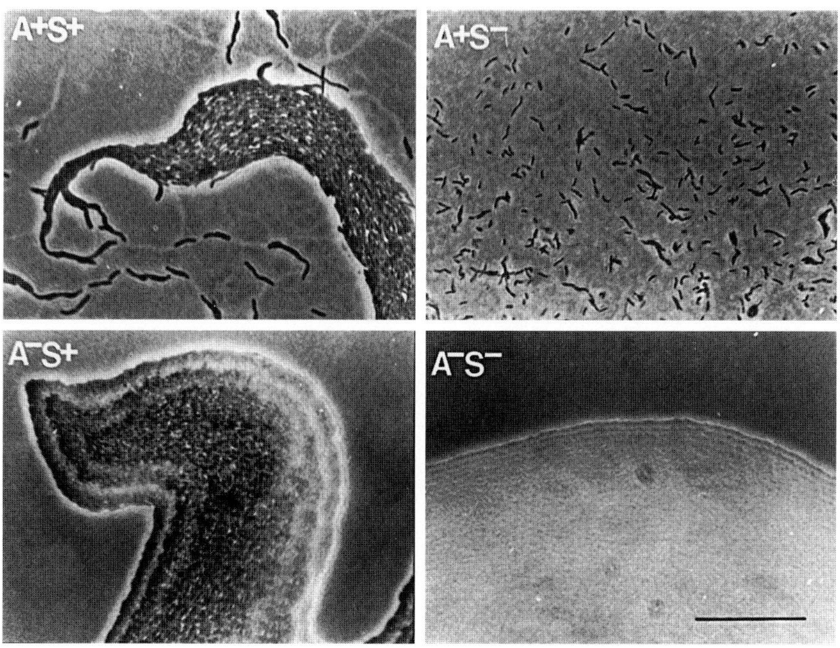

Figure 1. Effects of A- and S-motility mutations on cell behavior. *M. xanthus* cells were inoculated onto CTT agar (Hodgkin and Kaiser, 1977) and incubated at 32°C for 3 days. The colony edge was photographed under phase-contrast. Bar, 50 μm. Strains: A^+S^+, DK1622; A^-S^+, DK2608; A^+S^-, DK3468; A^-S^-, DK3460. (Courtesy of L. Shimkets [Shimkets, 1986a].)

kin and Kaiser, 1979b). In fact, mutations at the *mgl* locus are the only single mutations that completely abolish motility in *M. xanthus*. As such, *mgl* seems to be necessary for both A motility and S motility. Like A^-S^- mutants, *mgl* mutants form nonspreading colonies (Hodgkin and Kaiser, 1979b). *mgl* mutants are completely nonmotile.

Members of another class of mutants retain the ability to move but have lost the ability to control their movements. These "frizzy" (*frz*) mutants were isolated as part of an analysis of developmental mutants (Morrison and Zusman, 1979; Zusman, 1982). Cells of *frz* mutants failed to form fruiting bodies but instead aggregated into tangled, frizzy filaments (Fig. 2). Analyses of the behavior of individual cells revealed the reason for this phenotype. Wild-type cells of *M. xanthus* move slowly over agar surfaces at a rate of about one cell length every 2 min. Individual cells move in one direction for about 7 to 8 min (on average) and then reverse their direction of movement so that the front of the bacterium becomes the rear. Presumably, the cells respond to external stimuli to determine when they will reverse. In contrast, the *frz* mutants very rarely reverse their direction of movement (an average of once every 2 h for some mutants) (Blackhart and Zusman, 1985b; Zusman and McBride, 1991). This

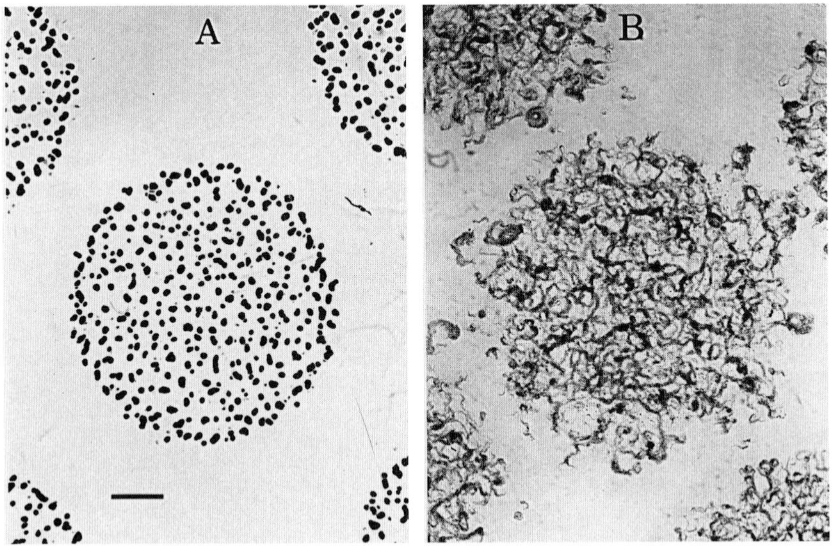

Figure 2. Morphology of frizzy mutants. Cultures of *M. xanthus* were spotted onto fruiting agar and allowed to incubate for 7 days at 28°C. (A) Strain DZF1 (wild type); (B) strain DZF1227 (*frzE*). Bar, 1 mm. Reprinted from Zusman (1982).

behavior accounts for their inability to form fruiting bodies and their tendency to form long filaments and swirls of cells.

Genetic, molecular, and behavioral studies of the mutants described above have taught us much about myxobacterial motility. The next sections of this chapter describe what we know about these mutants and what we have learned from them.

A AND S MOTILITIES

Mutations in an estimated 10 genes (*sgl* [social gliding], *tgl* [conditional transient gliding], and *dsp* loci) result in deficiencies in the S motility system that lead to reduced swarming movement (Hodgkin and Kaiser, 1979b). These mutants are all less cohesive than wild-type cells. Mutations in at least four of these loci (*sglA, sglB, sglG,* and *tgl*) result in loss of polar pili (Kaiser, 1979). Pili, which extend approximately one cell length, may be involved in cell-cell communication and may explain the need for cells to be within one cell length for S motility to be employed. Although S motility has been correlated with the presence of polar pili, *dsp* mutants are deficient in S motility but retain pili. Like the other S^- mutants, *dsp* mutants are less cohesive than wild-type cells, probably because of a failure of *dsp* mutants to form extracellular fibrils (Shimkets, 1986b).

Mutations in any of 25 genes (*agl, cgl,* and *lps* loci) result in a deficiency in A motility (Hodgkin and Kaiser, 1979a; Fink and Zissler, 1989). The primary defects of the *agl* (adventurous gliding) or *cgl* (conditional gliding) mutants have not been completely characterized. *lps* mutants were isolated with the aid of monoclonal antibodies that recognize the O antigen of *M. xanthus* lipopolysaccharide (LPS) (Fink and Zissler, 1989). These mutants were deficient in this antigen and in A motility. In a qualitative analysis of LPS levels, Fink and Zissler (1979) found that the O-antigen levels of *agl* and *cgl* mutants are similar to those of wild-type cells. This may indicate that the *lps* mutants represent a third class of A-motility mutants. LPS may play a role in motility or in the interaction of cells with their substrata or other cells. Alteration of LPS may also hinder proper insertion or assembly of motility components in the outer membrane.

Hodgkin and Kaiser (1977, 1979a, 1979b) found that some of the A- and S-motility mutants were stimulatable by contact with cells that carried the corresponding wild-type alleles. These included *cgl* mutants (A^-) and *tgl* mutants (S^-). Hodgkin and Kaiser hypothesized that these genes code for cell surface or extracellular components that stimulate gliding. Kalos and Zissler (1990) used the transposon Tn*phoA* to identify proteins that are exported across the cell membrane. This transposon contains the region of *phoA* that codes for the carboxy terminus of PhoA but lacks the region that codes for the amino-terminal signal secretion sequence. PhoA is active only after export from the cytoplasm. Therefore, the detection of phosphatase activity indicates that fusion in frame with a gene for an exported protein has occurred. Three of the phosphatase-positive Tn*phoA* insertions caused A-motility defects. These mapped within the known *cgl* genes and could be stimulated to move by contact with *cgl*$^+$ strains. This is further evidence that these *cgl* gene products are exported.

Cells with mutations in both A- and S-motility genes form nonspreading colonies and were originally described as completely nonmotile. However, time-lapse videomicroscopy indicates that at least some A$^-$S$^-$ mutants, including DK1252 (*cglF1 tgl-3*), are still able to move (Dworkin, personal communication). The cells move only short distances before reversing direction and therefore form colonies that do not spread. Further study is necessary to determine whether other A$^-$S$^-$ strains retain the ability to move.

Swarming and the Functions of the Two Motility Systems

Swarming of gliding bacteria involves the coordinated movement of many cells over a solid surface. Swarming of *M. xanthus* is dependent on cell density (Kaiser and Crosby, 1983). The swarming rate increases with increasing cell density until a saturating concentration is reached. Swarming is affected by mutations in both A and S motility systems. Recently, the swarming of *M. xanthus* on different surfaces was studied by using wild-type cells (A$^+$S$^+$ motile) and mutants carrying mutations in each of the motility systems (A$^-$S$^+$ and A$^+$S$^-$) (Shi and Zusman, in press). Wild-type cells swarm slowly on standard growth media (Casitone-yeast extract [CYE]) solidified with 1.5% agar. Swarming was much greater on a very moist surface (e.g., CYE with 0.3% agar). Efficient swarming on 0.3% agar required a functional S motility system and a functional *frz* system but did not require A motility. In contrast, efficient swarming on 1.5% agar required a functional A motility system. A$^+$S$^-$ cells migrated relatively rapidly over 1.5% agar, whereas A$^-$S$^+$ cells spread much less effectively. The rate of spreading of a fully motile strain was equal to or greater than the sum of the rates of A$^+$S$^-$ and A$^-$S$^+$ cells measured independently (Kaiser and Crosby, 1983; Shi and Zusman, in press). The two different motility systems thus allow cells to move efficiently over different surfaces. These results suggest a rationale for maintaining two different motility systems in *M. xanthus*.

Effect of A- and S-Motility Mutations on Development and Rippling

The most striking feature of the myxobacteria is their ability to form multicellular fruiting bodies. Fruiting body formation of *M. xanthus* occurs under nutrient-poor conditions. Aggregation of cells results in the formation of a raised mound containing approximately 10^6 cells. Within the mounds, rod-shaped cells develop into spherical myxospores. Strains defective in A or S motility are often developmentally deficient. This is especially evident with S^- mutants, many of which are completely unable to form fruiting bodies (Hodgkin and Kaiser, 1979b). Rippling, which is the periodic formation of ridges of cells (Reichenbach, 1966; Shimkets and Kaiser, 1982), is also disrupted by mutations in genes of either the A or S system of motility.

Do the A and S Motility Systems Code for Two "Motors"?

The exact functions of most of the A- and S-motility genes are not known. They could code for components of two gliding motors. Such apparent redundancy of motility functions may not be uncommon. *Vibrio parahaemolyticus* has two different flagellar systems that allow the bacteria to migrate under different conditions (McCarter and Silverman, 1990). Polar flagella allow the cells to swim through liquid, whereas large numbers of peritrichous flagella allow the cells to swarm over solid surfaces in multicellular rafts. The presence of two independent motors in *M. xanthus* could explain why nonmotile mutants are so rare, since mutations in each system would be necessary for this phenotype. However, the A- and S-motility genes that have been analyzed do not code for proteins with obvious functions in a motility apparatus, and at least some A^-S^- strains are still capable of movement. This may indicate that the two motility systems of *M. xanthus* do not code for components of two gliding motors. They may instead allow the functioning of one or more motors (coded for by other genes) on different surfaces. Further analyses of A- and S-motility genes and mutants should help us determine the role(s) of these genes in gliding motility.

mgl LOCUS

As the only site of mutations that completely abolish motility, the *mgl* locus is of interest in determining the mechanism of gliding motility and has been the subject of an intensive investigation. Stephens and Kaiser (1987) cloned a 2.5-kb fragment of DNA that complemented several *mgl* mutations. The DNA sequence revealed two open reading frames (ORFs) (Stephens et al., 1988). Most of the mutations that abolished gliding were in the downstream ORF, *mglA*. Antibodies raised against recombinant MglA were used to identify the 22-kDa protein in extracts of *M. xanthus*. MglA was also identified in extracts of *Stigmatella aurantiaca,* suggesting that it is highly conserved among the myxobacteria (Hartzell and Kaiser, 1991a). This is supported by the finding that the cloned *mgl* fragment from *M. xanthus* rescues motility in an *S. aurantiaca mgl* mutant (Hartree, personal communication). Some distantly related gliding bacteria (*Microscilla* spp. and *Flexibacter columnaris*) also produced material that cross-reacted with antisera raised against MglA, but whether these proteins are related to MglA or are involved in gliding is not known.

The essential role of *mglA* in gliding suggested that MglA might be a component of the gliding motor or a global regulator of motility genes. To begin to understand the function of MglA, its cellular location was determined (Hartzell and Kaiser, 1991a). Immunoelectron microscopy and immunoassay of cell fractions demonstrated that MglA is primarily distributed in the cytoplasm. There was no evidence of any organization that would indicate its localization to structures involved in gliding.

The predicted amino acid sequence of MglA indicates significant similarity to members of a class of GTP-binding proteins (Hartzell and Kaiser, 1991a). These proteins, such as p21^{ras} and sar1P, include membrane-associated and soluble proteins with diverse functions in transformation, protein secretion, and signal transduction (Bourne et al., 1991). The region of similarity with MglA spans the consensus site for GTP binding and hydrolysis. Although binding and hydrolysis of GTP by purified MglA has yet to be reported, the similarities between MglA and this class of G proteins suggests a role for MglA as a modulator of A and S components involved in gliding motility. However, MglA is probably not involved in regulating expression of A- and S-motility genes. A number of Tn5 *lac* insertions with A- and S-motility genes were isolated and used to measure the level of gene expression in wild-type or *mgl* backgrounds (Hartzell, unpublished data). Activity of β-galactosidase was nearly identical regardless of the background, indicating that loss of *mglA* function had little or no effect on expression of these genes.

Sequence analysis of the *mgl* region identified a second ORF (*mglB*) upstream of the *mglA* gene (Stephens et al., 1988). *mglB* and *mglA* are cotranscribed. Two mRNAs of 1.3 and 1.5 kb that spanned both ORFs were identified by Northern (RNA) blot analysis. To determine the function of *mglB*, mutants that carried deletions of portions of *mglB* but produced normal levels of *mgl* RNA were constructed. These mutants exhibited reduced motility and colony spread (Hartzell and Kaiser, 1991b). Enzyme-linked immunosorbent assays indicated that these *mglB* mutants contained lower levels of MglA protein (11 to 17%) than did wild-type cells. The level of MglA protein corresponded roughly to the spreading rate of the *mglB* mutants. These results indicate that *mglB* may affect translation of *mglA* or that MglB protein may interact with and perhaps stabilize MglA. The effects of the *mglB* deletions are apparently not polar effects, since heterozygous partial diploids (*mglB*/*mglA* with the wild-type alleles in *trans*) have normal motility.

The predicted protein product of *mglB* shows similarity to one of the calcium-binding sites of yeast calmodulin. The significance of this finding is uncertain, but calcium is required for gliding of both *M. xanthus* and *S. aurantiaca* (Burchard, 1984; Womack et al., 1989). Calcium seems to be necessary for both A and S motility, since the calcium chelator ethyleneglycoltetraacetic acid (EGTA) inhibits both (Hartzell, unpublished data). In addition to its immediate effects on gliding, calcium also affects the expression of *S. aurantiaca* genes involved in gliding (Womack et al., 1989).

frz MUTANTS

The *frz* mutants exhibit aberrant directional control and are unable to form fruiting bodies. Blackhart and Zusman (1985a) cloned and genetically analyzed a region of DNA that complemented the *frz* mutations. The *frz* mutations were recessive to the wild-type genes and were present in five complementation groups: *frzA*, *frzB*, *frzC*, *frzE*, and *frzF*. In addition, two Tn*5* mutations between *frzC* and *frzE* defined the *frzD* locus. *frzD* mutants did not exhibit the *frz* phenotype but instead formed nonspreading colonies. *frzD* mutations were dominant to wild-type alleles. In contrast to the majority of *frz* mutants (*frzA*, *frzB*, *frzC*, *frzE*, and *frzF*), which rarely reverse their direction of movement, *frzD* mutants reverse their direction of movement very frequently (about once every 2 min) and do not vary the interval between reversals.

The behavior patterns of the *frz* mutants are similar in some respects to those of chemotaxis mutants of swimming bacteria such as *E. coli* and *Salmonella typhimurium* (Macnab, 1987). *E. coli* cells sense chemicals in their environment and respond to changing concentrations of these chemicals by altering their behavior (Macnab, 1987). They either continue to swim smoothly (if moving in a favorable direction) or tumble erratically (if moving in an unfavorable direction). Each tumble reorients the cell for another swimming run. Cells tumble less frequently when they are moving in a favorable direction. Most chemotaxis mutants of *E. coli* tumble either much more frequently or less frequently than wild-type cells.

frz Gene Products Are Similar to Enteric *che* Gene Products

The behavioral similarities of *frz* mutants of *M. xanthus* and *che* mutants of *E. coli* provided the stimulus to subclone and sequence the *frz*

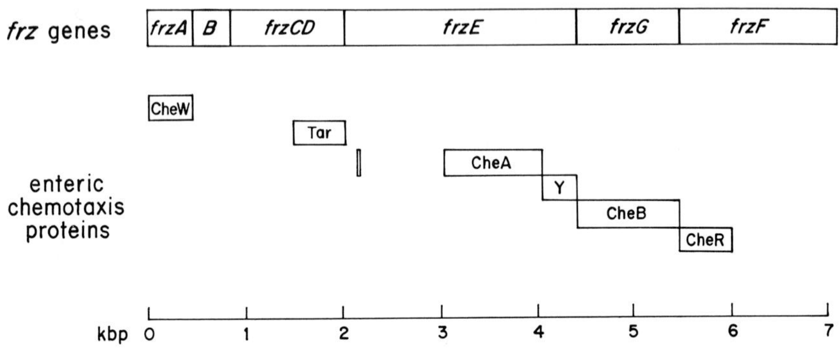

Figure 3. Similarities between enteric chemotaxis proteins and deduced *frz* proteins. Tar is an MCP. Reprinted from Zusman and McBride (1991).

genes (McBride et al., 1989; McCleary et al., 1990; McCleary and Zusman, 1990a). Sequence analyses revealed that five of the six *frz* genes are homologous to chemotaxis genes of flagellated enteric bacteria (Fig. 3). In particular, *frzA* is similar to *cheW, frzCD* (a single gene composed of the *frzC* and *frzD* loci) is similar to the genes for the enteric methyl-accepting chemotaxis proteins (MCPs), *frzE* is similar to both *cheA* and *cheY, frzF* is similar to *cheR,* and *frzG* (a region of DNA between *frzE* and *frzF* formerly referred to as the "gap" region) is similar to *cheB*. The deduced *frz* proteins exhibit 28 to 40% identity to their *che* counterparts at the amino acid level. Comparison of the *che* and *frz* systems suggests functions for many of the *frz* gene products.

The chemotaxis signal transduction pathway of enteric bacteria consists of a series of interacting proteins (Macnab, 1987; Parkinson, 1988). MCPs act as chemoreceptors. These membrane-spanning proteins detect chemicals in the environment and initiate a signal relay within the cytoplasm by interacting with CheW and CheA. CheA is an autokinase that phosphorylates itself following stimulation and then transfers the phosphate to CheY. CheY-phosphate is thought to interact with the flagellar motor and cause clockwise rotation, resulting in a tumble. Following stimulation by a chemoeffector, the cell readjusts to its original swimming behavior of runs and tumbles over a period of seconds to minutes. This is necessary to allow the cell to respond to new changes in its chemical environment. Adaptation is accomplished by covalent modification of the cytoplasmic portion of the transmembrane MCPs. These modifications are mediated by CheR and CheB. CheR catalyzes the methylation of specific glutamate residues on the MCPs. CheB catalyzes the demethylation of these same residues and also the deamidation of specific glutamine residues. Demethylation of these glutamine residues generates additional glutamate residues for methylation by CheR. All of these modifications are thought to alter the conformation and thus the signaling state of the cytoplasmic portion of the MCP.

Many of the *frz* proteins have been identified in extracts of *M. xanthus*. Antisera raised against recombinant forms of FrzA, FrzB, FrzCD, and FrzE have allowed identification and analyses of the biochemical properties of these proteins. The *frz* proteins exhibit biochemical similarities to and significant differences from their *che* homologs. The specific similarities and differences are outlined below. We begin by examining homologs to the components at the signal reception end of the enteric chemotaxis chain (the MCPs and the proteins that modify the MCPs) and proceed toward homologs to the components that interact with the enteric flagellar motor.

The *frzC* and *frzD* loci of *M. xanthus* lie within a single ORF that we now call *frzCD* (McBride et al., 1989). (The *frzD* mutants produce truncated forms of the FrzCD protein, which probably accounts for their phenotype.) The *frzCD* gene codes for a 43.6-kDa protein that is similar in sequence to the cytoplasmic portion of the enteric MCPs. The regions of strongest similarity include the signaling domain of the MCPs and the sites of modification by methylation. Three or four sites resembling the consensus sequences for methylatable glutamate residues of MCPs are found in FrzCD. FrzCD is reversibly methylated in vivo (McCleary et al., 1990). The methyl groups are easily removed by gentle base treatment, which is consistent with carboxymethylation of glutamate residues. FrzF, the homolog to the enteric methyl transferase CheR, is required for methylation of FrzCD (McCleary et al., 1990). The sequence similarity between FrzG and CheB suggests that FrzG is involved in demethylation and deamidation of FrzCD. This was a surprise, because mutations in *frzG* do not result in the *frz* phenotype. Instead, these mutants form fruiting bodies nearly as well as the wild-type strain, and their gliding behavior is similar to that of wild-type cells. *frzG* mutants are able to modify FrzCD (methylation and demethylation) (McBride et al., 1992). This may indicate that another enzyme in the cell can compensate for loss of FrzG and could explain why *frzG* mutants can form fruiting bodies. Methylation of FrzCD, like methylation of the MCPs, is responsive to various external stimuli. The time course of methylation or demethylation of FrzCD following stimulation is considerably longer than that observed for the enteric bacteria. This is consistent with the slow movement of myxobacterial cells. The swimming enteric bacteria must respond and adapt very quickly to stimuli in order to migrate effectively in a gradient. These cells must make frequent alterations in their movements, because they are continually being reoriented by Brownian motion. The myxobacteria, moving slowly over solid surfaces, are not under such stringent time constraints.

Mutations in *frzB*, like those in *frzF*, also affect the abilities of cells to modify FrzCD by methylation. The *frzB* region of DNA contains an ORF that codes for a protein of about 13.5 kDa. This protein does not bear any striking similarity to other proteins in the data bases. We have obtained antibodies against a recombinant form of FrzB and have identified the FrzB protein in extracts of *M. xanthus*. FrzB is required for proper functioning of the *frz* system. Mutants that do not produce FrzB but make FrzCD, FrzA, FrzE, and FrzF have the *frz* phenotype. Analyses of the motility behavior of *frz* mutants provides some indication of the function of FrzB. Wild-type cells of *M. xanthus* respond to additions of isoamyl alcohol by reversing repeatedly (McBride et al., 1992). Presumably, isoamyl alcohol interacts with some component near the signal reception end of the *frz* signal transduction chain to result in this behavior. Cells carrying deletions of the *frz* region or cells carrying mutations in *frzA*, *frzCD*, or *frzE* do not alter their motility following addition of isoamyl alcohol, indicating that these mutations affect components that are crucial for signal transmission. They probably constitute the central components of the signal transduction chain. In their absence, no signal is sent to the motility apparatus, and cells do not reverse their directions of movement. In contrast, cells with mutations in *frzB* or *frzF* still respond to additions of isoamyl alcohol, indicating that FrzB and FrzF may be peripherally associated with the signal transduction chain. This is consistent with the proposed role of these proteins in adaptation rather than signal transmission.

The results outlined above indicate many similarities between FrzCD and the enteric MCPs. FrzCD, however, bears no similarity to the N-terminal (transmembrane and periplasmic) portions of the MCPs. Analysis of the predicted sequence of FrzCD suggests that the protein is probably not an integral membrane protein. Consistent with this analysis, McBride et al. (1992) found that FrzCD is present primarily in the soluble fractions of cell extracts. FrzCD presumably functions near the signal reception end of the *frz* signal transduction chain, but we do not know how it perceives stimuli. It may detect signals within the cytoplasm, or it may interact with membrane-spanning proteins that function as receptors for extracellular stimuli.

In *E. coli*, the MCPs probably interact with CheW (Liu and Parkinson, 1991). The exact function of CheW is uncertain, but the protein is thought to mediate signal transmission between the MCPs and CheA. *M. xanthus* FrzA exhibits 28% identity to enteric CheW at the amino acid level. However, the predicted differences between these two proteins are quite dramatic. The predicted pIs of CheW and FrzA are 4.5 and 11.1, respectively, suggesting that these two proteins have different charges within their respective cells. This could dramatically affect how these proteins interact with other components of their signal transduction systems.

In *E. coli,* CheW interacts with CheA (Matsumura et al., 1990). CheA is a kinase that phosphorylates itself on a histidine residue and then transfers the phosphate to an aspartate residue of CheY. These two proteins are members of a conserved class of bacterial two-component regulatory proteins, most of which are involved in regulating gene expression. In the absence of CheY-phosphate, the flagella spin counterclockwise and propel the cell in a swimming run. Phospho-CheY is thought to interact with the flagellar switch, causing the flagella to reverse their rotation to clockwise, resulting in a cell tumble (Parkinson, 1988; Sockett et al., 1992). The central region of *M. xanthus* FrzE is similar to the central and C-terminal regions of CheA and includes a site resembling the histidine phosphorylation site of CheA (McCleary and Zusman, 1990a). The carboxy terminus of FrzE is similar to CheY and includes a site resembling the aspartate phosphorylation site of CheY. FrzE thus resembles VirA of *Bordetella pertussis,* which also incorporates both members of a two-component regulatory system in a single polypeptide (Stibitz et al., 1989). FrzE contains a stretch of 70 amino acids consisting of 70% alanine and proline that lies between the two putative phosphorylation sites. This sequence may form a flexible hinge, allowing the different domains of the protein to interact. Recombinant FrzE functions as an autokinase in vitro (McCleary and Zusman, 1990b). The phosphate is removed by incubation with acid or base, suggesting that it is an acyl phosphate, which is consistent with phosphorylation on the conserved aspartate residue. The rate of phosphorylation of FrzE is independent of the concentration of FrzE in solution. This result caused McCleary and Zusman (1990b) to speculate that FrzE first phosphorylates itself on the conserved histidine and then transfers the phosphate to the aspartate by an intramolecular reaction.

CheZ of *E. coli* enhances the dephosphorylation of CheY (Hess et al., 1987). We have not yet identified a homolog to CheZ in *M. xanthus.* FrzE-phosphate (like CheY-phosphate) is labile and loses its phosphate with time. Since the myxobacteria move so much more slowly than *E. coli,* it may not be necessary to enhance the rate of dephosphorylation of FrzE.

Model of the *frz* Signal Transduction Pathway

Figure 4 presents a model of the *frz* signal transduction system. In this model, a putative (unidentified) receptor protein(s) in the cell membrane recognizes external stimuli and conveys this information to FrzCD. The

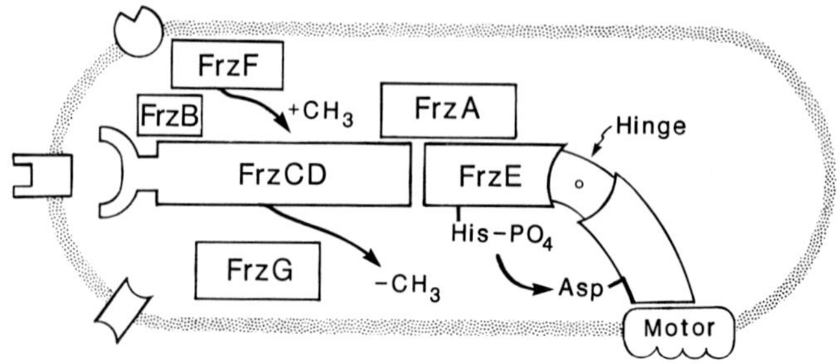

Figure 4. Model of the *frz* signal transduction system that incorporates all of the *frz* proteins as well as hypothetical receptor proteins and a hypothetical gliding motor to account for signal reception and transmission-controlling cell movements in *M. xanthus*. Reprinted from McBride et al. (1992).

transmembrane receptor molecules in this model are purely speculative, and equally plausible models lacking such receptors are possible. FrzCD interacts with FrzA and FrzE and presumably affects the phosphorylation state of FrzE. We hypothesize that it is FrzE (or phospho-FrzE) that interacts with some component of the gliding motility machinery and determines whether cells reverse their directions of movement. FrzF and FrzB are involved in methylation of FrzCD, and FrzG is involved in demethylation of FrzCD. These modifications may occur to mediate adaptation to changing levels of stimuli.

frz System of *M. xanthus* Controls Chemotaxis

The complex movements and behavior of myxobacterial cells suggested to many investigators that these cells were exhibiting chemotaxis. Attempts to demonstrate chemotaxis of myxobacteria and define the chemoeffectors have led to conflicting results (Ho and McCurdy, 1979; McVittie and Zahler, 1962; Shimkets et al., 1979; Dworkin and Eide, 1983). Dworkin and Eide (1983) postulated that chemotaxis based on a temporal sensing system such as that found in the enteric bacteria is unlikely to be present in the myxobacteria. This conclusion was based on their inability to demonstrate chemotaxis toward a wide variety of potential chemoeffectors and on the argument that a slow-moving bacterium would not be able to migrate effectively in a gradient by making temporal comparisons of the amount of soluble chemicals surrounding the cell.

McBride et al. (1992) employed an indirect approach to identifying potential chemoeffectors. Since the *frz* system controls myxobacterial directed movement, they searched for chemicals that interact with the *frz* system as potential chemoattractants or chemorepellents. Enteric MCPs become heavily methylated following binding of attractants and rapidly demethylated following binding of repellents (Macnab, 1987). Methylation of MCPs results in faster migration during sodium dodecyl sulfate-polyacrylamide gel electrophoresis (Nowlin et al., 1988). FrzCD is also reversibly methylated, and methylation alters its mobility in the same way (McCleary et al., 1990). Electrophoretic analysis of FrzCD extracted from cells exposed to different chemicals was used as an assay for potential chemoeffectors. Yeast extract, Casitone, and several defined chemicals, including lauric acid and lauryl alcohol, resulted in increased methylation of FrzCD (McBride et al., 1992). In contrast, several short-chain alcohols and dimethyl sulfoxide caused demethylation of FrzCD. These also resulted in a dramatic increase in the frequency with which cells reversed their directions of movement. The lipids and solvents mentioned above might interact specifically with components of the *frz* system, or they may act less specifically by partitioning into the cell membrane and interacting with many membrane proteins.

Shi et al. (in press) demonstrated that *M. xanthus* exhibits chemotaxis toward or away from some of the chemicals mentioned above. They employed two-chambered petri dishes to establish sharp chemical gradients and demonstrated movement of *M. xanthus* toward yeast extract or Casitone and away from dimethyl sulfoxide or short-chain alcohols. Positive and negative chemotactic movements were additive, and the movements toward attractants or away from repellents were correlated with methylation or demethylation of FrzCD. *frz* mutants were defective in chemotactic behavior. These results establish that *M. xanthus* does exhibit chemotaxis and that this chemotaxis is controlled by the *frz* system.

Other Forms of Myxobacterial Taxis

Besides chemotaxis, myxobacteria are capable of several other forms of directed motility. Some myxobacteria exhibit light-oriented movement (Aschner and Chorin-Kirsch, 1970). A more common form of directed movement among the myxobacteria is elasticotaxis, which is the directed movement of myxobacterial cells based on the structural features of the substratum (Stanier, 1942; Dworkin, 1983). Wild-type myxobacterial cells tend to follow stress lines on an agar surface and also to enter and preferentially glide on the slime trails deposited by members of other related or unrelated gliding bacteria (Burchard, 1982). Presumably, this is a passive form of taxis in which individual cells follow the path of least resistance. *frz* mutants also exhibit elasticotaxis. They follow stress lines on an agar surface, migrate preferentially in preexisting slime trails, and accumulate around glass beads placed on an agar surface (McBride and Zusman, unpublished data).

frz Genes Control Developmental and Vegetative Social Behaviors

The *frz* gene products are produced by both vegetative and developmental cells. Frz mutants sporulate normally, but they do not form fruiting bodies (Zusman, 1982). The *frz* genes are developmentally regulated (Weinberg and Zusman, 1989). The level of gene expression for some of the *frz* genes increases during the stage of development corresponding to the time of mound formation. That the *frz* system plays a role in development may also be deduced from the observation of cell movements during development. Several groups have reported the formation of swirls or rings of cells during myxobacterial fruiting body formation (Reichenbach, 1966; Vasquez et al., 1985; O'Connor and Zusman, 1989). The phenotypes of *frz* mutants (which form predominately swirls) may be an exaggeration of this behavior. Cells may respond to signal molecules produced by their neighbors to result in swirl formation. FrzCD becomes heavily methylated during aggregation of cells into mounds (an early stage of fruiting body development) (McBride and Zusman, submitted). It is likely that this is the result of increased levels of some aggregation signal that is produced by the cells and is detected by the *frz* system, causing the cells to aggregate. The nature of this signal(s) remains to be elucidated. Shi et al. (in press) extended the results described above and found that aggregation of cells to form fruiting bodies corresponded to increased methylation of FrzCD, whereas swarming of cells away from the colony center corresponded to decreased methylation of FrzCD. They also found that the *frz* genes were required for efficient swarming on CYE growth medium containing 0.3% agar. This demonstrates that the *frz* chemotaxis system plays an important role in both developmental and vegetative social behaviors of *M. xanthus*.

MECHANISM OF GLIDING MOTILITY

Analyses of motility mutants and *frz* mutants have not yet elucidated the mechanism of myxobacterial movement. It is tempting to speculate, however, that myxobacterial gliding motility and the gliding motilities of other bacteria are based on the same principles as flagellar motility. *che* proteins control the migration of swimming cells by interacting with flagellar motors. *frz* proteins control the directional motility of *M. xanthus* by an unknown mechanism. It is likely that one or more of the *frz* proteins interact with the gliding motor, and the most likely candidate for this role is FrzE. The regions of CheY that are believed to interact with the switch proteins have been mapped by genetic analyses (Matsumura et al., 1990; Sockett et al., 1992). Several of the critical amino acid residues are conserved between CheY and FrzE. This suggests that FrzE interacts with some protein that is similar to the switch proteins of enteric flagellar basal bodies. We do not know whether the gliding motor is similar to a flagellar basal body, but it seems likely that the switch on the gliding motor bears some similarity to the switch on the flagellar motor. Studies performed with other gliding bacteria suggest that rotary motors may be involved in some forms of gliding motility. Cells of cytophagas have been observed to attach one pole to glass slides and rotate, much as flagellated bacteria do when tethered by their flagella (Pate and Chang, 1979; Lapidus and Berg, 1982). These cells also propel latex beads along their surfaces or make the beads rotate in place. These rotary movements suggest the functioning of rotary motors similar to those that power the flagellar rotation of swimming bacteria (Pate, 1988). Ultrastructural examination of *Cytophaga johnsonae* and *F. columnaris* cells revealed structures in the cell membrane that appear to be similar to flagellar basal bodies (Pate and Chang, 1979). If some forms of gliding motility are related to swimming motility, it might be possible to obtain mutants of flagellated bacteria that can no longer swim but can move over surfaces. This may have been recently observed. Mutants of *Serratia marcescens* that flack flagellar filaments are able to spread over moist agar surfaces (O'Rear et al., 1992). They may employ some of the components of the flagellar motility machinery to spread. This movement may or may not be related to bacterial gliding motility. However, it is possible that mutations similar to those observed in the laboratory with *S. marcescens* occurred in many different lines of the eubacteria to result in the organisms that we recognize today as gliding bacteria.

The mechanism of gliding motility of the myxobacteria, or of any bacteria, remains unknown. It is likely that experiments conducted in the next few years will provide us with a much clearer picture of how myxobacteria move and how they control their movements in response to environmental stimuli.

REFERENCES

Aschner, M., and I. Chorin-Kirsh. 1970. Light oriented locomotion in certain Myxobacter species. *Arch. Mikrobiol.* **74:**308–314.

Blackhart, B. D., and D. R. Zusman. 1985a. Cloning and complementation analysis of the "frizzy genes" of *Myxococcus xanthus*. *Mol. Gen. Genet.* **198:**243–254.

Blackhart, B. D., and D. R. Zusman. 1985b. "Frizzy" genes of *Myxococcus xanthus* are involved in control of frequency of reversal of gliding motility. *Proc. Natl. Acad. Sci. USA* **82**:8767–8770.

Bourne, H. R., D. A. Sanders, and F. McCormick. 1991. The GTPase superfamily: conserved structure and molecular mechanism. *Nature* (London) **349**:117–127.

Burchard, A. C., R. P. Burchard, and J. A. Kloetzel. 1977. Intracellular, periodic structures in the gliding bacterium *Myxococcus xanthus*. *J. Bacteriol.* **132**:666–672.

Burchard, R. P. 1970. Gliding motility mutants of *Myxococcus xanthus*. *J. Bacteriol.* **104**: 940–947.

Burchard, R. P. 1981. Gliding motility of prokaryotes: ultrastructure, physiology, and genetics. *Annu. Rev. Microbiol.* **35**:497–529.

Burchard, R. P. 1982. Trail following by gliding bacteria. *J. Bacteriol.* **152**:495–501.

Burchard, R. P. 1984. Gliding motility and taxes, p. 139–161. *In* E. Rosenberg (ed.), *The Myxobacteria*. Springer-Verlag, New York.

Duxbury, T., B. A. Humphrey, and K. C. Marshall. 1980. Continuous observations of bacterial gliding motility in a dialysis microchamber: the effects of inhibitors. *Arch. Microbiol.* **124**:169–175.

Dworkin, M. Personal communication.

Dworkin, M. 1983. Tactic behavior of *Myxococcus xanthus*. *J. Bacteriol.* **154**:452–459.

Dworkin, M., and D. Eide. 1983. *Myxococcus xanthus* does not respond chemotactically to moderate concentration gradients. *J. Bacteriol.* **154**:437–442.

Dworkin, M., K. H. Keller, and D. Weisberg. 1983. Experimental observations consistent with a surface tension model of gliding motility of *Myxococcus xanthus*. *J. Bacteriol.* **155**: 1367–1371.

Fink, J. M., and J. F. Zissler. 1989. Defects in motility and development of *Myxococcus xanthus* lipopolysaccharide mutants. *J. Bacteriol.* **171**:2024–2028.

Glagoleva, T. N., A. N. Glagoleva, M. V. Gusev, and K. A. Nikitina. 1980. Proton motive force supports gliding in Cyanobacteria. *FEBS Lett.* **117**:49–53.

Grimm, K., and H. Kuhlwein. 1973. Untersuchungen an spontanen Mutanten von *Archangium violaceum* (Myxobacterales). I. Bewegliche und unbewegliche Zellen von *A. violaceum*. *Arch. Mikrobiol.* **89**:105–119.

Halfen, L. N., and R. W. Castenholz. 1971. Gliding motility in the blue-green alga *Oscillatoria princeps*. *J. Phycol.* **7**:133–145.

Hartree, D. Personal communication.

Hartzell, P. Unpublished data.

Hartzell, P., and D. Kaiser. 1991a. Function of MglA, a 22-kilodalton protein essential for gliding in *Myxococcus xanthus*. *J. Bacteriol.* **173**:7615–7624.

Hartzell, P., and D. Kaiser. 1991b. Upstream gene of the *mgl* operon controls the level of MglA protein in *Myxococcus xanthus*. *J. Bacteriol.* **173**:7625–7635.

Hess, J. F., K. Oosawa, P. Matsumura, and M. Simon. 1987. Protein phosphorylation is involved in bacterial chemotaxis. *Proc. Natl. Acad. Sci USA* **84**:7609–7613.

Ho, J., and H. D. McCurdy. 1979. Demonstration of positive chemotaxis to cyclic GMP and 5′ AMP in *Myxococcus xanthus* by means of a simple apparatus for generating practically stable concentration gradients. *Can. J. Microbiol.* **25**:1214–1218.

Hodgkin, J., and D. Kaiser. 1977. Cell-to-cell stimulation of movement in nonmotile mutants of *Myxococcus*. *Proc. Natl. Acad. Sci. USA* **74**:2938–2942.

Hodgkin, J., and D. Kaiser. 1979a. Genetics of gliding motility in *Myxococcus xanthus* (Myxobacteriales): genes controlling movement of single cells. *Mol. Gen. Genet.* **171**:167–176.

Hodgkin, J., and D. Kaiser. 1979b. Genetics of gliding motility in *Myxococcus xanthus:* two gene systems control movement. *Mol. Gen. Genet.* **171**:177–191.

Kaiser, D. 1979. Social gliding is correlated with the presence of pili in *Myxococcus xanthus*. *Proc. Natl. Acad. Sci. USA* **76**:5952–5956.

Kaiser, D., and C. Crosby. 1983. Cell movement and its coordination in swarms of *Myxococcus xanthus*. *Cell Motil.* **3:**227–245.

Kalos, M., and J. F. Zissler. 1990. Transposon tagging of genes for cell-cell interactions in *Myxococcus xanthus*. *Proc. Natl. Acad. Sci. USA* **87:**8316–8320.

Keller, K. H., M. Grady, and M. Dworkin. 1983. Surface tension gradients: feasible model for gliding motility of *Myxococcus xanthus*. *J. Bacteriol.* **155:**1358–1366.

Lapidus, I. R., and H. C. Berg. 1982. Gliding motility of *Cytophaga* sp. strain U67. *J. Bacteriol.* **151:**384–398.

Liu, J., and J. S. Parkinson. 1991. Genetic evidence for interaction between the CheW and Tsr proteins during chemoreceptor signalling by *Escherichia coli*. *J. Bacteriol.* **173:** 4941–4951.

Lunsdorf, H., and H. Reichenbach. 1989. Ultrastructural details of the apparatus of gliding motility of *Myxococcus fulvus*. *J. Gen. Microbiol.* **135:**1633–1641.

Macnab, R. M. 1987. Motility and chemotaxis, p. 732–759. *In* F. C. Neidhardt, J. L. Ingraham, K. B. Low, B. Magasanik, M. Schaechter, and H. E. Umbarger (ed.), *Escherichia coli and Salmonella typhimurium: Cellular and Molecular Biology,* vol. 1. American Society for Microbiology, Washington, D.C.

MacRae, T. H., and H. D. McCurdy. 1976. The isolation and characterization of gliding motility mutants in *Myxococcus xanthus*. *Can. J. Microbiol.* **22:**1282–1292.

Matsumura, P., S. Roman, K. Volz, and D. McNally. 1990. Signalling complexes in bacterial chemotaxis, p. 135–154. *In* J. P. Armitage and J. M. Lackie (ed.), *Biology of the Chemotactic Response*. Society of General Microbiology symposium 46. Cambridge University Press, Cambridge.

McBride, M. J., T. Kohler, and D. R. Zusman. 1992. Methylation of FrzCD, a methyl-accepting taxis protein of Myxococcus xanthus, is correlated with factors affecting cell behavior. *J. Bacteriol.* **174:**4246–4257.

McBride, M. J., R. A. Weinberg, and D. R. Zusman. 1989. "Frizzy" aggregation genes of the gliding bacterium *Myxococcus xanthus* show sequence similarities to the chemotaxis genes of enteric bacteria. *Proc. Natl. Acad. Sci. USA* **86:**424–428.

McBride, M. J., and D. R Zusman. Submitted for publication.

McBride, M. J., and D. R. Zusman. Unpublished data.

McCarter, L., and M. Silverman. 1990. Surface-induced swarmer cell differentiation of Vibrio parahaemolyticus. *Mol. Microbiol.* **4:**1057–1062.

McCleary, W. R., M. J. McBride, and D. R. Zusman. 1990. Developmental sensory transduction in *Myxococcus xanthus* involves methylation and demethylation of FrzCD. *J. Bacteriol.* **172:**4877–4887.

McCleary, W. R., and D. R. Zusman. 1990a. FrzE of *Myxococcus xanthus* is homologous to both CheA and CheY of *Salmonella typhimurium*. *Proc. Natl. Acad. Sci. USA* **87:** 5898–5902.

McCleary, W. R., and D. R. Zusman. 1990b. Purification and characterization of the *Myxococcus xanthus* FrzE protein shows that it has autophosphorylation activity. *J. Bacteriol.* **172:**6661–6668.

McVittie, A., and S. A. Zahler. 1962. Chemotaxis in *Myxococcus*. *Nature* (London) **194:** 1299–1300.

Morrison, C. E., and D. R. Zusman. 1979. *Myxococcus xanthus* mutants with temperature-sensitive stage-specific defects: evidence for independent pathways in development. *J. Bacteriol.* **140:**1036–1042.

Niklitschek, A. 1934. Das Problem der Oscillatorienbewegung. *Beih. Bot. Zentralbl. A* **52:** 205–254.

Nowlin, D. M., J. Bollinger, and G. L. Hazelbauer. 1988. Site-directed mutations altering methyl-accepting residues of a sensory transducer protein. *Proteins Struct. Funct. Genet.* **3**:102–112.

O'Connor, K. A., and D. R. Zusman. 1989. Patterns of cellular interactions during fruiting body formation in *Myxococcus xanthus. J. Bacteriol.* **171**:6013–6024.

O'Rear, J., L. Alberti, and R. M. Harshey. 1992. Mutations that impair swarming motility in *Serratia marcescens* 274 include but are not limited to those affecting chemotaxis or flagellar function. *J. Bacteriol.* **174**:6125–6137.

Parkinson, J. S. 1988. Protein phosphorylation in bacterial chemotaxis. *Cell* **53:1–2.**

Pate, J. L. 1988. Gliding motility in prokaryotic cells. *Can. J. Microbiol.* **34**:459–465.

Pate, J. L., and L.-Y. E. Chang. 1979. Evidence that gliding motility in prokaryotic cells is driven by rotary assemblies in the cell envelopes. *Curr. Microbiol.* **2**:59–64.

Reichenbach, H. 1966. *Myxococcus* spp. (Myxobacteriales). Schwarmentwicklung und Bildung von Protocysten. Publikationen zu wissenschaftlichen Filmen, vol. 1A, p. 557–578. Göttingen, Germany.

Reichenbach, H., W. Ludwig, and E. Stackebrandt. 1986. Lack of relationship between gliding cyanobacteria and filamentous gliding heterotrophic eubacteria: comparison of 16S rRNA catalogues of *Spirulina, Saprospira, Vitreoscilla, Leucothrix,* and *Herpetosiphon. Arch. Microbiol.* **145**:391–395.

Ridgway, H. F. 1977. Source of energy for gliding motility in *Flexibacter polymorphus:* effects of metabolic and respiratory inhibitors on gliding movement. *J. Bacteriol.* **131:** 544–556.

Rosenberg, E. (ed.). 1984. *Myxobacteria: Development and Cell Interactions.* Springer-Verlag, New York.

Shi, W., T. Köhler, and D. R. Zusman. Chemotaxis plays a role in the social behaviors of *Myxococcus xanthus. Mol. Microbiol.,* in press.

Shi, W., and D. R. Zusman. The two motility systems of *Myxococcus xanthus* show different selective advantages on various surfaces. *Proc. Natl. Acad. Sci. USA,* in press.

Shimkets, L. J. 1986a. Correlation of energy-dependent cell cohesion with social motility in *Myxococcus xanthus. J. Bacteriol.* **166**:837–841.

Shimkets, L. J. 1986b. Role of cell cohesion in *Myxococcus xanthus* fruiting body formation. *J. Bacteriol.* **166**:842–848.

Shimkets, L. J., M. Dworkin, and K. H. Keller. 1979. A method for establishing stable concentration gradients in agar suitable for studying chemotaxis on a solid surface. *Can. J. Microbiol.* **25**:1460–1467.

Shimkets, L. J., and D. Kaiser. 1982. Induction of coordinated cell movement in *Myxococcus xanthus. J. Bacteriol.* **152**:451–461.

Sockett, H., S. Yamaguchi, M. Kihara, V. M. Irikura, and R. M. Macnab. 1992. Molecular analysis of the flagellar switch protein FliM of *Salmonella typhimurium. J. Bacteriol.* **174:** 793–806.

Stanier, R. Y. 1942. A note on elastiocotaxis in myxobacteria. *J. Bacteriol.* **44**:405–412.

Stephens, K., P. Hartzell, and D. Kaiser. 1988. Gliding motility in *Myxococcus xanthus: mgl* locus, RNA, and predicted protein products. *J. Bacteriol.* **171**:819–830.

Stephens, K., and D. Kaiser. 1987. Genetics of gliding motility in *Myxococcus xanthus:* molecular cloning of the *mgl* locus. *Mol. Gen. Genet.* **207**:256–266.

Stibitz, S., W. Aaronson, D. Monack, and S. Falkow. 1989. Phase variation in *Bordetella pertussis* by frameshift mutation in a gene for a novel two-component system. *Nature* (London) **338**:266–269.

Sutherland, I. W., and S. Thomson. 1975. Comparison of polysaccharides produced by *Myxococcus* strains. *J. Gen. Microbiol.* **89**:124–132.

Thaxter, R. 1892. On the *Myxobacteriaceae,* a new order of Schizomycetes. *Bot. Gaz.* **17:** 389–406.

Vasquez, G. M., F. Qualls, and D. White. 1985. Morphogenesis of *Stigmatella aurantiaca* fruiting bodies. *J. Bacteriol.* **163:**515–521.

Weinberg, R. A., and D. Zusman. 1989. Evidence that the *Myxococcus xanthus frz* genes are developmentally regulated. *J. Bacteriol.* **171:**6174–6186.

Womack, B. J., D. F. Gilmore, and D. White. 1989. Calcium requirement for gliding motility in myxobacteria. *J. Bacteriol.* **171:**6093–6096.

Zusman, D. R. 1982. Frizzy mutants: a new class of aggregation-defective developmental mutants of *Myxococcus xanthus*. *J. Bacteriol.* **150:**1430–1437.

Zusman, D. R., and M. J. McBride. 1991. Sensory transduction in the gliding bacterium *Myxococcus xanthus*. *Mol. Microbiol.* **5:**2323–2329.

Myxobacteria II
Edited by Martin Dworkin and Dale Kaiser
© 1993 American Society for Microbiology, Washington, DC 20005

Chapter 14

Myxospore and Fruiting Body Morphogenesis

David White

Morphogenesis in the myxobacteria includes cellular morphogenesis (shape change during myxospore formation) and multicellular morphogenesis (formation of aggregates and fruiting bodies). Although morphogenesis has been described, mechanistic explanations are still very speculative. For example, in the area of myxospore formation, model construction would be helped by more information regarding the chemistry and metabolism of the cell well molecules. With regard to multicellular development, uncertainties about morphogenetic movements of cells during aggregate formation or fruiting body morphogenesis or about the cues that certainly guide the cells during these movements hamper our understanding of the process. In this respect, the chemistry and organization of the extracellular matrix molecules are important areas that remain to be explored. However, recently, increasing attention has been given to the morphogenesis of fruiting bodies, especially in the areas of intercellular signaling and morphogenetic movements (Vasquez et al., 1985; O'Connor and Zusman, 1989, Kim and Kaiser, 1990a, 1990b; Sager and Kaiser, in press, submitted). Also, computer models that analyze swarming behavior and aggregation in myxobacteria have been published (Pfistner, 1990; Stevens, 1990, 1991, in press).

MYXOSPORE MORPHOGENESIS

Reshaping a Cell

Cellular morphogenesis occurs when vegetative cells convert to myxospores. The shape changes accompanying myxospore formation can be

David White • Department of Biology, Indiana University, Bloomington, Indiana 47401.

relatively small, e.g., the shortening and thickening of rod-shaped cells that occur in *Stigmatella* spp., or more extensive, e.g., the conversion of rod-shaped cells to ovoid or nearly spherical cells that takes place in *Myxococcus* spp. Even minimal shape changes necessitate restructuring of the cell wall and thus present major survival problems, since the high turgor pressure makes it critical that the wall not develop weak spots during breakage and reformation of bonds. Similar considerations were previously pointed out by Koch in his "surface stress theory" concerning cell wall growth and morphogenesis (Koch, 1983, 1991). Some stages of cell wall morphogenesis in *Myxococcus* spp. occur at approximately the same time that the myxospore capsule (coat) is synthesized. As discussed below, this synchronicity may be a mechanism to provide osmotic stability during cell wall reshaping.

Glycerol Induction Is a Model System for Cellular Morphogenesis

Most studies of myxospore development have been of glycerol-induced *Myxococcus* myxospores because of the convenience of the system for experimental manipulation (Dworkin and Gibson, 1964). During glycerol-induced myxospore morphogenesis, several biochemical changes occur, including the cessation of net synthesis of DNA, RNA, protein, and phospholipid (Bacon and Rosenberg, 1967; Bacon et al., 1975; Rosenberg et al., 1967; Sadler and Dworkin, 1966). At the same time, a large increase in the activities of the glyoxylate cycle enzymes and a major shift in metabolism towards gluconeogenesis occur (Bacon et al., 1975; Orlowski et al., 1972). Over a 24-h period, the increase in polysaccharide is approximately 200%, and polysaccharide accounts for 20% of the dry weight of the cell. About 50% of the polysaccharide is in the soluble portion of the cell and probably represents reserve material. The shift towards gluconeogenesis is accompanied by a decrease in the respiratory rate (Bacon et al., 1975). Other physiological investigations done with glycerol-induced myxospores include a study of the in vivo inactivation of one of the glyoxylate cycle enzymes, isocitrate lyase, and an examination of the intracellular and extracellular proteases (Orlowski et al., 1972; Orlowski and White, 1974). Similar studies have not been done with myxospore formation in fruiting bodies. However, there have been other comparisons between glycerol-induced and fruiting body myxospores. For example, the respiratory rates of fruiting body myxospores are far lower than those of glycerol-induced myxospores (Dworkin, 1973; Dworkin and Niederpruem, 1964). The adenylate energy charge is surprisingly high in both glycerol-induced and fruiting body myxospores (Hanson and Dworkin, 1974; Smith and Dworkin, 1980). The germination of glycerol-induced myxospores is induced by P_i in buffer, but that of fruiting body myxospores is not (Ramsey and Dworkin, 1968; White, 1975). Some of the

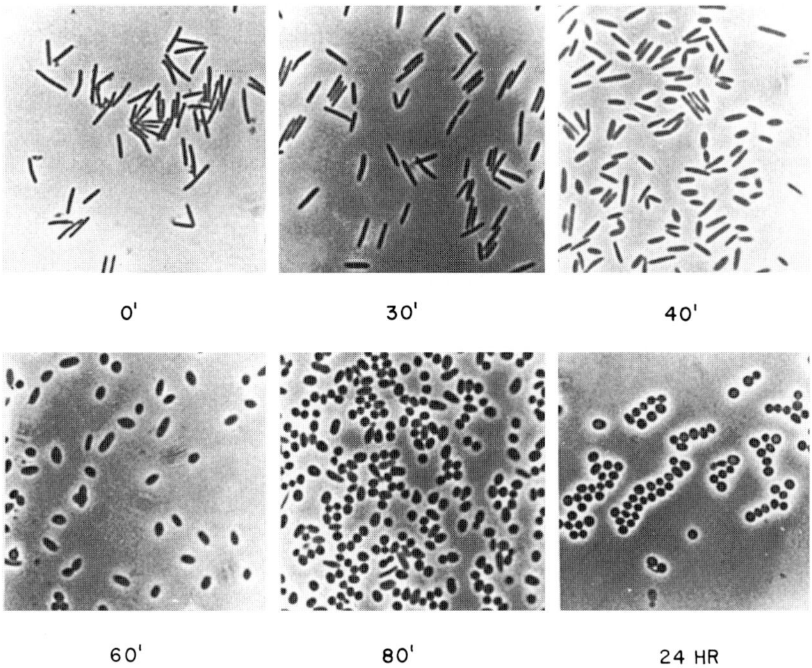

Figure 1. Morphological changes during glycerol-induced myxospore formation in *M. xanthus*. Glycerol at a final concentration of 0.5 M was added to a growing culture. At the times indicated, samples of cells were fixed in glutaraldehyde and photographed with a phase microscope. The average length of the vegetative cells at zero hour (0′) was 5 μm. Reprinted from White, 1975.

differences between fruiting body and glycerol-induced myxospores probably reflect incomplete maturation of glycerol-induced myxospores and the fact that they are formed in rich media, whereas fruiting body myxospores develop under starvation conditions. On the other hand, the shape changes appear identical, and glycerol-induced myxospore morphogenesis is a useful model for fruiting body myxospore morphogenesis, albeit not one that has received recent attention. During glycerol-induced myxospore formation, *Myxococcus xanthus* converts from rod-shaped cells approximately 5 by 0.75 μm to nearly spherical cells about 1.2 μm in diameter (Fig. 1). Additionally, a coat in the shape of the cell is synthesized external to the myxospore cell wall (Fig. 2 and 3). The coats of glycerol-induced myxospores are much thinner than those of fruiting body myxospores and lack protein S, which is present on the latter (Inouye et al., 1979).

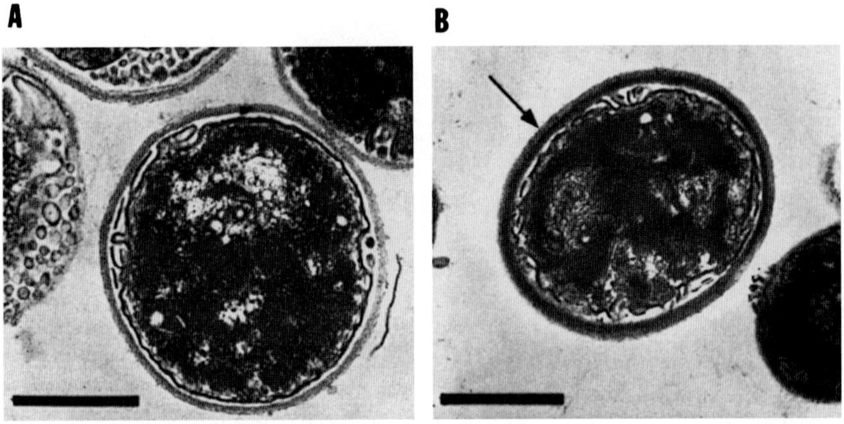

Figure 2. Thin sections of 8-h-old (A) and 24-h-old (B) glycerol-induced myxospores. Arrow indicates the extracellular coat. Marker is 0.5 μm. Reprinted from Kottel et al., 1975.

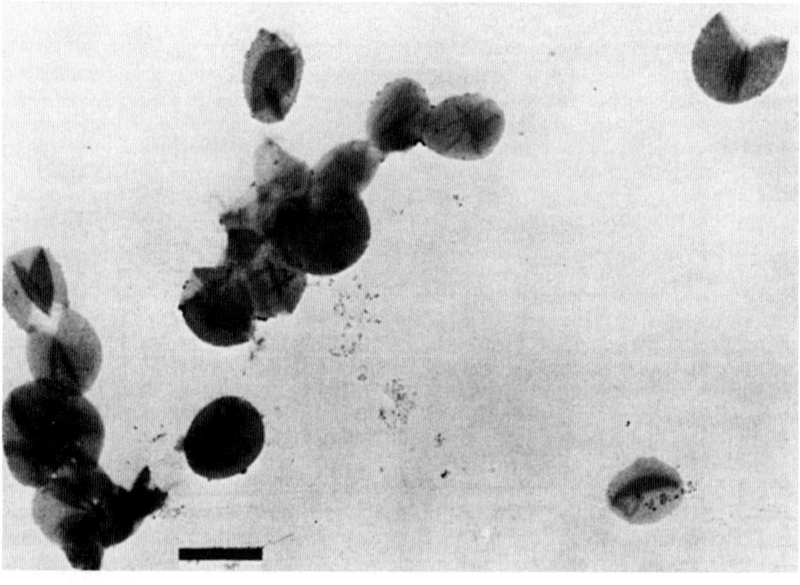

Figure 3. Negative stain of isolated, 24-h-old glycerol-induced myxospore coats. Marker is 2.0 μm. Reprinted from Kottel et al., 1975.

Myxospore Morphogenesis May Be Related to Patchlike Arrangement of Peptidoglycan Layer

Several years ago, it was reported that *M. xanthus* vegetative and myxospore cell walls contained peptidoglycan, a heteropolymer of amino sugars and amino acids that is responsible for maintaining the shapes of eubacteria with rigid cell walls (Mason and Powelson, 1958; White et al., 1968). In fact, in *Escherichia coli* and some other bacteria, the peptidoglycan has been isolated as a huge bag-shaped macromolecule approximately the size and shape of the cell (Weidel and Pelzer, 1965). It was reasonable to suggest that if the peptidoglycan in *M. xanthus* was also a bag-shaped molecule in the shape of the cell, then myxospore morphogenesis in *M. xanthus* should be accompanied by a breakage and reformation of covalent bonds within the peptidoglycan layer so that shape change could be accommodated. However, attempts to isolate the peptidoglycan from *M. xanthus* vegetative cells as a bag-shaped molecule of pure peptidoglycan were unsuccessful, raising doubt about the hypothesis that the peptidoglycan was solely responsible for maintaining the shape of the cell (White et al., 1968). Because the interpretation of these results is critical to an understanding of cellular morphogenesis, it is important to review how the experiments were performed. Cells were quickly pipetted from the growth medium into boiling water and heated for 15 min to inactivate wall lytic enzymes. When the boiled cells were treated with trypsin followed by sodium lauryl sulfate (or sodium lauryl sulfate followed by trypsin), they disintegrated, implying that the peptidoglycan existed in patches held together by material sensitive to trypsin and detergent. The patches of cell wall containing peptidoglycan could be sedimented by centrifugation at 78,410 $\times$ g for 60 min, which suggests that they were relatively large fragments. The chemical nature of the material that was suspected of holding the peptidoglycan patches together (the interpatch material) was not examined, but the material was thought to contain relatively large amounts of glycine, since this was solubilized by the trypsin. If indeed the peptidoglycan in *M. xanthus* vegetative cells exists as patches held together by some other material, perhaps proteinaceous, then the rigidity and shape of the cell wall are not due solely to peptidoglycan. (This does not mean that the rigid layer of the cell wall is not held together by covalent bonds throughout but rather that it may consist of material in addition to the peptidoglycan.) Therefore, the biochemistry of cell wall morphogenesis need not include simply alterations in the covalent structure of the peptidoglycan but may also involve changes in the putative interpatch material. It should be emphasized that the interpretation of a patchlike arrangement of the peptidoglycan rests on the assumption that autolysis did not occur prior to or during the pipetting of the cells into the boiling water and that boiling the cells irreversibly inactivated autolysins.

Changes in Cell Wall and External Layers During Myxospore Morphogenesis

When cells converted from ovoids to spheres during glycerol-induced myxospore formation, the combination of trypsin and detergent did not disintegrate the walls or fragment the peptidoglycan into small sedimentable pieces (White et al., 1968). It was concluded that (i) the peptidoglycan no longer existed as patches, or (ii) the interpatch material became resistant to trypsin and detergent, or (iii) additional layers external to the cell wall, e.g., a coat, were responsible for the resistance to trypsin and detergent. Indeed, it was later shown that coat components began to accumulate at approximately the time that the cells were converting from ovoids to spheres, i.e., when trypsin and detergent no longer disaggregated the cell walls (Filer et al., 1977b; Kottel et al., 1975). Johnson and White (1972) found that additional chemical changes occurred in the cell wall during morphogenesis. These changes included a significant increase in peptide cross-linking of the peptidoglycan chains and a large decrease in the amounts of a carbohydrate that appeared to be covalently bound to the peptidoglycan and contained only glucose. Those authors speculated that the glucose, along with the previously reported glycine, may have been part of the interpatch material postulated to hold the peptidoglycan patches together in vegetative cells.

Myxospore Coat

Kottel et al. (1975) isolated the myxospore coat synthesized by glycerol-induced myxospores as they converted to spheres (Fig. 2 and 3). In 8-h-old mature myxospores, approximately 75% of the dry weight of the coat was carbohydrate containing galactosamine and glucose, 14% was protein, 8% was glycine, and less than 1% was organic phosphorus. Filer et al. (1977b) reported that the galactosamine in the coat was N acetylated. It is likely that at least some of the glucose is in a polymer separate from the galactosamine, because glucose accumulation into coats occurred after galactosamine accumulation ceased (Kottel et al., 1975). Coats from fruiting body myxospores also contained carbohydrate consisting of only galactosamine, glucose, and glycine (they were not examined for protein or phosphorus), which suggests that the glycerol-induced coat represents one of the layers, presumably an inner layer, of the fruiting body myxospore coat (Kottel et al., 1975). Sutherland and Mackenzie (1977) extracted polysaccharide from whole glycerol-induced and fruiting body myxospores from *M. xanthus* and confirmed the presence of large amounts of galactosamine and glucose. The glucose was part of an α-1-3-glucan that could be separated from the galactosamine polysaccharide. The cellular locations of the glucan and the galactosamine-containing polysaccharide were not determined, but presumably, at least some of the polysaccharide was located in the myxospore coat.

A Model for Myxospore Morphogenesis

Any model that attempts to explain shape change on the basis of breakage and re-formation of covalent bonds in the cell wall must take into account the high intracellular turgor pressures known to exist in bacteria and must explain how the cell wall can be reshaped without producing fatal weaknesses as bonds are broken. One possible mechanism links cell wall morphogenesis to coat synthesis.

Since the coat contains large amounts of glucose and glycine, two molecules postulated to exist in the interpatch material in the vegetative cell wall, it is possible that the interpatch material evaginates during coat formation, bringing the peptidoglycan patches closer together and allowing them to be cross-linked by peptide bonds to each other, thus forming a layer of unbroken peptidoglycan (White, 1984). After cross-linking takes place, the bonds holding the evaginated interpatch material might be severed, releasing this material to be part of the developing coat. The cell would thus be protected, because according to the model, bonds connecting the interpatch material to the peptidoglycan would not be broken until after the cross-linking between the peptidoglycan patches had taken place. The model rationalizes the facts of peptidoglycan as patches in vegetative cell walls, increased cross-linking in the peptidoglycan from myxospores, the absence of evidence for a patchlike arrangement of the peptidoglycan in myxospores, and the presence of glucose and glycine in the myxospore coat.

Several predictions are made by the model. The chemical composition of the interpatch material, which is easily isolated after trypsin digestion of boiled and detergent-treated cell walls, should resemble that of the coat. Furthermore, radioactive tracer methods should indicate that the interpatch material serves as a metabolic precursor to the coat. One would also expect to find at the ovoid stage increased activities of enzymes that hydrolyze bonds holding the interpatch material to the peptidoglycan. Kottel and White (1974) did find in whole cells increased autolytic activity that began during the cell-shortening stage and peaked at the ovoid stage. The autolysins and their substrates were not studied. Dawson and Jones (1979) confirmed the results of Kottel and White and also reported a large increase in the metabolic turnover of a peptidoglycan component (diaminopimelic acid) during glycerol-induced myxospore formation, which suggests that bonds within the peptidoglycan were broken during morphogenesis. However, some data suggest that glycine in the coat is derived from a bactoprenol-linked intermediate rather than from an evagination of part of the cell envelope. Filer et al. (1977b) reported that glycine incorporation into the coat fraction was inhibited 75% by bacitracin, the antibiotic that interferes with the incorporation of bactoprenol-linked intermediates into cell walls. To add to the mystery, Sutherland and his coworkers made the observation that whereas phenol extraction of vegetative *Myxococcus* cells yielded lipopolysaccharide in the aqueous phase, extraction of fruiting body or glycerol-induced myxospores did not (Sutherland, 1976; Sutherland and Thomson, 1975), nor could lipopolysaccharides be found in the phenol phase. It was concluded that lipopolysaccharide was probably lost during myxospore formation (Sutherland and Mackenzie, 1977). This observation deserves further study because it suggests the possibilities that during myxospore morphogenesis, the lipopolysaccharide is degraded and some of its components, e.g., the oligosaccharides, are used for coat synthesis.

FRUITING BODY MORPHOGENESIS

An understanding of fruiting body morphogenesis requires a description of both morphogenetic movements and the extracellular polymers secreted by the cells during aggregate formation and fruiting body construction. The latter is probably important not only for the physical structure of fruiting bodies but also as tactic cues for morphogenetic cell movements. However, both areas of research pose technical difficulties, some actual and some potential. The observation of cellular movement is complicated by the high density of cells in aggregates and fruiting bodies that obscures individual cells. Analyses of extracellular polymers may be difficult to perform in some cases, because the extracellular matrix isolated from fruiting cultures can have a gellike consistency that complicates purification of components, and furthermore, cells and products of cell lysis are embedded in this material. However, protein-polysaccharide-lipid complexes have been isolated from the extracellular matrices of vegetative cultures grown in suspension, and exopolysaccharides have been isolated from vegetative cultures of several genera of myxobacteria (including *Myxococcus* and *Stigmatella*) grown on solid surfaces and from fruiting bodies (Gnosspelius, 1978; Sutherland, 1979; Sutherland and Thomson, 1975). Sutherand and Thomson (1975) recovered the mixture of cells and exopolysaccharides from vegetative cultures and fruiting bodies incubated on agar surfaces, separated the exopolysaccharides from the cells by vortex mixing followed by centrifugation, and precipitated the polysaccharides with acetone. The monosaccharides in the exopolysaccharides from vegetative cultures and from fruiting bodies were similar, prompting the conclusion that any differences between the polysaccharides from developmental and vegetative cultures may be in the fine structure rather than the individual components (also see chapter 3). It would be useful to know more about the heterogeneity, shapes, dimensions, viscosity, and gelling characteristics of the expolysaccharides (or proteins and proteoglycans) and the extent to which they are organized into supramolecular structures such as aggregates or fibrils.

A major impetus to studying fruiting body formation was the films published by Reichenbach et al., which showed the development of fruiting bodies (Reichenbach et al., 1965a, 1965b, 1975–1976). However, the movements of individual cells in the aggregates and fruiting bodies were not elucidated because of the failure or ordinary light microscopy to resolve single cells in a thick, densely packed population. At present, morphogenetic movements are inferred from scanning electron microscope photographs of *Chondromyces, Stigmatella,* and *Myxococcus* spp. (Grilione and Pangborn, 1975; Vasquez et al., 1985; O'Connor and Zusman, 1989) and from fluorescence confocal microscopy and time-lapse videography of *Myxococcus* spp. (Sager and Kaiser, in press, submitted).

Cell Patterns in *Stigmatella* Fruiting Populations

Figure 4 summarizes the stages of fruiting body formation in *Stigmatella* spp. on an agar surface. Aggregates form in areas of high cell density and grow in height. Eventually, stalks are formed and sporangioles differentiate. Stephens and White (1980b) reported that the stalks in 2- to 6-day-old mature fruiting bodies were cellular. Interestingly, stalk cells appeared larger than cells in the sporangioles, implying two different cell types. Grilione and Pangborn (1975) had earlier reported that the stalks of *Chondromyces* and *Stigmatella* spp. were mixtures of cells and slime in fruiting bodies that were 2 to 14 days old. Stalks in fruiting bodies of *Stigmatella* and *Chondromyces* spp. had previously been reported to be acellular, consisting of empty tubules that were continuous in *Chondromyces* spp. and discontinuous in *Stigmatella* spp., where they may have been the remnants of dead cells (McCurdy, 1969; Voelz and Reichenbach, 1969). It therefore appears that the stalks consist of cells in young fruiting bodies of *Chondromyces* and *Stigmatella* spp. but may become acellular because of cell lysis in older fruiting bodies.

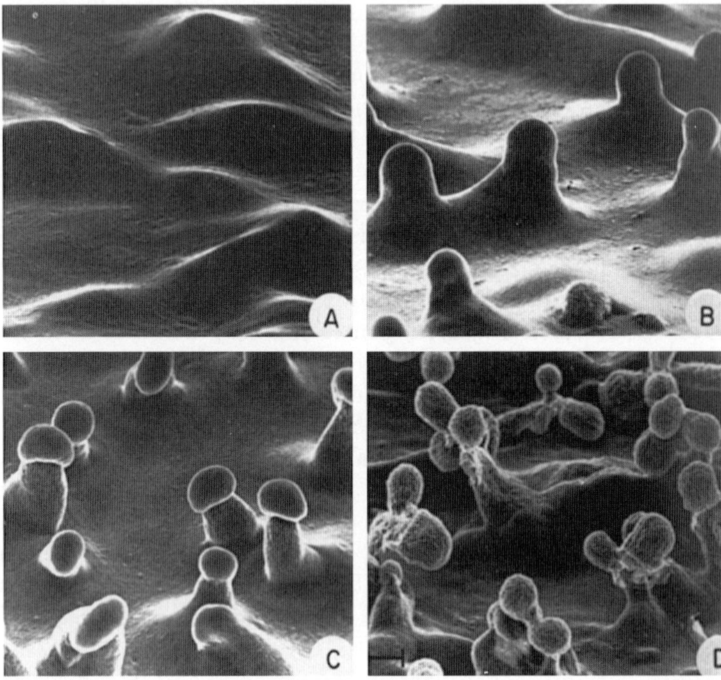

Figure 4. Stages of fruiting body formation in *S. aurantiaca*. (A) Early aggregates (9 h); (B) early stalks (1 h); (C) late stalks (15 h); (D) mature fruiting bodies (24 h). Bar, 20 μm. Reprinted from Qualls et al., 1978.

Vasquez et al. (1985) used scanning electron microscopy to study the surface patterns of cells. They reported that the cells in most incipient *Stigmatella* aggregates were organized in concentric circles or ellipses (Fig. 5A and B). At a later stage of development, a wide band of cells surrounded the developing aggregate and the mature fruiting body (Fig. 6). Similar bands of cells surrounding some *Stigmatella* fruiting bodies had previously been reported by Grilione and Pangborn (1975). The patterns of cells at the bases of the aggregates next to the circular bands of cells suggested that aggregates may grow taller as groups of cells leave the circular band and move in a spiral fashion up the aggregate (Fig. 7). Cell movements during stalk formation were different. During early stages of stalk formation, cells could be seen arranged perpendicularly at the base of the aggregate between the circular band of cells and the aggregate itself as the latter appeared to be lifted undisturbed off the surface (Fig. 8 and 9).

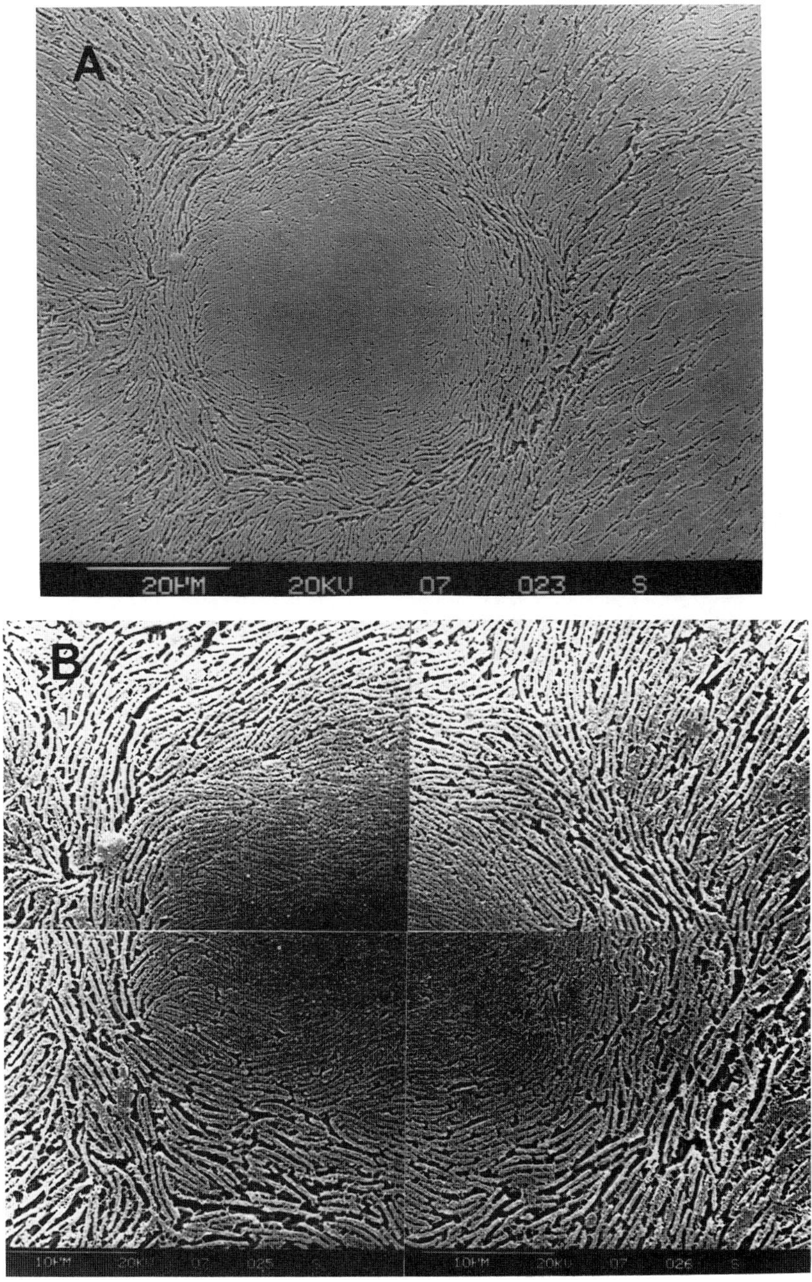

Figure 5. Early aggregate of *S. aurantiaca* showing circular or elliptical pattern of cells. (A) Low magnification; (B) high magnification, showing an elliptical pattern in the center (this photograph is a composite of four quadrants). Reprinted from Vasquez et al., 1985.

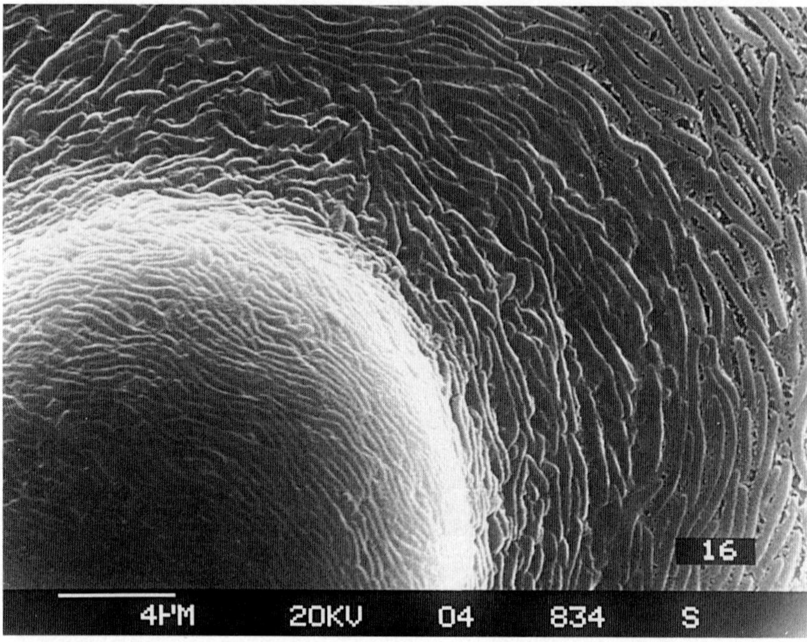

Figure 6. Top view of late aggregate of *S. aurantiaca* showing circular pattern of cells at base of aggregate and on aggregate. Reprinted from Vasquez et al., 1985.

Cell Patterns in *Myxococcus* Fruiting Populations

Figure 10 shows stages of fruit formation in *M. xanthus*. O'Connor and Zusman (1989) examined cell patterns during fruiting body formation in *M. xanthus* by using scanning electron microscopy. The formation of aggregation centers was correlated with spiral patterns of cells that persisted in the developing fruiting body (Fig. 11). Although spiral patterns were obvious at all stages of fruiting body development, other cell patterns were present, particularly when the interiors of sliced fruiting bodies were viewed. Sager and Kaiser (in press) viewed fruiting bodies on glass slides from underneath. They used fluorescence confocal and bright-field microscopy to probe the basal regions of *Myxococcus* fruiting bodies without destroying them. They discovered that the basal part of a nascent fruiting body consists of two regions, an outer hemispherical shell of densely packed cells arranged in an orderly pattern and an inner hemispherical region of less ordered cells that is approximately one-third as dense. Sager and Kaiser (submitted) proposed that myxospores begin their differentiation in the outer, denser region of a nascent fruiting body, because, accord-

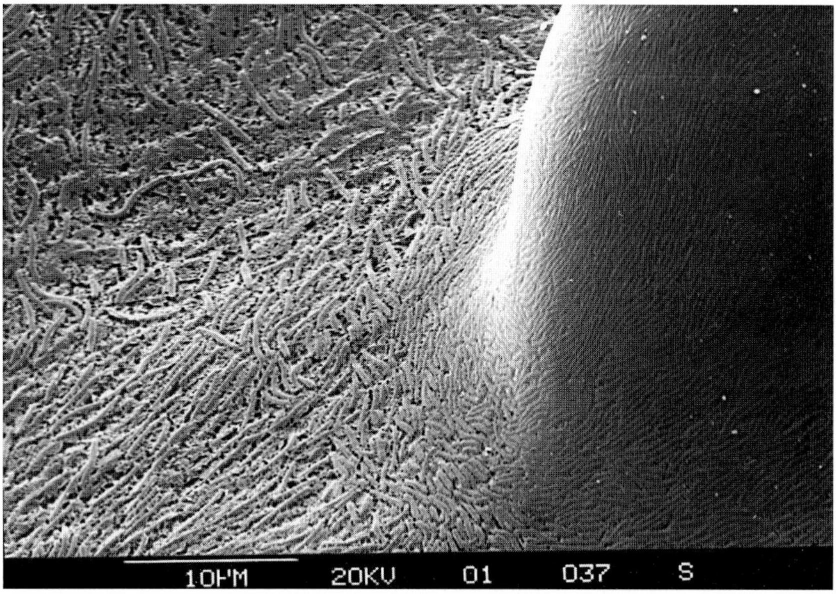

Figure 7. Aggregate of *S. aurantiaca* showing a narrow circular band of cells at base from which cells appear to have been moving into the aggregate as spiralling groups. Note the connections between cells at the periphery. Reprinted from Vasquez et al., 1985.

ing to Kim and Kaiser (1990a), high cell density favors C signaling. However, as these developing myxospores lost motility, they would be passively displaced toward the center of the fruiting body by the motion of cells that had not yet undergone morphogical change. Transport would explain a related observation that concerned the spatial expression of certain *lacZ* transcriptional fusions. Some transcriptional fusions reach their maximum β-galactosidase specific activity only in mature myxospores (Kroos et al., 1986). Sager and Kaiser (in press) observed that eight of these myxospore-specific fusions were more strongly expressed in the central, less dense area of the nascent fruiting body than in the outer region. Center-directed prespore transport would explain that otherwise unexpected observation.

Model for Fruiting Body Morphogenesis

A model to explain fruiting body formation in *Stigmatella* spp. on the basis of scanning electron microscopy has been proposed (Vasquez et al., 1985; White, 1992). The model is also based on two observations made with slide cultures. (i) Aggregates form in areas of high cell density;

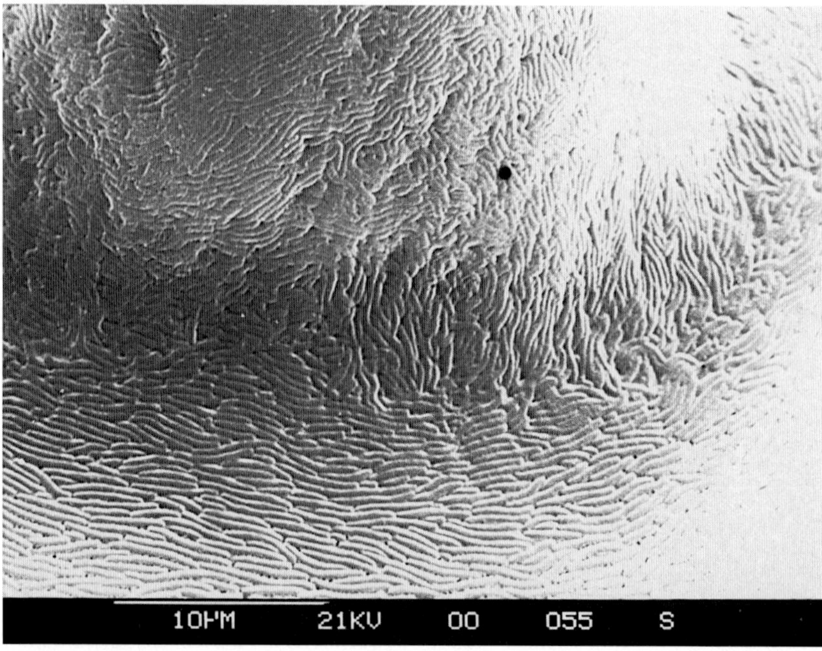

Figure 8. Beginning of stalk formation in *S. aurantiaca.* Cells between the aggregate and the substratum are arranged perpendicularly, as if they were entering underneath the aggregate. Note the spiral arrangement of cells in the aggregate. Reprinted from Vasquez et al., 1985.

i.e., there is no apparent need to attract cells in order to initiate centers of aggregation. (ii) Prior to the appearance of macroscopic aggregates on starvation media, microscopic transient aggregates form. In many of these, the cells move in circles or spirals but eventually leave the aggregate at a tangent by turning away from the circle. As a consequence, the microscopic aggregates recede. After several hours of nutrient depletion, an unknown event stabilizes the aggregates and they develop into fruiting bodies. One possible mechanism for stabilization might be the locking of the cells into a turning mode that traps them at the aggregation sites. This could be affected by a chemical, perhaps secreted locally at the site of the aggregate. Alternatively, the aggregate might be stabilized by a chemoattractant produced at the site of the aggregate. However, Dworkin and Eide (1983) were unable to demonstrate a chemotactic response in *M. xanthus* despite extensive attempts to do so. They suggested that *M. xanthus* may in fact not respond in any conventional way to chemical gradients. Therefore, I propose that enforced spiral or circular motility stabi-

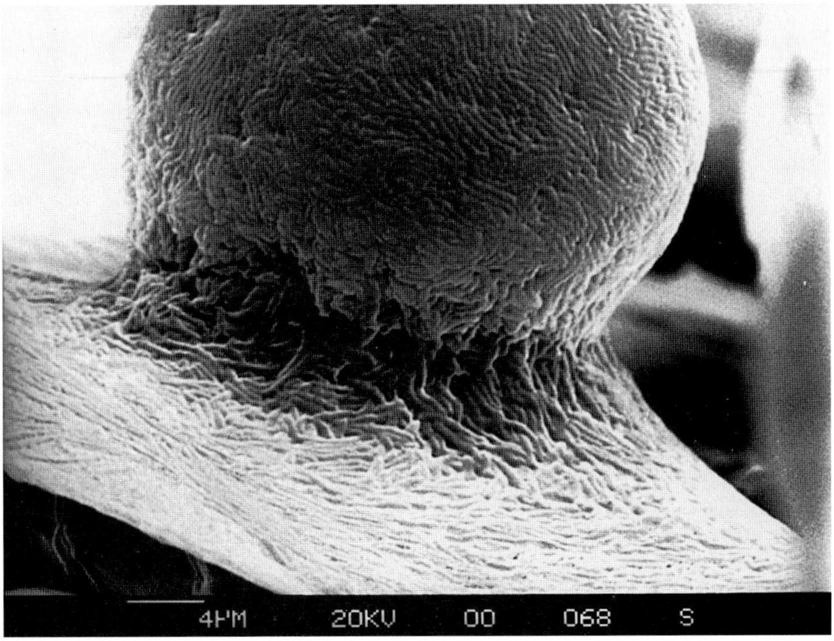

Figure 9. *S. aurantiaca* aggregate being lifted off the surface as the stalk forms. Note cells entering underneath aggregate. Reprinted from Vasquez et al., 1985.

lizes the aggregates and allows them to grow larger and that new cells are recruited to aggregates by existing slime trails leading to the aggregates. This would be expected, since as Reichenbach's films suggest, myxobacteria glide much faster when they can follow preexisting slime trails than when they must make their own slime trail. Eventually, this difference would result in major streams of cells entering the stabilized aggregates without the need for a chemoattractant. It is not known why myxobacteria move faster in preexisting slime trails. One can suppose that they are following tactic cues in the form of aligned polymers, perhaps polysaccharides, that may be dragged into alignment by the moving cells. *Stigmatella* aggregates appear to grow taller as cells at the base of aggregates move in an upward spiral into the aggregate (Fig. 7). The movement up the aggregate appears to be due to a change in pitch of circular movements at the periphery of the aggregate. The stalk appears to form as cells enter from beneath, while the aggregate is raised off the surface (Fig. 8 and 9). Stalk formation may be due to two factors: (i) the accumulation of matrix material (e.g., polysaccharide) at the base of the aggregate lifting the aggre-

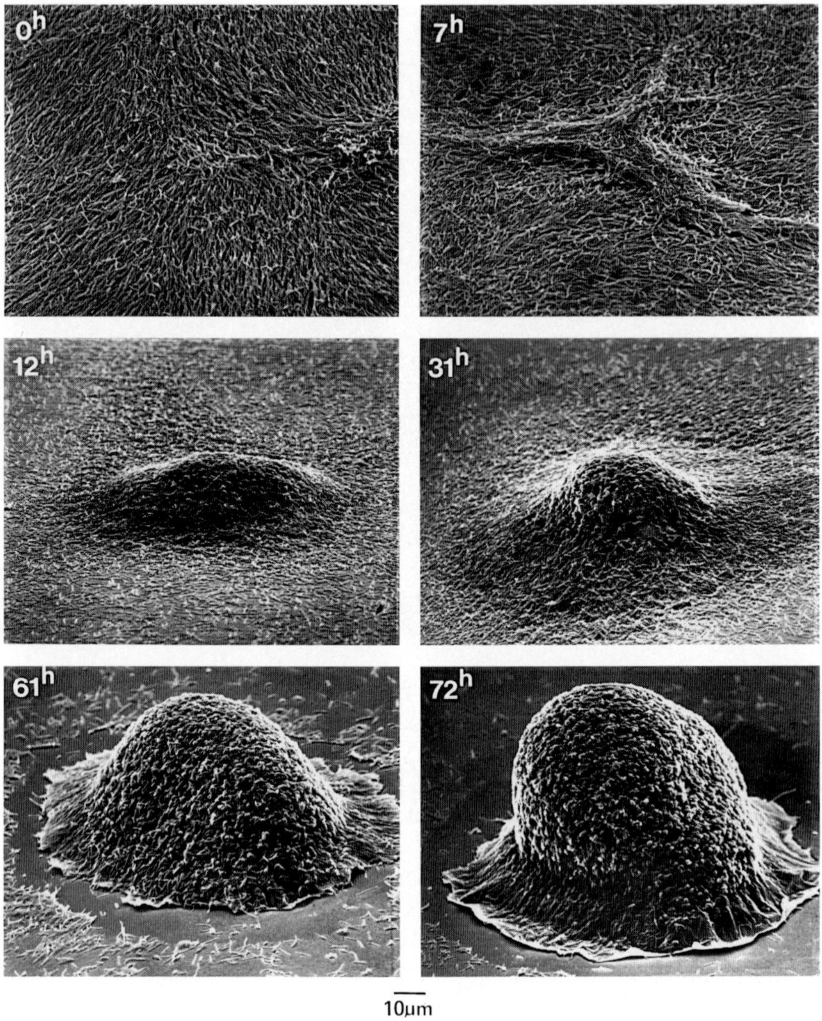

Figure 10. Scanning electron micrographs of *M. xanthus* fruiting body formation. Reprinted from Kuner and Kaiser, 1982.

gate off the surface, a suggestion made earlier by Dworkin (1972); and (ii) the accompanying movement of cells into the matrix that creates trails that other cells follow. Thus, I propose that circular motility at aggregation sites, trail following, and local accumulation of slime account for most phases of fruiting body morphogenesis in *Stigmatella* sp. Stevens (1990, 1991, in press) has written a computer program that simulates gliding and

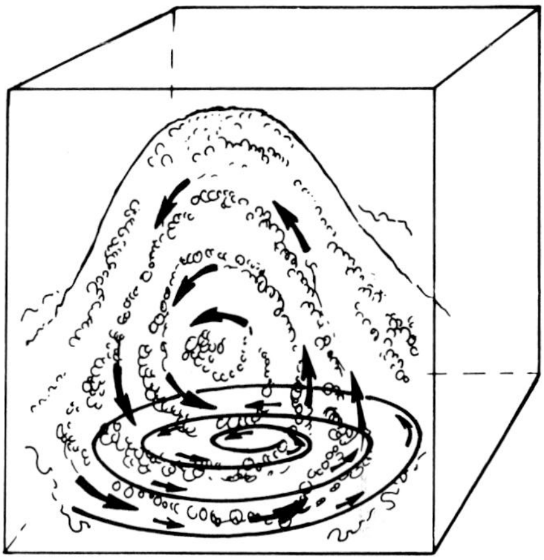

Figure 11. Interpretation of movement of *M. xanthus* during fruit morphogenesis based on scanning electron micrographs. Cells are interpreted as moving in spirals parallel to the substratum at the base of the young fruiting body and perpendicular to the substratum in the upper areas of the fruiting body. Reprinted from O'Connor and Zusman, 1989.

aggregation by myxobacteria. The program shows that trail following is important for pattern formation but cannot by itself account for aggregate formation. She found it necessary to postulate a density-dependent diffusing chemoattractant to stabilize the aggregation centers. Perhaps a density-dependent secreted chemical that locks the cells into a turning mode at the sites of aggregation, as proposed here, can serve the same purpose as a chemoattractant. The spiral cell patterns observed by O'Connor and Zusman (1989) with *Myxococcus xanthus* imply that the initial stages of fruiting body formation in *Myxococcus* spp. may also begin in areas of circular motility. However, it must be emphasized that we lack sufficient data to allow us to be comfortable about generalizations drawn about motility patterns during any stage of aggregate formation in *Myxococcus* or *Stigmatella* spp. or during later stages of morphogenesis of complex fruiting bodies such as those made by *Stigmatella* sp. Also, it is clear that myxobacteria can form aggregations of cells that need not involve circular or spiral motility, albeit the aggregates are not necessarily spherical. For example, in the dark, *Stigmatella* sp. forms elongated aggregates or ridges in which a circular pattern of cells is not evident (Stephens et al., 1982; Stephens and White, 1980a). These generally fail to develop into fruiting bodies, although myxospore formation takes place (Inouye et al., 1980). Also, under certain nutrient conditions, myxobacterial swarms move in waves of equidistant ridges called ripples, which for *Myxococcus* spp. consist of accumulations of moving cells aligned at approximately a 40° angle to the direction of wave movement (Shimkets and Kaiser, 1982). Finally, when cell patterns are viewed with the scanning electron microscope, one is never certain whether the patterns have been disturbed during the preparation of the sample or whether some underlying pattern is obscured by the dense population. Thus, the interpretation of scanning electron micrographs has its limitations, and we need to develop methods of following individual cell movements in live samples at high cell densities. The use of fluorescent probes during fruiting appears promising for monitoring the movements of such cells (Sager and Kaiser, in press).

Cellular Interactions and Morphogenesis

The results of many studies suggest that cellular interactions play an important role in morphogenesis in both *Myxococcus* and *Stigmatella* spp. In general, one suspects such interactions because (i) fruiting body formation requires a threshold concentration of cells (Wireman and Dworkin, 1975; Qualls et al., 1978), (ii) a fruiting population on agar induces the formation of fruiting bodies in a nearby population of cells of the same species (Lev, 1954), and (iii) scanning electron micrographs of fruiting populations reveal large numbers of threadlike connections between the cells (Fig. 7). The connections may be a mixture of true fibrils and amorphous matrix material that assumes a threadlike appearance upon dehydration. Specific examples of cell-cell interactions and their possible roles in morphogenesis are described below. Intercellular signaling is discussed more fully in chapter 12.

Morphogenetic Intercellular Signaling in *Stigmatella* sp.

Stigmatella aurantiaca secretes a lipoidal pheromone at the time stable aggregates appear (Stephens et al., 1982). The pheromone is assayed on low-cell-density preparations in the dark and has the following activities: (i) shortening of the lag time before the appearance of aggregates; (ii) increase in the numbers of aggregates that form with a given number of cells; and (iii) fruiting body formation in the absence of light, which is a requirement at low cell densities (White et al., 1980). Since the pheromone is lipoidal, it would not be expected to diffuse very far from the producing cells. It may be that cells receive the pheromone either when in actual contact with each other or by absorbing it from the site of deposition in slime trails. Because the pheromone is not detectable prior to the appearance of stable aggregates, its primary function may be to stabilize aggregates that form spontaneously because of slime trail following rather than to cause the formation of aggregates. The pheromone was not investigated with respect to a possible chemotactic role, its effect on the motility rates, or its influence on the turning patterns of individual cells. During a screen to determine whether adenine or guanine compounds also stimulated fruiting body formation, it was discovered that the addition of guanosine and guanosine nucleotides mimicked the effect of exogenously added pheromone (Stephens et al., 1982; Stephens and White, 1980a). This suggests that the pheromone signaling pathway may involve guanine nucleosides or nucleotides.

Development Cell Cohesion and Role of Calcium Ion in *Stigmatella* Morphogenesis

During aggregation, *Stigmatella* sp. develops an energy-dependent cohesion system (system B) that is induced specifically by calcium ion (Qualls and White, 1982; Gilmore and White, 1985). Induction requires RNA and protein synthesis, which may be necessary for cell envelope and extracellular matrix (glycocalyx) modifications that take place in calcium-induced cells (Chang and White, 1992). The modifications are (i) an increase in the production of extracellular fibrils, (ii) the synthesis of a 30-kDa protein located in the membrane fraction, (iii) translocation from the membranes to the glycocalyx of a 55-kDa protein that immunologically cross-reacts with a 65-kDa protein located on *M. xanthus* fibrils (Behmlander and Dworkin, 1991), and (iv) synthesis of nine polypeptides located in a cell extract fraction thought to be enriched for glycocalyx material. The nine polypeptides bind affinity-purified antibody directed against the 55-kDa protein. Calcium also induces the ability of *Stigmatella* sp. to glide and is required for gliding motility in cells already induced (Womack et al., 1989). These calcium-induced modifications of the cell envelope and glycocalyx may play a role in calcium-induced cohesion and motility. The requirement of metabolic energy for system B cohesion suggests that cohesion is not due simply to passive noncovalent bonding between ligands present on the surface of induced cells. Rather, the energy requirement suggests that (i) the ligands must be continuously synthesized and/or exported to the cell surface, (ii) an energy-requiring event that exposes the ligands at the cell surface in the proper orientation must occur, (iii) the bonding process itself requires energy, or (iv) a combination of any of the above must occur. Calcium requirements for the induction of gliding motility as well as cohesion and the mutual requirements for metabolic energy suggest that cohesion and motility might be functionally related. As discussed below, there is good evidence that extracellular fibrils and energy-dependent cohesion are required for social motility in *Myxococcus* spp. (Arnold and Shimkets, 1988a, 1988b; Behmlander and Dworkin, 1991).

Morphogenetic Intercellular Signaling in *Myxococcus* spp.

Aggregation in *M. xanthus* under nonpermissive conditions (low cell density or high nutrient levels) is rescued by adenine derivatives (Shimkets and Dworkin, 1981). The inhibition of purine biosynthesis by hadacidin also inhibits fruiting body formation, and the inhibition is reversed by exogenous adenosine. Since high concentrations of adenosine were produced during fruiting body formation, it was suggested that excreted adenosine serves as a cell density signal for fruiting body formation.

At least four other developmental signaling molecules secreted by *M. xanthus* are necessary for aggregation and sporulation. These are the A, B, C, and D signals (Hagen et al., 1978; Kroos et al., 1986; Kuspa et al., 1986; Kroos and Kaiser, 1987) (see chapter 4.) Two of the signals, A and C, have been isolated (Kim and Kaiser, 1990a, 1992; Kuspa et al., 1992; Plamann et al., 1992). The C signal, a 17-kDa protein, has three activities, namely, (i) stimulation of movement of C^- mutants from disordered cell patterns into nascent fruiting bodies, (ii) stimulation of myxospore formation in C^- mutants, and (iii) stimulation of *csg*-dependent gene expression in C^- mutants. On the basis of positive autoregulation of its own transcription and a relatively higher threshold concentration required for the stimulation of sporulation as opposed to aggregation, Kim and Kaiser (1992) suggested that the C signal was a developmental timer, triggering first aggregation and then sporulation. Plamann et al. (1992) and Kuspa et al. (1992) reported that A-signal activity is most probably due to proteolytic generation of amino acids and peptides that stimulate aggregation and myxospore formation in an unknown manner. Shimkets and Rafiee (1990) have shown that the C signal is part of the extracellular matrix of the cells.

Role of Social Motility and Cohesion in Fruiting Body Morphogenesis in *Myxococcus* spp.

M. xanthus has two multigene systems for motility (Hodgkin and Kaiser, 1979a, 1979b). One of these, the A system, is responsible for single-cell motility, and the other, the S system, is responsible for group (social) motility. Approximately 67% of A^+S^- and 10% of A^-S^+ strains are defective in fruiting body formation, which suggests the relative importance of the S system for fruiting body formation (Hodgkin and Kaiser, 1979b; Shimkets, 1990). *M. xanthus* also has an energy-dependent cohesion system that is important for functioning of the S motility system (Shimkets, 1986a, 1986b; Arnold and Shimkets, 1988a, 1988b; Behmlander and Dworkin, 1991). Cohesion is mediated by surface fibrils and requires Mg^{2+} and Ca^{2+}. Whether cohesion plays an additional role in fruiting body morphogenesis aside from its involvement in social motility is not known. Certainly, close apposition of cells, if not cohesion per se, is necessary for the transfer of some morphogenetic signals. This conclusion follows from experiments showing that developmental mutants could complement each other if they were mixed in pairs (McVittie et al., 1962) but not when they were separated by a semipermeable membrane that allowed the passage of diffusible molecules but not of cells (Hagen et al., 1978). Recently, Kim and Kaiser (1990b) demonstrated that a morphogenetic signal for fruiting body formation, the C signal, stimulated C-signal-dependent gene expression in recipient cells only if donor and recipient cells were closely aligned along their lengths. Thus, motility is important not only for the morphogenetic movements that help shape the fruiting bodies but also for positioning cells for the transmission of morphogenetic signals. As discussed below, motility mutants are frequently developmentally defective.

Myxococcus Mutants Defective in Morphogenesis

Many *Myxococcus* mutants that are defective in myxospore morphogenesis, fruiting body morphogenesis, or both have been isolated. Some of these are motility mutants. As stated above, approximately 10% of the strains defective in single-cell motility (A^-) and 67% of the strains defective in social motility (S^-) fail to form fruiting bodies (Hodgkin and Kaiser, 1979b). Other motility mutants defective in fruiting body formation include the *frz* mutants (Blackhart and Zusman, 1985a, 1985b, 1986). The *frz* mutants are motile, but they reverse their directions of motility more frequently than does the wild type. They fail to form normal aggregates but do sporulate. If one assumes that prolonged forward movement is necessary to form circular and spiral patterns and that these patterns are required for normal fruiting body morphogenesis, then a mutation that results in frequent reversals of movement would be expected to interfere with morphogenesis. Fink and Zissler (1989) isolated Mx1 phage-resistant mutants lacking an epitope present in the lipopolysaccharide core from wild-type cells that were developmentally defective. Two mutants were blocked in both aggregation and myxospore formation, whereas one aggregated but did not form spore-filled fruiting bodies (Fink and Zissler, 1989). The mutants were motile (A^+S^+) but were characterized by small colonies with decreased swarming. In contrast, mutants defective in the synthesis of the O antigen, which is generated by Tn*5* transposon mutagenesis or resistance to Mx8 phage, aggregated and formed normal-looking spore-filled fruiting bodies, albeit after a significant delay for many strains. Interestingly, most of the mutants defective in O-antigen synthesis were A^-S^+; i.e., they lacked single-cell motility. The phenotypes of the lipopolysaccharide mutants imply a role for lipopolysaccharide in motility and fruiting body formation. This is not surprising when one considers that the lipopolysaccharide is a major component of the cell outer envelope and therefore might be involved in interactions with extracellular matrices and other cells. Other developmental mutants have apparently normal motility. These include the *tag* mutants, which are defective in aggregation but form myxospores (Torti and Zusman, 1981; O'Connor and Zusman, 1990). The *tag* mutants develop normally at 28°C but do not aggregate properly at 34°C. This is true even for mutants generated by Tn*5* insertions into the *tag* genes, which suggests that *M. xanthus* has two genetically separable aggregation systems, one that operates at 28°C and does not require the *tag* genes and one that operates at 34°C and is dependent on the *tag* genes.

PERSPECTIVES

Myxospore Formation

It is unfortunate that glycerol-induced morphogenesis is no longer receiving much attention as a subject of research. It is a tractable system for studying cell morphogenesis and affords an opportunity for exploring the interface between biochemistry and morphology. An understanding of the events that take place during shape change would require a full exploration of the autolysins that are active during myxospore morphogenesis as well as the biochemical relationship between cell wall morphogenesis and coat synthesis. At the same time, the research could provide additional information on cell wall and cell surface molecules that would be helpful in understanding other areas of *Myxococcus* development, such as molecular interactions of the outer envelope with the secreted matrix. Glycerol-induced myxospore formation also provides an opportunity for studying interesting metabolic shifts and the regulation of enzyme activities during cell differentiation.

Fruiting Body Morphogenesis

It appears that multicellular morphogenesis is due primarily to morphogenetic movements of individual cells or groups of cells, notwithstanding possible important structural roles for secreted polymers, e.g., in stalk formation. From this perspective, it is important to learn more about cell motility and what controls its directionality (see chapter 11). In addition, much work needs to be done in identifying the molecules in the extracellular matrix, particularly in the slime trails, and in determining whether they are organized structurally or exist as a disordered array. It may be that matrix polymers bind to cell surface molecules, the consequences of such binding being a change in the rate and/or directionality of gliding motility. Under these circumstances, one might expect that ordered arrays of ligand sites in matrix molecules would be important for productive binding to the cell surface.

REFERENCES

Arnold, J. W., and L. J. Shimkets. 1988a. Cell surface properties correlated with cohesion in *Myxococcus xanthus*. *J. Bacteriol.* **170:**5771–5777.

Arnold, J. W., and L. J. Shimkets. 1988b. Inhibition of cell-cell interactions in *Myxococcus xanthus* by Congo red. *J. Bacteriol.* **170:**5765–5770.

Bacon, K., D. Clutter, R. H. Kottel, M. Orlowski, and D. White. 1975. Carbohydrate accumulation during myxospore formation in *Myxococcus xanthus*. *J. Bacteriol.* **124:**1635–1636.

Bacon, K., and E. Rosenberg. 1967. Ribonucleic acid synthesis during morphogenesis in *Myxococcus xanthus*. *J. Bacteriol.* **94:**1883–1889.

Behmlander, R. M., and M. Dworkin. 1991. Extracellular fibrils and contact-mediated cell interactions in *Myxococcus xanthus*. *J. Bacteriol.* **173:**7810–7821.

Blackhart, B. D., and D. R. Zusman. 1985a. "Frizzy" genes of *Myxococcus xanthus* are involved in control of the frequency of reversal of gliding motility. *Proc. Natl. Acad. Sci. USA* **82:**8767–8771.

Blackhart, B. D., and D. R. Zusman. 1985b. Cloning and complementation analysis of the "frizzy" genes of *Myxococcus xanthus*. *Mol. Gen. Genet.* **198:**243–254.

Blackhart, B. D., and D. R. Zusman. 1986. Analysis of the products of the *Myxococcus xanthus frz* genes. *J. Bacteriol.* **166:**673–678.

Chang, B. Y., and D. White. 1992. Cell-surface modifications induced by calcium ion in the myxobacterium *Stigmatella aurantiaca*. *J. Bacteriol.* **174:**5780–5787.

Dawson, H., and M. V. Jones. 1979. Cell wall turnover during myxospore formation in *Myxococcus xanthus*. *J. Gen. Microbiol.* **112:**143–148.

Dworkin, M. 1972. The myxobacteria: new directions in studies of procaryotic development. *Crit. Rev. Microbiol.* **1:**435–452.

Dworkin, M. 1973. Cell-cell interactions in the myxobacteria, p. 125–142. *In* J. M. Ashworth and J. E. Smith (ed.), *Microbial Differentiation*. Society for General Microbiology symposium 23. Cambridge University Press, Cambridge.

Dworkin, M., and D. Eide. 1983. *Myxococcus xanthus* does not respond chemotactically to moderate concentration gradients. *J. Bacteriol.* **154:**437–442.

Dworkin, M., and S. M. Gibson. 1964. A system for studying microbial morphogenesis: rapid formation of microcysts in *Myxococcus xanthus*. *Science* **246:**243–244.

Dworkin, M., and D. J. Niederpruem. 1964. Electron transport system in vegetative cells and microcysts of *Myxococcus xanthus*. *J. Bacteriol.* **87:**316–322.

Filer, D., S. H. Kindler, and E. Rosenberg. 1977a. Myxospore coat synthesis in *Myxococcus xanthus:* enzymes associated with uridine 5′diphosphate-*N*-acetylgalactosamine formation during myxospore development. *J. Bacteriol.* **131:**745–750.

Filer, D., D. White, S. H. Kindler, and E. Rosenberg. 1977b. Myxospore coat synthesis in *Myxococcus xanthus:* in vivo incorporation of acetate and glycine. *J. Bacteriol.* **131:** 751–758.

Fink, J. M., and J. F. Zissler. 1989. Defects in motility and development of *Myxococcus xanthus* lipopolysaccharide mutants. *J. Bacteriol.* **171:**2042–2048.

Gilmore, D. F., and D. White. 1985. Energy-dependent cell cohesion in myxobacteria. *J. Bacteriol.* **161:**113–117.

Gnosspelius, G. 1978. Purification and properties of an extracellular protease from *Myxococcus virescens*. *J. Bacteriol.* **133:**17–25.

Grilione, P. L., and J. Pangborn. 1975. Scanning electron microscopy of fruiting body formation by myxobacteria. *J. Bacteriol.* **124:**1558–1565.

Hagen, D. C., A. P. Bretscher, and D. Kaiser. 1978. Synergism between morphogenetic mutants of *Myxococcus xanthus*. *Dev. Biol.* **64:**284–296.

Hanson, C. W., and M. Dworkin. 1974. Intracellular and extracellular nucleotides and related compounds during the development of *Myxococcus xanthus*. *J. Bacteriol.* **118:**486–496.

Hodgkin, J., and D. Kaiser. 1979a. Genetics of gliding motility in *Myxococcus xanthus* (Myxobacterales): genes controlling movements of single cells. *Mol. Gen. Genet.* **171:**167–176.

Hodgkin, J., and D. Kaiser. 1979b. Genetics of gliding motility in *Myxococcus xanthus* (Myxobacterales): two gene systems control movement. *Mol. Gen. Genet.* **172:**177–191.

Inouye, M., S. Inouye, and D. R. Zusman. 1979. Biosynthesis and self-assembly of protein S, a development specific protein of *Myxococcus xanthus*. *Proc. Natl. Acad. Sci. USA* **76:**209–213.

Inouye, S., D. White, and M. Inouye. 1980. Development of *Stigmatella aurantiaca:* effects of light and gene expression. *J. Bacteriol.* **141:**1360–1365.

Johnson, R. J., and D. White. 1972. Myxospore formation in *Myxococcus xanthus:* chemical changes in the cell wall during cellular morphogenesis. *J. Bacteriol.* **112:**849–855.

Kim, S. K., and D. Kaiser. 1990a. Purification and properties of C-factor, an intercellular signaling protein. *Proc. Natl. Acad. Sci. USA* **87:**3635–3639.

Kim, S. K., and D. Kaiser. 1990b. Cell alignment required in differentiation of *Myxococcus xanthus*. *Science* **249:**926–928.

Kim, S. K., and K. Kaiser. 1992. C-factor has distinct aggregation and sporulation thresholds during *Myxococcus* development. *J. Bacteriol.* **173:**1722–1728.

Koch, A. L. 1983. The surface stress theory of microbial morphogenesis. *Adv. Microb. Physiol.* **24:**301–366.

Koch, A. L. 1991. Effective growth by the simplest means: the bacterial way. *ASM News* **57:**633–637.

Kottel, R., and D. White. 1974. Autolytic activity associated with myxospore formation in *M. xanthus*. *Arch. Microbiol.* **95:**91–95.

Kottel, R. H., K. Bacon, D. Clutter, and D. White. 1975. Coats from *Myxococcus xanthus:* characterization and synthesis during myxospore formation. *J. Bacteriol.* **124:**550–557.

Kroos, L., and D. Kaiser. 1987. Expression of many developmentally regulated genes depends on a sequence of cell interactions. *Genes Dev.* **1:**840–854.

Kroos, L., A. Kuspa, and D. Kaiser. 1986. A global analysis of developmentally regulated genes in *Myxococcus xanthus*. *Dev. Biol.* **117:**252–266.

Kuner, J. M., and D. Kaiser. 1982. Fruiting body morphogenesis in submerged cultures of *Myxococcus xanthus*. *J. Bacteriol.* **151:**458–461.

Kuspa, A., L. Kroos, and D. Kaiser. 1986. Intercellular signalling is required for developmental gene expression in *Myxococcus xanthus*. *Dev. Biol.* **117:**267–276.

Kuspa, A., L. Plamannn, and D. Kaiser. 1992. Identification of heat-stable A-factor from *Myxococcus xanthus*. *J. Bacteriol.* **174:**3319–3326.

Lev, M. 1954. Demonstration of a diffusible fruiting factor in myxobacteria. *Nature* (London) **173:**501.

Mason, D. J., and D. Powelson. 1958. The cell wall of *Myxococcus xanthus*. *Biochim. Biophys. Acta* **29:**1–7.

McCurdy, H. D., Jr. 1969. Light and electron microscope studies on the fruiting bodies of *Chondromyces crocatus*. *Arch. Mikrobiol.* **65:**380–390.

McVittie, A., F. Messik, and S. A. Zahler. 1962. Developmental biology of *Myxococcus*. *J. Bacteriol.* **84:**546–551.

O'Connor, K. A., and D. R. Zusman. 1989. Patterns of cellular interactions during fruiting body formation in *Myxococcus xanthus*. *J. Bacteriol.* **171:**6013–6024.

O'Connor, K. A., and D. R. Zusman. 1990. Genetic analysis of the *tag* mutants of *Myxococcus xanthus* provides evidence for two developmental aggregation systems. *J. Bacteriol.* **172:**3868–3878.

Orlowski, M., P. Martin, D. White, and M. C. Wong. 1972. Changes in activity of glyoxylate cycle enzymes during myxospore development in *Myxococcus xanthus*. *J. Bacteriol.* **111:** 784–790.

Orlowski, M., and D. White. 1974. Intracellular proteolytic activity in developing myxospores of *Myxococcus xanthus*. *Arch. Microbiol.* **97:**347–357.

Pfistner, B. 1990. A one dimensional model for the swarming behavior of myxobacteria, p. 556–563. *In* W. Alt and G. Hoffmann (ed.), *Lecture Notes in Biomathematics,* vol. 89. Springer Verlag, Berlin.

Plamann, L., A. Kuspa, and D. Kaiser. 1992. Proteins that rescue A-signal-defective mutants of *Myxococcus xanthus*. *J. Bacteriol.* **174:**3311–3318.

Qualls, G. T., K. Stephens, and D. White. 1978. Morphogenetic movements and multicellular development in the fruiting myxobacterium, *Stigmatella aurantiaca*. *Dev. Biol.* **66:** 270–274.

Qualls, G. T., and D. White. 1982. Developmental cell cohesion in *Stigmatella aurantiaca*. *Arch. Microbiol.* **131:**334–337.

Ramsey, W. S., and M. Dworkin. 1968. Microcyst germination in *Myxococcus xanthus*. *J. Bacteriol.* **95:**2249–2257.

Reichenbach, H., H. K. Galle, and H. H. Heunert. 1975–1976. *Stigmatella aurantiaca* spp. (Myxobacteriales [sic]): Schwarmentwicklung und Fruchtkorperbildung. Encyclop. Cinematogr. film E2421. Institut für den Wissenschaftlichen Film, Gottingen, Germany.

Reichenbach, H., H. H. Heunert, and H. Kuczka. 1965a. *Chondromyces apiculatus* (Myxobacterales): Schwarmentwicklung und Morphogenese. Encyclop. Cinematogr. film E779. Institut für den Wissenschaftlichen Film, Gottingen, Germany.

Reichenbach, H., H. H. Heunert, and H. Kuczka. 1965b. Schwarmentwicklung and Morphogenese bei Myxobackterien: *Archangium, Myxococcus, Chondrococcus* und *Chondromyces*. Encyclop. Cinematogr. film C893. Institut für den Wissenschaftlichen Film, Gottingen, Germany.

Rosenberg, E., M. Katarski, and P. Gottlieb. 1967. Deoxyribonucleic acid synthesis during exponential growth and microcyst formation in *Myxococcus xanthus*. *J. Bacteriol.* **93:** 1402–1408.

Sadler, W., and M. Dworkin. 1966. Induction of cellular morphogenesis in *Myxococcus xanthus*. II. Macromolecular synthesis and mechanism of inducer action. *J. Bacteriol.* **91:**1520–1525.

Sager, B., and D. Kaiser. Two cell-density domains within the *Myxococcus xanthus* fruiting body. *Proc. Natl. Acad. Sci. USA,* in press.

Sager, B., and D. Kaiser. Spatial-restriction of cellular differentiation. Submitted for publication.

Shimkets, L. J. 1986a. Role of cell cohesion in *Myxococcus xanthus* fruiting body formation. *J. Bacteriol.* **166:**842–848.

Shimkets, L. J. 1986b. Correlation of energy-dependent cell cohesion with social motility in *Myxococcus xanthus*. *J. Bacteriol.* **166:**837–841.

Shimkets, L. J. 1990. Social and developmental biology of the myxobacteria. *Microbiol. Rev.* **54:**473–501.

Shimkets, L. J., and M. Dworkin. 1981. Excreted adenosine is a cell density signal for the initiation of fruiting body formation in *Myxococcus xanthus*. *Dev. Biol.* **84:**51–60.

Shimkets, L. J., and D. Kaiser. 1982. Induction of coordinated cell movement in *Myxococcus xanthus*. *J. Bacteriol.* **152:**451–461.

Shimkets, L. J., and H. Rafiee. 1990. CsgA, an extracellular protein essential for *Myxococcus xanthus* development. *J. Bacteriol.* **172:**5299–5306.

Smith, B. A., and M. Dworkin. 1980. Adenylate energy charge during fruiting body formation by *Myxococcus xanthus*. *J. Bacteriol.* **142:**1007–1009.

Stephens, K., G. D. Hegeman, and D. White. 1982. Pheromone produced by the myxobacterium *Stigmatella aurantiaca*. *J. Bacteriol.* **149:**739–747.

Stephens, K., and D. White. 1980a. Morphogenetic effects of light and guanine derivatives on the fruiting myxobacterium *Stigmatella aurantiaca*. *J. Bacteriol.* **144:**322–326.

Stephens, K., and D. White. 1980b. Scanning electron micrographs of fruiting bodies of the myxobacterium *Stigmatella aurantiaca* lacking a coat and revealing a cellular stalk. *FEMS Microbiol. Lett.* **9:**189–192.

Stevens, A. 1990. Simulations of the gliding behaviour and aggregation of myxobacteria. Biological motion, p. 548–555. *In* W. Alt and G. Hoffmann (ed.), *Lecture Notes in Biomathematics,* vol. 89. Springer Verlag, Berlin.

Stevens, A. 1991. A model for gliding and aggregation of myxobacteria, p. 269–276. *In* A. Holden, M. Markus, and H. G. Othmer (ed.), *Nonlinear Wave Processes in Excitable Media*. Plenum Press, New York.

Stevens, A. Aggregation of myxobacteria—a many particle system. *In First European Conference of Mathematics Applied to Biology and Medicine,* in press. Springer Verlag, Berlin.

Sutherland, I. W. 1976. Novel surface polymer changes in development of *Myxococcus* spp. *Nature* (London) **259:**46–47.

Sutherland, I. W. 1979. Polysaccharides produced by *Cystobacter, Archangium, Sorangium* and *Stigmatella* species. *J. Gen. Microbiol.* **111:**211–216.

Sutherland, I. W., and C. L. Mackenzie. 1977. Glucan common to the microcyst walls of cyst-forming bacteria. *J. Bacteriol.* **129:**599–605.

Sutherland, I. W., and S. Thomson. 1975. Comparison of polysaccharides produced by *Myxococcus* strains. *J. Gen. Microbiol.* **89:**124–132.

Torti, S., and D. Zusman. 1981. Genetic characterization of aggregation-defective developmental mutants of *Myxococcus xanthus*. *J. Bacteriol.* **147:**768–775.

Vasquez, G. M., F. Qualls, and D. White. 1985. Morphogenesis of *Stigmatella aurantiaca* fruiting bodies. *J. Bacteriol.* **163:**515–521.

Voelz, H., and H. Reichenbach. 1969. Fine structure of fruiting bodies of *Stigmatella aurantiaca* (Myxobacterales). *J. Bacteriol.* **99:**856–866.

Weidel, W., and H. Pelzer. 1965. Bagshaped macromolecules—a new outlook on bacterial cell walls, p. 193–232. *In* F. F. Nord (ed.), *Advances in Enzymology*. John Wiley, New York.

White, D. 1975. Myxospores of *Myxococcus xanthus,* p. 44–51. *In* P. Gerhardt, H. L. Sadoff, and R. N. Costilow (ed.), *Spores VI*. American Society for Microbiology, Washington, D.C.

White, D. 1984. Structure and function of myxobacteria cells and fruiting bodies, p. 51–67. *In* E. Rosenberg (ed.), *Myxobacteria: Development and Cell Interactions*. Springer-Verlag, New York.

White, D. 1992. Morphogenesis in myxobacteria, p. 7–28. *In* E. F. Rossomando and S. Alexander (ed.), *Morphogenesis: an Analysis of the Development of Biological Form*. Marcel Dekker, Inc., New York.

White, D., M. Dworkin, and D. J. Tipper. 1968. Peptidoglycan of *Myxococcus xanthus:* structure and relation to morphogenesis. *J. Bacteriol.* **95:**2186–2197.

White, D., W. Shropshire, Jr., and K. Stephens. 1980. Photocontrol of development in *Stigmatella aurantiaca*. *J. Bacteriol.* **142:**1023–1024.

Wireman, J. W., and M. Dworkin. 1975. Morphogenesis and developmental interactions in myxobacteria. *Science* **189:**516–523.

Womack, B. J., D. F. Gilmore, and D. White. 1989. Calcium requirement for gliding motility in myxobacteria. *J. Bacteriol.* **171:**6093–6096.

Myxobacteria II
Edited by Martin Dworkin and Dale Kaiser
© 1993 American Society for Microbiology, Washington, DC 20005

Chapter 15

Stigmatella aurantiaca, an Organism for Studying the Genetic Determination of Morphogenesis

Hans Ulrich Schairer

Differentiation, which occurs in most organisms, consists of alterations in cellular morphology and/or the expression of cellular functions. The process of differentiation can be governed by intracellular events subject to an internal timer that is independent of most environmental factors, by extracellular signals, or by a combination of both. An important focus for such studies is the temporal and spatial (in multicellular systems) regulation of gene expression.

In prokaryotes, differentiation can be relatively simple, as in the case of the formation of flagella and pili in *Caulobacter crescentus* or of photosynthetic membranes associated with the cell membranes of photosynthetic bacteria. On the other hand, more-complex developmental systems include the formation of endospores in bacilli, the development of heterocysts in cyanobacteria, the formation of aerial mycelia in streptomycetes, and the development of spores and fruiting bodies in myxobacteria. One of the main problems to be addressed in the study of any developmental process is how a linear sequence of nucleotides encoded in the developmental genes is translated into new morphological shapes, both cellular and multicellular. The myxobacteria are particularly suitable for these studies.

Whereas *Myxococcus xanthus,* for which methods of bacterial genetics are available, forms morphologically simple spore-filled mounds in the course of starvation on solid surfaces, *Stigmatella aurantiaca* forms

---

Hans Ulrich Schairer · Zentrum für Molekulare Biologie, Universität Heidelberg, Heidelberg, Germany.

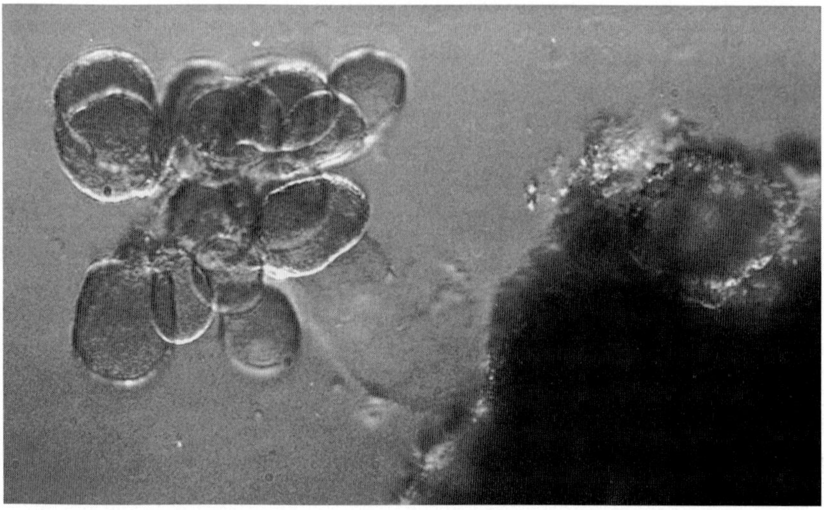

Figure 1. Fruiting body of *S. aurantiaca* (magnification, ×255.) Courtesy of Hans Reichenbach, GBF Braunschweig.

elaborate fruiting bodies with stalks and multiple sporangioles (Fig. 1). Thus, *S. aurantiaca* is better suited for studying the genetic determination of form than is *M. xanthus,* because *S. aurantiaca* offers the opportunity of asking questions about forces and factors that influence the formation of elaborate multicellular structures. In the past, a drawback to using *S. aurantiaca* for the study of morphogenesis of fruiting bodies has been that procedures for performing genetic experiments were not available. As discussed in this chapter, this situation is rapidly changing.

FRUITING BODY MORPHOGENESIS *S. AURANTIACA*

Conditions for Synchronous Fruiting

Fruiting body formation in *S. aurantiaca* has been studied in detail by D. White and his coworkers (Qualls et al., 1978a; Vasquez et al., 1985; White, 1981). They showed that *S. aurantiaca* grown dispersed in cell suspension will develop synchronously when deposited in drops on a solid agar surface devoid of added nutrients. Efficient fruiting body development requires incandescent light. If cells are incubated in the dark, the development of typical fruiting bodies is highly reduced. Instead, such cells tend to aggregate into clumps and ridges that contain myxospores (Inouye et al., 1980; Qualls et al., 1978b; White et al., 1980b). Different combinations of cations can have distinct effects on the morphologies of the fruiting bodies (White et al., 1980a). However, one must be careful about how the cells are cultured. Prolonged subculturing (i.e., 4 to 6 weeks) in cell suspension can lead to a diminished capacity to form fruiting bodies. In order to maintain a fruiting strain in liquid culture, it is necessary to periodically return to dried fruits for an inoculum. One should keep a parallel stock culture on solid medium that supports both growth and fruiting body development.

Stages of Fruiting Body Formation

White and coworkers described several stages in the morphogenesis of fruiting bodies (see chapter 14). After 8 to 10 h of starvation, aggregates form. At about 14 h after nutrient deprivation, mushroom-like structures with stalks supporting single globular immature sporangioles form (Qualls et al., 1978a; White, 1981). After 14 to 20 h, the immature sporangiole has been transformed into several smaller spore-filled sporangioles. Scanning electron microscopy by Vasquez et al. (1985) led to the suggestion that aggregate formation was initiated in areas of circular or spiral motility (see Fig. 5A and B, chapter 14). The circular pattern of cells remained as a band around the base of each aggregate (see Fig. 6, chapter 14). It was suggested that cells are recruited into the growing aggregate from the band of cells around the base (see Fig. 7 and 8, chapter 14). Stalk formation coincided with a change in the patterns of cells at the bases of the aggregates. The cell patterns were interpreted as representing vertical movement of cells underneath the aggregate as it lifted off the surface. The vertically moving cells and the slime produced may lift the aggregate, from which the future sporangioles arise (see Fig. 9, chapter 14) (Vasquez et al., 1985). Electron micrographs from negatively stained thin longitudinal sections of mature fruiting bodies suggest that the base of the stalk is surrounded by concentric circular streams of cells. In the stalk, cells are arranged parallel to the longitudinal axis. In the sporangioles, the spores seem to originate from different packs of cells arranged in parallel and piled up in the sporangiole that move into different directions during sporulation (Lünsdorf and Heidelbach, personal communication; Heidelbach, 1992).

Pheromone Is Produced during Fruiting Body Formation

Karlson and Lüscher (1959) defined pheromones as a class of biologically active substances that are secreted by an organism into the environment and that act to elicit a specific behavioral or developmental response from other organisms of the same species. This type of substance is secreted by *S. aurantiaca* during fruiting. Addition of preparations containing this compound reduces the time needed for aggregate formation at reduced cell densities and increases the number of aggregates formed (Stephens et al., 1982). Continuous removal of the pheromone from starving cells by dialysis prevents formation of aggregates (Plaga, personal communication). For the production of pheromone, 2×10^{12} to 4×10^{12} cells are spotted onto filter paper in 5-μl portions containing 4×10^8 cells each and incubated in the presence of light. Since the cells adhere tightly to the filter paper, the pheromone produced can be extracted continuously by descending chromatography. The pheromone is extracted from the eluate by steam distillation. The total biological activity of the extract is found in the distillate but none stays in the waste of the distillation. In subsequent steps, the pheromone is purified further by reversed-phase chromatography (Plaga, personal communication). Dialysis has shown that the molecular weight of the compound is below 500. It is soluble in apolar solvents such as cyclopentane. Because of the minute amounts of the pheromone available, no reliable structural data have been obtained.

It was mentioned above that light is a prerequisite for reliably efficient fruiting body formation (Qualls et al., 1978b; White et al., 1980b), but the pheromone or guanosine nucleotides partially substitute for light (Stephens et al., 1982; Stephens and White, 1980). A combination of both substances synergistically stimulates fruiting in the dark (Stephens et al., 1982).

Sporulation May Be Uncoupled from Fruiting Body Formation

Dworkin and Gibson (1964) discovered that *M. xanthus* cells in suspension can be induced to form myxospores by the addition of glycerol to a final concentration of 0.5 M. In contrast to that in *M. xanthus,* sporulation in *S. aurantiaca* can also be induced by certain monovalent cations (Reichenbach and Dworkin, 1970), certain antibiotics, various aliphatic or aromatic alcohols, and other organic compounds, of which indole and 3-methylindole seem to be the most potent (Gerth, 1975; Gerth and Reichenbach, 1978; Gerth et al., in press). In a detailed and thorough investigation, Gerth and Reichenbach (1978) determined the physiological parameters of *S. aurantiaca* spore formation. They showed that in the narrow range between 40 and 41°C, the vegetative cells of strain Sga1 converted into typical myxospores after the temperature shift from 30°C. Reduction of the temperature back to 30°C resulted in transition of the sporelike cells to vegetative rods. Heat induction and glycerol induction were shown to be complementary. If cells in growth medium were first incubated at 41°C for a defined time and subsequently cooled down to 30°C during simultaneous addition of glycerol, the change of cell shape always occurred ca. 60 min after the initial switch to 41°C (Gerth and Reichenbach, 1978). We believe that heat shock, starvation, and different types of stress modulate a regulatory circuit that is part of the developmental pathway. This belief is substantiated by the observation that a number of genes labeled by a promoter probe are induced by starvation, methylindole treatment, and heat shock. Interestingly, Gerth isolated mutants of strain Sga1 that were resistant to the induction of sporulation by certain organic substances. He was able to show that resistance to one substance may be independent of resistance to other compounds. Different mutants affect the induction of sporulation by different substance classes. He suggested that different "receptors" were affected in the different mutant classes (Gerth, 1978; Gerth et al., in press).

GENETICS OF *S. AURANTIACA*

Genetic Methods Available for *M. xanthus*

Elaborate bacterial-genetic methods have been established for *M. xanthus* in the course of the last 15 years. Of particular importance are Tn*5* insertion mutagenesis using abortive *Escherichia coli* phage P1 transduction, generalized transduction with the myxobacterial phages Mx4 and Mx8, conjugation systems using RP4 and some of its derivatives, transcomplementation systems using the Mx8 integration system, electroporation, and, recently, an Mx4 phage-derived plasmid. These methods contributed substantially to the elucidation of the involvement of certain genes in *Myxococcus* development and gliding.

Recent Developments in Establishing Genetic Methods for *S. aurantiaca*

For the induction of insertional mutants, RP4 derivatives harboring variants of Tn*5* were transferred from *E. coli* to *S. aurantiaca* in the course of mating experiments on membrane filters (Glomp et al., 1988). Tn*5*-containing recombinants were isolated by selection for kanamycin resistance. Kanamycin-resistant mutants were obtained with a frequency of 10^{-5} to 10^{-6} per recipient *S. aurantiaca* cell and were analyzed by Southern hybridization after restriction of the chromosomal DNA. Although no circular RP4 DNA could be demonstrated in the transconjugants, RP4-derived sequences of variable lengths were inserted in the chromosomes of 40% of the kanamycin-resistant *S. aurantiaca* clones (Glomp et al., 1988). Instead of RP4, small multicopy plasmids (Simon et al., 1983) harboring the *mob* region of RP4 and Tn*5* can be used. The *tra* genes necessary for the transfer were inserted in the chromosome of the corresponding *E. coli* strain. Kanamycin-resistant transconjugants were obtained with a frequency of about 10^{-4} to 10^{-5}, and 10 to 20% of them contained plasmid DNA inserted into the *S. aurantiaca* chromosome.

For monitoring promotor activity of genes expressed during development, the promoter probe Tn*5lac* (Kroos and Kaiser, 1984) was inserted into a conjugable plasmid. When the promoter probe was used, the yield of the kanamycin-resistant transconjugants was reduced to about 5% of the yield obtained with wild-type Tn*5*. Hoping that the transposition frequency of the promoter probe would increase if nonessential sequences were deleted from Tn*5lac,* they reduced the size of the transposon from 12 to 7.5 kbp. Though a 4.5-kbp segment was deleted, no significant increase in transposition was observed. This smaller transposon, Tn*5lacZ,* was used in further experiments, because larger DNA fragments (up to 15 kbp) with the transposon inserted can be cloned in λ replacement vectors (Pospiech et al., 1993).

DEVELOPMENT- AND MOTILITY-SPECIFIC POLYPEPTIDES AND GENES

Developmentally Regulated Putative Outer Membrane Receptor

By using a Tn*5* promoter probe, Tn*5lacZ* insertional mutants have been generated. More than 600 mutants have been screened for the induction of β-galactosidase activity by light, heat shock, indole, or starvation (Pospiech, 1992; Pospiech et al., 1993). Several mutant types have been found: mutants that express β-galactosidase constitutively; mutants in which β-galactosidase activity increases during starvation or after the addition of indole, a compound that induces sporulation in liquid culture (Gerth, 1975; Gerth et al., in press); and mutants in which β-galactosidase levels increase during heat shock. Interestingly, in all the mutants in which β-galactosidase is induced by indole, it is also induced by starvation. Furthermore, mutants defective in fruiting body formation were isolated, and most of them do not form β-galactosidase under the conditions tested. These mutants fall into three classes. Members of the first class form anomalous fruiting bodies, those of the second class aggregate into clumps, and those of the third class do not aggregate at all (Pospiech, 1992; Pospiech et al., 1993). It is possible to phenotypically complement a mutant from the second class, strain AP182, with mutants of the third class. A mushroom-like structure, which corresponds to an intermediate stage in fruiting body formation, is obtained by this complementation.

The transposon with adjacent *S. aurantiaca* DNA sequences was cloned from strain AP182, which aggregated but did not fruit, and introduced into the wild type by using a conjugable plasmid. The clones in which wild-type sequences were replaced by the mutant gene showed the mutant AP182 phenotype. About 700 bp downstream of the transposon, an open reading frame encoded a polypeptide with significant homology to an outer membrane polypeptide encoded by the *nodC* gene of *Rhizobium* spp. The similarity of the amino acid sequence deduced from the nucleotide sequence of the putative gene to that from the *Rhizobium* sp. is about 62%. The *Stigmatella* protein also resembled the NodC protein in having a similar molecular mass (45.5 kDa) and conserved hydrophobic domains. Preliminary Northern (RNA) blot data suggest that the gene is transcribed after about 15 h of development and is part of an operon (Pospiech, 1992).

A Spore-Specific Polypeptide

A spore-specific polypeptide, SP21 (molecular weight, 21,000), has been isolated from spores induced by methylindole (Heidelbach, 1992; Heidelbach et al., 1993). The polypeptide was sedimented with the spore membranes, extracted by a detergent, and purified further by ion-exchange chromatography and electrophoresis. The polypeptide is not detectable in vegetative cells by specific antisera and first appears about 2 h after induction of sporulation. By immunoelectron microscopy, it was shown that the polypeptide is associated with the membrane and the cytoplasm and that it forms aggregates. In fruiting bodies, most of the cross-reacting material was found in the spaces between the spores. It was very striking that the polypeptide was also induced by heat shock or oxygen deprivation. Ten minutes after the increase of temperature and about 2 h after deprivation of oxygen, enough SP21 had been accumulated to be detected by a Western (immunoblot) analysis. Thus, this spore protein seemed to be a stress protein induced by heat, chemical stress (indole), and starvation. This corresponds quite well to the results obtained by Gerth and Reichenbach (1978), who demonstrated that ca. 60 min were needed after induction of sporulation before the shape of the cells changed. Sixty minutes was the sum of various lengths of heat shock and glycerol treatments (see above).

By using antisera against SP21 and degenerate oligonucleotides deduced from a sequence of seven amino acids in the N-terminal domain, the gene *hspA* (heat shock polypeptide), which encodes SP21, was isolated from a lambda λgt11 expression gene library of *Hpa*II-restricted *S. aurantiaca* chromosomal DNA. Comparison of the protein sequence encoded by *hspA* revealed that SP21 belongs to a family of small stress proteins from arabidopsis and soybean. The sequences are ca. 60% similar. The C-terminal sequence is very characteristic for the members of the alpha-crystallin family. It is generally accepted that these proteins aggregate as a consequence of their C-terminal sequences (Lindquist and Craig, 1988).

This would explain the aggregation observed by electron microscopy and the sedimentation with membranes of the spores (Heidelbach, 1992; Heidelbach et al., submitted).

Fruiting Body Formation in *S. aurantiaca* Depends on *csgA* Function

In *M. xanthus,* the *csgA* polypeptide plays a central role in coordination of the developmental process (Kim and Kaiser, 1990a, 1990b; Li et al., 1992). By using the *M. xanthus csgA* gene as a probe, the *S. aurantiaca* gene was isolated. In the direct neighborhood, the *fprA* gene (Shimkets, 1990) was also located. The *csgA* gene was conserved about 70% compared with the *M. xanthus* gene. The amino acid sequence of the *csgA* gene-derived polypeptide has an identity of only 54% with that of *M. xanthus*. Inactivation of the gene by insertional mutagenesis affects fruiting body formation (Butterfaß, 1992).

mgl Gene of *S. aurantiaca*

A mutation in the *mglA* gene leads to a nonmotility phenotype in *M. xanthus* and secondarily to the inability to transmit and receive the *csgA* signal (Hartzell and Kaiser, 1991a, 1991b; Hodgkin and Kaiser, 1979; Kim and Kaiser, 1990c, 1990d). By using the *mgl* gene of *M. xanthus* as a probe, the *S. aurantiaca* gene was isolated from a lambda *dash* gene library. The *S. aurantiaca* gene was shown to restore motility in an *mgl* mutant of *M. xanthus*. The restored motility was slightly reduced compared with that of the *M. xanthus* wild-type strain. A similar nonmotile mutant was constructed in *S. aurantiaca* by insertion of transposon Tn*5* into the cloned gene followed by the introduction of the interrupted gene into a wild-type strain of *S. aurantiaca* by using a conjugable plasmid. Only those kanamycin-resistant transconjugants in which the wild-type gene was replaced (via a double crossover) by the mutant gene (ca. 3% of the kanamycin-resistant clones) were nonmotile. The nucleotide sequences of the *S. aurantiaca* genes *mglA* and *mglB* are at least 90% identical to the sequences of the corresponding *M. xanthus* genes. Interestingly, sequence homology is highly reduced outside of the two open reading frames, but evidence from Southern hybridization indicates that other genes within 5 kb of the *mgl* genes are conserved in the two species with regard to both sequence homology and location (Hartree, personal communication).

RNA POLYMERASE AND ITS SIGMA FACTORS

Potential Role for Sigma Factors in Myxobacterial Development

Gene expression in prokaryotes is mainly regulated by a tight control of transcription. The specificity of transcription is highly dependent on the interaction of the polymerase and the promoters of the genes. The sigma subunit determines the promoter specificity of the polymerase that does not recognize the promoter without that subunit. Thus, a gene family is selectively transcribed by the core polymerase, which is associated with the gene family-specific sigma factor. In *Bacillus subtilis,* the progress of differentiation is tightly correlated with the appearance of specific sigma factors involved in the transcription of developmentally regulated gene families (Losick and Stragier, 1992). Assuming that similar mechanisms are involved in the coordination of developmental genes in myxobacteria, S. Inouye's group (Apelian and Inouye, 1990; Inouye, 1990) and my group (Skladny et al., 1992, submitted) started to characterize sigma factors of *M. xanthus* and *S. aurantiaca,* respectively. A prerequisite for studying specific sigma factors in transcription assays is the availability of purified core polymerase.

Purification of RNA Polymerase from *S. aurantiaca*

The RNA polymerase was purified by chromatography on heparin Sepharose, gel filtration on S-200 Sepharose, and final ion-exchange chromatography on Q-Sepharose. From 50 g (wet weight) of cells, about 45 mg of purified polymerase was obtained. In sodium dodecyl sulfate-polyacrylamide gel electrophoresis, only three polypeptide bands, with molecular masses of about 150, 100, and 45 kDa, were detected. By using highly specific antisera against the β and β′ subunits of *E. coli* RNA polymerase, it was unequivocally shown that the 150-kDa band contained the β and β′ subunits. Antiserum against the *S. aurantiaca* polymerase labeled the β and β′ subunits of the *E. coli* RNA polymerase. The 100-kDa band was identified as a sigma factor by using antiserum against a fusion protein composed of the MS2 polymerase and a central part of the main *M. xanthus* sigma factor. By analogy to other prokaryotic RNA polymerases, the 45-kDa band was assumed to be the α subunit (Heidelbach et al., 1992).

sigA and *sigB* Genes in *S. aurantiaca*

By using gene probes derived from the genes *rpoD* of *E. coli* and *sigA* of *M. xanthus* (Inouye, 1990), various homologous DNA fragments have been detected by Southern analysis of restricted *S. aurantiaca* DNA. A gene probe spanning a central region of the *sigA* gene that encoded a highly conserved sequence of about 210 amino acids was used to detect the corresponding *S. aurantiaca* genes in lambda-derived gene libraries. This central region of the gene is thought to code for domains of the sigma factor that bind to the core polymerase and to DNA sequences of the promoters (Helman and Chamberlin, 1988). The gene *sigA,* for the main sigma factor of *S. aurantiaca,* is about 87% homologous to that of *M. xanthus*. The homology of the adjacent DNA sequences is strikingly lower (Skladny et al., submitted).

By using the *sigA* gene probe, about 14 other *Sal*I fragments have been detected by Southern analysis of *Sal*I-restricted chromosomal DNA of *S. aurantiaca*. Similar results were obtained when two oligonucleotides derived from the most conserved regions of the *S. aurantiaca sigA* gene were used.

By using one of the oligonucleotides derived from the conserved regions of the *S. aurantiaca sigA* gene as a probe, a second sigma gene, *sigB,* was detected in a lambda gt11 library. The gene *sigB* encodes a polypeptide of 296 amino acids with a predicted molecular mass of 33,297 kDa. Alignment with other sigma sequences reveals that it is a member of the heat shock sigma cluster (Lonetto et al., 1992). Homology is about 95% with the *M. xanthus* SigB polypeptide and about 62% with *E. coli* RpoH (Skladny et al., 1992). Inactivation of *sigB* by insertional mutagenesis does not affect fruiting body morphology or heat shock-dependent expression of the stress protein SP21 (see above). A merodiploid strain containing a complete *sigB* gene and a truncated *sigB* gene fused in frame to the β-galactosidase gene were used to show that β-galactosidase is produced during indole-induced sporulation or in the late stage of fruiting body formation. This unequivocally demonstrates that *sigB* is activated during development.

SIZE AND PHYSICAL MAP OF *S. AURANTIACA* CHROMOSOME

Genome Size

The complex biochemical and physiological processes that lead to cell-cell interactions and finally to the formation of fruiting bodies as well as the capability to produce a rich repertoire of secondary metabolites should be reflected in the complexity of the genome. Therefore, one would expect the myxobacteria to have genomes substantially larger than those of most other bacteria. For determining the length of the genome, chromosomal DNA of *S. aurantiaca* was restricted by using the rarely cutting enzymes *Spe*I and *Ase*I, and the fragments were separated by pulsed-field gel electrophoresis. For *Spe*I, 21 bands were obtained, and for *Ase*I, 26 bands were obtained. DNA bands harboring more than one fragment as judged from band intensity and sharpness were analyzed further by reciprocal restriction with an appropriate enzyme followed by a second pulsed-field gel electrophoresis. Analysis of the restriction fragment pattern obtained by adding the apparent fragment lengths resulted in the exact number of fragments hidden in one band. Thus, 33 and 25 DNA fragments were obtained after *Ase*I and *Spe*I restriction, respectively, totalling about 9,350 kbp. This result agrees quite well with the data obtained by Shimkets' group for *M. xanthus* (Chen et al., 1990). Restriction analysis of various *Stigmatella* strains revealed that they correlated with five different restriction patterns, perhaps because of altered methylation patterns or restriction fragment size variation (Neumann et al., 1992).

It is tempting to speculate that the large chromosomes of the myxobacteria are needed for the complex life cycle and for fruiting body formation, since genome complexity increases with structural diversity. Screening a set of ca. 600 Tn*5lacZ* insertional mutants showed that about 20% of the genes labeled are activated during starvation and that less than 4% of the mutants are affected in fruiting body formation (Pospiech et al., 1993). But very little is known about the physiology of *S. aurantiaca*. Which genes of the anabolic or secondary metabolism are activated as a consequence of a shift from complex to nonnutrient medium? Which stress genes, which may play a role in development, are induced? Therefore, it is premature to draw conclusions, from the results given above about the fraction of the genome involved in development.

Preliminary Physical Map

A *Spe*I and *Ase*I linking clone library was established with the aim of arranging the various restriction fragments on a physical gene map. Some of the clones were analyzed by hybridizing them to Southern blots of *Spe*I- or *Ase*I-restricted *S. aurantiaca* DNA. In addition, DNA fragments of a chromosomal *Spe*I or *Ase*I digest were used as probes in a Southern analysis of the corresponding reciprocal restriction digest. Furthermore, a Smith-Birnstiel analysis with defined probes of various fragments was performed after a time-dependent limited restriction with a second enzyme. The combination of these methods resulted in a preliminary proposal of a circular physical map on which more than 40 genes were localized by using probes from *S. aurantiaca, M. xanthus, Streptomyces* spp., and *E. coli* (Neumann, 1992; Neumann et al., submitted).

CONCLUSIONS

Because of its complex fruiting body morphology, *S. aurantiaca* is a valuable organism for the analysis of how genetic information is translated into multicellular structures. Although the analyses are in the preliminary stages, the results thus far are encouraging and suggest that all of the molecular genetic manipulations that are used with *Myxococcus* spp.

can also be performed with *S. aurantiaca*. Comparison of the data derived from *M. xanthus* and *S. aurantiaca,* particularly DNA sequences and the genetic map, may teach us how the diversification of the myxobacteria occurred. Analyses of mutants can illuminate which genes are involved in the determination of fruiting body morphology. A study of these genes and the regulation of their expression and (very importantly) identification of the gene products can stimulate the formulation of biochemical and biophysical models for morphogenesis.

Acknowledgments. I thank David White for revision of the manuscript and for *Stigmatella* strains; P. Hartzel, D. Kaiser, S. Inouye, H. Reichenbach, and L. Shimkets for strains and gene probes; S. Nazareth and K. Coutinho for typing the manuscript; and M. Dworkin, who encouraged us to continue the work with *S. aurantiaca*.

The work of this laboratory was supported by Bundesministerium für Forschung und Technologie grant BCT 0381/5, by Deutsche Forschungsgemeinschaft grant SFB229, and by the Fonds der chemischen Industrie.

REFERENCES

Apelian D., and S. Inouye. 1990. Development-specific sigma-factor essential for late-stage differentiation of *Myxococcus xanthus*. *Genes Dev.* **4:**1396–1403.

Butterfaß, H.-J. 1992. Isolierung und Charakterisierung des *csgA*-Gens aus *Stigmatella aurantiaca*. Diploma thesis. University of Heidelberg, Heidelberg, Germany.

Chen, H. W., I. M. Keseler, and L. J. Shimkets. 1990. The genome size of *Myxococcus xanthus* determined by pulsed-field gel electrophoresis. *J. Bacteriol.* **172:**4206–4213.

Dworkin, M., and S. M. Gibson. 1964. A system for studying microbial morphogenesis: rapid formation of microcysts in *Myxococcus xanthus*. *Science* **146:**243–244.

Gerth, K. 1975. Untersuchungen über den molekularen Mechanismus der Induktion von Myxosporen bei *Stigmatella aurantiaca* Sga1 (Myxobacteralcs). Ph.D. thesis. University of Freiburg, Freiburg, Germany.

Gerth, K., R. Metzger, and H. Reichenbach. 1993. Induction of myxospores in *Stigmatella aurantiaca* (myxobacteria). II. Inducers and inhibitors of myxospore formation, and mutants with a changed sporulation behavior. *J. Gen. Microbiol.*, **139,** in press.

Gerth, K., and H. Reichenbach. 1978. Induction of myxospore formation in *Stigmatella aurantiaca* (Myxobacterales). *Arch. Microbiol.* **117:**173–182.

Glomp, I., P. Saulnier, J. Guespin-Michel, and H. U. Schairer. 1988. Transfer of IncP plasmids into *Stigmatella aurantiaca* leading to insertional mutants affected in spore development. *Mol. Gen. Genet.* **214:**213–217.

Hartree, D. Personal communication.

Hartzell, P., and D. Kaiser. 1991a. Function of *mglA,* a 22-kilokalton protein essential for gliding in *Myxococcus xanthus*. *J. Bacteriol.* **173:**7615–7624.

Hartzell, P., and D. Kaiser. 1991b. Upstream gene of the *mgl* operon controls the level of *mglA* protein in *Myxococcus xanthus*. *J. Bacteriol.* **173:**7625–7635.

Heidelbach, M. 1992. Expression eines Hitzeschockgens während der Streβantwort und der Entwicklung von *Stigmatella aurantiaca*. Ph.D. thesis. University of Heidelberg, Heidelberg, Germany.

Heidelbach, M., H. Skladny, and H. U. Schairer. 1992. Purification of the DNA-dependent RNA polymerase from the myxobacterium *Stigmatella aurantiaca*. *J. Bacteriol.* **174:** 2733–2735.

Heidelbach, M., H. Skladny, and H. U. Schairer. 1993. Purification and characterization of SP21, a development-specific protein of the myxobacterium *Stigmatella aurantiaca*. *J. Bacteriol.* **175:**905–908.

Heidelbach, M., H. Skladny, and H. U. Schairer. Heat shock and development induce synthesis of a stress-responsive protein belonging to the alpha-crystallin family in the myxobacterium Stigmatella aurantiaca. Submitted for publication.

Helman, J. D., and M. J. Chamberlin. 1988. Structure and function of bacterial sigma factors. *Annu. Rev. Biochem.* **57:**839–872.

Hodgkin, J., and D. Kaiser. 1979. Genetics of gliding motility in *Myxococcus xanthus* (Myxobacterales): two gene systems control movement. *Mol. Gen. Genet.* **171:**177–191.

Inouye, S. 1990. Cloning and DNA sequence of the gene coding for the major sigma factor from *Myxococcus xanthus*. *J. Bacteriol.* **172:**80–85.

Inouye, S., D. White, and M. Inouye. 1980. Development of *Stigmatella aurantiaca:* effects of light on gene expression. *J. Bacteriol.* **141:**1360–1365.

Karlson, P., and M. Lüscher. 1959. "Pheromones": a new term for a class of biologically active substances. *Nature* (London) **183:**55–56.

Kim, S. K., and D. Kaiser. 1990a. Purification and properties of *Myxococcus xanthus* C-factor, and intercellular signaling protein. *Proc. Natl. Acad. Sci. USA* **87:**3635–3639.

Kim, S. K., and D. Kaiser. 1990b. C-factor: a cell-cell signaling protein required for fruiting body morphogenesis of *M. xanthus*. *Cell* **61:**19–26.

Kim, S. K., and D. Kaiser. 1990c. Cell motility is required for the transmission of C-factor, an intercellular signal that coordinates fruiting body morphogenesis of *M. xanthus*. *Genes Dev.* **4:**896–904.

Kim, S. K., and D. Kaiser. 1990d. Cell alignment in differentiation of *Myxococcus xanthus*. *Science* **249:**926–928.

Kroos, L., and D. Kaiser. 1984. Construction of Tn*5lac*, a transposon that fuses *lacZ* expression to exogenous promoters, and its introduction into *Myxococcus xanthus*. *Proc. Natl. Acad. Sci. USA* **81:**5816–5820.

Li, S.-F., B.-U. Lee, and L. J. Shimkets. 1992. *csgA* expression entrains *Myxococcus xanthus* development. *Genes Dev.* **6:**401–410.

Lindquist, S., and E. A. Craig. 1988. The heat-shock proteins. *Annu. Rev. Genet.* **22:**631–677.

Lonetto, M., M. Gribskov, and C. A. Gross. 1992. The σ^{70} family: sequence conservation and evolutionary relationships. *J. Bacteriol.* **174:**3843–3849.

Losick, R., and P. Stragier. 1992. Crisscross regulation of cell-type-specific gene expression during development in *B. subtilis*. *Nature* (London) **355:**601–604.

Lünsdorf, H., and M. Heidelbach. Personal communication.

Neumann, B. 1992. Genomstruktur des Myxobakteriums *Stigmatella aurantiaca DW4/3.1*. Ph.D. thesis. University of Heidelberg, Heidelberg, Germany.

Neumann, B., A. Pospiech, and H. U. Schairer. 1992. Size and stability of the genomes of the myxobacteria *Stigmatella aurantiaca* and *Stigmatella erecta*. *J. Bacteriol.* **174:** 6307–6310.

Neumann, B., A. Pospiech, and H. U. Schairer. The physical and genetic map of the *Stigmatella aurantiaca* DW 4/3.1 chromosome. Submitted for publication.

Plaga, W. Personal communication.

Pospiech, A. 1992. Isolierung und Charakterisierung entwicklungsbiologischer Mutanten des Myxobakteriums *Stigmatella aurantiaca*. Ph.D. thesis. University of Heidelberg, Heidelberg, Germany.

Pospiech, A., B. Neumann, B. Silakowski, and H. U. Schairer. 1993. Detection of developmentally regulated genes of the myxobacterium *Stigmatella aurantiaca* with the transposon Tn*5lacZ*. *Arch. Microbiol.*, **159:**201–206.

Qualls, G. T., K. Stephens, and D. White. 1978a. Morphogenetic movements and multicellular development in the fruiting myxobacterium. *Stigmatella aurantiaca. Dev. Biol.* **66:** 270–274.

Qualls, G. T., K. Stephens, and D. White. 1978b. Light-stimulated morphogenesis in the fruiting myxobacterium, *Stigmatella aurantiaca. Science* **201:**444–445.

Reichenbach, H., and M. Dworkin. 1970. Induction of myxospore formation in *Stigmatella aurantiaca* (Myxobacterales) by monovalent cations. *J. Bacteriol.* **101:**325–326.

Shimkets, L. J. 1990. The *Myxococcus xanthus fprA* gene causes increased flaving biosynthesis in *Escherichia coli. J. Bacteriol.* **172:**24–30.

Simon, R., U. Piefer, and A. Pühler. 1983. A broad host range mobilization system for *in vivo* genetic engineering: transposon mutagenesis in Gram negative bacteria. Bio/Technology **1:** 784–791.

Skladny, H., M. Heidelbach, and H. U. Schairer. Cloning and DNA sequence of the gene coding for the major sigma factor of the myxobacterium *Stigmatella aurantiaca.* Submitted for publication.

Skladny, H., M. Heidelbach, and H. U. Schairer. 1992. Cloning and DNA sequence of *sigB* gene of *Stigmatella aurantiaca. Nucleic Acids Res.* **20:**6416.

Stephens, K., G. D. Hegeman, and D. White. 1982. Phermone produced by the myxobacterium *Stigmatella aurantiaca. J. Bacteriol.* **149:** 739–747.

Stephens, K., and D. White. 1980. Morphogenetic effects of light and guanine derivatives on the fruiting myxobacterium *Stigmatella aurantiaca. J. Bacteriol.* **144:**322–326.

Stragier, P. 1992. Dances with sigmas. *EMBO J.* **10:**3559–3566.

Vasquez, G. M., M. Qualls, and D. White. 1985. Morphogenesis of *Stigmatella aurantiaca* fruiting bodies. *J. Bacteriol.* **163:**515–521.

White, D. 1981. Cell interactions and the control of development in *Myxobacteria* populations. *Int. Rev. Cytol.* **71:**203–277.

White, D., J. A. Johnson, and K. Stephens. 1980a. Effects of specific cations on aggregation and fruiting body morphology in the myxobacterium *Stigmatella aurantiaca. J. Bacteriol.* **144:**400–405.

White, D., W. Shropshire, Jr., and K. Stephens. 1980b. Photocontrol of development by *Stigmatella aurantiaca. J. Bacteriol.* **142:**1023–1024.

Myxobacteria II
Edited by Martin Dworkin and Dale Kaiser
© 1993 American Society for Microbiology, Washington, DC 20005

Chapter 16

Production of Bioactive Secondary Metabolites

Hans Reichenbach and Gerhard Höfle

Until recently, the myxobacteria have been notable mainly for their impressive morphogenetic potential. In the past decade, however, it has been shown that they are also outstanding as producers of bioactive secondary metabolites, which should make them very interesting for the pharmaceutical and chemical industries. Many myxobacterial metabolites are completely novel chemical structures, and some show rare and even unique mechanisms of action. Up to now, 50 different basic structures and 300 structural variants have been isolated from myxobacteria and fully characterized chemically. Among these structures are only three that had been known before from other microorganisms. The 350 compounds constitute 3.5% of the presently known 10,000 secondary metabolites of microbial origin and bring the myxobacteria into third place among the prokaryotes as producers of bioactive substances, leading the pseudomonads, following the bacilli, and, of course, coming at a great distance behind the actinomycetes. This is true even though only very few and usually small laboratories have participated worldwide in research on secondary metabolites from myxobacteria. Progress in the field has been reviewed a number of times (Reichenbach et al., 1984; Rosenberg and Varon, 1984; Reichenbach et al., 1988a; Reichenbach and Höfle, 1989, in press; Höfle and Reichenbach, 1990).

It has been known for many years that myxobacteria produce antibiotic activities. However, for a long time, nobody was able or even seri-

Hans Reichenbach • Microbial Secondary Metabolites, Gesellschaft für Biotechnologische Forschung, D-38124 Braunschweig, Germany. **Gerhard Höfle** • Chemistry of Natural Products, Gesellschaft für Biotechnologische Forschung, D-38124 Braunschweig, Germany.

ously tried to elucidate the chemical structures of the antibiotics, partly because until the late 1960s, myxobacteria were believed to be extremely difficult to handle and could not readily be cultivated in liquid media. Also, at that time, other, easily accessible microorganisms still supplied many new compounds. However, myxobacteria had already been screened for antimicrobial substances early in the antibiotic era and had been found positive (Oxford, 1947; Norén, 1953; Katô, 1955; Norén and Raper, 1962). Even antiviral activities have been described (Singh et al., 1973). In 1973, inhibitory substances from myxobacteria were chemically characterized for the first time. They were produced by *Myxococcus xanthus,* inhibited the germination of fungal spores, turned out to be ordinary cellular fatty acids, and thus were trivial (Norén and Odham, 1973). This is an obvious pitfall, as gram-positive bacteria tend to be very sensitive to fatty acids. Finally, in 1977, the chemical structure of a real myxobacterial antibiotic, viz., that of the antifungal compound ambruticin, was elucidated (Fig. 1; Ringel et al., 1977; Connor et al., 1977). As mentioned above, a substantial number of structures of myxobacterial secondary metabolites have since been published. They will be discussed in more detail below.

At this point, to avoid a completely wrong impression, the reader must be warned of some taxonomic confusion in the past. For some time, the cytophagas and *Cytophaga*-like bacteria were classified among the myxobacteria, yet we know today that they are totally unrelated and belong to a different main branch, or phylum, of the bacterial phylogenetic tree. Further, the bacteria now classified as genus *Lysobacter* were regarded as true myxobacteria for more than a decade after their discovery. Some were even assigned directly to the myxobacterial genus *Sorangium* (in the chemical literature, this name is often distorted to *Sporangium*). Again, the two are not related at all; genus *Lysobacter* is close to the xanthomonads and thus belongs to the gamma branch of the proteobacteria (formerly the purple bacteria). The *Cytophaga*-like bacteria and the lysobacters produce quite a few antibiotics, among them some rather inter-

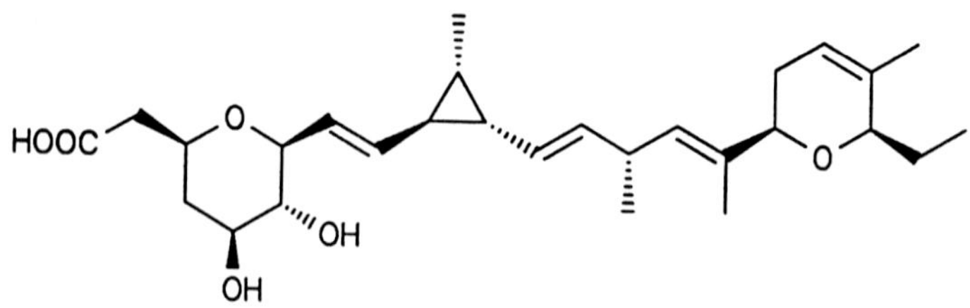

Figure 1. Ambruticin.

esting compounds. The phenazine antibiotic myxin was even commercialized and sold for some time as a broad-spectrum antibiotic for veterinary applications. Myxin was obtained from "*Sorangium*" strain 3 C, actually a strain of *Lysobacter antibioticus* (for further details, see Reichenbach [1992]). Thus, if in the older literature an antibiotic is attributed to a myxobacterium, careful examination of the taxonomic position of the producer is suggested. The latest case of confusion in this respect is that of the glidobactins. It has been stated that these antitumor compounds are produced by the myxobacterium *Polyangium brachysporum,* a new species (Oka et al., 1988), yet the strain distributed by the American Type Culture Collection is not a gliding but a flagellated bacterium and is definitely not a myxobacterium (personal inspection by H. Reichenbach). In fact, this conclusion was already suggested by the scanty description of the organism and its physiological behavior.

The elucidation of every magnificent new structure begins with a humble screening. Not much has been published about screening procedures with myxobacteria, so we shall rely here mainly on our own experience.

As a first step, strains must be isolated. We get them from soil samples and other suitable material that has been collected all over the world. In this way, the strains obtained are genetically more diverse than if they were isolated from a restricted area only. Presently, between 300 and 500 new strains are isolated annually in our laboratory and introduced in our screening. All in all, we may have screened about 3,000 strains of myxobacteria, many of them encountered repeatedly with different sets of screening tests.

We perform our screenings exclusively with shake flask cultures. Newly isolated myxobacterial strains almost always grow in nodules, flakes, and clumps when first inoculated into a liquid medium, but this is acceptable for a screening. We use standard growth media, usually peptone based. A nonionic adsorber resin, e.g., 2 to 3% Amberlite XAD-16 or XAD-1180 (Rohm and Haas), is added to the culture from the beginning. The resin is well tolerated by almost all organisms and has several advantages. It adsorbs all excreted substances if they are not extremely hydrophilic, and thereby concentrates and stabilizes the produced metabolites, so that one sampling at the end of the growth of the culture is sufficient and taking a series of samples along the growth curve is no longer necessary. In addition to the concentrated methanolic eluate from the resin, an acetone extract of the separated cell mass and a sample of the lyophilized culture supernatant are then introduced into the tests.

Most of our substances have been discovered with rather simple screening assays. These are usually growth inhibition tests with a collection of gram-positive and gram-negative bacteria, yeast cells, and filamentous fungi. One can expect roughly 50% of all myxobacterial strains to produce some activity in these comprehensive tests (Reichenbach et al., 1984). In some groups, specifically the cellulose-degrading myxobacteria of the genus *Sorangium,* that figure goes up to nearly 100%. Very interesting substances have been found in a screening with animal cell cultures, mainly with the robust mouse fibroblast strain L929 (Sasse et al., 1993). Here, not only growth inhibition but also more-subtle effects, e.g., changes in cell morphology, can be seen. Occasionally, tests with the brine shrimp *Artemia salina* have been performed. They could indicate substances that interact, e.g., with moulting or with the nervous system, but so far, nothing useful has been found in this way. Also, antiviral tests, e.g., in cooperation with the German Primate Center in Göttingen (G. Hunsmann, Deutsches Primatenzentrum), a test for substances that specifically block human immunodeficiency virus type 1 propagation in cell cultures (Jurkiewicz et al., 1992), have been tried. In this test, 11% of the 340 screened strains of myxobacteria produced inhibitory substances. The three activities that were studied further indeed led to three novel compounds, but all three were electron transport inhibitors in complex I (Jansen et al., 1991, 1992; Kunze et al., 1992a; Trowitzsch-Kienast et al., 1992). A few compounds have been found in enzyme inhibition tests, e.g., with elastase and angiotensin-converting enzyme. Finally, new structures have also been discovered through striking chemical signals when a strain was studied for some other antibiotic. In this way, e.g., the aurachins were first seen in culture extracts of *Stigmatella aurantiaca* Sg a15. They are efficient electron transport inhibitors but were present only in minute quantities along with two other electron transport inhibitors, stigmatellin and myxalamid, which were both produced in much larger quantities (Kunze et al., 1987).

What kinds of activities can be expected from myxobacteria? The range of activities is indeed very wide. Antifungal activities are very common, and antibacterial activities, mostly against gram-positive bacteria but occasionally against gram-negative bacteria or both, are also often seen. More rarely, insecticidal activities and substances that act more or less specifically on animal cells appear. Of course, a strain may be active against different types of organisms, but these activities are often due to different substances. Whether that is the case can usually quickly be found out by studying a bioautogram after thin-layer chromatography.

When the decision has been made to investigate a certain activity further, the first task is to produce sufficient material for structure elucidation. For this, a stable producer strain must be available, so clone selection is required, since the wild strains used in the screening usually are genetic mosaics. Cloning, however, may present a technical problem with myxobacteria, in particular with representatives of the suborder *Sorangineae*, because, first, the organisms tend to grow in flakes and clumps, so the homogeneous cell suspension needed for diluting and plating is not easily obtained, and second, single cells may not readily grow into colonies, and substantial efforts may have to be invested in the development of a suitable plating medium. The problem has not yet been solved in all cases. Further, a useful production medium has to be found, which can usually be achieved within a couple of weeks. Some myxobacteria, in particular *Myxococcus* species, are not very flexible in their metabolism, and there are not so many possible medium variations. Still, many myxobacteria can be cultivated on technical substrates, like corn steep powder, maize gluten, soya and peanut meals, or skim milk (Gerth et al., 1983, 1984), so there is some choice. The influences of different substrates on production may be substantial. As the initial yields of myxobacterial substances normally are low (between 0.5 and 20 mg/liter), fermentations on a larger scale, i.e., between 50 and 5,000 liters, are necessary to produce the required material. Myxobacteria can be grown without difficulty in bioreactors, and scaleup is also not a problem. Of course, the processing of such large volumes of fermentation broth is not trivial, and efficient methods have to be found for the isolation of the desired substances. Fermentation in the presence of an Amberlite adsorber resin is often helpful. Today, preparative high-pressure liquid chromatography (HPLC) greatly facilitates purification of the substances. As a rule, it takes between 4 and 9 months to prepare the quantity of pure substance required for structure elucidation. It may then take another 3 to 6 weeks to determine the structure, usually by application of nuclear magnetic resonance and mass spectroscopy and occasionally by using X-ray crystallography. Only then can it be known with certainty whether a new compound has been discovered and whether the whole procedure was worthwhile.

If a compound is new, further studies are indicated. Strain, fermentation, and downstream processing must be improved so that gram amounts of the substance for extended testing can be produced. By feeding ^{13}C-labeled potential precursors to the organisms, a component analysis of the substance is performed. Studies on the enzymology of biosynthesis and its regulation are very complex and time-consuming and therefore are justified only if a substance is of practical interest or in some way unusual. The mechanism of action is investigated, and this work may involve rather sophisticated biochemistry. The mechanism of action is not yet known for all substances.

What kinds of compounds have been found in myxobacteria? First, we can see that the substances produced by myxobacteria are chemically very diverse (see Fig. 1 to 20). There are peptides, macrocyclic and polyenic compounds, polyethers, aromatics, quinoids, heterocyclics, and alkaloids either pure or in various combinations. The occurrence of unsubstituted phenyl rings in quite a few myxobacterial compounds is conspicuous and unusual. Macrocyclic lactones and lactams are particularly prominent among myxobacterial secondary metabolites. Also, many myxobacterial compounds contain oxazole and thiazole rings. These facts are connected with the biosynthetic roots of myxobacterial secondary metabolites. The compounds usually arise from amino acid or lipid metabolism; in fact, many of them are polyketides. In contrast, modified sugars have been found only rarely and only as components of larger structures. No compound that derives from nucleotide metabolism has been discovered in myxobacteria so far. As is often observed with antibiotic producers, a strain may produce a large family of related substances. In some cases, we isolated and characterized between 20 and 50 different structural variants. A complete list of all published structures is given in Table 1.

Table 1. Bioactive secondary metabolites from myxobacteria

Compound	Yr of publication of structure[a]	Activity spectrum	Status
Ambruticin	1977	Fungi	New
Myxothiazol	1980	Fungi	New
Althiomycin	1982	Bacteria	Known
Myxovirescin	1982	Bacteria	New
Pyrrolnitrin	1982	Fungi	Known
Myxalamid	1983	Fungi	New
Myxopyronin	1983	Bacteria	New
Stigmatellin	1984	Fungi	New
Angiolam	1985	Bacteria	New
Corallopyronin	1985	Bacteria	Variant
Sorangicin	1985	Bacteria	New
Aurachin	1987	Fungi	New
Saframycin	1988	Bacteria, animal cells	Known
Myxochelin A	1989	Bacteria	Variant
Myxovalargin	1989	Bacteria	New
Phenoxan	1991	Fungi	New
Stipiamide-phenalamid	1991	Fungi	Variant
Nannochelin	1992	Bacteria	Variant
Thiangazol	1992	Fungi	New
Rhizopodin	1993	Animal cells	New
Soraphen	1993	Fungi	New

[a] Only fully published structures are considered. References are given in the text. The year of publication refers to publication for the myxobacterial compound if the substance was discovered earlier in another organism.

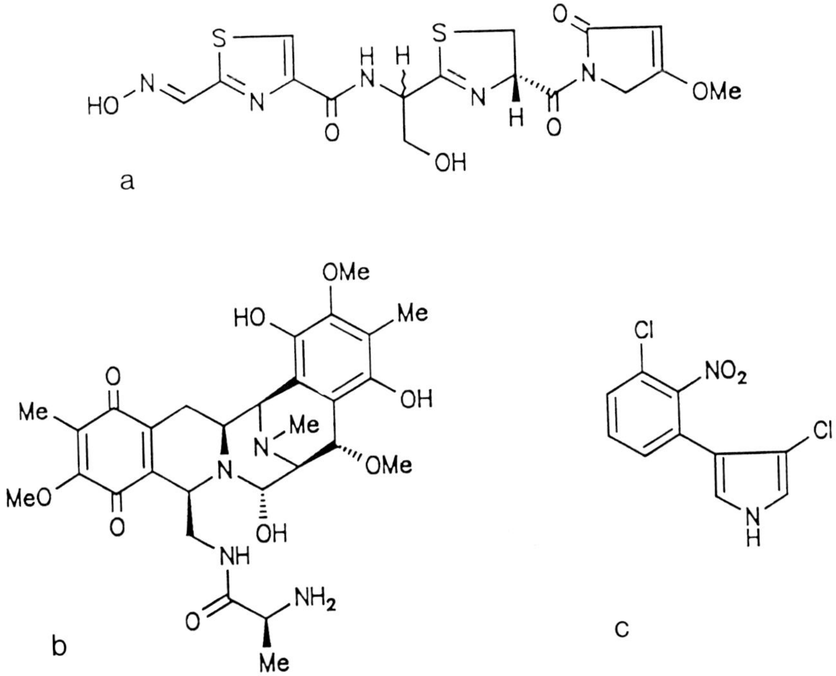

Figure 2. Myxobacterial antibiotics described earlier from other organisms. (a) Althiomycin; (b) saframycin Mx1; (c) pyrrolnitrin. Me, methyl group.

Second, we have the pleasant situation that almost all myxobacterial secondary metabolites are new compounds. So far, there have been only three exceptions (Fig. 2). Althiomycin (Kunze et al., 1982) was first discovered in some streptomycetes (Sakakibara et al., 1974; the identity with matamycin [Sensi et al., 1959] has been corroborated by Kunze et al. [1982]). Pyrrolnitrin and aminopyrrolnitrin (Gerth et al., 1982b), which also are of considerable practical interest, although as synthetic analogs, were originally found in pseudomonads (Arima et al., 1964; Chang et al., 1981), and the saframycins (Irschik et al., 1988; Trowitzsch-Kienast et al., 1988a) were initially described in *Streptomyces lavendulae* (Arai et al., 1977, 1980). The myxobacterial substance is one of the highly efficient saframycin variants with an OH group on C-21 (which can easily be replaced experimentally by a cyano group).

In several cases, compounds that are structurally related to myxobacterial substances are known to occur in other organisms. Thus, the myxalamids (Gerth et al., 1983; Jansen et al., 1983, 1984) and the phenalamids (Trowitzsch-Kienast et al., 1992; Höfle et al., 1992) or stipiamides (Kim et al., 1991) are partly identical to the molecules of piericidin from *Streptomyces pactum* (Hall et al., 1966; Takahashi et al. 1966; Fig. 3), and the striking structure of thiangazol (Jansen et al., 1992) clearly repeats part of the structure of tantazole B from the blue-green bacterium *Scytonema mirabile* (Carmeli et al., 1990; Fig. 4). However, many myxobacterial substances have completely novel structures that have no precedents.

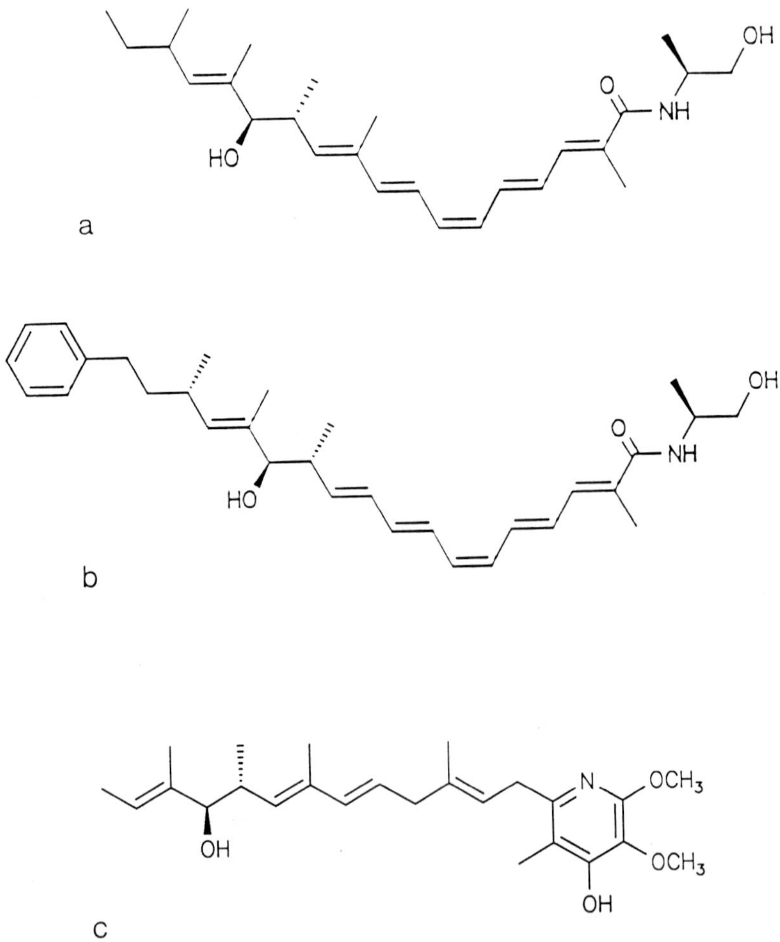

Figure 3. Myxalamid A (a), phenalamid or stipiamide (b), and structurally related piericidin A (c) from streptomycetes.

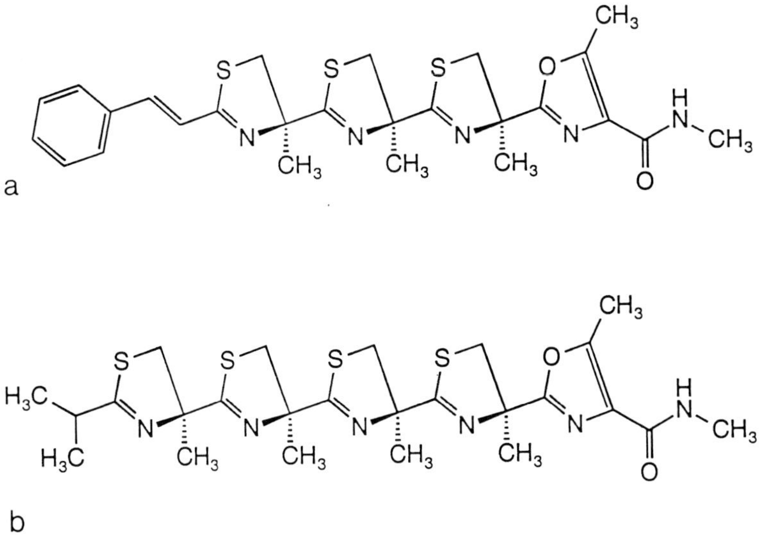

Figure 4. Thiangazol (a) from myxobacteria and structurally related tantazole (b) from cyanobacteria.

Let us now take a closer look at bioactive metabolites from myxobacteria.

The myxobacteria have yielded an impressive collection of electron transport inhibitors. So far, 11 basically different, new electron transport inhibitors have been found. The first antibiotic discovered and characterized in our laboratories and the second myxobacterial antibiotic for which a structure was proposed, myxothiazol, was of this type (Gerth et al., 1980; Trowitzsch et al., 1980). Some of those myxobacterial compounds block the respiratory chain in the cytochrome bc_1 complex or in complex III (myxothiazol, stigmatellin, and the aurachins C and D: Fig. 5), while others block the NADH:ubiquinone oxidoreductase or complex I (the myxalamids and phenalamids [Fig. 3], thiangazol [Fig. 4], phenoxan, and the aurachins A and B [Fig. 6]). As mentioned above, phenoxan, thiangazol, and the phenalamids were discovered because they interfere with human immunodeficiency virus propagation in cell cultures. Several of those compounds also act on *b*-type cytochromes in the photosystems of higher plants and bacteria. Thus, aurachin D and stigmatellin are the most potent inhibitors of the cytochrome b_6-f complex in spinach chloroplasts known today, with 50% inhibitory values below 60 nM (Oettmeier et al., 1985, 1990). They also block the reducing side of photosystem II.

The electron transport inhibitors often show good antifungal and in-

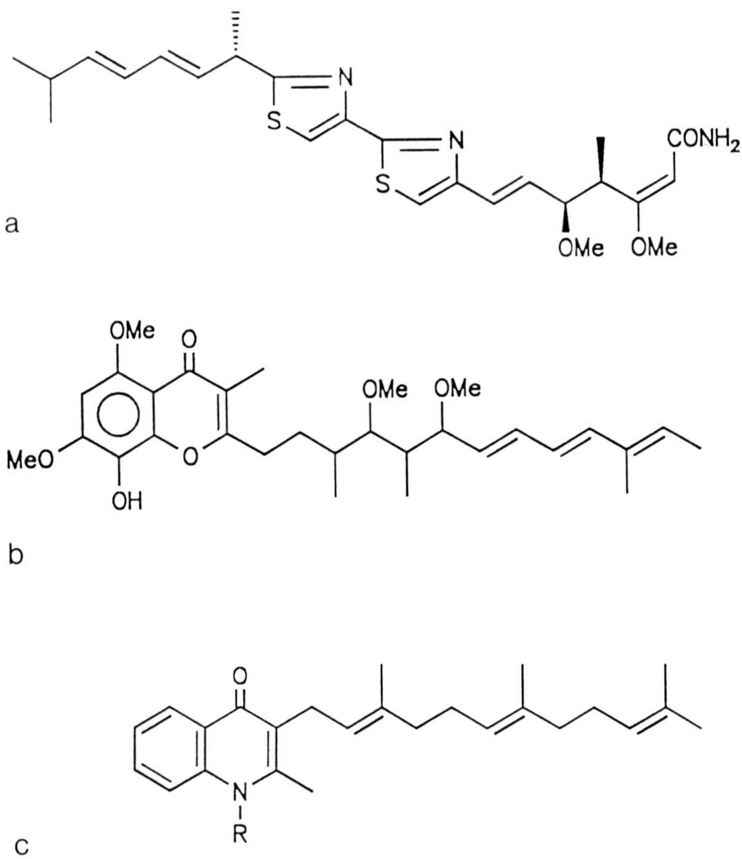

Figure 5. Myxobacterial electron transport inhibitors (complex III). (a) Myxothiazol A; (b) stigmatellin A; (c) aurachins C (R = OH) and D (R = H). Me, methyl group.

secticidal activities, but bacteria may also be inhibited (Table 2). All electron transport inhibitors are, of course, highly toxic for mammals, especially since some of them are extremely efficient drugs. Thus, the 50% lethal doses (LD_{50}s) for the mouse were 2 mg/kg of body weight intraperitoneally for myxothiazol (Gerth et al., 1980) and 2 mg/kg subcutaneously (s.c.) and 30 mg/kg perorally for stigmatellin (Kunze et al., 1984). The LD_{100} of myxalamid was 3 mg/kg s.c. and 10 mg/kg perorally (Gerth et al., 1983). The K_d for complex III inhibition by myxothiazol and stigmatellin is below 10^{-10} M (von Jagow and Link, 1986). The complex III inhibitors myxothiazol and stigmatellin have become useful biochemical probes. They contributed substantially to the elucidation of the sophisticated bio-

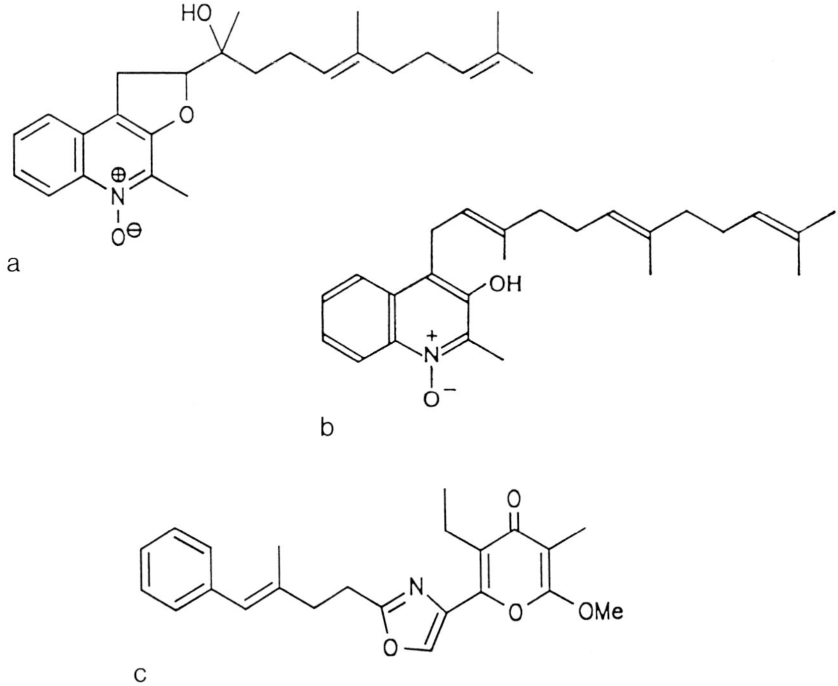

Figure 6. Myxobacterial electron transport inhibitors (complex I). (a) Aurachin A; (b) aurachin B; (c) phenoxan. Me, methyl group.

Table 2. Inhibitory spectra of myxobacterial electron transport inhibitors

Inhibitor	Reference	MIC (μg/ml)[a]			
		Gram-positive bacteria	Gram-negative bacteria	Filamentous fungi	Yeasts
Complex III					
Myxothiazol	Gerth et al., 1980; Thierbach and Michaelis, 1982	–	–	(0.02–2)	(0.3)
Stigmatellin	Kunze et al., 1984	(2.5–5)	–	+	(0.1–2.5)
Aurachins C and D	Kunze et al., 1987	(0.05–0.8)	–	±	(0.2–10)
Complex I					
Myxalamid	Kunze et al., 1992b	±	±	+	+
Aurachins A and B	Kunze et al., 1987	(0.2–5)	–	–	–
Phenoxan	Kunze et al., 1992a	–	–	+	–

[a] Evaluation +, positive; ±, moderately active and variable; –, negative. Where known, MICs are given.

chemistry of this essential part of the energy-producing machinery of the cell (see, e.g., von Jagow and Engel [1981], von Jagow et al. [1984], von Jagow and Ohnishi [1985], and von Jagow and Link [1986]). Both inhibitors are available from drug companies (Boehringer Mannheim, Fluka, Sigma). Conceivably, the myxobacterial complex I inhibitors may play a similar role in the elucidation of the even more complicated biochemistry of complex I. They are presently being tested for their relevance in this context.

Myxothiazol (Fig. 5) was originally isolated from *Myxococcus fulvus* Mx f16 (Gerth et al., 1980) but has since been found in many other myxobacteria (Table 3). The antibiotic is produced during growth phase and remains essentially intracellular or attached to the cells. It contains two thiazole rings (Trowitzsch et al., 1980). Its stereochemistry (Trowitzsch et al., 1981) and its building blocks (Trowitzsch-Kienast et al., 1986) have

Table 3. Occurrence of bioactive myxobacterial secondary metabolites in different taxa

Secondary metabolite	Source(s)[a]
Althiomycin	*Cb. fuscus,*[b] *M. virescens, M. xanthus, Streptomyces althioticus,*[b] *Streptomyces matensis*
Ambruticin	*So. cellulosum*[b]
Angiolam	*An. disciformis*[b]
Aurachin	*Sg. aurantiaca,*[b] *Sg. erecta*
Corallopyronin	*Cc. coralloides*[b]
Myxalamid	*Cb. velatus, Cc. coralloides, M. stipitatus, M. virescens, M. xanthus,*[b] *Sg. aurantiaca, Sg. erecta*
Myxochelin A	*An. disciformis*[b]
Myxopyronin	*M. fulvus*[b]
Myxothiazol	*An. disciformis, Cb. velatus, M. fulvus,*[b] *M. virescens, Sg. aurantiaca, Sg. erecta*
Myxovalargin	*Ar. gephyra, Cc. coralloides, M. fulvus,*[b] *M. xanthus*
Myxovirescin	*M. fulvus, M. virescens,*[b] *M. xanthus*
Nannochelin	*N. aggregans, N. exedens*[b]
Phenalamid	*M. stipitatus*[b]
Phenoxan	*Pl. fumosum*[b]
Pyrrolnitrin	*Cb. ferrugineus, Cb. minor, Cc. exiguus, Melittangium boletus, M. fulvus,*[b] *Ps. aureofaciens, Ps. fluorescens, Ps. pyrrocinia,*[b] *Ps. pyrrolnitrica*
Rhizopodin	*M. stipitatus*[b]
Saframycin	*M. fulvus, M. virescens, M. xanthus,*[b] *Streptomyces lavendulae*[b]
Sorangicin	*So. cellulosum*[b]
Soraphen	*So. cellulosum*[b]
Stigmatellin	*Cb. velatus, Cb. minor, Sg. aurantiaca*[b]
Thiangazol	*Pl. fumosum*[b]

[a] Abbreviations for genera: *An., Angiococcus; Ar., Archangium; Cb., Cystobacter; Cc., Corallococcus; M., Myxococcus; Pl., Polyangium; Ps., Pseudomonas; Sg., Stigmatella;* and *So., Sorangium.*
[b] Organism in which the compound was originally discovered.

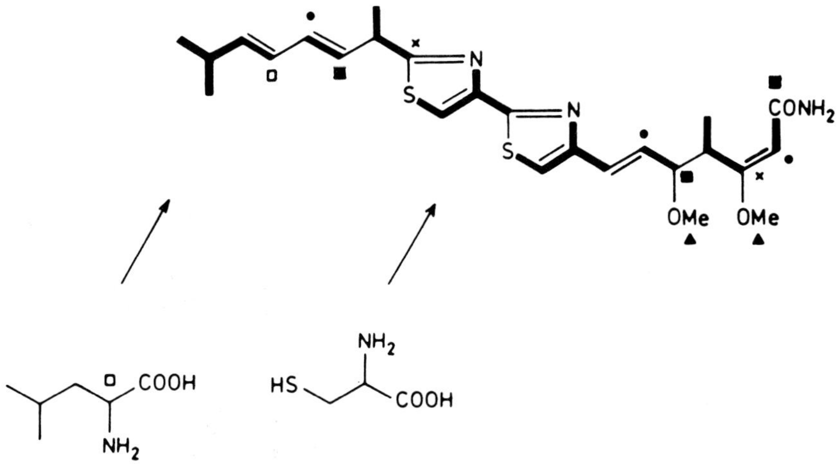

Figure 7. Building blocks of myxothiazol: acetate, propionate, leucine, cysteine, and methyl groups (Me) from methionine. (Reprinted from Trowitzsch-Kienast et al. [1986] with permission.)

been elucidated. It is constructed from acetate, propionate, cysteine, leucine, and methyl groups from methionine and thus is a typical example of combined biosynthetic pathways (Fig. 7). It was discovered early on that myxothiazol blocks electron transfer in the cytochrome bc_1 complex of beef heart submitochondrial particles selectively and with high efficiency (Thierbach and Reichenbach, 1981a, 1981b). Subsequent studies revealed still more details. Thus, it was shown that the new inhibitor blocks the outer ubiquinol oxidizing center (Q_o), i.e., the low-potential heme center (cytochrome b_l or b_{566}), whereas the inner Q_i center, the antimycin-sensitive, high-potential heme center (cytochrome b_h or b_{562}) is still open to ubiquinol (von Jagow et al., 1984; von Jagow and Link, 1986). When antimycin and myxothiazol were applied at the same time, however, all reduction of cytochrome *b* in the complex was prevented (von Jagow and Engel, 1981). The efficiency of myxothiazol in this experimental system is so high that the heme b_{566} content can be titrated directly; one molecule of the antibiotic is bound per cytochrome b_{566} (Thierbach and Reichenbach, 1981b). Regarding the wide distribution of respiratory enzymes among bacteria, it is somewhat astonishing that bacteria are not inhibited by myxothiazol. A study of *Paracoccus denitrificans,* whose respiratory chain closely resembles that of mitochondria, showed, however, that myxothiazol was indeed active (Thierbach and Reichenbach, 1983): NADH oxidation by membrane particles was strongly depressed (50% inhibition at 0.25 mol of myxothiazol per mol of heme *b* and 100%

at 0.5 mol of myxothiazol per mol of heme *b;* the latter value can, however, only be extrapolated because of some side reaction). Further, the study of reduction kinetics by dual-wavelength measurements demonstrated that myxothiazol temporarily prevented any reduction of *c*-type but not *b*-type cytochromes. Continued growth of the bacterium in the presence of myxothiazol might be explained by alternative respiratory chains, e.g., via cytochrome *o*.

The biochemical data on myxothiazol are corroborated by genetic and molecular biological results. Resistance loci in baker's yeast, *Saccharomyces cerevisiae,* are found in the mitochondrial apocytochrome *b* gene, either in exon (or box) 1 (locus *myx-1*) or in exon 6 (locus *myx-2;* Thierbach and Michaelis, 1982). In a later study, a third resistance locus was identified in exon 5 (di Rago et al., 1989). When the respective gene segments were sequenced, it turned out that myxothiazol resistance in yeast cells is conferred by replacement of single amino acids, i.e., Phe-129 or Gly-137 in exon 1, Asn-256 in exon 5, or Leu-275 in exon 6 (di Rago et al., 1989; Tron and Lemesle-Meunier, 1990). These amino acids are all located near the outer side of the inner membrane of the mitochondrion.

More than 100 articles dealing with various aspects of myxothiazol and myxothiazol effects in diverse systems have been published. Many of the relevant results have just been reviewed (Reichenbach and Höfle, in press) and thus need not be repeated here.

Stigmatellin (Fig. 5) has been discovered in *S. aurantiaca* Sg a15 (Kunze et al., 1984). It is only rarely found among myxobacteria (Table 3). The compound contains a chromone system and remotely resembles a menaquinone molecule (Höfle et al., 1984). Stigmatellin is produced during growth with yields of up to 12 mg/liter and remains intracellular. Good yields are obtained in a lean peptone medium supplemented with glucose (0.5%), maltose, starch, or dextrin 10. A suitable fermentation medium can be prepared from maize gluten. As all macromolecular syntheses are stopped immediately by the antibiotic and as the inhibitory effect of the antibiotic on yeast cells can be completely abolished by the addition of glucose to the medium, interference with respiratory energy metabolism was inferred (Kunze et al., 1984). This is indeed the case (Thierbach et al., 1984). Stigmatellin and myxothiazol act in very similar but not identical ways. In addition to the outer heme center of the cytochrome bc_1 complex, where it appears to compete with myxothiazol for the same binding site (Thierbach et al., 1984), stigmatellin also binds tightly to the sulfur iron cluster (the Rieske protein) of complex III (von Jagow and Ohnishi, 1985) and blocks electron flow from cytochrome c_1 back to the Rieske protein, a rather exceptional effect (degli Esposti et al., 1986). Studies on the structure-activity relationship that used a series of chemical derivatives demonstrated, e.g., that the keto and hydroxy functions on the chromone are essential for good in vitro and in vivo activities (Thierbach et al., 1984) and that the structure of the side chain is critical in binding to the Rieske protein (Ohnishi et al., 1988). When six independently derived stigmatellin-resistant yeast mutants were mapped, all six showed a replacement at one single site in the mitochondrial apocytochrome *b* gene, viz., in Ile-147 in exon 3 (di Rago et al., 1989).

The aurachins (Fig. 5 and 6) were discovered as satellite antibiotics in the *S. aurantica* strain that contained stigmatellin and appear to be rare metabolites (Kunze et al., 1987). They are a family of at least four different compounds that all share a quinoline–γ-quinolone system and an isoprenoid side chain, both of which are relatively unusual features for bacterial secondary metabolites. There is a remote resemblance to the well-known complex III inhibitor HQNO (2-heptyl-4-hydroxyquinoline *N*-oxide) from *Pseudomonas aeruginosa*. The aurachins are essentially intracellular (90%), and individual compounds appear at different times during fermentation. For instance, aurachin C is seen only transiently during growth phase, while aurachin A is produced only in stationary phase. Further, the oxygen supply during fermentation determines the proportion of the variants. Thus, under oxygen limitation, a high yield of aurachin B can be expected. All four aurachins act on the respiratory chain but at different sites (Pridzun et al., in preparation). While aurachins A and B are complex I inhibitors, aurachins C and D act with high efficiency on complex III. The four compounds also differ in their inhibitory spectra (Table 2).

The myxalamids (Fig. 3) were first obtained from *M. xanthus* Mx x12 (Gerth et al., 1983) but have since been found in many other myxobacteria, including the stigmatellin- and aurachin-producing *S. aurantiaca* strain (Kunze et al., 1984). They appear to be very common in myxobacteria and are often produced in substantial quantities (more than 100 mg/liter). Their synthesis responds sensitively to culture conditions. With Mx x12, myxalamid yields rose by a factor of 5 when the organism was grown on maize gluten or single-cell protein (Probion Hoechst) compared with a simple peptone medium and could be further increased (to 120 mg/liter) by the addition of 0.3% sodium acetate. With Sg a15, the yields in a lean peptone medium varied between 25 and 70 mg/liter depending on the type of carbohydrate added (0.5%), with glucose, maltose, starch, dextrin 10, and pyruvate giving the best results. Even better results (50 to 120 mg/liter) were obtained with technical substrates, again with maize gluten as the most effective medium. In both organisms, synthesis occurred during growth. The myxalamids are intracellular (95%), but when the adsorber resin XAD-1180 (1%) was present during fermentation, the antibiotic was transferred quantitatively to the resin, at least in the case of Mx x12. The myxalamids are a family of polyenic compounds with an alaninol amide head group (Jansen et al., 1983). Their absolute configuration has been determined by chemical methods and X-ray analysis (Jansen et al., 1984). Recently, a further structural variant with a benzene ring at the left end has been discovered (Kim et al., 1991; Trowitzsch-Kienast et al., 1992). These compounds, the phenalamids or stipiamids (Fig. 3), occur also as several structural variants (the name phenalamid was first published in a patent application [Höfle et al., 1992] and may have precedence over stipiamide). The phenalamids were isolated from a strain of *Myxococcus stipitatus* and in fact are the compounds responsible for the strong, yellow fluorescence shown by swarm colonies of that species in UV light of 366 nm (Lampky and Brockman, 1977). The phenalamids are synthesized during growth and early stationary phases. When fermentation occurs in the presence of adsorber resin, the substance can be recovered practically quantitatively from the resin. A component analysis showed that the benzene ring comes from phenylalanine, the head group comes from alanine, and the chain is a pure polyketide composed of acetate and propionate units (Trowitzsch-Kienast et al., 1992). Myxalamid inhibits NADH oxidation in beefheart submitochondrial particles with a 50% inhibitory dose of 170 pmol/mg of protein, or 1.0 to 1.7 mol/mol of flavin mononucleotide (which is an essential constituent of complex I and thus a better reference point for inhibitory activity; Gerth et al., 1983). Inhibition by piericidin A in this experimental system occurred at 20 to 60 pmol/mg of protein and was thus three to seven times better than that with myxalamid B. Difference spectroscopy showed a specific block in the NADH: ubiquinol oxidoreductase. As is to be expected, myxalamid does not affect NADH oxidation by yeast mitochondria. The structural variants and the all-*trans* isomers all show biological activity.

Phenoxan (Fig. 6) was the first bioactive metabolite isolated from a *Polyangium* strain (Kunze et al., 1992a). It appears to be rather common among organisms of the *Polyangium spumosum-Polyangium fumosum* group. The antibiotic can be produced by fermentation on technical substrates, like skim milk powder or single-cell protein, with yields of up to 10 mg/liter. It is recovered from the cell mass by extraction with acetone. Phenoxan is a new γ-pyrone compound with an oxazole and an unsubstituted phenyl ring (Jansen et al., 1991). Structurally related substances from streptomycetes (aureothin, spectinabilin) are known; the pyrone system resembles the quinone part of ubiquinone. The antibiotic inhibits NADH oxidation in beefheart submitochondrial particles with a 50% inhibitory dose of about 30 ng/mg of protein. Difference spectra and the reduction kinetics of cytochrome *b* indicate a specific block in complex I (Kunze et al., 1992a).

Thiangazol (Fig. 4) is another example of a *Polyangium* compound (Kunze et al., unpublished data). In contrast to phenoxan, it is, however, a rare metabolite. The molecule consists of a very peculiar sequence of oxazole and thiazole rings with a benzene ring at one end (Jansen et al., 1992). Also, thiangazol is intracellular and can be obtained from the cell mass by acetone extraction. Its mechanism of action is similar to that of phenoxan (Kunze, personal communication).

Among the 50 known myxobacterial antibiotics are four new and specific inhibitors of eubacterial RNA polymerase (one not yet published; Fig. 8). This number is remarkable because such inhibition is a very rare mechanism of action: among the thousands of microbial compounds, a mere handful act on that important enzyme.

The first antibiotic of this type discovered in myxobacteria was myxopyronin (Irschik et al., 1983a). It is produced by *M. fulvus* Mx f50 and appears to be rare. Myxopyronin can be produced in lean peptone media with yields of up to 26 mg/liter. It is excreted and can be recovered by extraction of the culture broth with ethyl acetate. Myxopyronin is active against most gram-positive bacteria tested (MIC, 0.3 to 50 μg/ml), but a few gram-negative bacteria are also inhibited (MIC, 0.8 to 1.3 μg/ml). The producer strain is also moderately sensitive to its own antibiotic (MIC, 25 to 50 μg/ml, depending on the structural variant). Myxopyronin is an α-pyrone compound that occurs in two structural variants whose left side chains differ in length by one carbon atom (Kohl et al., 1983). Its building blocks have been identified as glycine, acetate, and methyl groups from methionine (Kohl et al., 1984). Labeled [^{13}C]acetate is readily taken up by the organism and incorporated into the antibiotic, whose synthesis is not inhibited below a concentration of 0.1% acetate and 0.05% glycine. The synthesis of myxopyronin is more complicated than it may seem (Fig. 9). The molecule actually consists of two polyacetate chains, one starting with glycine and the other starting with acetate, which are then connected head to head in the pyrone ring. While three of the methyl groups come from methionine, the fourth stems from C-2 of acetate. This is relatively unusual. The carboxyl group on the nitrogen appears to be introduced by a direct, biotin-catalyzed carboxylation.

Myxopyronin specifically blocks eubacterial DNA-dependent RNA polymerase (Irschik et al., 1983a). The *Escherichia coli* enzyme is inhibited in vitro with a 50% inhibitory concentration of 0.6 to 0.8 μg/U, whereas in vivo, the bacterium is insensitive to the antibiotic. The discrepancy may be explained by a penetration barrier in gram-negative bacteria. Even at high doses (250 μg/ml), a low, residual RNA synthesis remains

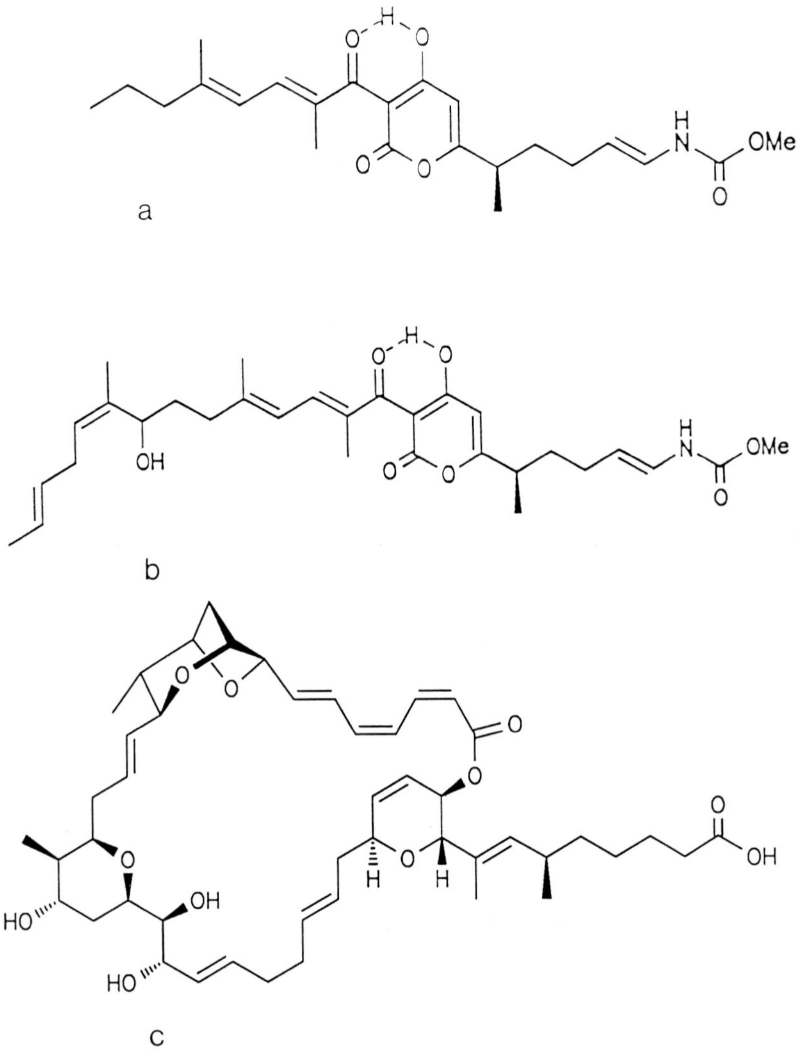

Figure 8. Myxobacterial inhibitors of bacterial RNA polymerase. (a) Myxopyronin A; (b) corallopyronin A; (c) sorangicin A. Me, methyl group.

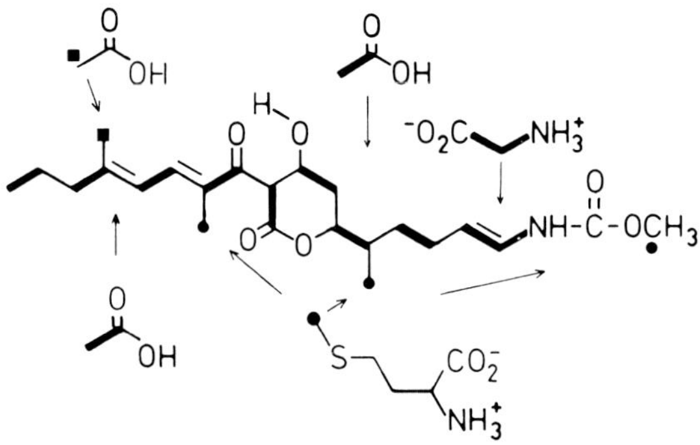

Figure 9. Building blocks of myxopyronin: glycine, acetate, and methyl groups from methionine. (Reprinted from Kohl et al. [1984] with permission.)

both in vivo and in vitro. Wheat germ RNA polymerase II does not respond to myxopyronin. The mechanism of inhibition is in all probability identical to that of corallopyronin (see below).

Corallopyronin (Fig. 8) was discovered in *Corallococcus (Myxococcus) coralloides* Cc c127 (Irschik et al., 1985). It shows up relatively often in strains of that species. The antibiotic can be produced with yields of up to 10 mg/liter by fermentation in a lean peptone medium enriched with 0.1% starch. The fact that the organism cannot utilize glucose and maltose yet metabolizes starch is somewhat astonishing. In this sense, its behavior is similar to that of *M. xanthus* (Dworkin, 1962). The explanation is that starch is degraded to trisaccharides, which are readily taken up and processed (Irschik and Reichenbach, 1985a). The compound is recovered by extraction of the culture supernatant. The corallopyronins (there are several structural variants) differ from the myxopyronins only in the left side chain, which itself varies in the two myxopyronins. In the corallopyronins, this side chain is considerably longer and is also modified by oxygen function and isomerization (Jansen et al., 1985). The corallopyronins and the myxopyronins have similar inhibitory spectra (Irschik et al., 1985), and also in this case, the producing strain is sensitive to its own metabolite (MIC, 50 μg/ml). Variations in the side chain have a substantial effect on the efficiency of the compound (MICs different by a factor of 5 to 15). In general, the corallopyronins with their longer side chains are the more efficient compounds, a tendency that has already been observed in the myxopyronins, where compound B, with a longer side chain, is superior to compound A. In addition, the stereoisomer A′ of corallopyronin A shows a clear increase in MICs.

Like the myxopyronins, the corallopyronins specifically block eubacterial RNA polymerase (Irschik et al., 1985), but in contrast to myxopyronin, corallopyronin can inhibit RNA synthesis by 100%. In *Staphylococcus aureus* cultures, all RNA synthesis stops within 10 min of the addition of 1 μg of corallopyronin A per ml. In vitro, *E. coli* RNA polymerase responds to the antibiotic immediately with a much reduced rate of RNA synthesis. This seems to indicate that corallopyronin, like streptolydigin and in contrast to rifamycin, stops RNA synthesis even after chain initiation.

The third RNA polymerase inhibitor from myxobacteria is sorangicin (Fig. 8), which is produced by *Sorangium cellulosum* So ce12 (Irschik et al., 1987). The compound has occasionally been observed in other strains of that organism but appears to be relatively rare. Sorangicin can be produced by fermentation in a glucose-mineral salts medium (CK6 l.m.) supplemented with 0.05% peptone, which leads to more homogeneous growth and improves the yield. The yield was initially low (around 1 mg/liter) but could be substantially increased to almost 1 g/liter (Irschik, personal communication). Fermentation lasts about 90 h. Sorangicin is excreted into the medium and can be obtained by extraction with ethyl acetate or by treatment of the supernatant with the adsorber resin XAD-1180 (Jansen et al., 1989a). Sorangicin is a relatively large macrocyclic lactone (the molecular weight of sorangicin A is 806) with an acidic side chain, 15 asymmetric carbon atoms, three internal six-membered ether rings, and one five-membered ether ring (Jansen et al., 1985, 1989a). The absolute configuration of the molecule has been determined (Jansen et al., 1989a). Despite its complex structure, sorangicin arises from a single polyacetate chain to which four methyl groups from methionine are added (Fig. 10). The starter molecule appears to be a malonyl coenzyme A unit in the side chain (Jansen et al., 1989c). In addition to the main compound, sorangicin A, an appreciable number of natural variants have been characterized, including two 21-*O*-β-D-glucosides and a 21-*O*-β-D-6′-deoxyglucoside, the sorangiosides (Jansen et al., 1989b). Because of its potential as an antibacterial drug, sorangicin has been studied intensively for some time. A large number of derivatives have been made, and the structure-activity relationship has been investigated (Jansen et al., 1990; Schummer, 1989; Schummer et al., in press). While no substantial improvement in the efficacy of the antibiotic could be achieved, certain derivatives showed equal activities, thus identifying sites in the molecule that could be modified, e.g., some esters and amides and certain substitutions of the hydroxyl on C-25 (Jansen et al., 1990), without reducing the potency of the drug.

Sorangicin inhibits growth of gram-positive bacteria very efficiently (MIC, between 8 and 125 ng/ml; Irschik et al., 1987). Gram-negative bacte-

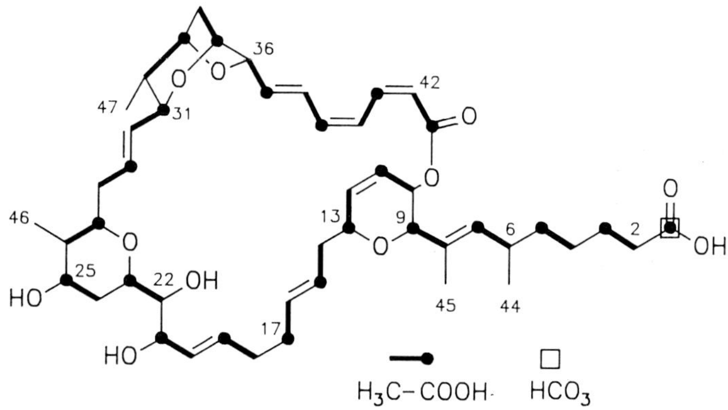

Figure 10. Building blocks of sorangicin: acetate and CO_2. (Reprinted from Jansen et al. [1989c] with permission.)

ria, including *Pseudomonas* spp., also respond but usually at much higher doses (MIC, 100 ng/ml to 35 μg/ml), obviously because of reduced penetration. The myxobacteria themselves are relatively sensitive to the antibiotic (MIC, 0.1 to 3 μg/ml), which parallels the exceptional sensitivity of those organisms to actinomycin D (Dworkin, 1969). The producer strain was rather resistant (MIC, 50 μg/ml), and even more resistant strains could easily be isolated, but other, nonproducing *S. cellulosum* strains showed the same high sensitivities as other myxobacteria. The sorangiosides (glycosides of sorangicin) are less active by a factor of 1,000 and more. In *S. aureus* cultures, all RNA synthesis comes to an end within 3 min of the addition of sorangicin A (50 ng/ml). In vitro, *E. coli* RNA polymerase responds to 20 ng of sorangicin A per ml with a 50% reduction in activity, and total inhibition is observed at 100 ng/ml, but only if the antibiotic is added some minutes after the reaction has started. This suggests that sorangicin blocks eubacterial RNA polymerase in the same way rifampin does, i.e., by interfering with chain initiation (Irschik et al., 1987). In fact, it was found that resistance to sorangicin is due to mutation of the RNA polymerase gene, with the consequence that the enzyme becomes resistant in vitro to the antibiotic (Römmele et al., 1990). There is high but not complete cross-resistance between rifampin and sorangicin; i.e., strains that have been selected for resistance to rifampin are often but not always resistant to sorangicin and vice versa. Apparently, the binding regions for the two antibiotics on the RNA polymerase overlap to a considerable degree. Further, it was found that sorangicin interferes with hepatitis B virus propagation in cell cultures, probably by blocking the viral polymerase (Galle and Theilmann, 1990).

An effort has also been made to produce sorangicin with alginate-immobilized cells (Becker, 1990). The cultures could be kept sterile for 700 h, at which time the bacteria finally degenerated within the beads. After the beads were loaded, the bacteria were allowed to colonize them for 70 h. After that, fresh medium was continuously pumped into the bioreactor, and the antibiotic yield in the effluent was determined. Production of sorangicin started 40 h after the medium exchange was initiated and continued until almost the end of the experiment. The time-volume production rate was up to three times better than with suspended cells. For *Cystobacter* and *Corallococcus* strains, an improvement of antibiotic production by 25 to 50% was reported when the cells were immobilized in κ-carrageenan or on nylon pads (Yasouri and Foster, 1991).

At least three myxobacterial antibiotics interfere with prokaryotic or eukaryotic protein synthesis. Myxovalargin (Fig. 11) is the best-studied inhibitor of this type. It is an example of a myxobacterial peptide antibiotic. Several more are known by now (all unpublished). The antibiotic was discovered in *M. fulvus* Mx f65 (Irschik et al., 1983b). It has also occasionally been observed in other myxobacteria, e.g., in *M. xanthus, C. coralloides,* and *Archangium gephyra* strains, but appears to be relatively rare. Myxovalargin inhibits gram-positive (MIC, 0.3 to 5 μg/ml) and gram-negative (MIC, 6 to 100 μg/ml) bacteria but is inactive against yeasts and molds. The producer strain was initially relatively sensitive (MIC, 4 to 6 μg/ml), but more-resistant strains could be selected (MIC, 20 to 50 μg/ml). Myxovalargin is unusual in that the MIC depends very drastically on cell density in the test culture; the MICs rise by a factor of 2 to 5 when the cell density increases from 10^5 to 10^7 cells per ml. Mice infected with

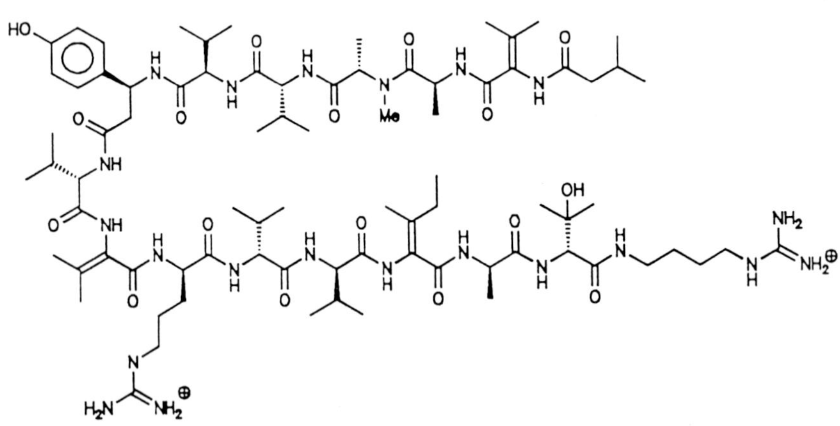

Figure 11. Myxovalargin A, a peptide antibiotic from myxobacteria. Me, methyl group.

S. aureus and *E. coli* could be cured with relatively low doses of myxovalargin (e.g., 2 × 0.4 mg/kg s.c.), but for reasons that will be explained below, the drug is also rather toxic for animals. For the mouse, the LD_{50} is 10 mg/kg s.c. and the LD_{100} is 30 mg/kg. Myxovalargin can be produced in a 1% peptone medium, with yields of up to 30 mg/liter. It is excreted during growth phase and can be recovered by butanol extraction of the culture supernatant. A typical fermentation takes 2 days.

In cultures of *S. aureus*, all protein synthesis stops at once when 3 μg of myxovalargin per ml is added (Irschik et al., 1983b). In an *E. coli* in vitro system, protein synthesis is reduced 50% by 1 μg and 100% by 10 μg of myxovalargin per ml (Irschik and Reichenbach, 1985b), but protein synthesis responds only very gradually if the antibiotic is added 5 min after the reaction has been started. Eukaryotic in vitro protein synthesis can also be inhibited but only to a moderate extent and at much higher doses (around 300 μg/ml; wheat germ and rabbit reticulocyte systems). Myxovalargin appears to act on the A site of the bacterial ribosome and to block the transpeptidation reaction by interfering with the binding of aminoacyl-tRNA much as lincomycin and spectinomycin do. It binds tightly enough to ribosomes to be removed from the incubation mixture by centrifugation. The specific effects of myxovalargin just described are, however, seen only when the antibiotic is applied at low concentrations around the MIC (Irschik and Reichenbach, 1985b). In *Bacillus subtilis* cultures, RNA synthesis stops instantaneously upon the addition of 20 μg of myxovalargin per ml (rather than the MIC of 0.5 μg/ml). Further, UV-absorbing material is released from the cell, and respiration breaks down. This suggests that at high doses, myxovalargin destroys the cell membrane. In fact, erythrocytes respond with hemolysis upon treatment with myxovalargin. Yeasts and molds show no membrane effects, probably because of their compact cell walls.

Myxovalargin is a linear tetradecapeptide blocked at both ends with nonamino acids with the following sequence (from the amino end): isovaleric acid–dehydro-D-Val–D-Ala–*N*-methyl-L-Ala–D-Val–D-β-Tyr–L-Val–dehydro-D-Val–D-Arg–D-Val–D-Val–dehydro-DL-Ile–D-Ala–hydroxy-D-Val–agmatine (decarboxylated Arg; Steinmetz et al., 1989). Several structural variants have been found. Because of unusual components, a normal degradation of the molecule was impossible, and elucidation of the structure took a long time (Steinmetz et al., 1989). The biosynthesis has also been studied and appears to occur on a huge, 280,000-Da multienzyme complex (Schulz, 1984). Inhibitors of protein synthesis do not interfere with myxovalargin synthesis. Curiously, similar large, sodium dodecyl sulfate-resistant proteins could also be demonstrated in other *M. fulvus* strains from which peptide antibiotics are not known. The origins and times of incorporation of the various components into the antibiotic were determined. Low but definite myxovalargin production (2 mg/liter) could be achieved in a fully defined amino acid-pyruvate medium, but efforts to synthesize myxovalargin with an in vitro system were not successful.

Another myxobacterial inhibitor that acts on bacterial protein synthesis is angiolam (Fig. 12). The compound was discovered in *Angiococcus disciformis* An d30 (Kunze et al., 1985). The same strain also produces 6 mg of myxothiazol per liter (which remains in the cell mass). Angiolam appears to be very rare among myxobacteria. The fermentation is best performed in the presence of an adsorber resin, e.g., 1% Amberlite XAD-1180, in a medium based on single-cell protein, at 30°C, and a partial O_2 pressure between 90 and 30% saturation. Fermentation lasts about 40 h. Yields of up to 6 mg/liter could be obtained, and 90% of the material was bound to the resin. Angiolam is moderately active on some gram-positive bacteria (MIC, 0.8 μg/ml). Among the gram-negative bacteria, only mutant strains with increased permeability (MIC, 2.5 μg/ml) and spheroplasts were sensitive. This clearly indicates that the antibiotic has difficulty penetrating cells. Yeasts and molds did not respond. No acute toxicity for mice was observed at doses of up to 300 mg/kg s.c. The inhibitory effect is bacteriostatic. All protein synthesis stopped immediately upon the addition of angiolam to cultures of *Bacillus thuringiensis,* while other macromolecular syntheses continue for some time. The exact biochemical mechanism of the block has not been worked out. Angiolam is a highly oxidized macrocyclic lactone-lactam ring with a side chain (Kohl et al., 1985) whose absolute configuration was determined by X-ray analysis. The building blocks are alanine, acetate, and propionate; propionate is responsible for all the methyl groups (Fig. 12). A few structural variants are also produced in minor quantities.

Only a few myxobacterial compounds are known to interact with DNA or interfere with DNA synthesis. The best-known example is saframycin, which was, however, originally discovered in a streptomycete (Arai et al., 1977, 1980). Also, all studies on the mechanism of action of saframycin have been performed with the streptomycete substances. The myxobacterial saframycins have been found in *M. xanthus* Mx x48 (Irschik et al., 1988). While they are occasionally also observed in other strains of myxobacteria, they are relatively rare (Table 3). Saframycin can be produced in a lean peptone medium and is excreted into the medium, where it can be collected on an adsorber resin that is added after or, preferably, during fermentation. The producer strain is sensitive to its own antibiotic (MIC, 15 μg/ml), and highly resistant strains have not yet been obtained. Yields are between 5 and 20 mg/liter. Saframycin Mx1 is a very efficient inhibitor of many gram-positive-bacteria (MIC, 1 to 500

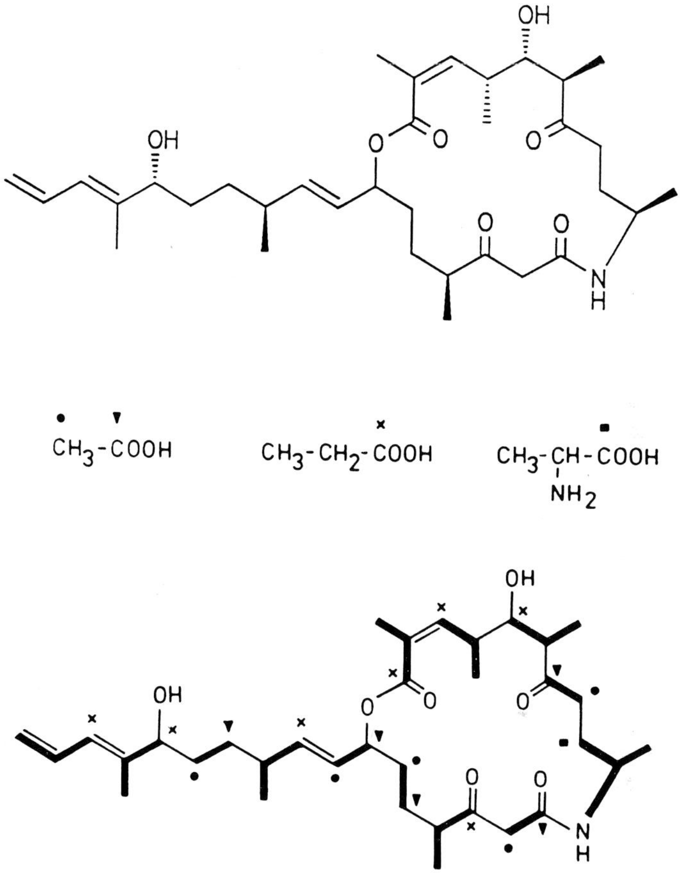

Figure 12. Angiolam and its building blocks: alanine, acetate, and propionate. (Reprinted from Kohl et al. [1985] with permission.)

ng/ml) but is less efficient with gram-negative bacteria, including various myxobacteria (MIC, 1 to 20 μg/ml). Also, archaebacteria of the genus *Halobacterium* are very sensitive. Saframycin Mx1 does not act on yeasts and molds, but animal cells respond to very low doses. Like some of the streptomycete compounds, the myxobacterial saframycins show interesting antitumor effects. Chemically, the saframycins can be regarded as modified tyrosine diketopiperazines. The aromatic rings are transformed into two quinone-hydroquinone systems (Fig. 2). The myxobacterial compounds carry an OH group on C-21 (Trowitzsch et al., 1988a). This hydroxyl is essential for the efficiency of the compound and can easily be replaced in a chemical reaction by a cyano group. The cyano compound is equally efficient but more stable. The absolute configuration has also been determined (Trowitzsch et al., 1988a). The saframycins decompose in the light and at pHs above 7. The myxobacterial saframycins differ from the various streptomycete compounds in that they comprise a different combination of substituents. The side chain ends with alanine instead of pyruvic acid (this can be achieved in the streptomycete only by mutasynthesis [Arai et al., 1985]). Further, the OH group on C-21 is present only on saframycin S, and the methoxy group on C-14 is present only on saframycin C.

In *S. aureus* cultures, saframycin Mx1 at a dose of 0.5 μg/ml stops DNA, RNA, and protein syntheses instantaneously (Irschik et al., 1988). However, dose-response curves show that protein synthesis is somewhat less sensitive than DNA and RNA syntheses, and in an *E. coli* in vitro system, protein synthesis is not inhibited at all by up to 50 μg of saframycin Mx1 per ml. The *E. coli* RNA polymerase reaction is, however, inhibited, and oxidized and reduced Mx1 saframycins are equally effective. These results clearly indicate that myxobacterial saframycin acts in the same way as the highly efficient streptomycete saframycins A and S (Kishi et al., 1984). The various streptomycete compounds show big differences in efficiency, which can be explained by three different reactions with DNA. While all saframycins weakly bind to DNA through electrostatic forces after having attached themselves in the minor groove of the double helix (single-stranded DNA is not attacked), the highly efficient variants may then form a covalent bond between C-21 and the amino group of a guanine. Further, in the presence of oxygen, reduced saframycins produce single-strand breaks by creating oxygen and hydroxyl radicals (see, e.g., Ishiguro et al. [1981], Lown et al. [1982], and Hill and Remers [1991]).

An example of a myxobacterial compound with a novel mechanism of action is soraphen (Fig. 13; Reichenbach et al., 1988b; Gesellschaft für Biotechnologische Forschung, Wissenschaftlicher Ergebnisbericht 1988:

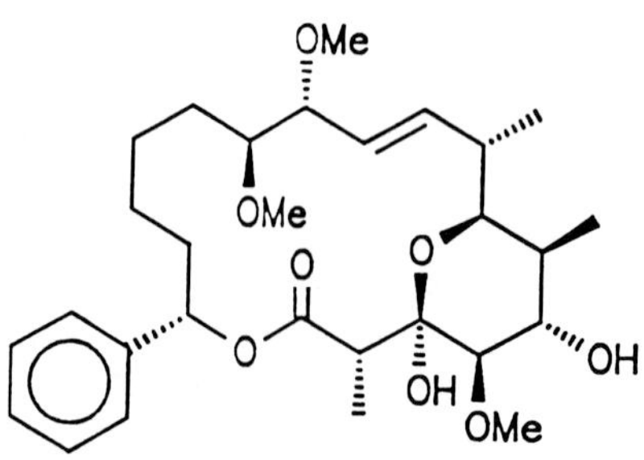

Figure 13. Soraphen $A_{1\alpha}$. Me, methyl group.

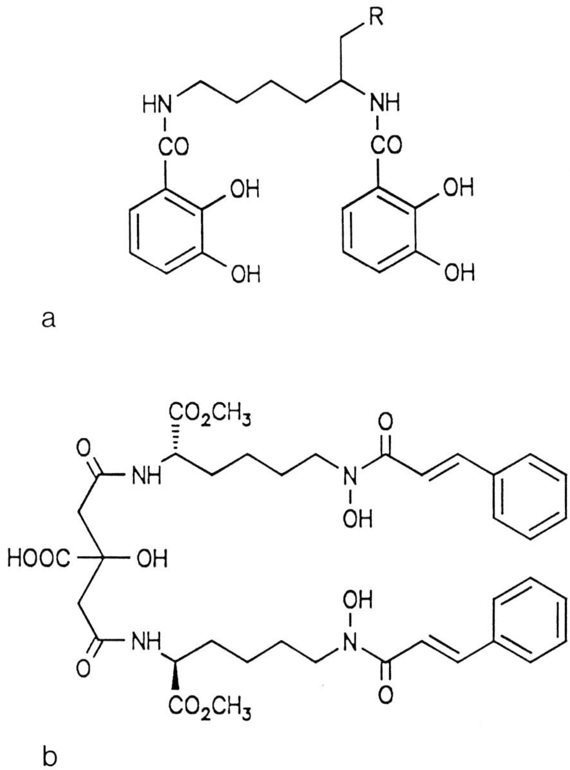

Figure 14. Myxobacterial siderophores. (a) Myxochelin A (R = OH); (b) nannochelin A.

30). Soraphen is a macrocyclic lactone with an internal ether ring and an unsubstituted phenyl ring in a short side chain. It was discovered in *S. cellulosum* So ce26 but has occasionally also been observed in other strains of that species, although never in another species (Gerth, personal communication). Soraphen is a potent antifungal inhibitor. It specifically blocks fungal acetyl coenzyme A carboxylase and is completely inactive against the corresponding enzyme in the chloroplasts of higher plants (Pridzun, 1992). It also inhibits acetyl coenzyme A carboxylase from rat liver.

Three compounds isolated from myxobacterial cultures very probably serve as siderophores for the producers (Fig. 14). The first of these is myxochelin A (and structurally closely related myxochelin B). Myxochelin A was discovered in *A. disciformis* An d30 (which also synthesizes angiolam and myxothiazol; Kunze et al., 1989). It has also been observed in another strain of *A. disciformis* but not in any of 28 other myxobacteria and thus may be species or genus specific. The compound is produced at relatively high yields (44 mg/liter) when the organism is cultivated in an iron-depleted peptone medium (the optimum is 10^{-7} M $FeCl_3$). Myxochelin is excreted into the medium and can be recovered by extraction with ethyl acetate at pH 4.0. Its isolation is much facilitated by a transient permutation of myxochelin into its water-soluble Cu(II) complex, which allows removal of contaminating material. In the agar diffusion assay, myxochelin is slightly active against gram-positive bacteria (the MIC for *S. aureus* is 25 μg/ml), probably because it sequesters iron or other essential ions in the medium. Myxochelin is a classical catechol siderophore. A similar compound, azotochelin, has been reported in *Azotobacter vinelandii* (Corbin and Bulen, 1969).

The second example, nannochelin (Fig. 14), comes from *Nannocystis exedens* Na e485, which belongs to the second suborder of the myxobacteria (Kunze et al., 1992b). The compound has also been demonstrated in many other strains of *N. exedens* and *Nannocystis aggregans* but so far in no other myxobacterium. It can be produced by fermentation in a lean peptone medium for 90 h and is recovered from culture broth by adding the adsorber resin XAD-1180 (*Nannocystis* strains do not grow in the presence of the resin [Kunze, personal communication]). Yields are more than 30 mg/liter. Iron deficiency does not stimulate biosynthesis. Nannochelin, too, shows weak antimicrobial activity against gram-positive bacteria and a few yeasts and molds. The MICs for *Brevibacterium ammoniagenes* are 1.5 μg/ml (nannochelins A and B) and 0.4 μg/ml (nannochelin C). Nannochelin is a citrate-hydroxamate-type siderophore like schizokinen *(Bacillus megaterium),* arthrobactin *(Arthrobacter pascens),* and aerobactin (*Aerobacter* strains; for a review, see Neilands and Ratledge [1982]). Three structural variants of nannochelin differ in the methyl ester groups on the carboxyls of the hydroxylysine part of the molecule.

While this has not yet been rigorously proven, the compounds mentioned above almost certainly function as siderophores for the respective myxobacteria. Myxochelin synthesis by *Angiococcus* spp. indeed responds to the iron content of the medium, as is characteristic for siderophores of many organisms. A siderophore effect has in fact been demonstrated for both compounds in certain enterobacteria, where the myxobacterial substances can replace the indigenous compounds and obviously are recognized by the special transport systems developed in those bacteria (Rabsch and Winkelmann, 1991). Thus, the enterobactin transporter FepA appears to recognize and transport myxochelin. Nannochelin is used by only a few organisms and is recognized by an undefined aero-

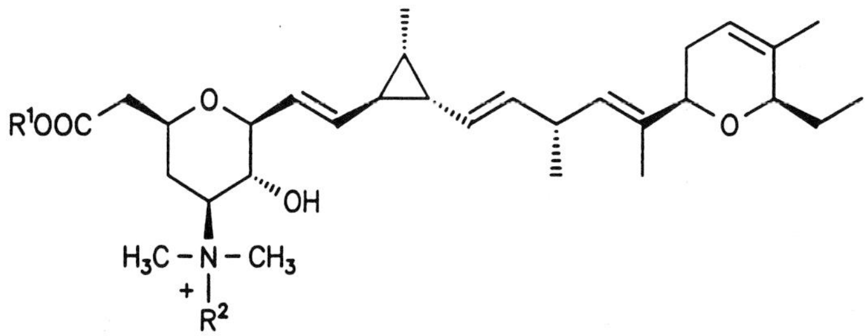

Figure 15. New ambruticins VS-2 (R^1 and R^2 = CH_3) and VS-3 (R^1 and R^2 = H).

bactin receptor that is 2,000 Da larger than the usual aerobactin receptor, Iut (Rabsch et al., 1991). Very probably, further compounds involved in iron transport can be discovered in other myxobacteria.

For a number of myxobacterial secondary metabolites, the mechanism of action has been explained only hypothetically or not at all. Ambruticin has already been mentioned as the first myxobacterial antibiotic whose chemical structure has been determined (Fig. 1). Recently, a series of homologs with a (substituted) NH_3^+ group on one of the ether rings has been described (Fig. 15; Höfle et al., 1991). Ambruticin was discovered in a strain of *S. cellulosum* (then named *Polyangium cellulosum* [Ringel et al., 1977]); the novel ambruticins came from *S. cellulosum* So ce10. Many more strains of that organism but no other myxobacteria have since been found to produce these antibiotics (Gerth, personal communication). The ambruticins are obtained by fermentation in a peptone (or soybean concentrate)-starch medium at 32°C (Ringel et al., 1977). The highest titers are reached at about 96 h. The antibiotic is excreted and can be recovered by extraction of the broth with ethyl acetate. Alternatively, fermentation can be performed in the presence of an adsorber resin, e.g., XAD-1180, and the antibiotic can be recovered by elution of the resin with 4 bed volumes of methanol (Höfle et al., 1991). The ambruticins are purified by column chromatography on silica gel. In the original study, the main compound, ambruticin S, constituted 70 to 90% of a mixture of homologs. The total yield in our fermentations was between 50 and 90 mg/liter.

The ambruticins are active against fungi, including many human pathogens, although often with unfavorable MICs (Ringel et al., 1977; Ringel, 1978; Shadomy et al., 1978a). A weak inhibitory effect on a few gram-positive bacteria (MIC, 12 to 50 μg/ml) has also been reported (Ringel et al., 1977). The acute toxicity for the mouse is very low (LD_{50}, 315 mg/kg intravenously and more than 1,000 mg/kg perorally). In vitro experiments showed good activity against dermatophytes (MICs were often between 5 and 150 ng/ml, but many strains are more resistant [Ringel, 1978; Shadomy et al., 1978a]). Also, many clinical isolates respond well, in particular *Histoplasma capsulatum, Coccidioides immitis, Blastomyces dermatitidis,* and *Mucor* species (MIC, 10 to 250 ng/ml). The efficiency of ambruticin is strongly influenced by the pH (the antibiotic becomes much less active at a pH above 7) and by the presence of (human) serum. Guinea pigs infected with *Trichophyton mentagrophytes* could be completely cured by oral treatment with 40 mg/kg twice daily for 10 days. Topical treatment gave even better results (Ringel, 1978). Mice infected with *Coccidioides immitis* have been cured with ambruticin (Levine et al., 1978). Also *H. capsulatum* responded quite well, although at a much higher dose than was required with amphotericin B (150 versus 25 mg/kg and day [Shadomy et al., 1978b]).

Depending on the organism, resistance to ambruticin may or may not develop very quickly between two transfers (Ringel, 1978). At growth temperature, but not at 4°C, the effect of the antibiotic is fungicidal (Ringel, 1978). In *Microsporum fulvum,* the antibiotic does not cause leakage of cell contents. The incorporation of precursors into RNA and protein is immediately stopped by 0.2 μg of ambruticin per ml. This could suggest interference with mRNA synthesis. A study of ambruticin effects in *Candida parapsilosis* showed that amino acid incorporation is not blocked if 10 mM acetate or succinate is present in the medium. Glucose, maltose, and glycerol had no such effect. The conclusion was that the antibiotic inhibits carbohydrate utilization and thereby impedes energy generation and transport processes (Simpkin, 1985).

Chemically, ambruticin is very interesting because it represents a new class of antibiotics. A substantial number of derivatives have been prepared (Connor and von Strandtmann, 1979a, 1979b; Connor et al., 1979), and after 10 years of struggle, a total synthesis has also been achieved (Kende et al., 1990).

Myxovirescin (Fig. 16) has been discovered in *Myxococcus virescens* Mx v48 (Gerth et al., 1982a). It has also been observed in many other strains of that species, many *M. xanthus* strains, and a few *M. fulvus* strains and appears to be a rather common myxobacterial antibiotic. Myxovirescin fermentation may be performed in a corn steep-soy meal medium at 32°C and a constant pH of 7.4 with an aeration rate of 0.2 volume of air per volume of medium per min. Fermentation is finished after 60 h. The antibiotic is produced toward the end of growth phase, when the oxygen level in the culture begins to rise again. It is excreted into the medium, from which it is recovered by ethyl acetate extraction, and is purified by preparative HPLC. Alternatively, fermentation can be performed in the presence of the adsorber resin XAD-1180; the antibiotic

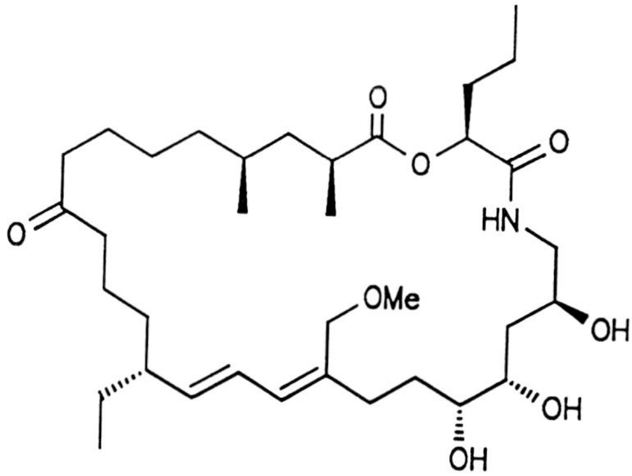

Figure 16. Myxovirescin A_1. Me, methyl group.

is then harvested on the resin (Trowitzsch-Kienast et al., 1989). Yields are in the range of 10 to 30 mg/liter. Myxovirescin is produced as a mixture of about 20 different structural variants, in which myxovirescin A is far in excess.

Substantial efforts have been made to understand the controlling factors of biosynthesis and to optimize fermentation and recovery (Steinmetz et al., 1984; Gerth and Reichenbach, 1986; Hecht et al., 1987, 1990). Determination of intracellular NH_3-NH_4^+ concentrations showed that with rising levels in the medium, intracellular levels became surprisingly high, up to 140 mM (Gerth and Reichenbach, 1986). This obviously must have serious consequences for intracellular pH and enzyme activities. Ammonium accumulation in the medium cannot be avoided, because the organism does not utilize carbohydrate. Therefore, other strategies were developed to control the NH_3-NH_4^+ level in the medium. One elegant possibility, although only on an experimental scale, is to extract the ammonia on-line from the culture. This can be achieved with a hollow-fiber module consisting of a hydrophobic membrane that allows the passage of gases only. The effects are dramatic: cell dry mass increases by a factor of 2.6, and the antibiotic yield increases by a factor of 8 (Hecht et al., 1990). Another possibility is a gradual feeding of peptone to a batch culture to prevent any unnecessary metabolism of amino acids (Nigam et al., 1984). In fact, it became quite clear in these experiments that there was virtually no production during exponential growth and that some growth limitation was required in order to induce antibiotic synthesis. This could be achieved by keeping the peptone content of the medium at a certain level. In a 7.5-liter batch with an initial peptone concentration of 1.6 g/liter, optimum production was obtained with a feeding rate of 1.5 g of peptone per h. The respiratory quotient during the production phase was close to 0.8. The relation between specific production rate and specific peptone uptake rate showed a very sharp optimum between 0.15 and 0.20 g of peptone used per g of cell mass (dry weight) per h. In chemostat cultures, continuous removal of the antibiotic from the culture increased the volumetric production rate from 0.12 to 0.55 mg/liter and h, i.e., by a factor of 4.5 (Hecht et al., 1987). The removal was achieved by passing the culture through a vortex chamber filled with the adsorber resin XAD-1180. While in the beginning about 25% of the produced myxovirescin remained inside the cells, the antibiotic content in the cells and in the medium fell almost to zero after the adsorption chamber was connected to the bioreactor. A method of recovering myxovirescin from larger culture volumes (4,700 liters) was developed (Steinmetz et al., 1984). At the end of the fermentation, 15 liters of adsorber resin XAD-1180 was added to the culture with continuous stirring, and then another 4 liters of fresh resin was added 4 h later. Five hours after the adsorption procedure was started, all myxovirescin was on the resin, which could be separated from the cells and culture broth on a sieve (0.25-mm pore size). The resin was packed into a column and eluted with methanol-water (1:1) and pure methanol.

The producer strain Mx v48 was initially rather sensitive to its own antibiotic (MIC, 0.5 μg/ml; 50% growth inhibition at 0.25 μg/ml), but more-resistant mutants (MICs, 30 to 40 μg/ml) could be obtained by treatment with *N*-methyl-*N*′-nitro-*N*-nitrosoguanidine (NTG) (Gerth et al., 1982a). Strains with improved yields were obtained by selection of mutants that grew with elevated (0.5%) levels of $(NH_4)_2SO_4$ in the medium. Myxovirescin A is especially active against gram-negative bacteria and in particular against enterobacteria (MIC, ~1 μg/ml). At higher concentrations, even pseudomonads are inhibited (MIC, 30 μg/ml). Gram-positive bacteria are less sensitive (MIC, 30 μg/ml), and yeasts and molds do not respond. Dose-response curves for the various bacteria differ in a characteristic way (Gerth et al., 1982a). In gram-negative bacteria, there is little inhibition until a threshold value is reached, but then maximum inhibition is reached within a narrow dose interval. In gram-positive bacteria, e.g., *S. aureus*, there is a broad dose range starting at rather low concentrations within which a gradually increasing growth inhibition is observed. For *E. coli*, myxovirescin is lethal within 30 min. The inhibitory effect is counteracted by Ca^{2+} ions; at 0.1% $CaCl_2 \cdot 2H_2O$ (7 mM), the organism becomes essentially resistant. Mg^{2+} and Ba^{2+} have no such effect, nor is Ca^{2+} protective for *S. aureus*. The addition of EDTA, on the other hand, potentiates the effect of myxovirescin on *E. coli* and *P. aeruginosa*. In the latter, the MIC is reduced from 30 to 7 μg/ml by 0.01% EDTA. *E. coli* can be further protected against myxovirescin by (bovine) serum. The interfering component in the serum appears to be the phospholipids. This may explain why myxovirescin is ineffective in curing experimental infections in animals.

The mechanism of action of myxovirescin is not yet established. Incorporation experiments with *E. coli* appear to indicate a specific effect on peptidoglycan synthesis (Gerth et al., 1982a).

Chemically, myxovirescin is a relatively polar macrocyclic lactone-lactam peptolide antibiotic (Trowitzsch et al., 1982). Myxovirescin A occurs as two epimers, A_1 and A_2, whose constitutions and absolute configurations have been determined (Trowitzsch et al., 1985). Further, about 20 natural structural variants have been isolated from one strain and fully characterized (Trowitzsch-Kienast et al., 1989). The molecule consists of glycine (as the starter molecule), acetate, and methyl groups from methionine (Trowitzsch et al., 1983). One C-2 unit has a relatively unusual origin (Trowitzsch-Kienast et al., 1989). It comes from methylmalonate, which is obviously produced from succinate by methylmalonate mutase (Fig. 17). Partial and total syntheses of myxovirescin have also been achieved (Borgschulte, 1986; Borgschulte et al., 1990; Williams and McGill, 1990; Seebach et al., 1991).

Myxovirescin has a long and tortuous history. Its first appearance in the literature is very probably as antibiotic TA isolated from *M. xanthus* TA (Rosenberg et al., 1973; Vaks et al., 1973). While no chemical structure has yet been proposed for antibiotic TA and the early chemical data (UV, infrared spectrum, molecular weight, molar extinction coefficient) do not fit the myxovirescin data particularly well or at all, later chemical results (Rosenberg et al., 1982) make the identity with myxovirescin A_1 almost certain (Trowitzsch et al., 1985), in spite of differences in the proposed molecular formulas and partial structures that appear to have arisen from erroneous interpretation of the data. More recently, Japanese investiga-

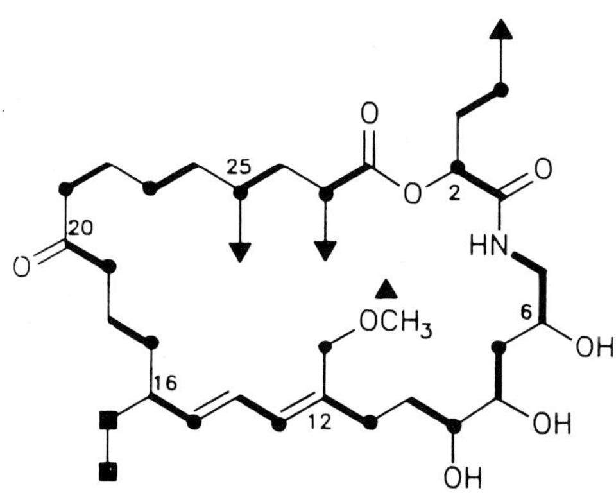

Figure 17. Building blocks of myxovirescin: glycine, acetate (●—), succinate-methylmalonate (■—■), and methyl groups from methionine. (Reprinted from Trowitzsch-Kienast et al. [1989] with permission.)

tors isolated myxovirescin A and 26,27-dehydromyxovirescin A, which is identical to our myxovirescin B, from an *M. xanthus* strain (Onishi et al., 1984). Another Japanese group described myxovirescin A_1 and previously published structural variants of myxovirescin under an entirely unjustified new name, megovalicin, which they obtained from an *M. xanthus* or *M. virescens* strain for which they created a new species, *Myxococcus flavescens* (Miyashiro et al., 1988; Takayama et al., 1988). Curiously, in this new species, all of the antibiotic was intracellular, but as with Mx v48, biosynthesis took place in late growth and early stationary phases. Despite statements to the contrary, the new compounds are no more active against *P. aeruginosa* than myxovirescin A is, as can be seen in the original publication (Gerth et al., 1982a).

There are many parallels in the biologies of antibiotic TA and myxovirescin (Rosenberg et al., 1973; Vaks et al., 1973). Thus, the biosynthesis of antibiotic TA is sensitive to elevated peptone concentrations (almost complete inhibition at 2% Casitone), it occurs in late log and early stationary phases, and the substance is found in the culture liquid. Antibiotic TA is more active against gram-negative than gram-positive bacteria, and its effect on *E. coli* is bactericidal: the bacterium begins to lyse after 30 min. The killing effect is, however, suspended when the bacteria are treated in the presence of chloramphenicol, i.e., if they are not growing. In contrast to myxovirescin, antibiotic TA did not act against *Serratia marcescens* and *Pseudomonas fluorescens,* but this failure may be a matter of dose. As in the case of myxovirescin, RNA and DNA syntheses do not respond for a considerable time, and there appears to be a block in cell wall synthesis. Later studies corroborated that antibiotic synthesis is inhibited by peptone concentrations above 1% (Fytlovitch et al., 1983). Synthesis could, however, be stimulated 100% by the addition of alanine and serine, an effect that is not seen in Mx v48. The labeled amino acids were incorporated into the antibiotic, obviously after transformation into acetate. Glycine was also incorporated, apparently directly, which again would fit the structure of myxovirescin. The production of antibiotic TA was scaled up to 750 liters (Mizrahi et al., 1976). The fermentation lasted 4 days, and the substance was harvested either by extracting the cell-free culture supernatant with chloroform or by adsorbing the antibiotic to kaolin, which was then extracted with 20% methanol in chloroform. Recovery was around 90%. An investigation into the mechanism of action of antibiotic TA suggests that the inhibitor interferes with the polymerization reaction during peptidoglycan synthesis (Zafriri et al., 1981).

Some of the most interesting and potentially important properties of antibiotic TA are its abilities to bind tightly to a variety of tissues and surfaces and to retain its antibacterial activity while in the bound form

(Rosenberg et al., 1984). Antibiotic TA binds to dental hard tissues (Manor et al., 1985), and this binding has been used to treat severe gingivitis and to reduce plaque in humans (Manor et al., 1989).

Pyrrolnitrin (Fig. 2) is an intensively studied compound, but all studies have been performed with material from pseudomonads or chemical syntheses. Here, we shall deal only with the myxobacteria.

In myxobacteria, pyrrolnitrin was discovered in *M. fulvus* Mx f147 (Gerth et al., 1982b). It has since also been found in strains of *Corallococcus exiguus, Cystobacter ferrugineus, Cystobacter minor,* and *Melittangium lichenicola* and thus appears to be not uncommon in myxobacteria. The antibiotic can be produced by fermentation in a lean peptone medium at 32°C. Harvest is after 4 days, and yield is 4 mg/liter (not optimized). Since 70% of the substance is in the supernatant and 30% is within the cells, the two fractions are separated and extracted independently with ethyl acetate and acetone, respectively. Purification is by column chromatography on silica gel. Two compounds are obtained: pyrrolnitrin and its immediate precursor, aminopyrrolnitrin. The production behavior of the myxobacteria differs somewhat from that of the pseudomonads. While the former synthesize the antibiotic strictly in growth phase and excrete a substantial proportion into the medium, the latter begin production only after growth has stopped, and all of their antibiotic is contained in the cells.

Pyrrolnitrin shows good activity against yeasts and molds and especially against phycomycetes (MIC, 0.1 to 0.2 μg/ml). Its effect is fungistatic. Bacterial growth may also be inhibited but at much higher doses (MIC, at least 50 to 100 μg/ml). Reports on the mechanism of action vary considerably. Pyrrolnitrin clearly inhibits respiration and electron transport in several systems, but this may not be the primary effect. In comparison with other electron transport inhibitors, the antibiotic is not particularly efficient, and other effects, e.g., swelling of mitochondria, are observed at much lower concentrations (Wong and Airall, 1970; Wong et al., 1971; Kawai et al., 1983).

For althiomycin also (Fig. 2), we restrict our discussion to myxobacteria. The compound has been found in *Cystobacter fuscus* Cb 685 but has also been observed in strains of *M. virescens* and *M. xanthus* (Kunze et al., 1982). It appears, however, relatively rarely in myxobacteria. Fermentation can be performed in 1% peptone medium. The antibiotic is excreted into the medium during growth phase. A maximum yield of 0.7 mg/liter is achieved after 2 to 3 days. Recovery is by ethyl acetate extraction of the cell-free broth. Althiomycin inhibits certain gram-positive and gram-negative bacteria (MIC, 0.7 to 6 μg/ml) and seems to interfere with

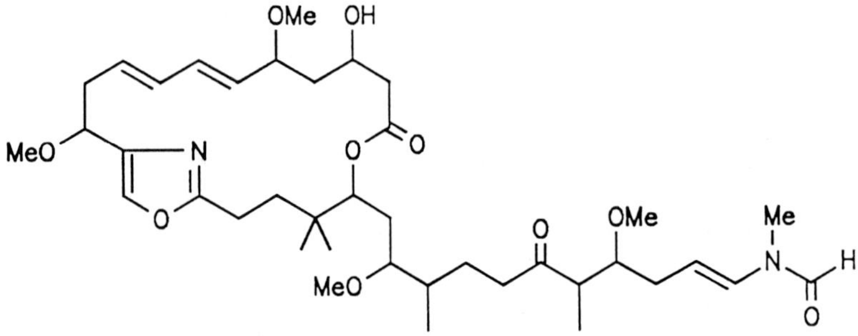

Figure 18. Rhizopodin. Me, methyl group.

protein synthesis, perhaps by blocking the puromycin reaction (Fujimoto et al., 1970).

A very fascinating recent discovery is rhizopodin (Fig. 18). The compound was tracked down after it was observed that culture extracts from *M. stipitatus* Mx f164 induced mouse fibroblasts to form long, branching, rhizopodium-like extensions (Fig. 19; Sasse et al., 1993). The responsible agent was isolated and chemically characterized. Rhizopodin is a macrocyclic lactone with an oxazole ring and a side chain that ends with a formamide group. Somewhat related structures (for which, however, no such effects are described) are known to occur in cyanobacteria, nudibranch molluscs, and sponges. Rhizopodin fermentations are performed in a lean peptone medium in the presence of the adsorber resin XAD-16. Synthesis occurs during growth phase, and at harvest, 5 days later, all of the antibiotic is on the resin. The compound is recovered by elution with methanol. Yields are very low (0.5 to 1 mg/liter). Rhizopodin does not inhibit bacteria and is only weakly active against a few yeasts and some molds. Animal cells, however, are very sensitive (inhibitory concentration, 12 to 30 ng/ml). Adherently growing cell lines respond within hours by producing the long runners mentioned above. This effect is irreversible, and the cells eventually die, but there are always some cells in the culture that are not affected and continue to grow. Cell lines that grow in suspension do not form the rhizopodium-like extensions but instead develop bubbles on their surfaces. Such cell lines are also killed, however. The observed effects suggest that rhizopodin may interfere in some way with the cellular protein phosphorylation system.

In addition to bioactive substances or, more exactly, substances for which a biological effect has been recognized, secondary metabolites that appear to be biologically inert have been isolated from myxobacteria and chemically characterized. One example is the isoquinoline alkaloid iso-

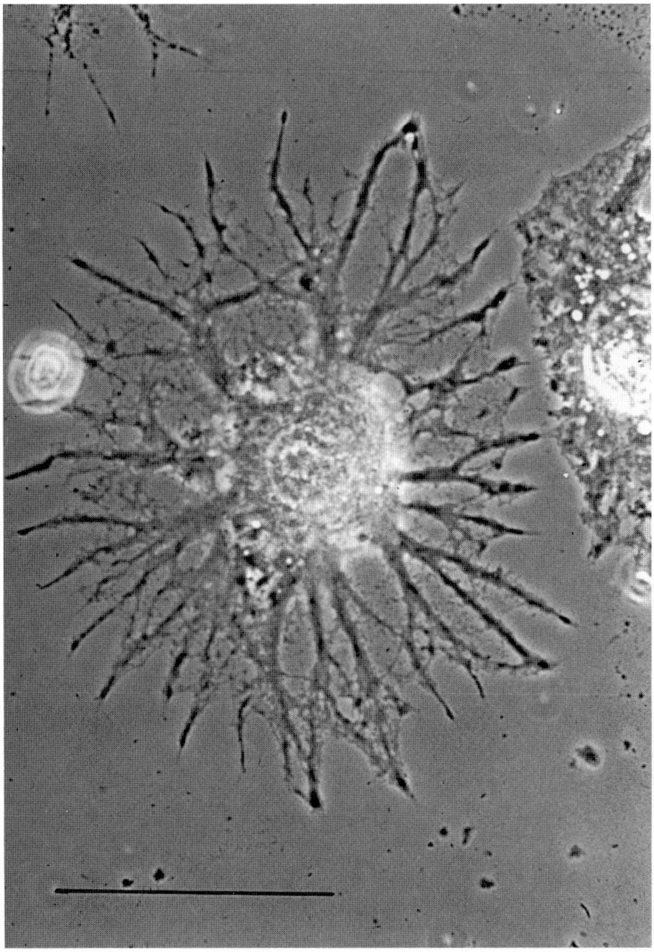

Figure 19. Effect of rhizopodin on L929 mouse fibroblasts. Bar = 100 μm. (Courtesy of F. Sasse, Gesellschaft für Biotechnologische Forschung.)

lated from *M. xanthus* Mx x48, the producer of saframycin (Fig. 20; Trowitzsch-Kienast et al., 1988b). The compound is structurally related to saframycin and may actually arise through a side reaction.

More intriguing, perhaps, is a collection of indole compounds isolated from various myxobacteria that probably arise from tryptophan metabolism. Certain indole derivatives have been found to be very efficient inducers or inhibitors of myxospore formation in *S. aurantiaca* (Gerth, 1975; Gerth et al., 1993).

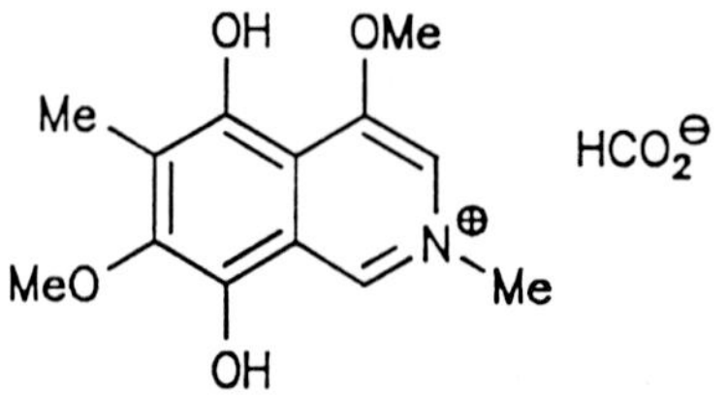

Figure 20. Isoquinoline compound from *M. xanthus*. Me, methyl group.

A few antibiotic activities described in myxobacteria in recent years and in some detail still are not substantiated by a chemical structure. The case of antibiotic TA has already been discussed in connection with myxovirescin. *C. (M.) coralloides* D produces a substance that inhibits gram-positive bacteria (Arias et al., 1976, 1979a, 1979b). It is bactericidal for *S. aureus* and acts only on growing cells. The compound was produced in a Casitone (0.5%)-glucose (1%) medium; is synthesized only during growth phase, with a maximum yield after 2 to 3 days; and is excreted into the medium, from which it can be extracted at pH 9 with chloroform. It is unstable under acidic conditions and migrates toward the anode in electrophoresis at pHs from 6.5 up. By mass spectroscopy, a (hypothetical) molecular weight of 260 was determined. While the production behavior of the organism, the inhibitory spectrum of the substance, and its UV absorption suggest that the substance could be identical with corallopyronin, other data, such as its infrared spectrum, behavior as an acid, and molecular weight, are not in agreement.

In the same strain of *C. coralloides,* two pigments with antibiotic properties have been described (Arias et al., 1983). The pigments are extracted from the cell mass with acetone and are inhibitory for gram-positive bacteria. Their absorption spectra resemble those of a carotenoid and a keto-carotenoid, e.g., myxobacton. Perhaps the inhibitory effect was due to fatty acids, which may have been liberated during the test by saponification of their ester bonds to the carotenoid-glucoside.

Regarding the microbiological aspects of myxobacterial secondary metabolism, myxobacteria in general do not behave differently from other producers of secondary metabolites.

The potential to synthesize a certain compounds is strain specific. This is of considerable consequence, because there are only about 40 species of myxobacteria but a very large number of strains. In a series of strains of the same species, individual strains may produce quite different compounds (Table 3) or nothing at all. Of course, biosynthesis also depends on the cultivation conditions, so that one can never be sure about the synthetic capabilities of a strain. However, phylogenetically related strains should not be too different in their production requirements, and under identical or similar growth conditions, one may expect the same spectrum of substances to be produced if the potential is there. As a result of our screening, we feel that there cannot be much doubt that synthesis potential is strain specific.

The same compound may show up in strains of different species, genera, or families (Table 3). So far, only the border between suborders has never been crossed; i.e., no substance found in a representative of one suborder has ever been rediscovered in a strain of the other suborder. This probably means only that this happens very rarely, for compounds that had been discovered before in totally unrelated bacteria do occur in myxobacteria (Fig. 2).

All this does not mean, however, that there is no connection at all between taxonomy and biosynthetic potential. Thus, certain compounds have been isolated so far only from strains of certain species. The phenalamids and stipiamides were discovered independently in two strains of *M. stipitatus*. The fact that the phenalamids are responsible for the yellow fluorescence of *M. stipitatus* strains suggests that all *M. stipitatus* strains and only they synthesize phenalamids. This conclusion may not be completely sound, however, because we now classify *Myxococcus* strains of the *M. fulvus-M. stipitatus* group as *M. stipitatus* if they show yellow fluorescence, and this may not be correct. There may be *M. stipitatus* strains without fluorescence, and *M. fulvus* strains with fluorescence could also exist.

There is a long list of substances found so far only in *S. cellulosum* strains, e.g., ambruticin, sorangicin, and soraphen. In fact, no compound discovered in *S. cellulosum* has ever been found in another organism. However, unresolved taxonomy may again fool us, for the 900 strains in our collection classified as *S. cellulosum* are certainly not all members of one species, although they probably do belong to one genus. Further, the metabolism of cellulose-degrading *Sorangium* spp. differs substantially from that of other myxobacteria, so it would be understandable if their secondary metabolism also differed.

Undeniably, certain genera and species are rich and other are poor in secondary metabolites. Thus, *S. cellulosum* is an extremely rich source of novel compounds. So far, we have obtained 18 new basic structures from that species alone. Also, while the *Myxococcus* species are relatively rich in secondary metabolites, *Corallococcus* and *Nannocystis* species are relatively poor. However, no myxobacterial species is completely devoid of secondary metabolites. Also, our experience is that whenever a new taxon is harnessed for screening and production, new substances are obtained as well, as in *Polyangium, Chondromyces,* and *Nannocystis* spp.

A single strain may produce a large family of chemically related substances, a fact well known for, e.g., streptomycetes. Examples from myxobacteria are the 20 myxovirescins isolated from one strain and the 50 soraphens also isolated from one strain. Myxothiazol is produced without variants by the original strain, *M. fulvus* Mx f16, but from another *M. fulvus* strain, 35 structural variants have been characterized.

A strain may synthesize several structurally unrelated compounds. This fact, too, is well known from other producers of antibiotics. Examples from myxobacteria have been mentioned above, e.g., *S. aurantiaca* Sg a15, which synthesizes three different electron transport inhibitors, viz., stigmatellin, myxalamid, and the aurachins, the last with four substantially different structural variants.

The production phase in myxobacterial fermentations usually coincides with growth phase. Rarely, synthesis begins only in late log phase and extends into stationary phase, e.g., with myxovirescin, even when there is a sustained stationary phase, which often is not the case. However, some growth limitation may be needed in order to get good production, for as is often observed in antibiotic fermentations, a maximum growth rate normally has a negative effect on antibiotic synthesis. Thus, myxothiazol is produced with reasonable yields only when the oxygen supply is limited (Gerth et al., 1980), or as mentioned above, myxovirescin yields respond to the peptone concentration in the medium with a sharp maximum. Also, stigmatellin production is clearly better if a lean rather than a rich peptone medium is used (Kunze et al., 1984). The specific problem of ammonia accumulation in fermentations with *Myxococcus* species has already been discussed above in connection with myxovirescin, and the possibility of production with immobilized cells has been discussed in connection with sorangicin. Myxobacteria grow relatively slowly, and fermentations take between 2 and 14 days, with a concomitantly rising risk of contamination. A substantial biomass can be reached: 10^9 to 10^{10} cells per ml, and 1 to 3 g of dry mass per liter.

While initial yields are almost always rather low, considerable improvements in yield are made possible by the application of familiar techniques. Thus, by a random screen of mutants, we have improved yields from around 1 mg/liter to 1 g/liter and more. Mutagenization can be done by UV or NTG treatment. Also, a directed screening of wild strains often provides superior producers. For example, in one case, the yields of the original strain were about 0.5 mg/liter, but wild strains that started at 70 to 150 mg/liter were found. This suggests that our low yields may often be simply a consequence of the low number of strains that have been screened, but this lack cannot easily be remedied because the isolation of myxobacteria is relatively tedious and time-consuming. Also, better-producing wild strains have not always been found, and some antibiotics are very rare.

Besides undirected mutagenesis, modern techniques of molecular genetics are also available for yield improvements with myxobacteria. Several methods of gene transfer have been developed for myxobacteria: phage transduction (for a review see Kaiser [1984]), conjugation with *E. coli* using Inc plasmids as vectors (Breton et al., 1985; Jaoua et al., 1990), and transformation by electroporation (Kuspa and Kaiser, 1989). Genes involved in antibiotic synthesis can be identified by disruption through transposon insertion (Saulnier, 1989) and can thus be directly manipulated. One difficulty with genetics, however, is that all studies thus far have been performed only with certain strains of *M. xanthus* and *S. aurantiaca,* but yield increases have to be achieved with taxonomically very diverse producer strains. Considerable efforts may be required before such strains can be handled properly (Saulnier et al., 1988; Jaoua et al., 1992). The regulation of production of antibiotic TA (see above) has been subjected to genetic analysis. The genes involved in the production of TA have been mapped (Varon et al., 1992), and the transposon Tn*5 lac* has been used as a reporter gene to examine the expression of genes required for TA production (Tolchinsky et al., 1992).

What may the function of myxobacterial secondary metabolites for producer strains be? There is little doubt that the compounds serve some useful purpose and that some selection pressure ensures their conservation. Their synthesis always requires an appreciable number of highly specialized enzymes. From other producing organisms, e.g., streptomycetes, it is known that the genes involved in antibiotic syntheses are clustered at one or at most two sites on the genome. The situation is probably not much different with the myxobacteria. This high degree of organization on the genetic level can certainly not be maintained without a constant correction through selection. On the other hand, these many different compounds will probably not all be used to the same end.

The iron chelators myxochelin and nannochelin (Fig. 14) are almost certainly myxobacterial siderophores. With all other compounds, however, evidence for a function is circumstantial at best. Many myxobacteria seem to live in nature as "micropredators" by enzymatically lysing other microorganisms. An old hypothesis suggested that enzymatic attack would be facilitated or made possible if the prey was first killed or paralyzed with the aid of antibiotics (Oxford, 1947; Norén, 1953). Since growing bacteria are more resistant to enzymatic attack, probably because enzymes do not easily enter intact cells and all damages are quickly repaired, this idea is indeed attractive, but there are also many confusing arguments. Thus, many strains seem to do quite well without antibiotics, and no antibiotic will hamstring every kind of microorganism, but we do not know what compounds are produced under natural conditions, and in order to be beneficial, a compound need not be 100% efficient. More difficult to understand in this context is the role of intracellular substances, but as some proportion of the population may always be lysing, even those compounds may diffuse into the environment. Further, we do not know whether these compounds remain intracellular under all conditions, and they might exert their inhibiting effect in direct cell contact between the myxobacterium and its prey.

The classic idea of antibiotics as defense agents may also not be unreasonable for some of the myxobacterial inhibitors. Young fruiting bodies are, as a rule, free of contaminants, so that one can quickly get pure cultures if one succeeds in removing part of the fruiting body without carrying along some of the substrate (unfortunately, this is possible only with *Myxococcus* and *Chondromyces* fruiting bodies). The inner parts of the swarm colonies, on the other hand, are practically always contaminated. Further, it is interesting that *Sorangium* strains are virtually always producing some bioactive material and very often are producing antifungal compounds. These cellulose-degrading organisms live on decaying wood and other cellulose fabric, where their chief competitors most likely are fungi.

Finally, there may be some connection between morphogenesis and secondary metabolites. This has been discussed also for other producers, although without clear evidence. The connection could as well be indirect. Some organisms with morphogenetic potentials have larger genomes, synthesize extra biochemicals such as pigments or special structural material, and certainly have sophisticated regulatory circuits, so their metabolisms may be more flexible and it may be more easier for them to stow away the additional genetic information necessary for secondary metabolism. On plates, production of antibiotic TA appears to coincide with fruiting body formation, but nonfruiting mutants synthesize the antibiotic as well (Rosenberg et al., 1973). The case of uncontaminated fruiting bodies just mentioned could indicate that inhibitory substances are synthesized during fruiting. In our laboratories, all studies on myxobacterial metabolites are performed with liquid cultures in the absence of any morphogenetic events, but in streptomycetes, the connection between antibiotic synthesis and morphogenesis is seen only in plate cultures, so the issue of a connection between morphogenesis and secondary metabolites is open. It appears that involvement of a secondary metabolite in the mechanism of fruiting or myxospore formation would be likely only when every strain of a species is producing the same compound.

The impact of the new myxobacterial secondary metabolites can already be felt. Several myxobacterial electron transport inhibitors, especially myxothiazol and stigmatellin, have become valuable tools in biochemical research. Some compounds appear to have a potential for application, e.g., as fungicides, insecticides, or antitumor agents. For chemists who deal with synthesis, the novel and often difficult structures are a challenge for partial or total synthesis. The synthetic work on myxalamid (Cox and Whiting, 1991a, 1991b), myxopyronin A (Funk, 1986), and ambruticin (Just and Potvin, 1980; Barnes et al., 1981) may be added to the examples already given. A large number of derivatives of soraphen have been prepared (Böhlendorf, 1991). Above all, the myxobacteria are now firmly established as producers of new, interesting, bioactive secondary metabolites. More compounds can definitely be expected from that source. In 1992, in our laboratories alone, seven new compounds were elucidated, among them a new inhibitor of bacterial RNA polymerase, two compounds that interact with DNA synthesis, one inhibitor of protein synthesis, and a new electron transport inhibitor.

REFERENCES

Arai, T., K. Takahashi, and A. Kubo. 1977. New antibiotics, saframycins A, B, C, D and E. *J. Antibiot.* **30:**1015–1018.

Arai, T., K. Takahashi, S. Nakahara, and A. Kubo. 1980. The structure of a novel antitumor antibiotic, saframycin A. *Experientia* **36:**1025–1027.

Arai, T., K. Yazawa, K. Takahashi, A. Maeda, and Y. Mikami. 1985. Directed biosynthesis of new saframycin derivatives with resting cells of *Streptomyces lavendulae. Antimicrob. Agents Chemother.* **28:**5–11.

Arias, J. M., R. Carmona, and E. Montoya. 1983. Pigments with antibiotic activity from *Myxococcus coralloides. Microbios* **38:**27–32.

Arias, J. M., C. Rodriguez, and E. Montoya. 1976. Estudio de las condiciones optimas para la produccion de antibioticos por *Myxococcus coralloides. Laboratorio* **42:**317–328.

Arias, J. M., C. Rodriguez, and E. Montoya. 1979a. Biological activity of an antibiotic produced by *Myxococcus coralloides. Microbios* **24:**123–131.

Arias, J. M., C. Rodriguez, and E. Montoya. 1979b. Purification and partial characterization of an antibiotic produced by *Myxococcus coralloides. J. Antibiot.* **32:**205–211.

Arima, K., H. Imanaka, M. Kousaka, A. Fukuda, and G. Tamura. 1964. Pyrrolnitrin, a new antibiotic substance, produced by *Pseudomonas. Agric. Biol. Chem.* **28:**575–576.

Barnes, N.J., A. H. Davidson, L. R. Hughes, G. Procter, and V. Rajcoomar. 1981. Synthetic studies on ambruticin: determination of the absolute stereochemistry by a chiral synthesis of the cyclopropane ring. *Tetrahedron Lett.* **22:**1751–1754.

Becker, K. 1990. Antibiotica—Produktion mit trägerfixierten Myxobakterien. Doctoral thesis. Technical University, Braunschweig, Germany.

Böhlendorf, B. 1991. Chemische Eigenschaften und Derivatisierung des antifungischen Makrolids Soraphen A. Doctoral thesis. Technical University, Braunschweig, Germany.

Borgschulte, K. 1986. Darstellung semisynthetischer Myxovirescine und Untersuchung der Struktur-Wirkungsbeziehung. Doctoral thesis. Technical University, Braunschweig, Germany.

Borgschulte K., W. Trowitzsch-Kienast, and G. Höfle. 1990. Semisynthetische Myxovirescine: Austausch des α-Hydroxycarbonsäure-Segments und Modifikation der Ringgröße. *Justus Liebigs Ann. Chem.* **1990:**69–81.

Breton, A. M., S. Jaoua, and J. Guespin-Michel. 1985. Transfer of plasmid RP4 to *Myxococcus xanthus* and evidence for its integration into the chromosome. *J. Bacteriol.* **161:** 523–528.

Carmeli, S., R. E. Moore, and G. M. L. Patterson. 1990. Tantazoles: unusual cytotoxic alkaloids from the blue-green alga *Scytonema mirabile*. *J. Am. Chem. Soc.* **112:**8195–8197.

Chang, C. J., H. G. Floss, D. J. Hook, J. A. Mabe, P. E. Manni, L. L. Martin, K. Schröder, and T. L. Shieh. 1981. The biosynthesis of the antibiotic pyrrolnitrin by *Pseudomonas aureofaciens*. *J. Antibiot.* **34:**555–566.

Connor, D. T., R. C. Greenough, and M. von Strandtmann. 1977. W-7783, a unique antifungal antibiotic. *J. Org. Chem.* **42:**3664–3669.

Connor, D. T., S, Klutchko, and M. von Strandtmann. 1979. Antifungal agents. 3. Chemical modification of antibiotics from *Polyangium cellulosum* var. *fulvum*. A divinylcyclopropane-cycloheptadiene rearrangement. *J. Antibiot.* **32:**368–370.

Connor, D. T., and M. von Strandtmann. 1979a. Antifungal agents. 4. Chemical modification of antibiotics from *Polyangium cellulosum* var. *fulvum*. Ester and amide analogues of ambruticin. *J. Med. Chem.* **22:**1144–1147.

Connor, D. T., and M. von Strandtmann. 1979b. Antifungal agents. 5. Chemical modification of antibiotics from *Polyangium cellulosum* var. *fulvum*. Alcohol, ketone, aldehyde, and oxime analogues of ambruticin. *J. Med. Chem.* **22:**1055–1059.

Corbin, J. L., and W. A. Bulen. 1969. The isolation and identification of 2,3-dihydroxybenzoic acid and 2-N,6-N-di-(2,3-dihydroxybenzoyl)-L-lysine formed by iron-deficient *Azotobacter vinelandii*. *Biochemistry* **8:**757–762.

Cox, C. M., and D. A. Whiting. 1991a. Synthetic studies on electron transport inhibitors. Part 1. Chiral synthesis of a synthon for myxalamide D, piericidin A, and the actinopyrones. *J. Chem. Soc. Perkin Trans.* **1991:**1901–1905.

Cox, C. M., and D. A. Whiting. 1991b. Synthetic studies on electron transport inhibitors. Part 2. Approaches to the synthesis of myxalamide D. *J. Chem. Soc. Perkin Trans.* **1991:** 1907–1911.

degli Esposti, M., A. L. Tsai, G. Palmer, and G. Lenaz. 1986. On the oxidation pathways of the mitochondrial bc_1 complex from beef heart. Effects of various inhibitors. *Eur. J. Biochem.* **160:**547–555.

di Rago, J. P., Y. J. Coppée, and A. M. Colson. 1989. Molecular basis for resistance to myxothiazol, mucidin (strobulin A), and stigmatellin. *J. Biol. Chem.* **264:**14543–14548.

Dworkin, M. 1962. Nutritional requirements for vegetative growth of *Myxococcus xanthus*. *J. Bacteriol.* **84:**250–257.

Dworkin, M. 1969. Sensitivity of gliding bacteria to actinomycin D. *J. Bacteriol.* **98:**851–852.

Fujimoto, H., T. Kinoshita, H. Suzuki, and H. Umezawa. 1970. Studies on the mode of action of althiomycin. *J. Antibiot.* **23:**271–275.

Funk, E. 1986. Totalsynthese des Antibiotikums Myxopyronin A und analoger Verbindungen. Doctoral thesis. Technical University, Braunschweig, Germany.

Fytlovitch, S., P. D. Nathan, D. Zafriri, and E. Rosenberg. 1983. Amino acid precursors of *Myxococcus xanthus* antibiotic TA. *J. Antibiot.* **36:**1525–1530.

Galle, P. R., and L. Theilmann. 1990. Inhibition of hepatitis B virus polymerase activity by various agents. *Arzneim. Forsch. Drug Res.* **40:**1380–1382.

Gerth, K. Personal communication.

Gerth, K. 1975. Untersuchungen über den molekularen Mechanismus der Induktion von Myxosporen bei *Stigmatella aurantiaca* Sg a1 (Myxobacterales). Doctoral thesis, University Freiburg, Freiburg, Germany.

Gerth, K., H. Irschik, H. Reichenbach, and W. Trowitzsch. 1980. Myxothiazol, an antibiotic from *Myxococcus fulvus* (Myxobacterales). I. Cultivation, isolation, physico-chemical and biological properties. *J. Antibiot.* **33:**1474–1479.

Gerth, K., H. Irschik, H. Reichenbach, and W. Trowitzsch. 1982a. The myxovirescins, a family of antibiotics from *Myxococcus virescens* (Myxobacterales). *J. Antibiot.* **35:** 1454–1459.

Gerth, K., R. Jansen, G. Reifenstahl, G. Höfle, H. Irschik, B. Kunze, H. Reichenbach, and G. Thierbach. 1983. The myxalamids, new antibiotics from *Myxococcus xanthus* (Myxobacterales). I. Production, physico-chemical and biological properties and mechanism of action. *J. Antibiot.* **36:**1150–1156.

Gerth, K., R. Metzger, and H. Reichenbach. 1993. Induction of myxospores in *Stigmatella aurantiaca* (myxobacteria). Inducers and inhibitors of myxospore formation, and mutants with a changed sporulation behaviour. *J. Gen. Microbiol.* **139:**865–871.

Gerth, K., and H. Reichenbach. 1986. Determination of bacterial ammonia pools using *Myxococcus virescens* as an example. *Anal. Biochem.* **152:**78–82.

Gerth, K., W. Trowitzsch, G. Piehl, R. Schultze, and J. Lehmann. 1984. Inexpensive media for mass cultivation of myxobacteria. *Appl. Microbiol. Biotechnol.* **19:**23–28.

Gerth, K., W. Trowitzsch, V. Wray, G. Höfle, H. Irschik, and H. Reichenbach. 1982b. Pyrrolnitrin from *Myxococcus fulvus* (Myxobacterales). *J. Antibiot.* **35:**1101–1103.

Hall, C., M. Wu, F. L. Crane, H. Takahashi, S. Tamura, and K. Folkers. 1966. Piericidin A: a new inhibitor of mitochondrial electron transport. *Biochem. Biophys. Res. Commun.* **25:**373–377.

Hecht, V., L. Bischoff, and K. Gerth. 1990. Hollow fiber supported gas membrane for in-situ removal of ammonium during an antibiotic fermentation. *Biotechnol. Bioeng.* **35:** 1042–1050.

Hecht, V., J. Vorlop, H., Kalbitz, K. Gerth, and J. Lehmann. 1987. Vortex chamber for in situ recovery of the antibiotic myxovirescin A in continuous culture. *Biotechnol. Bioeng.* **29:**222–227.

Hill, G. C., and W. A. Remers. 1991. Computer simulation of the binding of saframycin A to d(GATGCATC)$_2$. *J. Med. Chem.* **34:**1990–1998.

Höfle, G., N. Bedorf, R. Jansen, H. Steinmetz, W. Trowitzsch-Kienast, H. Reichenbach, E. Forche, K. Gerth, H. Irschik, B. Kunze, F. Sasse, G. Hunsmann, and E. Jurkiewicz. July 1992. German patent application DE 40 41 686 A1.

Höfle, G., B. Kunze, C. Zorzin, and H. Reichenbach. 1984. Stigmatellin A und B—zwei neue Antibiotika aus *Stigmatella aurantiaca* (Myxobacterales). *Justus Liebigs Ann. Chem.* **1984:**1883–1904.

Höfle, G., and H. Reichenbach. 1990. Biologically active substances from microorganisms—an interdisciplinary research project at the GBF. *Sci. Annu. Rep. Ges. Biotechnol. Forsch. Braunschweig* **1990:**5–22.

Höfle, G., H. Steinmetz, K. Gerth, and H. Reichenbach. 1991. Ambruticins VS: new members of the antifungal ambruticin family from *Sorangium cellulosum*. *Justus Liebigs Ann. Chem.* **1991:**941–945.

Irschik, H. Personal communication.

Irschik, H., K. Gerth, G. Höfle, W. Kohl, and H. Reichenbach. 1983a. The myxopyronins, new inhibitors of bacterial RNA synthesis from *Myxococcus fulvus* (Myxobacterales). *J. Antibiot.* **36:**1651–1658.

Irschik, H., K. Gerth, T. Kemmer, H. Steinmetz, and H. Reichenbach. 1983b. The myxovalargins, new peptide antibiotics from *Myxococcus fulvus* (Myxobacterales). I. Cultivation, isolation, and some chemical and biological properties. *J. Antibiot.* **36:**6–12.

Irschik, H., R. Jansen, K. Gerth, G. Höfle, and H. Reichenbach. 1987. The sorangicins, novel and powerful inhibitors of eubacterial RNA polymerase isolated from myxobacteria. *J. Antibiot.* **40:**7–13.

Irschik, H., R. Jansen, G. Höfle, K. Gerth, and H. Reichenbach. 1985. The corallopyronins, new inhibitors of bacterial RNA synthesis from myxobacteria. *J. Antibiot.* **38:**145–152.

Irschik, H., and H. Reichenbach. 1985a. An unusual pattern of carbohydrate utilization in *Corallococcus (Myxococcus) coralloides* (Myxobacterales). *Arch. Microbiol.* **142:**40–44.

Irschik, H., and H. Reichenbach. 1985b. The mechanism of action of myxovalargin A, a peptide antibiotic from *Myxococcus fulvus*. *J. Antibiot.* **38:**1237–1245.

Irschik, H., W. Trowitzsch-Kienast, K. Gerth, G. Höfle, and H. Reichenbach. 1988. Saframycin Mx1, a new natural saframycin isolated from a myxobacterium. *J. Antibiot.* **41:** 993–998.

Ishiguro, K., K. Takahashi, K. Yazawa, S. Sakiyama, and T. Arai. 1981. Binding of saframycin A, a heterocyclic quinone anti-tumor antibiotic to DNA as revealed by the use of the antibiotic labeled with [^{14}C]tyrosine or [^{14}C]cyanide. *J. Biol. Chem.* **256:**2162–2167.

Jansen, R., H. Irschik, H. Reichenbach, and G. Höfle. 1985. Corallopyronin A, B and C—drei neue Antibiotika aus *Corallococcus coralloides* Cc c127 (Myxobacterales). *Justus Liebigs Ann. Chem.* **1985:**822–836.

Jansen, R., H. Irschik, H. Reichenbach, D. Schomburg, V. Wray, and G. Höfle. 1989a. Sorangicin A, ein hochwirksames Antibiotikum mit neuartiger Makrolid-Polyether-Struktur aus *Sorangium cellulosum,* So ce12: spektroskopische Strukturaufklärung, Kristall- und Lösungsstruktur. *Justus Liebigs Ann. Chem.* **1989:**111–119.

Jansen, R., H. Irschik, H. Reichenbach, V. Wray, and G. Höfle. 1989b. Natürliche Strukturvarianten von Sorangicin A aus *Sorangium cellulosum,* So ce12. *Justus Liebigs Ann. Chem.* **1989:**213–222.

Jansen, R., H. Irschik, H. Reichenbach, V. Wray, and G. Höfle. 1989c. Biosynthese von Sorangicin A in *Sorangium cellulosum,* So ce12. *Justus Liebigs Ann. Chem.* **1989:**309–313.

Jansen, R., B. Kunze, H. Reichenbach, E. Jurkiewicz, G. Hunsmann, and G. Höfle. 1992. Thiangazole: a novel inhibitor of HIV-1 from *Polyangium* spec. *Justus Liebigs Ann. Chem.* **1992:**357–359.

Jansen, R., B. Kunze, V. Wray, H. Reichenbach, E. Jurkiewicz, G. Hunsmann, and G. Höfle. 1991. Phenoxan: a novel inhibitor of HIV-1 infection in cell cultures from *Polyangium* sp., strain Pl VO19 (myxobacteria). *Justus Liebigs Ann. Chem.* **1991:**707–708.

Jansen, R., G. Reifenstahl, K. Gerth, H. Reichenbach, and G. Höfle. 1983. Myxalamide A, B, C und D, eine Gruppe homologer Antibiotika aus *Myxococcus xanthus* Mx x12 (Myxobacterales). *Justus Liebigs Ann. Chem.* **1983:**1081–1095.

Jansen, R., D. Schummer, H. Irschik, and G. Höfle. 1990. Chemical modification of sorangicin A and structure-activity relationship. I. Carboxyl and hydroxyl group derivatives. *Justus Liebigs Ann. Chem.* **1990:**975–988.

Jansen, R., W. S. Sheldrick, and G. Höfle. 1984. Die absolute Konfiguration der Myxalamide A und B. *Justus Liebigs Ann. Chem.* **1984:**78–84.

Jaoua, S., B. Letouvet-Pawlak, C. Monnier, and J. F. Guespin-Michel. 1990. Mechanism of integration of the broad-host-range plasmid RP4 into the chromosome of *Myxococcus xanthus*. *Plasmid* **23:**183–193.

Jaoua, S., S. Neff, and T. Schupp. 1992. Transfer of mobilizable plasmids to *Sorangium cellulosum* and evidence for their integration into the chromosome. *Plasmid* **28:**157–165.

Jurkiewicz, E., R. Jansen, B. Kunze, W. Trowitzsch-Kienast, E. Forche, H. Reichenbach, G. Höfle, and G. Hunsmann. 1992. Three new potent HIV-1 inhibitors from myxobacteria. *Antiviral Chem. Chemother.* **3:**189–193.

Just, G., and P. Potvin. 1980. Synthesis of ring A of ambruticin and proof of its absolute stereochemistry. *Can. J. Chem.* **58:**2173–2177.

Kaiser, D. 1984. Genetics of myxobacteria, p. 163–184. *In* E. Rosenberg (ed.), *Myxobacteria: Development and Cell Interactions*. Springer Verlag, New York.

Katô, H. 1955. Notes on myxobacteria. II. Antibacterial strains of *Myxococcus fulvus*. *Ecol. Rev.* **14:**25–28.

Kawai, K., H. Shiojiri, R. Watanabe, and Y. Nozawa. 1983. Effect of pyrrolnitrin on mitochondrial reactions: induction of swelling. *Res. Commun. Chem. Pathol. Pharmacol.* **39:** 311–320.

Kende, A. S., Y. Fujii, and J. S. Mendoza. 1990. Total synthesis of natural ambruticin. *J. Am. Chem. Soc.* **112:**9645–9646.

Kim, Y. J., K. Furihata, S. Yamanaka, R. Fudo, and H. Seto. 1991. Isolation and structural elucidation of stipiamide, a new antibiotic effective to multidrug-resistant cancer cells. *J. Antibiot.* **44:**553–556.

Kishi, K., K. Yazawa, K. Takahashi, Y. Mikami, and T. Arai. 1984. Structure-activity relationships of saframycins. *J. Antibiot.* **37:**847–852.

Kohl, W., H. Irschik, H. Reichenbach, and G. Höfle. 1983. Myxopyronin A und B—zwei neue Antibiotika aus *Myxococcus fulvus* Stamm Mx f50. *Justus Liebigs Ann. Chem.* **1983:** 1656–1667.

Kohl, W., H. Irschik, H. Reichenbach, and G. Höfle. 1984. Die Biosynthese des Antibiotikums Myxopyronin A aus *Myxococcus fulvus* Stamm Mx f50. *Justus Liebigs Ann. Chem.* **1984:**1088–1093.

Kohl, W., B. Witte, B. Kunze, V. Wray, D. Schomburg, H. Reichenbach, and G. Höfle. 1985. Angiolam A—ein neues Antibiotikum aus *Angiococcus disciformis* (Myxobacterales). *Justus Liebigs Ann. Chem.* **1985:**2088–2097.

Kunze, B. Personal communication.

Kunze, B., N. Bedorf, W. Kohl, G. Höfle, and H. Reichenbach. 1989. Myxochelin A, a new iron-chelating compound from *Angiococcus disciformis* (Myxobacterales). Production, isolation, physico-chemical and biological properties. *J. Antibiot.* **42:**14–17.

Kunze, B., G. Höfle, and H. Reichenbach. 1987. The aurachins, new quinoline antibiotics from myxobacteria: production, physico-chemical and biological properties. *J. Antibiot.* **40:**258–265.

Kunze, B., R. Jansen, G. Höfle, and H. Reichenbach. Manuscript in preparation.

Kunze, B., R. Jansen, L. Pridzun, E. Jurkiewicz, G. Hunsmann, G. Höfle, and H. Reichenbach. 1992a. Phenoxan, a new oxazole-pyrone from myxobacteria: production, antimicrobial activity and its inhibition of the electron transport in complex I of the respiratory chain. *J. Antibiot.* **45:**1549–1552.

Kunze, B., T. Kemmer, G. Höfle, and H. Reichenbach. 1984. Stigmatellin, a new antibiotic from *Stigmatella aurantiaca* (Myxobacterales). I. Production, physico-chemical and biological properties. *J. Antibiot.* **37:**454–461.

Kunze, B., W. Kohl, G. Höfle, and H. Reichenbach. 1985. Production, isolation, physico-chemical and biological properties of angiolam A, a new antibiotic from *Angiococcus disciformis* (Myxobacterales). *J. Antibiot.* **38:**1649–1654.

Kunze, B., H. Reichenbach, H. Augustiniak, and G. Höfle. 1982. Isolation and identification of althiomycin from *Cystobacter fuscus* (Myxobacterales). *J. Antibiot.* **35:**635–636.

Kunze, B., W. Trowitzsch-Kienast, G. Höfle, and H. Reichenbach. 1992b. Nannochelins A, B and C, new iron-chelating compounds from *Nannocystis exedens* (myxobacteria). Production, isolation, physico-chemical and biological properties. *J. Antibiot.* **45:**147–150.

Kuspa, A., and D. Kaiser. 1989. Genes required for developmental signalling in *Myxococcus xanthus:* three *asg* loci. *J. Bacteriol.* **171:**2762–2772.

Lampky, J. R., and E. R. Brockman. 1977. Fluorescence of *Myxococcus stipitatus*. *Int. J. Syst. Bacteriol.* **27:**161.

Levine, H. B., S. M. Ringel, and J. M. Cobb. 1978. Therapeutic properties of oral ambruticin (W 7783) in experimental pulmonary coccidioidomycosis of mice. *Chest* **73:**202–206.

Lown, J. W., A. V. Joshua, and J. S. Lee. 1982. Molecular mechanisms of binding and single-strand scission of deoxyribonucleic acid by the antitumor antibiotics saframycins A and C. *Biochemistry* **21:**419–428.

Manor, A., I. Eli, M. Varon, H. Judes, and E. Rosenberg. 1989. Effect of adhesive antibiotic TA on plaque and gingivitis in man. *J. Clin. Periodontol.* **16:**621–624.

Manor, A., M. Varon, and E. Rosenberg. 1985. Adsorption of antibiotic TA to dental hard tissues. *J. Dent. Res.* **64:**1371–1373.

Miyashiro, S., S. Yamanaka, S. Takayama, and H. Shibai. 1988. Novel macrocyclic antibiotics: megovalicins A, B, C, D, G and H. I. Screening of antibiotic-producing myxobacteria and production of megovalicins. *J. Antibiot.* **41:**433–438.

Mizrahi, A., J. Arnan, G. Miller, Z. Liron, M. Manai, Y. Badus, and E. Rosenberg. 1976. Scaling-up of process for production of antibiotic of *Myxococcus xanthus* TA. *J. Appl. Chem. Biotechnol.* **26:**160–166.

Neilands, J. B., and C. Ratledge. 1982. Microbial iron transport compounds, p. 565–574. *In* A. I. Laskin and H. A. Lechevalier (ed.), *CRC Handbook of Microbiology,* 2nd ed., vol. IV. CRC Press, Inc., Boca Raton, Fla.

Nigam, J. N., J. Lehmann, K. Gerth, H. Piehl, R. Schulze, and W. Trowitzsch. 1984. Feeding strategy for the production of the new antibiotic myxovirescin A from *Myxococcus virescens* (Myxobacteriales). *Appl. Microbiol. Biotechnol.* **19:**157–160.

Norén, B. 1953. On the production of antibiotics by myxobacteria. *Sven. Bot. Tidskr.* **47:** 402–410.

Norén, B., and G. Odham. 1973. Antagonistic effects of *Myxococcus xanthus* on fungi. II. Isolation and characterization of inhibitory lipid factors. *Lipids* **8:**573–583.

Norén, B., and K. B. Raper. 1962. Antibiotic activity of myxobacteria in relation to their bacteriolytic capacity. *J. Bacteriol.* **84:**157–162.

Oettmeier, W., R. Dostatni, and C. Majewski. 1990. The aurachins, naturally occurring inhibitors of photosynthetic electron flow through photosystem II and the cytochrome b_6/f-complex. *Z. Naturforsch.* **45c:**322–328.

Oettmeier, W., D. Godde, B. Kunze, and G. Höfle. 1985. Stigmatellin. A dual type inhibitor of photosynthetic electron transport. *Biochim. Biophys. Acta* **807:**216–219.

Ohnishi, T., U. Brandt, and G. von Jagow. 1988. Studies on the effect of stigmatellin derivatives on cytochrome *b* and the Rieske iron-sulfur cluster of cytochrome *c* reductase from bovine heart mitochondria. *Eur. J. Biochem.* **176:**385–389.

Oka, M., Y. Nishiyama, S. Ohta, H. Kamei, M. Konishi, T. Miyaki, T. Oki, and H. Kawaguchi. 1988. Glidobactins A, B and C, new antitumor antibiotics. I. Production, isolation, chemical properties and biological activity. *J. Antibiot.* **41:**1331–1337.

Onishi, N., K. Izaki, and H. Takahashi. 1984. A macrocyclic antibiotic M-230 B produced by *Myxococcus xanthus*. Isolation and characterization. *J. Antibiot.* **37:**13–19.

Oxford, A. E. 1947. Observations concerning the growth and metabolic activities of myxococci in a simple protein-free liquid medium. *J. Bacteriol.* **53:**129–138.

Pridzun, L. 1992. Untersuchungen zum Wirkungsmechanismus von Soraphen A. Ph.D. thesis. Technical University, Braunschweig, Germany.

Rabsch, W., G. Höfle, H. Reichenbach, and R. Reissbrodt. 1991. Iron supply of *E. coli* and *Salmonella* strains by siderophores from myxobacteria, abstr. 42, p. 35. VAAM/DGHM Meet. 1991, Freiburg. *Bioforum* **14**(1/2):42(35).

Rabsch, W., and G. Winkelmann. 1991. The specificity of bacterial siderophore receptors probed by bioassays. *Biol. Metals* **4:**244–250.

Reichenbach, H. 1992. The genus *Lysobacter,* p. 3256–3275. *In* A. Balows, H. G. Trüper, M. Dworkin, W. Harder, and K. H. Schleifer (ed.), *The Prokaryotes,* 2nd ed. Springer Verlag, New York.

Reichenbach, H., K. Gerth, H. Irschik, B. Kunze, and G. Höfle. 1988a. Myxobacteria, a source of new antibiotics. *Trends Biotechnol.* **6:**115–121.

Reichenbach, H., K. Gerth, H. Irschik, B. Kunze, G. Höfle, A. Augustiniak, R. Jansen, T. Kemmer, W. Kohl, H. Steinmetz, and W. Trowitzsch. 1984. Results of a screening for antibiotics with gliding bacteria, p. 15–20. *In Third European Congress on Biotechnology,* vol. 1. Verlag Chemie, Weinheim, Germany.

Reichenbach, H., and G. Höfle. 1989. The gliding bacteria: a treasury of secondary metabolites, p. 79–100. *In* M. E. Bushell and E. Gräfe (ed.), *Bioactive Metabolites from Microorganisms.* Elsevier, Amsterdam.

Reichenbach, H., and G. Höfle. Biologically active secondary metabolites from myxobacteria. *In* B. R. Glick (ed.), *Biotechnology Advances.* Pergamon Press, Oxford, in press.

Reichenbach, H., W. Trowitzsch-Kienast, J. Steinmetz, B. Kunze, H. Irschik, N. Bedorf, G. Höfle, R. Jansen, H. Augustiniak, K. Gerth, F. Sasse, and E. Forche. September 1988b. German patent application DE 37 07 955 C1.

Ringel, S. M. 1978. In vitro and in vivo studies of ambruticin (W 7783): new class of antifungal antibiotics. *Antimicrob. Agents Chemother.* **13:**762–769.

Ringel, S. M., R. C. Greenough, S. Roemer, D. Connor, A. L. Gutt, B. Blair, G. Kanter, and M. von Strandtmann. 1977. Ambruticin (W 7783), a new antifungal antibiotic. *J. Antibiot.* **30:**371–375.

Römmele, G., G. Wirz, R. Solf, K. Vosbeck, J. Gruner, and W. Wehrli. 1990. Resistance of *Escherichia coli* to rifampicin and sorangicin A—a comparison. *J. Antibiot.* **43:**88–91.

Rosenberg, E., S. Fytlovitch, S. Carmeli, and Y. Kashman. 1982. Chemical properties of *Myxococcus xanthus* antibiotic TA. *J. Antibiot.* **35:**788–793.

Rosenberg, E., J. M. Porter, P. N. Nathan, A. Manor, and M. Varon. 1984. Antibiotic TA: an adherent antibiotic. *Biotechnology* **2:**796–799.

Rosenberg, E., B. Vaks, and A. Zuckerberg. 1973. Bactericidal action of an antibiotic produced by *Myxococcus xanthus. Antimicrob. Agents Chemother.* **4:**507–513.

Rosenberg, E., and M. Varon. 1984. Antibiotics and lytic enzymes, p. 109–125. *In* E. Rosenberg (ed.), *Myxobacteria: Development and Cell Interactions.* Springer Verlag, New York.

Sakakibara, H., H. Naganawa, M. Ohno, K. Maeda, and H. Umezawa. 1974. The structure of althiomycin. *J. Antibiot.* **27:**897–899.

Sasse, F., H. Steinmetz, G. Höfle, and H. Reichenbach. 1993. Rhizopodin, a new compound from *Myxococcus stipitatus* (myxobacteria), causes formation of rhizopodia-like structures in animal cell cultures. Production, isolation, physico-chemical and biological properties. *J. Antibiot.* **46:**581–588.

Saulnier, P. 1989. Dévelopment d'outils génétiques utilisables chez des myxobactéries—application à l'étude génétique de la production de myxothiazol chez *Myxococcus fulvus.* Ph.D. dissertation. Université de Technologie, Compiègne, France.

Saulnier, P., J. Hanquier, S. Jaoua, H. Reichenbach, and J. F. Guespin-Michel. 1988. Utilization of IncP-1 plasmids as vectors for transposon mutagenesis in myxobacteria. *J. Gen. Microbiol.* **14:**195–227.

Schulz, R. 1984. Untersuchungen zur Biosynthese von Myxovalargin, einem Antibiotikum aus *Myxococcus fulvus* (Myxobacterales). Doctoral thesis. Technical University, Braunschweig, Germany.

Schummer, D. 1989. Darstellung semisynthetischer Derivate und Strukturvarianten vom Polyether-Makrolid-Antibiotikum Sorangicin A. Doctoral thesis. Technical University, Braunschweig, Germany.

Schummer, D., B. Böhlendorf, M. Kiffe, and G. Höfle. Chemical modification of the antifungal macrolide soraphen $A_{1\alpha}$: cleavage and reconstruction of the lactone bond. *In Proc. 3rd Internat. Symp. Chem. Synthesis Antibiot. Related Microb. Prod.,* Kloster Banz, Sept. 1992. VCH Verlag, Weinheim, Germany, in press.

Seebach, D., M. A. Maestro, M. Sefkow, A. Neidlein, F. Sternfeld, G. Adam, and T. Sommerfeld. 1991. A highly convergent total synthesis of (+)-myxovirescine M_2. *Helv. Chim. Acta* **74:**2112–2118.

Sensi, P., R. Ballotta, and G. G. Gallo. 1959. Matamycin, a new antibiotic. II. Isolation and characterization. *Antibiot. Chemother.* **9:**76–80.

Shadomy, S., D. M. Dixon, A. Espinel-Ingroff, G. E. Wagner, H. P. Yu, and H. J. Shadomy. 1978a. In vitro studies with ambruticin, a new antifungal antibiotic. *Antimicrob. Agents Chemother.* **14:**99–104.

Shadomy, S., C. J. Utz, and S. White. 1978b. In vivo studies with ambruticin in murine histoplasmosis. *Antimicrob. Agents Chemother.* **14:**95–98.

Simpkin, K. G. 1985. Inhibition of growth and uptake of amino-acids in the yeast *Candida parapsilosis* by ambruticin. *J. Med. Vet. Mycol.* **23:**425–432.

Singh, N. B., O. P. Babbar, and J. N. S. Yadava. 1973. Studies on antiviral property of fruiting myxobacteria *(Myxococcus virescens). Indian J. Exp. Biol.* **11:**252–253.

Steinmetz, H., N. Bedorf, W. Trowitzsch, R. Krützfeld, K. Gerth, H. Reichenbach, and G. Höfle. 1984. Large scale isolation of myxovirescins, p. 643–647. *In Third European Congress on Biotechnology,* vol. 1. Verlag Chemie, Weinheim, Germany.

Steinmetz, H., H. Irschik, H. Reichenbach, and G. Höfle. 1989. Structure elucidation of the peptide antibiotics myxovalargin A–D, p. 13–18. *In* W. A. König and W. Voelter (ed.), *Chemistry of Peptides and Proteins,* vol. 4. Attempto Verlag, Tübingen, Germany.

Takahasi, N., A. Suzuki, and S. Tamura. 1966. Chemical structure of piericidin A. III. Structures of piericidin A and octahydropiericidin A. *Agric. Biol. Chem.* **30:**1–12.

Takayama, S., S. Yamanaka, S. Miyashiro, Y. Yokokawa, and H. Shibai. 1988. Novel macrocyclic antibiotics: megovalicins A, B, C, D, G and H. II. Isolation and chemical structures of megovalicins. *J. Antibiot.* **41:**439–445.

Thierbach, G., B. Kunze, H. Reichenbach, and G. Höfle. 1984. The mode of action of stigmatellin, a new inhibitor of the cytochrome *b-c₁* segment of the respiratory chain. *Biochim. Biophys. Acta* **765:**227–235.

Thierbach, G., and G. Michaelis. 1982. Mitochondrial and nuclear myxothiazol resistance in *Saccharomyces cerevisiae. Mol. Gen. Genet.* **186:**501–506.

Thierbach, G., and H. Reichenbach. 1981a. Myxothiazol, a new antibiotic interfering with respiration. *Antimicrob. Agents Chemother.* **19:**504–507.

Thierbach, G., and H. Reichenbach. 1981b. Myxothiazol, a new inhibitor of the cytochrome b-c_1 segment of the respiratory chain. *Biochim. Biophys. Acta* **638:**282–289.

Thierbach, G., and H. Reichenbach. 1983. The effect of the new antibiotic myxothiazol on the respiration of *Paracoccus denitrificans. Arch. Microbiol.* **134:**104–107.

Tolchinsky, S., N. Fuchs, M. Varon, and E. Rosenberg. 1992. Use of Tn*5lac* to study expression of genes required for production of the antibiotic TA. *Antimicrob. Agents Chemother.* **36:**2322–2327.

Tron, T., and D. Lemesle-Meunier. 1990. Two substitutions at the same position in the mitochondrial cytochrome *b* gene of *S. cerevisiae* induce a mitochondrial myxothiazol resistance and impair the respiratory growth of the mutated strains albeit maintaining a good electron transfer activity. *Curr. Genet.* **18:**413–419.

Trowitzsch, W., K. Borgschulte, V. Wray, D. Schomburg, and G. Höfle. 1985. Konstitution und absolute Konfiguration der epimeren Myxovirescien A_1 and A_2. *Justus Liebigs Ann. Chem.* **1985:**1629–1652.

Trowitzsch, W., K. Gerth, V. Wray, and G. Höfle. 1983. On the biosynthesis of the antibiotic myxovirescin A_1 by *Myxococcus virescens. J. Chem. Soc. Chem. Commun.* **1983:** 1174–1175.

Trowitzsch, W., G. Höfle, and W. S. Sheldrick. 1981. The stereochemistry of myxothiazol. *Tetrahedron Lett.* **22:**3829–3832.

Trowitzsch, W., G. Reifenstahl, V. Wray, and K. Gerth. 1980. Myxothiazol, an antibiotic from *Myxococcus fulvus* (Myxobacterales). II. Structure elucidation. *J. Antibiot.* **33:** 1480–1490.

Trowitzsch, W., V. Wray, K. Gerth, and G. Höfle. 1982. Structure of myxovirescin A, a new macrocyclic antibiotic from gliding bacteria. *J. Chem. Soc. Chem. Commun.* **1982:** 1340–1342.

Trowitzsch-Kienast, W., E. Forche, V. Wray, E. Jurkiewicz, G. Hunsmann, and G. Höfle. 1992. Phenalamide, neue HIV-1-Inhibitoren aus *Myxococcus stipitatus* Mx s40. *Justus Liebigs Ann. Chem.* **1992:**659–664.

Trowitzsch-Kienast, W., H. Irschik, H. Reichenbach, V. Wray, and G. Höfle. 1988a. Isolierung und Strukturaufklärung der Saframycine Mx 1 und Mx 2, neue antitumoraktive Antibiotika aus *Myxococcus xanthus*. *Justus Liebigs Ann. Chem.* **1988:**475–481.

Trowitzsch-Kienast, W., H. Irschik, V. Wray, H. Reichenbach, and G. Höfle. 1988b. Ein neues Isochinolin-Alkaloid aus *Myxococcus xanthus*. *Justus Liebigs Ann. Chem.* **1988:** 483–484.

Trowitzsch-Kienast, W., K. Schober, V. Wray, K. Gerth, H. Reichenbach, and G. Höfle. 1989. Zur Konstitution der Myxovirescine B-T und Biogenese des Myxovirescins A. *Justus Liebigs Ann. Chem.* **1989:**345–355.

Trowitzsch-Kienast, W., V. Wray, K. Gerth, H. Reichenbach, and G. Höfle. 1986. Biosynthese des Myxothiazols in *Myxococcus fulvus* Mx f16. *Justus Liebigs Ann. Chem.* **1986: 93–98.**

Vaks, B., A. Zuckerberg, and E. Rosenberg. 1973. Purification and partial characterization of an antibiotic produced by *Myxococcus xanthus*. *Can. J. Microbiol.* **20:**155–161.

Varon, M., N. Fuchs, M. Monosov, S. Tolchinsky, and E. Rosenberg. 1992. Mutation and mapping of genes involved in antibiotic TA production in *Myxococcus xanthus*. *Antimicrob. Agents Chemother.* **36:**2316–2321.

von Jagow, G., and W. D. Engel. 1981. Complete inhibition of electron transfer from ubiquinol to cytochrome *b* by the combined action of antimycin and myxothiazol. *FEBS Lett.* **136:**19–24.

von Jagow, G., and T. A. Link. 1986. Use of specific inhibitiors on the mitochondrial bc_1 complex. *Methods Enzymol.* **126:**253–271.

von Jagow, G., P. O. Ljungdahl, P. Graf, T. Ohnishi, and B. L. Trumpower. 1984. An inhibitor of mitochondrial respiration which binds to cytochrome *b* and displaces quinone from the iron-sulfur protein of the cytochrome bc_1 complex. *J. Biol. Chem.* **259:**6318–6326.

von Jagow, G., and T. Ohnishi. 1985. The chromone inhibitor stigmatellin—binding to the ubiquinol oxidation center at the C-side of the mitochondrial membrane. *FEBS Lett.* **185:** 311–315.

Williams, D. R., and J. M. McGill. 1990. Total synthesis of myxovirescin B. *J. Org. Chem.* **55:**3457–3459.

Wong, D. T., and J. M. Airall. 1970. The mode of action of antifungal agents: effect of pyrrolnitrin on mitochondrial electron transport. *J. Antibiot.* **23:**55–62.

Wong, D. T., J. S. Horng, and R. S. Gordee. 1971. Respiratory chain of a pathogenic fungus, *Microsporum gypseum:* effect of the antifungal agent pyrrolnitrin. *J. Bacteriol.* **106:** 168–173.

Yasouri, F. N., and H. A. Foster. 1991. Improved yield of antibiotics by immobilization of cells of myxobacteria. *Biotechnol. Tech.* **5:**149–152.

Zafriri, D., E. Rosenberg, and D. Mirelman. 1981. Mode of action of *Myxococcus xanthus* antibiotic TA. *Antimicrob. Agents Chemother.* **19:**349–351.

INDEX